Vegetative Powers

INTERNATIONAL ARCHIVES OF THE HISTORY OF IDEAS

ARCHIVES INTERNATIONALES D'HISTOIRE DES IDÉES

234

VEGETATIVE POWERS

Fabrizio Baldassarri
Andreas Blank

More information about this series at http://www.springer.com/series/5640

Fabrizio Baldassarri • Andreas Blank
Editors

Vegetative Powers

The Roots of Life in Ancient, Medieval and Early Modern Natural Philosophy

Springer

Editors
Fabrizio Baldassarri
Institute for Research in the Humanities
University of Bucharest
Bucharest, Romania

Andreas Blank
Institute of Philosophy, Universitätsstr
Alpen-Adria University Klagenfurt
Klagenfurt, Austria

ISSN 0066-6610 ISSN 2215-0307 (electronic)
International Archives of the History of Ideas Archives internationales d'histoire des idées
ISBN 978-3-030-69708-2 ISBN 978-3-030-69709-9 (eBook)
https://doi.org/10.1007/978-3-030-69709-9

This Springer imprint is published by the registered company Springer Nature Switzerland AG
The registered company address is: Gewerbestrasse 11, 6330 Cham, Switzerland

L.

Auaricia viciorũ ancilla atq; ministra.

Luxuriosꝰ homo vt bestia.

Gulosus vt planta.

Piger vt lapis.

¶ Qui vero sese ab extremis abercuerit/ qui immobilis equa lance constiterit in medio sederitq; in homine: hic merito veraciter homo/ hic studiosꝰ/ probus/ sapiens/ felix beatusq; censeatur. Circũstant itaq; vnicũ hominẽ/ precipua mala sex: que velut truculẽtissime fere ex abditis / pestifera in eum arma mouent illiusq; felicitati insidiantur. His vero cadem instrumentaria Auaricie labes famulatur. Vniuersas ceca habẽdi rabies (velut impijssima ancilla) fouet/ alit/ nutrit. Largitur siquidẽ hec diuitias: proter quas aut nosipsos honotibus dignos arbitramur (Superbe inquam eleuamur/ Inhonorãtibus succẽsemus/ paribꝰ inuidemus) aut misera carnis blandimẽta fluxasq; voluptates insipienter prosequimur. ¶ Efficit quippe luxuries: & immoderatus fede libidinis amor/ hominẽ haud bestijs absimilem illumq; ab humana sede; in brutorum animantiũ gradum/ infeliciter exturbat. quãdoquidem nil prestantius brutis animãtibus adsit: q̃ seminis profusio sueq; speciei propagatio. Gula siue corporee alimonie irrepressa auiditas: hũc a primo proprioue loco/ deiicit in tertium: plantisq; persimilem efficit. que licet totiꝰ sensationis & voluptatis probentur expertes: officia tamen alimentationis exercent. Porro extrema Acedia: hominem in imũ vltimumq; gradum extrudit: facitq; mineralibus persimilẽ. ¶ Sicut enim mineralia que in extremo sedent gradu: haud aliud aliquid habent/ q̃ ipsum esse nullaq; naturali operatione excerceri aut per seipsa dimoueri illis indultum est: ita & quoscũq; portentosũ Acedie mõstrum obsederit: assiduo ferme somno cõsopescunt/ ab actu omni & operatione remittũtur. Immoti vt lapides perstãt: tanq̃ si simplex eẽ sine vlla preclara viatq; laudabilium operationum facultate: natura mater illis esset elargita. Perseuerant quippe isti/ plusq̃ omnes/ in nature muneribus ingrati: nullam naturalium suarum virium excercentes. in ipsa ratione penitus irrationabiles. In sensu/ insensibiles: in vita exanimes/ emortui/ infecundi. In ipsa deniq; substantia: ociosi/ desides/ torpẽtes. Et hec clarius/ ex subiecta figura dinoscuntur

Charles de Bovelles, *Liber de sapiente, in Que hoc volumine continentur: Liber de intellectu ; Liber de sensu ; Liber de Nichilo ; Ars oppositorum ; Liber de generatione ; Liber de sapiente ; Liber de duodecim numeris ; Epistole complures*, [Paris], [Henri Estienne], [1510], f. 117 v°. (Courtesy of Bibliothèque Nationale de France, BP16_101472, www.gallica.fr)

Acknowledgments

This volume is the result of a combination of different projects and the suggestions of many people. Fabrizio Baldassarri came up with a first idea in Israel in 2016, organized a panel at the annual meeting of RSA 2017 in Chicago and a symposium at the conference of the BSHP 2017 in Sheffield. Andreas Blank joined the project in 2017, when we started to plan a volume, inviting several contributors to explore this notion in seventeenth century philosophy, sharing ideas, and later submitting a proposal happily welcomed by the editors of this series.

In 2018, thanks to the help of professors Gianluigi Baldo, Pierdaniele Giaretta and Fabio Zampieri, we have received generous funds from the University of Padua and the Department of Cardiac, Thoracic and Vascular Sciences of the University of Padua to set up a two-day conference on vegetative powers. For this task, we have also benefitted from a grant of the Society for the Social History of Medicine (SSHM) and the support of the Società Italiana di Storia della Scienza (SISS), all of whom we need to thank heartily. Without all this important help, we would have been unable to have the workshop, which took place in Padua on September 12–13, 2018.[1] On this occasion, the lovely encounters, rich presentations, and productive discussions led us to the conviction that the discussions over vegetative powers could not be merely restricted to the seventeenth century, and deserved a bigger picture. Subsequently, the enterprise of the volume grew unexpectedly—we would have loved to include more contributions to fill as much gaps as possible, but in the end, this has not been possible for many reasons.

The volume we have put together profited from the help of professors at the University of Padua, discussions at seminars, conferences, and the discussions with colleagues at Herzog August Bibliothek in Wolfenbüttel, at ICUB-Humanities in Bucharest, but also the passion and expertise of all contributors, the experience of the editors, and the support of a grant of the Romanian National Authority for Scientific Research and Innovation (CNCS – UEFISCDI), project number

[1] See the website for more information: https://vegetativepowers.wordpress.com/ (last access 15/11/2019).

PN-III-P1-1.1-PD-2016-1496, “The Overlooked History of Vegetal Life. From the Vegetative Soul to Metabolism in Early Modern Philosophy and Biomedicine”. We would especially thank Guido Giglioni, Hiro Hirai, Sarah Hutton, Christoph Lüthy, Ohad Nachtomy, Carla Rita Palmerino, and Theo Verbeek, with whom we discussed the subject matter which comprises this volume at different stages.

Finally, a special thank goes to Ilaria Carrino, whose patience and lovely support have been a fundamental root of life for Fabrizio Baldassarri.

November 2019 Fabrizio Baldassarri

Andreas Blank

Contents

About the Authors

Igor Agostini is Professor of Philosophy at the University of Salento, where he is also Head of the Philosophy Programs and Director of the *Centro Dipartimentale di Studi su Descartes e il Seicento – Ettore Lojacono*. Visiting Fellow in 2013 and 2014 at the Philosophy Department of Princeton University; Invited Professor at the ENS of Paris in 2015, at the Universidade Federal de Uberlândia in 2014 and in 2018, at the Université de Caen in 2019. He is the author of some sixty publications on early modern philosophy and theology. His three main books constitute a trilogy on the problem of the knowledge of God in the early modern age: *L'infinità di Dio. Il dibattito da Suárez a Caterus. 1597–1641* (Editori Riuniti, 2008); *L'idea di Dio in Descartes. Dalle* Meditationes *alle* Responsiones (Mondadori Education, 2010); *La démonstration de l'existence de Dieu. Les conclusions des cinq voies de Thomas d'Aquin et la preuve* a priori *dans le thomisme du XVIIe siècle* (Brepols, 2016). He collaborated on the Italian edition of all Descartes's writings (Bompiani, 2005–2009) and on the French edition of Descartes's *Méditations. Objections et Réponses* by J.-M. Beyssade and D. Kambouchner (Gallimard, 2018). He is currently working on the *Nouvel Index Scolastico-cartésien*.

Raphaële Andrault is researcher at the CNRS (IHRIM-ENS de Lyon). She deals with the history of early modern philosophy, and life sciences and medicine, focusing especially on the mind-body problem, the notions of life mechanism and organism and the history of anatomy through the sixteenth to the eighteenth centuries. She has published monographs on early modern philosophy and life sciences, and in particular a critical edition of Steno's *Discours sur l'anatomie du cerveau* (Garnier 2009), *La vie selon la raison. Physiologie et métaphysique chez Spinoza et Leibinz* (Champion, 2014), and *La raison des corps. Mécanisme et sciences médicales* (Vrin, 2016). She co-edited with Mogens Laerke, *Steno and the Philosophers* (Brill, 2018). She has published in *Perspectives on Science*, *Studia Leibnitiana*, *British Journal for the History of Philosophy*, *Early Science and Medicine*, and *Dix-septième siècle*. She is now conducting a research project in Lyon on pain in early modern culture.

Fabrizio Baldassarri, PhD, is currently PI of the research project "The Overlooked History of Vegetal Life. From the Vegetative Soul to Metabolism in Early Modern Philosophy and Biomedicine" at ICUB-Humanities in Bucharest, and was post-doc fellow at HAB in Wolfenbüttel, recipient of the Kristeller-Popkin fellowship of the JHP, which he spent at Utrecht University, post-doc fellow at Bar-Ilan University in Tel Aviv, and DAAD post-doc fellow at Gotha Research Centre. He has taught medical humanities at the University of Padua. He has organized sessions and panels at several annual meetings, organized conferences such as "Manipulating Flora", which is now an edited special issue for *Early Science and Medicine* (Brill, 2018), and VivaMente2020, and widely published on Descartes' philosophy and medicine, and the study of botany in early modern times, see in particular: "«[P]er experientiam scilicet, vel deductionem». Descartes' Battle for *Scientia* in the Early 1630s", in *Historia Philosophica*, 15 (2017): 115–133; "Descartes' Bio-Medical Study of Plants: Vegetative Activities, Soul, and Power", in *Early Science and Medicine* 23 (2018): 509–529; "The Mechanical Life of Plants: Descartes on Botany", in *British Journal for the History of Science* 52 (2019): 41–63; and "Descartes and the Dutch: Botanical Experimentation in the Early Modern Period", in *Perspectives on Science*, forthcoming. In 2021, he will have a Marie Skłodowska Curie fellowship at Ca' Foscari Venice and Bloomington Indiana University.

Andreas Blank holds a Research Position funded by the Austrian Science Fund at the Alpen-Adria Universität Klagenfurt, Austria. Previously, he has been Visiting Associate Professor at the University of Hamburg and Bard College Berlin, as well as Visiting Fellow at the Center for Philosophy of Science at the University of Pittsburgh and the Cohn Institute for the History and Philosophy of Science, Tel Aviv University. He has published some 60 articles on early modern philosophy and science in edited volumes and journals such as *British Journal for the History of Philosophy*, *History of Philosophy Quarterly*, *Journal of the History of Ideas*, *Intellectual History Review*, *History of European Ideas*, *The Monist*, *Studia Leibnitiana*, *Perspectives on Science*, *Annals of Science*, *Science in Context*, and *Early Science and Medicine*.

Justin Begley is a Postdoctoral Researcher in the Humanities Division at the University of Helsinki. He has held short-term fellowships at the Folger Shakespeare Library in Washington DC, the Institute for Research in the Humanities at the University of Bucharest, and, most recently, The Royal Society with a Lisa Jardine Grant. Justin has published in journals such as *Intellectual History Review*, *The Review of English Studies*, *Notes & Queries*, and *The Seventeenth Century*. Currently, he is transforming the DPhil thesis that he completed at the University of Oxford in 2017, "Margaret Cavendish, The Last Natural Philosopher," into a monograph, and is working on an edition of the Cavendish medical book that is held by the University of Nottingham. Along with his work on Cavendish, Justin has written on Pierre Gassendi, Thomas Tryon, Nehemiah Grew, and Kenelm Digby, among others, and has special interests in the history of substance theories, the reception of Aristotelian thought, and early modern developments in comparative anatomy and botany.

Paola Bernardini graduated in History of Medieval Philosophy at the University of Siena in 1997, and then she received her PhD in Philosophy at the University of Florence in 2003. Since then, she has been working at the University of Siena, with research and teaching positions in the field of Medieval Philosophy. Her research focuses on edited and unedited texts of natural philosophy, as far as the reception of Aristotle's De anima in the XIIIth century is concerned. In particular, she has studied the commentaries produced in the Faculties of Arts and the origins of the so-called ' Latin averroism'. Amongst her publications, *Anonymi Magistri Artium Quaestiones super librum de anima* (ms. Siena, Biblioteca Comunale degli Intronati, L.III.21,ff.134ra-174va), 2009 (with a substantial Introduction); *Temporibus autem meis. Theologians' Errors with Regard to the Human Soul*, in *Roger Bacon's* Communia naturalium. *A 13th Century Philosopher's* Workshop, ed. by P. Bernardini-A. Rodolfi, SISMEL, Firenze 2014, pp. 139–158; "From the First to the Second Averroism. The Attribution of Art. 113: *Quod homo est homo praeter animam rationalem* (Paris, 1277)", *Recherches de Théologie et Philosophie Médiévales* 2017 (I), pp. 37–73.

Fabrizio Bigotti is Senior Research Fellow at the Institut für Geschichte der Medizin, Julius-Maximilians University of Würzburg, and Honorary Fellow at the Centre for Medical History of the University of Exeter. As an intellectual historian, with a specialisation in the history of science, medicine and technology in the early modern period, his work focuses primarily on the history of quantification and the reconstruction of scientific instruments, with broader interests in the role that classical and medieval philosophy played in the development of early modern ideas on logic, method, theory of matter, taxonomy, anatomy and physiology. He is the founding director of the 'Centre for the Study of Medicine and the Body in the Renaissance' (CSMBR, Pisa), and the co-editor of the Series 'Palgrave Studies in Medieval and Early Modern Medicine' (PSMEMM). His recent publications include *Physiology of the Soul. Mind, Body and Matter in the Galenic Tradition of the Late-Renaissance (1550–1630)* (Brepols, 2019) and a co-edited volume, with Jonathan Barry, *Santorio Santori and the Emergence of Quantified medicine (1614–1790)*, forthcoming (Palgrave MacMillan, 2021).

Amber D. Carpenter (PhD, London) is Associate Professor at Yale-NUS College. She publishes in Ancient Greek philosophy, especially the ethics, epistemology and metaphysics of Plato. Her monograph *Indian Buddhist Philosophy* (Acumen/Routledge) appeared in 2014, and she continues to publish in the area. She has held research fellowships with the Einstein Forum (Potsdam, 'Metaphysics as Ethics in Indian Buddhist philosophy'), the Templeton Religious Trust ('Ethical Ambitions and their Formation of Character', *The Beacon Project*), the University of Melbourne, and the University of York, and currently leads an international research project on Buddhist-Platonist dialogues. She continues to publish in Greek and Indian Buddhist philosophy separately, while also pursuing questions that bring the two into critical conversation, focusing usually on the ethical implications and underpinnings of metaphysical and epistemological arguments. As the culmination of a British Academy supported project with *The Integrity Project*, she is co-editing a volume of *Portraits of Integrity* (Bloomsbury 2020).

Cristiano Casalini is Chair of Jesuit Pedagogy and Educational History at Lynch School of Education and Human Development and research scholar at the Institute of Advanced Jesuit Studies at Boston College. Among his publications are *Aristotle in Coimbra. The Cursus Conimbricensis and Education at the College of Arts* (Routledge, 2017), and – with Claude Pavur, S.J. – *Jesuit Pedagogy (1540–1616): A Reader* (Institute of Jesuit Sources, 2018).

Amalia Cerrito, PhD student at the University of Pisa and at the University of Florence, currently works on the dynamics of generation in Albertus Magnus' natural philosophy and metaphysics. She has published papers on the interaction between natural philosophy and biblical exegesis in Albertus Magnus: "Botany as science and exegetical tool in Albert the Great," *Aisthesis* 11(1) (2018): 97–107; "Paternitas divina and naturalis: Albert the Great on Matthew 6:9 and Luke 11:12", *Divus Thomas* 122 (2) (2019): 59–78; and on the techniques of commentary/paraphrase in Albertus Magnus' *De vegetabilibus*: Alberto Magno e il *De plantis*: ricostruire la botanica perduta di Aristotele, in *Raccolta di Saggi in onore di Marco Arosio*, ed. by M. Martorana *et al.* (Rome: IF Press, 2019), 9–28.

Tobias Cheung is Associate Professor for the History of Knowledge and Culture at the Institute of Cultural History and Theory of the Humboldt University in Berlin. Cheung is author of *The Organization of the Living: The Role of Cuvier's, Leibniz's and Kant's Notions of Organismic Order in the History of Biology* (2000), *Charles Bonnet's System Theory and Philosophy of Organic Order* (2005), *Res vivens. Agent Models of Organic Order 1600–1800* (2008), *Organisms. Agents Between Inner and Outer Worlds 1780–1860* (2014), and editor of *Transitions and Borders between Animals, Humans and Machines 1600–1800* (2010).

Klaus Corcilius is Professor of philosophy at the University of Tübingen and specialises in ancient philosophy. His recent publications include: *Aristoteles. De motu animalium.* New Greek Text and German translation with introduction and commentary (with O. Primavesi) (Meiner, Hamburg 2018), and "Ideal Intellectual Cognition in Timaeus 37a2-c5", *Oxford Studies in Ancient Philosophy* LIV (2018/1), 51–105.

François Duchesneau is Professor Emeritus of the history and philosophy of science at the Université de Montréal. His publications include: *Leibniz et la méthode de la science* (Paris: PUF, 1993); *La Dynamique de Leibniz* (Paris: Vrin, 1994); *Philosophie de la biologie* (Paris: PUF, 1997); *Les Modèles du vivant de Descartes à Leibniz* (Paris: Vrin, 1998); *Leibniz. Le vivant et l'organisme* (Paris: Vrin, 2010); *La Physiologie des Lumières* (Paris: Classiques Garnier, 2012); (with J. E. H. Smith) *The Leibniz-Stahl Controversy* (New Haven: Yale University Press, 2016);

Organisme et corps organique de Leibniz à Kant (Paris: Vrin, 2018). His present research bears on the concept of organism and the laws of vital organization in the modern life sciences up to the twentieth century.

Michael Fatigati is a PhD Candidate in Classical Arabic Philosophy at the University of Toronto's Centre for Medieval Studies, and the Collaborative Program in Ancient and Medieval Philosophy. His dissertation focuses on emotions in Avicenna's philosophical psychology—their nature, and their significance for knowledge and morality—and includes a translation of the lengthy discussion emotions in Avicenna's *Healing: Rhetoric*. Previously, he completed an MA in Greek and Latin thought at Villanova University, and he has experience teaching in interdisciplinary Great Texts programs. His prior publications include work on the medieval Arabic scientific community, as well as in Scholastic philosophy of mind.

Matteo Favaretti Camposampiero, PhD, is Associate Professor at the Department of Philosophy and Cultural Heritage of the Ca' Foscari University of Venice (Italy). His research is devoted to early modern philosophy from Descartes to Kant with a special interest in metaphysics, logic, rational theology, the philosophy of biology and medicine, the mind-body problem, and the theories of knowledge, language, and modalities. His publications include two books, *Filum cogitandi: Leibniz e la conoscenza simbolica* (Milan: Mimesis, 2007) and *Conoscenza simbolica: Pensiero e linguaggio in Christian Wolff e nella prima età moderna* (Hildesheim: Olms, 2009), six co-edited volumes, and several articles and essays primarily focused on Leibniz, Wolff, their sources, and their influence. Recent journal articles: 'Machines of Nature and Machines of Art: Christian Wolf's Reception of Leibniz', *Rivista di Storia della Filosofia*, 74 (2019), 431–452; 'Mereology and Mathematics: Christian Wolff's Foundational Programme', *British Journal for the History of Philosophy*, 27 (2019), 1151–1172; 'Infinite Regress: Wolff's Cosmology and the Background of Kant's Antinomies', *Kant-Studien* (forthcoming).

Guido Giglioni is Associate Professor of History of Philosophy at the University of Macerata, Italy. His research is focused on the interplay of life and imagination in the early modern period, on which he has written and edited several contributions. He has published two books, on *Jan Baptista van Helmont* (Milan, 2000) and *Francis Bacon* (Rome, 2011).

Sarah Hutton is Honorary Visiting Professor in the Department of Philosophy at the University of York. She studied at New Hall, University of Cambridge and the Warburg Institute, University of London. She has held posts at the University of Hertfordshire, Middlesex University and Aberystwyth University. She has published extensively on early modern century intellectual history, with special interests in the Cambridge Platonists, and women in early modern philosophy and science. Her publications include *Anne Conway, a Woman Philosopher* (CUP, 2004), *Newton and Newtonianism*, edited with James Force (Springer, 2004), and *British Philosophy in the Seventeenth Century* (OUP, 2015).

Bálint Kékedi is PhD in philosophy at the University of Aberdeen, through the doctoral programme "The Emergence of a Scientific Culture". His dissertation – "Descartes, the Sheep, and the Wolf: A Study in the Autonomy of Cartesian Automata" – focuses on the mechanical models of animals' perceptual cognition in Descartes' texts, arguing for a greater autonomy of complex material systems than what is generally recognised in the prevailing Cartesian scholarship. He graduated in philosophy at the Eötvös Loránd University in Budapest. Afterwards, he pursued his studies at the University of Sydney, dealing with the mechanistic philosophy of the Early Modern period, with a particular interest in the materialist wing of Cartesianism, as well as the history of automata. Currently, he is an independent scholar, working full-time outside academia.

Martin Klein, Ph.D. in philosophy (2016), Humboldt University of Berlin (supervisors: Dominik Perler, Martin Pickavé and Martin Lenz), is a Lecturer at the Department of Philosophy at the University of Würzburg. He studied philosophy, history and theology in Berlin, Toronto, Groningen and Rome (2005–16) and was a Research Fellow at the Topoi Excellence Cluster Berlin (2012–18), a Mellon Fellow at PIMS, Toronto (Manuscript Programme 2016–18) and a Visiting Researcher at Sorbonne University and EPHE, Paris (2018). Amongst his publications: *Philosophie des Geistes im Spätmittelalter. Intellekt, Materie und Intentionalität bei Johannes Buridan* (Studien und Texte zur Geistesgeschichte des Mittelalters, 124) Leiden: Brill 2019. *Consciousness and Memory in Medieval and Early Modern Philosophy*, special issue of *Science and Politics* 12/2 (2018) (co-ed. with N.O. Kupferblum and O.I. Toth). "John Buridan on the Singularity of Sense Perception," in: *Medieval Perceptual Puzzles. Theories of Sense Perception in the 13th and 14th Century* (Investigating Medieval Philosophy, 13), ed. E. Baltuta (Leiden: Brill 2019), 364–388. "Digestive Problems: John Buridan on Human Nutrition," in: *Nutrition and Nutritive Soul in Aristotle and Aristotelianism* (Topics in Ancient Philosophy), eds. G. Korobili and R. Lo Presti (Berlin: de Gruyter 2021).

Laura Madella, BA in Philology at University of Parma, received her PhD in the History of Education from Roma Tre University, discussing the idea of self-education developed by the Spanish heretic Juan de Valdés. She has been a fellow at Boston College (US), working on Jesuit Catalogs's historical records, and she currently serves as research assistant at the University of Parma. Her studies focus on texts and documents of early modern Europe related to schooling and educational frameworks, and especially on the moral content and context of *regimina sanitatis* in the sixteenth and seventeenth centuries, as well as the evolution of pedagogical theories and practice through eighteenth century Italian academies. She has recently published a monograph, *Sull'Alphabeto christiano di Juan de Valdés* (2018), and several articles and literary translations.

Elisabeth Moreau is a Visiting Postdoctoral Fellow at Princeton University and a Research Associate at the Université libre de Bruxelles (Belgium). Trained in History and Philosophy of Science, she works on medicine, alchemy and natural philosophy in late Renaissance Europe. Her PhD dissertation (2018) is centred on the emergence of atomistic and corpuscular theories in Galenic physiology between 1567 and 1634. Her current postdoctoral project is focused on the body's assimilation of food and drugs from the perspective of theories of matter in early modern medicine. For her doctoral training, she held fellowships at the Science History Institute (Philadelphia) and FNRS (Belgian National Fund for Scientific Research), and has been a visiting researcher at UPenn and the University of Cambridge.

Marilena Panarelli obtained a PhD degree in Philosophy at the University of Salento in co-tutorship with the University of Cologne. In her dissertation, she conducted a doctrinal study of Albert the Great's *De vegetabilibus* and edited the critical edition of book VII (ansae 1–2) of Henry of Herford's *Catena aurea entium*. Her main research interests concern medieval botanical traditions, with particular attention to the work of Albert and the Dominicans. In this regard, she has just published an edition of the commentary on the *De plantis* by the Dominican friar John Krosbein. She currently has scientific collaborations with the University of Cologne and the University of Cluj.

Martin Pickavé (PhD University of Cologne) is Professor of Philosophy and a Canada Research Chair in Medieval Philosophy in the Department of Philosophy and the Centre for Medieval Studies at the University of Toronto, where he also serves at the Chair of the Department of Philosophy (since 2015). He specializes in later medieval philosophy, especially later medieval philosophy of mind, metaphysics, and philosophy of action. He is currently working on two monographs: one on medieval theories of the emotions, another on medieval debates about free will. With Russell L. Friedman (KU Leuven) he is working on a *Companion to Cognitive Theory in the Later Middle Ages*. His publications include many journal articles and book chapters and the following books: *A Companion to James of Viterbo* (Brill, 2018, co-edited with Antoine Côté), *Emotion and Cognitive Life in Medieval and Early Modern Philosophy* (Oxford University Press, 2012; co-edited with Lisa Shapiro), and *Heinrich von Gent über Metaphysik als erste Wissenschaft* (Brill, 2007).

Dániel Schmal graduated in classical philology (1995) and philosophy (1998) from Eötvös Loránd University, Budapest. He was a Young Research Fellow at the *Research Institute for Classical Philology of the Hungarian Academy of Sciences* (1995–1998), and Research Fellow at the *Institute for the Philosophical Research* of the same Academy (2002–2005). More recently, he has been an Associate Professor in the Institute of Philosophy at *Pázmány Péter Catholic University*. He took his PhD at Eötvös Loránd University, Budapest. He has held scholarships in Amsterdam, Leuven and Rotterdam, and Paris. Area of specialization: early modern epistemology and moral philosophy. His English and French publications include articles on St. Augustine, Descartes, Malebranche and Leibniz. He has translated Descartes's *Notae in programma quoddam* (Notes on a Certain Broadsheet), John Locke's *Second Tract on Government*, Gottfried Wilhelm Leibniz's *De causa Dei*, and Nicolas Malebranche's *Entretiens sur la métaphysique et sur la religion*.

Robert Vinkesteijn has recently obtained his PhD at Utrecht University (2020), with highest distinction, with a dissertation on Galen, entitled "Philosophical Perspectives on Galen of Pergamum: Four Case-Studies on Human Nature and the Relation between Body and Soul" (which will be published in Brill's Philosophia Antiqua series). He has also recently published an article in *Phronesis* (2020), "Mixing body and Soul: Galen on the Substance of Soul in QAM and De Propriis Placitis." He currently works as a lecturer at the Philosophy Department at Leiden University.

Chapter 1
Missing a Soul That Endows Bodies with Life: An Introduction

Fabrizio Baldassarri and Andreas Blank

Abstract In the history of ideas, innumerable attempts to explain life and to define living activities have invoked the notion of the soul. Yet this theoretical entity seems to be an unfathomable thing. Difficulties beset the mere definition of it, and controversies span from whether the soul is a material body or an immaterial form, an immortal or a mortal thing, a subject of experiential or of theoretical knowledge, to the question of whether it is the subject of a specific discipline or rather of a *scientia de anima* that, in the Aristotelian tradition, was also regarded as a part of natural philosophy—the field of philosophy that deals with natural particulars. Not only was the soul thought, from a theological angle, to define the unity and unicity of living beings (and in some cases, even their individuality), but it was also seen as the source of several bodily activities such as nutrition and growth in living beings. And since some of the beings that grow and nourish themselves also sense and think, the relation of the origin of vegetative powers to the origin of perceptual and intellectual powers has always been an object of debate. In this sense, the soul has not just been a topic of one specific field, but has rather drawn upon many disciplines, combining the work of philosophers, religious thinkers, physicians, naturalists, chymists, and so on, thereby blurring the boundaries between different fields of knowledge. Botanical treatises of the Renaissance, for example, begin with the definition of the soul of plants and its primary role in bringing forth vegetal activities. Indeed, the presence of the soul was thought to be involved also in the life of animals, human beings, and, in some cases, it was even ascribed to minerals and stones, not to speak of the world-soul, the latent idea that the entire cosmos is animated with a soul. Furthermore, the theory of the materiality of the soul was developed since Antiquity and found new momentum in early modern times, in the reappraisal of Epicurean ideas by thinkers such as Guillaume Lamy, Pierre Gassendi, and Julian Offray de La Mettrie. Yet, with the Cartesian Henricus Regius, the understanding of vegetative,

F. Baldassarri (✉)
Institute for Research in the Humanities, University of Bucharest, Bucharest, Romania
e-mail: fabrizio.baldassarri@gmail.com

A. Blank
Institute of Philosophy, Alpen-Adria University Klagenfurt, Klagenfurt, Austria
e-mail: andreas.blank@aau.at

F. Baldassarri, A. Blank (eds.), *Vegetative Powers*, International Archives of the History of Ideas Archives internationales d'histoire des idées 234,
https://doi.org/10.1007/978-3-030-69709-9_1

sensitive, and rational souls as material importantly surfaces as a peculiar attempt to mechanize the living functions of bodies.

In the history of ideas, innumerable attempts to explain life and to define living activities have invoked the notion of the soul. Yet this theoretical entity seems to be an unfathomable thing. Difficulties beset the mere definition of it, and controversies span from whether the soul is a material body or an immaterial form, an immortal or a mortal thing, a subject of experiential or of theoretical knowledge, to the question of whether it is the subject of a specific discipline or rather of a *scientia de anima* that, in the Aristotelian tradition, was also regarded as a part of natural philosophy—the field of philosophy that deals with natural particulars.[1] Not only was the soul thought, from a theological angle, to define the unity and unicity of living beings (and in some cases, even their individuality), but it was also seen as the source of several bodily activities such as nutrition and growth in living beings. And since some of the beings that grow and nourish themselves also sense and think, the relation of the origin of vegetative powers to the origin of perceptual and intellectual powers has always been an object of debate. In this sense, the soul has not just been a topic of one specific field, but has rather drawn upon many disciplines, combining the work of philosophers, religious thinkers, physicians, naturalists, chymists, and so on, thereby blurring the boundaries between different fields of knowledge. Botanical treatises of the Renaissance, for example, begin with the definition of the soul of plants and its primary role in bringing forth vegetal activities. Indeed, the presence of the soul was thought to be involved also in the life of animals, human beings, and, in some cases, it was even ascribed to minerals and stones, not to speak of the world-soul, the latent idea that the entire cosmos is animated with a soul.[2] Furthermore, the theory of the materiality of the soul was developed since Antiquity and found new momentum in early modern times, in the reappraisal of Epicurean ideas by thinkers such as Guillaume Lamy, Pierre Gassendi, and Julian Offray de La Mettrie.[3] Yet, with the Cartesian Henricus Regius, the understanding of vegetative, sensitive, and rational souls as material importantly surfaces as a peculiar attempt to mechanize the living functions of bodies.

Such an intricate object of study has attracted a huge amount of attention from scholars. Even recently, several volumes have been devoted to the history of the notion of the soul, shedding important light on the historical explorations of this notion as well as on the relationship between historical positions and contemporary philosophical, neuro-biological, neuro-psychological, and plant-biological studies.[4] Still, a specific aspect of this complex notion appears to have attracted much less

[1] See Park and Kessler 2007, Park 2007, Kessler 2007, Serjeantson 2011. On questions related to life see Jonas 1966; Grmek 1990; and Landecker 2017.

[2] See Wilberding 2021.

[3] See Wolfe and Esveld 2014.

[4] For a reconstruction of the recent investigations on the soul in the history of philosophy, see Baldassarri 2017, and Salatowsky 2006, Heinämaa and Reuter 2009, Perler 2009, Corcilius and Perler 2014, Perler 2015.

attention.[5] This issue is the vegetative soul, or what was understood in the Aristotelian tradition to endow living bodies with the basic functions of life: namely, nutrition, growth, and reproduction. With this volume, we intend to provide a series of detailed accounts of theories of the vegetative powers of organic bodies, a set of operations traditionally related to the vegetative soul and its cognate concepts, which generally specify the basic and first acts of life.

Our initial plan was to put together a rather limited series of articles about vegetative powers in seventeenth-century natural philosophy. Soon, however, it became clear that the debates about powers such as nutrition, growth, generation, respiration, and sleep in the classical age took place against the background of a thorough familiarity with the corresponding debates in ancient, medieval, and Renaissance texts. And these texts range across what we would now regard as disciplinary boundaries, from theology and natural philosophy to medicinal theory and "chymistry" (the early experimental investigation of the causal powers of plant-based and mineral drugs). Hence, this volume plans to offer detailed coverage of the historical background of the early modern controversies. As it also turned out, many of the more speculative issues—such as the question of the natural sensibility of plants—remained on the agenda throughout the eighteenth century.[6] This is why we decided to substantially cover the developments preceding the late eighteenth-century emergence of the discipline of biology in the modern sense.

The scope of the volume is thus motivated by a certain perspective on what took place in early modern discussions about vegetative powers. We do not regard the events in the seventeenth century as a process in which something old and obsolete—Platonic, Aristotelian, Galenic conceptions of vegetative life and their many medieval Arabic, medieval Latin and Renaissance interpretations—was simply replaced by something more intelligible—reductionist, corpuscularian, chymical explanations of the basic functions of life. Rather, we would like to emphasize the persistent relevance of these older traditions for modernity—both in the sense that the mechanical philosophers had to show in detail that their explanations of vital functions were more persuasive than the explanations proposed by their predecessors, and in the sense that the explanatory gaps that the mechanical philosophies left especially in the life sciences led thinkers to consult the older traditions as sources of inspiration, even though often in a highly eclectic way. The chapters of this volume reveal the extent to which conceptual change in natural philosophy was the result of a tight web of interconnected controversies, observations, and experiments, involving the interaction between concepts stemming from philosophy with medical and physiological studies. As a result, the diagnosis of problems inherent in previous theories as well as the insight into the conceptual resources of these

---

[5] Mix 2018 gives a general overview of the history of the vegetative soul; Korobili and Lo Presti 2021 focus mainly on development in antiquity. On the Middle Ages, see Paravicini Bagliani 2009; Bakker, de Boer, and Leijenhorst 2012. Pioneering studies on early modern theories of plant generation include Giglioni 1999, 2016; Blank 2009, 2010, 2012. See also Hirai 2011 on the medical side, and Baldassarri 2020 on the relationship between plants and medicine.

[6] Nowadays, this is a subject of contemporary plant biology, see for example Marder 2014; Gagliano et al. 2014; cf. also Pollan 2013.

theories led to ever new, and sometimes eclectic combinations of theoretical assumptions. Rather than developing as a ladder, the history of the vegetative soul develops as a shrub of interconnected issues that give life to original interpretations.

What the present volume thus tries to make visible is the degree in which seventeenth- and eighteenth-century debates about vegetative powers can be seen as contributions to a centuries-old field of controversies that is characterized by a set of recurring questions and an astonishingly creative and subtle variety of answers. The length of the volume should not be taken as a sign of thematic comprehensiveness, though. In fact, there are many positions in the history of the analysis of vegetative powers that are not covered here at all.[7] This includes many ancient authors, such as Homer, Vergil, Cicero, or Pliny, as well as pantheism or paganism, Eastern Orthodox lines of thought, the folk traditions derived from Paracelsian ideas, the medical work of ibn Al Nafis, Andreas Vesalius, Cristóbal Acosta, and even Carolus Linnaeus, and the early modern eclectic or spiritual philosophers. But the volume should, at the very least, give a reliable survey of the questions that were discussed and a yet-to-be-completed map of answers—a map that could help to locate those answers that are not dealt with here in a space of theoretical alternatives.

We believe that even this less than comprehensive approach could fill a substantial lacuna in recent scholarly work. This is so because early modern theories of the higher functions of the soul such as sensation and intellection, as well as theories of the biological reproduction of animals, have recently received detailed attention from commentators.[8] By contrast, early modern theories of the lower functions of the soul such as nutrition, growth, plant generation, and sleep have not yet been studied in the degree of detail that is warranted by the availability of rich and extensive primary sources. Filling this gap will illustrate the ways in which the Aristotelian, Platonic and Galenic traditions shaped the emergent life sciences, including attempts at mechanizing vital powers, which were informed by the problems connected with these traditional approaches.

As well-known, for the largest part of European intellectual history, questions concerning the nature of vegetative powers were framed as questions dealing with the nature of the vegetative soul. A first occurrence of the locution is to be found in Aristotle's *De anima*, where a sketch or outline of the nature of the soul is offered.

[7] These positions include, just to name a few of them, Pythagorean, Epicurean and Stoic lines; Platonism and Neoplatonism and the vegetative soul; Islamic medicine and natural philosophy; theories of nutrition in Catholic natural philosophers such as the Coimbra Commentators and Francisco Suarez; medical conceptions of vegetative activities in Marcello Malpighi, Girolamo Sbaraglia, and Georg Wolfgang Wedel; theories of immaterial substantial forms in Julius Caesar Scaliger and Daniel Sennert's accounts of vital powers; the anatomical studies of Michael Servetus and Realdus Columbus; the Marburg hermetic school, in the work of early modern German physicians Heinrich Petrænd and Johannes Hartmann; the transformations of vegetative powers in natural philosophers like Francis Bacon, Pierre Gassendi, and John Locke; the scholastic Cartesianism of Juan Caramuel; the role of vegetation in the vitalist conceptions of matter developed by thinkers such as Joan Baptista van Helmont and Francis Glisson; and the methodological role of vital spirit in the natural philosophy of Isaac Newton—which is to say that there is enough material at least for yet another hefty volume.

[8] Some prominent publications in this field include Smith 2006; Nachtomy and Smith 2014, Blank 2016, Bucheneau, Lo Presti 2017.

Here the vegetative or nutritive soul is characterized as "an originative power through which they [i.e., plants] increase and decrease in all spatial direction" (Aristotle 1991, 23; on Aristotle's vegetative powers, see **Corcilius** in this volume).[9] In Aristotle, the vegetative soul is the principle of vegetative powers, through which living bodies (starting from plants, following the scale of beings) nurture, grow, reproduce, and, in short, continue to live. While the vegetative or nutritive soul develops as "the most primitive and widely distributed power of the soul, being indeed that one in virtue of which all are said to have life" (Aristotle 1991, 26), 'to vegetate' arises as a synonym of 'to live'. Yet, questions emerge, concerning (a) a definition of the vegetative soul, which for Aristotle was the form of the corporeal body, (b) the role of soul in vegetation, and also (c) the place to locate it.

While Aristotle separated the powers of nutrition and reproduction from the powers of sensation and appetite, and from any cognitive capacity, the relationship between the vegetative powers and higher soul-powers was differently shaped in the thought of Plato and the Platonic tradition. In *Timaeus*, Plato outlined a view according to which plants are endowed with natural sensation and desire.[10] Still, he also suggested that these operations are not genuinely cognitive functions, but rather functions grounded in the cognition of an intelligence outside of the plants. To be precise, while vegetative functions are individuated in plants, the intelligence of other functions is only derivative, flowing from the intelligence ordering the cosmos (see **Carpenter** in this volume).

Later, Galen of Pergamum integrated an Aristotelian account of matter and form with Platonic intuitions concerning the desiderative nature of plant-souls. Furthermore, appropriating Plato's tripartite soul (with some Hippocratic element and Hellenistic medical knowledge) Galen located the vegetative powers in the liver, with plants serving as his model (1) for discussing the vegetative soul as operating in living nature, and (2) for explaining some bodily functions and organs. Ultimately, he developed an analogy between plants and human bodies that had great success in later periods (see **Vinkesteijn** in this volume).

Yet reflections on vegetative powers did not progress linearly. As anticipated, although it is possible to isolate singular threads, the debate over vegetation generally follows a shrub-like diagram, in which different interpretative branches intertwine and variously combine in a fecund, and burgeoning aggregation of fibers-ideas. One example may be found in Avicenna, who incorporated Aristotelian and Galenic aspects into his work, framing the knowledge of vegetative faculties within an Aristotelian structure, but using Galen as the main point of reference for details related to the biological study of animals and plants, ultimately discussing the ways sensitive and intellective functions presuppose, and grant perfection to, vegetative functions (see **Fatigati** in this volume). By contrast, the usage of Aristotle during the middle ages appears to be more unitary. Thomas Aquinas, for example, repeated the claim that the principle of life is the vegetative soul that provides living bodies with conservation, differentiation, and reproduction, therefore remaining an orthodox Aristotelian (see **Pickavé** in this volume).

---

[9] See Polansky 2007.

[10] See Johansen 2020. On Neoplatonism, see Wilberding 2015.

Nevertheless, investigations followed different paths, with some medieval thinkers asking: is nutrition a kind of substantial change that combines the corruption of food with the production of new parts of living beings? In response, Thomas Aquinas and John Buridan discussed 'transmateriation', while Francis of Marchia spoke of 'generation' of parts of the body (see **Klein** in this volume). The relationship between vegetative activities and the higher faculties of humankind developed as a question concerning the autonomy of the vegetative power of the body (a question that will surface later in the seventeenth century), and dealing with the finality of the active principle already at work in the embryo. A further line consisted of the investigation of the naturalistic powers of vegetation strictly speaking. In Pseudo-Aristotle's *De plantis*, the author devotes the first part of the text to questions of the vegetative soul in plants, the fittest place to explore these powers, and their obscure life, which characterizes Albert the Great's investigations of plants, the relationship between plant and matter and the place of plants on the scale of nature (see **Cerrito** in this volume); this topic eventually resurfaces in late periods.[11]

From the middle ages onwards, the relations between material, vegetative, sensitive, and intellective powers raise a variety of *ontological* and theological questions—not only with respect to the existence of vegetative souls. One such question is whether functions such as nutrition involved lower-level substantial forms of organs or substantial forms of their parts—substantial forms that, in order to be analyzed, were understood to be dominated by the substantial form informing the entire plant or animal. The possible alternative to the hypothesis of a plurality of substantial forms in living beings—prominent in the Arabic tradition and its reception in the Latin West[12]—was the hypothesis of a single substantial form from which all vital functions of a living being are brought forth. While dealing with the nature of plants, medieval thinkers explored the faculties of vegetal bodies, mostly following Aristotelian texts, but ultimately concluding that vegetative powers in animals could not be understood independently of sensitive and cognitive powers (see **Panarelli** in this volume).

A further question was what accounts for the species membership of human beings. Regarding the intellective soul as conferring species membership was a fraught issue, since Catholic theology regarded the intellective soul as being created by God, not as being generated by human parents. If so, unlike animals of other species, humans seem incapable of generating offspring that belong to the same species. To put it differently, before the advent of a divinely created soul, their offspring seems to belong to a species other than the human species. This problem led thinkers to entertain the possibility that naturally generated substantial forms, which are the origin of vegetative and sensitive powers, could account for species membership. If this is the case, then there would be a plurality of substantial forms in humans without, however, the form that was traditionally taken to be the highest accounting for species membership—an argumentative strategy developed by

[11] On the reception of pseudo-Aristotle *De plantis*, see Giglioni and Ferrini, 2020.

[12] See Michael 1992.

Masters of Arts such as John de Sècheville, amongst others, and later judged as heretical (see **Bernardini** in the volume). A third ontological question connected with vegetative powers concerns the challenge of substance monism that was derived from some early modern interpretations of Averroes's theory of a unique active intellect. If vegetative powers can be understood as an outcome of the activities of substantial forms, then the analysis of phenomena such as nutrition, growth, and generation could provide arguments for substance pluralism—both a pluralism of substances within the world and a pluralism of substances within organic bodies. In the Renaissance, Nicolaus Taurellus developed this concern against Andrea Cesalpino's view, which he claimed to be incompatible with the Christian doctrine of creation (see **Blank** in this volume).

While dealing with the traditional definition of vegetation as a basic form of life, in the sixteenth and seventeenth centuries scholars began to combine the Aristotelian view of the essential difference between vegetative and cognitive functions with explorations of the causal link between vegetative powers and other bodily activities. Plants slowly emerged as a silent witness to life more generally. In the medical tradition, one finds a wealth of observations concerning how impairments of vegetative functions correlate with impairments of sensitive and intellective functions, thereby raising the question of whether these correlations express only causal interactions between otherwise modular powers or of whether in some respects they share a common nature. In the simplest sense, plants were used as architectural models, as the structure of trees was frequently compared to some parts of animal bodies. Yet, following the Galenic tenet that animal bodies incorporate plant-like structures, where physiological similarities between plant and animal physiology are used as an explanation for the sameness of the desiderative power of plant souls and animal souls, a field of comparative physiology slowly developed. In this field, different branches intertwine. While the Platonic, Galenic, and Aristotelian traditions dominate Giambattista Della Porta's *signatura rerum*, Juan Huarte de San Juan's study of sexual and desiderative capacities of plants, and William Harvey's embryological work respectively, the combination of these influences paved the way for modern botanical studies and modern anatomical studies of living bodies (see **Bigotti** in this volume). Another case concerns the debate about the nature of living beings that seem to be intermediaries between plants and animals, such as zoophytes, sponges, and corals. In the second half of the seventeenth century, these borderline cases reveal the difficulties in defining a 'vital phenomenon', as natural philosophers like Nehemiah Grew, Ralph Cudworth, and John Ray conceived of continuity between physical phenomena such as gravitation and vegetal life (see **Andrault** in this volume).

Moreover, the uses of plants as biological *models* surface in different traditions, and not only in strictly medical disciplines. Indeed, similarities in observable behavior between animals and plants—for instance, the similarity between sleep in animals and the closing of leaves in plants—raise the question of what can be learned about animal physiology from plant physiology, and vice versa. An important example is the idea that influencing the vegetative functions through a well-regulated diet could be beneficial for the emotional and ethical aspects of human life. This idea became influential in late scholastic moral philosophy, when the Jesuit cultivation of a vegetative soul through a sober diet emerges as a philosophical and medical issue (see **Casalini and Madella** in this volume). Along a different (but not entirely opposed) line, Daniel Sennert's chymical study of plants draws analogies between plant generation and grafting. In doing so, it makes plant physiology a common denominator of his medical, alchemical, and theological reflections. This can be seen as a way to develop the concept of a formative force that combines Aristotle with Galen, and to reconcile the assumption that such a force exists with the Paracelsian account of generation and life (see **Moreau** in this volume). In all of these interpretations, vegetative powers grounded definitions of life and provided the basis for comprehension of the higher faculties, sensation, self-motion, and cognition.

The important role that vegetation played in grounding the definition of life can be also detected in René Descartes, who unified vegetative and sensitive activities under his mechanical physiology (as famously represented by the phenomenon of the *Mimosa pudica*).[13] In Descartes one may find two approaches: the first is the metaphysical reconstruction of the vegetative powers as not depending on the soul, a proposition that will never be necessarily true (see **Agostini** in this volume);[14] the second is the physiological and anatomical study of vegetative powers of living bodies (animals and plants) as a mechanical set of activities coherently working in the organism. Using plants as models goes beyond the Cartesian doctrine of the mechanics of life. Still, early Cartesian scholars tried to follow this line with difficulties (see **Baldassarri** in this volume).

While the Cartesian idea of unified mechanical accounts of vegetation and sensation remained influential into the early seventeenth century, this was developed in a variety of ways, unearthing alternative interpretations of plants and vegetation. An important case is one related to vegetative perception, an elusive notion discussed by early modern scholars such as Marsilio Ficino, Cesalpino, Johannes Kepler, and William Harvey, who laid the ground for new ideas of digestion and the irritable self, ultimately ascribing to plants (and thus all living bodies) a sense of natural discernment (see **Giglioni** in this volume). Another case along a similar line of study is Francis Glisson, who investigated perceptions in plants and concluded that vegetative organization required a sort of perception, but his original perspective also contains contradictions and obscurities (see **Schmal** in this volume). A related discussion is of the relation between vegetative powers and instinctive behavior, which Marin Cureau de la Chambre underscored as he blurred the line between living and non-living bodies (see **Kékedi** in this volume).

A further response to the Cartesian, reductionist account of life consisted of the adoption of Neoplatonic ontologies to justify the view that all phenomena of life follow laws that cannot be reduced to the laws of mechanics or the laws of physics. In this case, an investigation of the relation between vegetative powers and the

[13] See Baldassarri 2018, Giglioni 2018. On strange, curious plants in the Renaissance and early modern time, see Brancher 2015.

[14] For the limitation of Descartes' metaphysical understanding of life, see Byers 2006. Cf. Des Chene 2001.

powers of *matter* importantly surfaces. These questions became particularly pressing because vegetative powers were widely regarded as being supported by "instrumental" powers, such as attraction, repulsion, retention, and excretion, which also occur outside of living beings. One particularly pressing issue was to determine what the concurrence of a complex series of such material causes could not explain. Such an explanatory gap is what motivated the postulation of the existence of vegetative souls in the first instance, but this is also one of the most prominent issues in which some of the early moderns began to depart from the ancient and medieval traditions. In fact, one of the hallmarks of the emerging mechanical philosophy of the seventeenth century is reductionist accounts of vegetative powers that regard the workings of plant parts as being nothing other than the workings of their material constituents. Such reductionist programs could be formulated in mechanistic terms—where the idea was that all explanatory concepts had to be fully intelligible according to the standards of Cartesian epistemology—or they could be formulated more broadly in physical terms—where the idea was that some physical concepts can be descriptive of phenomena whose inner nature remains hidden to the observer.[15] Anti-reductionist programs, however, did not always take the form of a defense of the existence of vegetative souls (or, for that matter, a world soul). In the case of Cambridge Platonists such as Thomas More and Cudworth, these scholars revived the functions of the Aristotelian vegetable soul and developed the notions of spirits of nature and plastic nature, which appears to be a combination of Platonic, Aristotelian, and Stoic tenets (see **Hutton** in this volume). Another response consisted in ascribing properties of life—such as perception and activity—to matter itself, a fascinating position that, moving from Margaret Cavendish's interpretation of vegetable life (which however importantly relies on a combination of Galenic and Aristotelian sources), anticipated later theories of vitalism (see **Begley** in this volume).

Debates over the vegetative soul continued well into the eighteenth century. Gottfried Wilhelm Leibniz and Georg Erns Stahl, for example, developed alternative interpretations of this soul, one derived from Leibniz's organic mechanism, and the other from Stahl's theory of a vital principle; and no matter how much these authors thought of their theories as incompatible alternatives, elements of both approaches were linked together by later eighteenth-century physiologists (see **Duchesneau** in this volume). Something similar can be observed in the school of Christian Wolff: while he claimed that the term 'vegetative soul' is an empty one that has no function for a mechanical understanding of life, his prominent disciple Michael Christoph Hanov suggested that life is a product of a vegetative force (see **Favaretti Camposampiero** in this volume). At the turn of the nineteenth century, Marie François Xavier Bichat developed a more complex understanding of life. His theory is illuminating in the sense that he chalked up a distinction between organic life—which is an internal (vegetal) life, common to all living nature—and animal life, which is a feature of animal functions, and gives birth to more complex natures (see **Cheung** in this volume).

[15] On the hidden life of plants in the early modern period, see Des Chene 2000.

From late antiquity to the early modern period, the vegetative soul and its cognate concepts have thus played a substantial role in understandings of life, living functions, and living bodies, sometimes blurring the line between living and non-living nature, and, at other moments, resulting in a strong restriction of life to a mechanical system of operations and powers. What surfaces in this volume is the necessity to combine diverse traditions in order to understand vegetation in philosophical, physiological, medical, and naturalistic contexts. In the history of ideas, the vegetal processes of incessant proliferation, generation, and organic growth help shed light on the notion of life, and result in a variety of diverging analyses of a concept that turned out to be crucial for European culture. Moreover, operating as the roots of life, vegetative powers fuel life throughout the body, ultimately determining diverse activities.

In sum, what we present here is, we hope, a significant, though initial and partial, contribution to the field. One that provides evidence showing that a less-comprehensively explored subject—namely, the vegetative soul, its faculties and powers, and vegetation as the root of life—is a vital topic in the history of European sciences, philosophy, medicine, life sciences, and botany, and that is a prominent issue in the history of knowledge.

References

Aristotle. 1991. *On the soul*. In *The complete works of Aristotle*, ed. J. Barnes. Princeton: Princeton University Press.

Baldassarri, Fabrizio. 2017. Manipulating souls. *Alvearium* 10: 98–107.

———. 2018. Descartes' bio-medical study of plants: Vegetative activities, soul, and power. *Early Science and Medicine* 23: 509–529.

———. 2020. Botany and medicine. In *Encyclopedia of early modern philosophy and the sciences*, ed. Charles Wolfe and Dana Jalobeanu. Cham: Springer.

Blank, Andreas. 2009. Material souls and imagination in late Aristotelian embryology. *Annals of Science* 67: 187–204.

———. 2010. Julius Caesar Scaliger on plant generation and the question of species constancy. *Early Science and Medicine* 15: 266–286.

———. 2012. Julius Caesar Scaliger on plants, species, and the ordained power of God. *Science in Context* 25: 503–523.

———, ed. 2016. *Animals: New essays*. Philosophia Verlag: Munich.

Brancher, Dominique. 2015. *Quand l'esprit vient aux plantes: Botanique sensible et subversion libertine*. Geneva: Droz.

Bakker, Paul J.J.M., Sander W. de Boer, and Cees Leijenhorst, eds. 2012. *Psychology and other disciplines. A case of cross-disciplinary interaction (1250–1750)*. Leiden/Boston: Brill.

Byers, Sarah. 2006. Life as 'self-motion': Descartes and 'the Aristotelians' on the soul as the life of the Body. *Review of Metaphysics* 59: 723–755.

Bucheneau, Stephanie, and Roberto Lo Presti, eds. 2017. *Human and animal cognition in early modern philosophy and medicine*. Pittsburgh: University of Pittsburgh Press.

Corcilius, Klaus, and Dominik Perler, eds. 2014. *Partitioning the soul. Debates from Plato to Leibniz*. Berlin/Boston: De Gruyter.

Des Chene, Dennis. 2000. *Life's form: Late Aristotelian conceptions of the soul*. Ithaca: Cornell University Press.

———. 2001. *Spirits & clocks. Machine and organism in Descartes*. Ithaca: Cornell University Press.
Gagliano, Monica, Michael Renton, Martial Depczynski, and Stefano Mancuso. 2014. Experience teaches plants to learn faster and forget slower in environments where it matters. *Oecologia* 175: 63–72.
Giglioni, Guido. 1999. Girolamo Cardano e Giulio Cesare Scaligero. Il dibattito sul ruolo dell'anima vegetativa. In *Girolamo Cardano. Le opere, le fonti, la vita*, ed. Marialuisa Baldi and Guido Canziani, 313–340. Milano: FrancoAngeli.
———. 2016. Sentient nature and the great paradox of early modern philosophy: How William Harvey and Francis Glisson reinterpreted Aristotelian ΦΥΣΙΣ. In *Adelino Cardoso, Marta Mendonça and Manuel Silvério Marques*, ed. Natureza, causalidade e formas de corporeidade, 9–28. Lisbon: Húmus.
———. 2018. Touch me not: Sense and sensibility in early modern botany. *Early Science and Medicine* 23: 420–443.
Giglioni, Guido, and Maria Fernanda Ferrini, eds. 2020. Περι φυτων. *Greek Botanical Treatises in the West and the East*. Macerata: EUM.
Grmek, Mirko. 1990. *La première revolution biologique. Réflexion sur la physiologie et la médecine du XVIIe siècle*. Paris: Payot.
Heinämaa, Sara, and Martina Reuter, eds. 2009. *Psychology and philosophy. Inquiries into the soul from late scholasticism to contemporary thought*. Boston: Springer.
Hirai, Hiro. 2011. *Medical humanism and natural philosophy: Renaissance debates on matter, life and the soul*. Leiden/Boston: Brill.
Johansen, Thomas K. 2020. Soul, life, and nutrition in the *Timaeus*. In *Heat, pneuma, and soul in ancient philosophy and science*, ed. Hynek Bartoš and Colin G. King, 121–139. Cambridge: Cambridge University Press.
Jonas, Hans. 1966. *The phenomenon of life: Toward a philosophical biology*. New York: Harper & Row.
Kessler, Eckhard. 2007. The intellective soul. In *The Cambridge history of renaissance philosophy*, ed. Charles B. Schmitt and Quentin Skinner, 485–534. Cambridge: Cambridge University Press.
Korobili, Giouli, and Roberto Lo Presti. 2021. *Nutrition and nutritive soul in Aristotle and the Aristotelianism*. Berlin/Boston: De Gruyter.
Landecker, Hannah. 2017. Metabolism, autonomy, and individuality. In *Biological individuality: Integrating scientific, philosophical, and historical perspectives*, ed. Scott Lidgard and Lynn K. Nyhart, 225–248. Chicago: The University of Chicago Press.
Marder, Michael. 2014. *The Philosopher's plant: An intellectual herbarium*. New York: Columbia University Press.
Michael, Emily. 1992. Averroes and the plurality of forms. *Franciscan Studies* 52: 155–182.
Nachtomy, Ohad, and Justin E.H. Smith, eds. 2014. *The life sciences in early modern philosophy*. Oxford: Oxford University Press.
Paravicini Bagliani, Agostino, ed. 2009. *Le monde végétal. Médecine, botanique, symbolique*. Florence: Sismel-Ed. del Galluzzo.
Park, Katharine, and Eckhard Kessler. 2007. The concept of psychology. In *The Cambridge history of Renaissance philosophy*, ed. Charles B. Schmitt and Quentin Skinner, 455–463. Cambridge: Cambridge University Press.
Park, Katharine. 2007. The organic soul. In *The Cambridge history of Renaissance philosophy*, ed. Charles B. Schmitt and Quentin Skinner, 464–484. Cambridge: Cambridge University Press.
Perler, Dominik, ed. 2009. *Transformations of the soul. Aristotelian psychology 1250–1650*. Leiden/Boston: Brill.
———, ed. 2015. *The faculties: A History*. Oxford: Oxford University Press.
Polansky, Roland. 2007. *Aristotle's De Anima*. Cambridge: Cambridge University Press.
Pollan, Michael. 2013. The intelligent plant: Scientists debate a new way of understanding flora. *The New Yorker*, 23/30 December, 92–105.

Salatowsky, Sacha. 2006. *De anima: Die Rezeption der aristotelischen Psychologie im 16. und 17. Jahrhundert*. Amsterdam/Philadelphia: Grüner.
Serjeantson, Richard. 2011. The soul. In *The Oxford handbook of philosophy in early modern Europe*, ed. Desmond M. Clarke and Catherine Wilson, 119–137. Oxford: Oxford University Press.
Smith, Justin E.H., ed. 2006. *The question of animal generation in early modern philosophy*. Cambridge: Cambridge University Press.
Wilberding, James. 2015. Neoplatonists on the causes of vegetative life. In *Causation and creation in late antiquity*, ed. Anna Marmodoro and Brian D. Prince, 171–185. Cambridge: Cambridge University Press.
———, ed. 2021. *The World-Soul: A History*. Oxford: Oxford University Press.
Wolfe, Charles, and Michela van Esveld. 2014. The material soul: Strategies for naturalising the soul in an early modern Epicurean context. In *Conjunctions of mind, soul and body from Plato to the enlightenment*, ed. Danijela Kambaskovic, 371–421. Dordrecht: Springer.

Chapter 2
Soul, Parts of the Soul, and the Definition of the Vegetative Capacity in Aristotle's *De anima*

Klaus Corcilius

Abstract The aim of this chapter is to explain Aristotle's definition of the vegetative part of the soul in the *De anima* from a methodological point of view. I discuss Aristotle's conception of the soul and his conception of "parts of the soul" before I turn to his definition of the vegetative part of the soul in *De anima* II 4. I argue that the definition of the vegetative capacity is deliberately abstract so as to cover its various activities under one common heading. I also argue that the *De anima* deliberately avoids a positive and determinate description of the activity of the soul that is properly responsible for the vegetative functions, and that Aristotle offers such a positive and determinate description in his works dedicated to the actions and affections "common to body and soul".

Nutrition, growth, and reproduction – in one word: vegetation – are basic life-activities. For Aristotle, these activities are necessary and sufficient for the possession of life.[1] He moreover attributes them – all of them – to one and the same "part" of the soul. In what follows, I will refer to this as the *vegetative part of the soul* (*to threptikon*, more literally "the part capable of nourishing").

My aim in this chapter is to explain what these statements mean. I will first briefly expound what "soul" is for Aristotle (i), what a part of the soul is for him, and why it makes sense for him to adopt the (originally Platonic) language of "parts of the soul" in the first place (ii). I will then consider how Aristotle can regard the

[1] See De an. II 4, 416b17-25, see also De an. II 1, 412a14-15 (without mention of reproduction). See Johansen 2012 for a convincing explanation as to why Aristotle, unlike Plato, thought it necessary to postulate a soul as responsible for the vegetative activities of living things (116–127). On Plato's *Timaeus*, see Carpenter, Chap. 3 in this volume.

K. Corcilius (✉)
University of Tübingen, Tübingen, Germany
e-mail: klaus.corcilius@uni-tuebingen.de

F. Baldassarri, A. Blank (eds.), *Vegetative Powers*, International Archives of the History of Ideas Archives internationales d'histoire des idées 234,
https://doi.org/10.1007/978-3-030-69709-9_2

vegetative part of the soul as one single part, despite the various activities he attributes to it (iii). I will discuss the issues in turn and end with an interpretation of the definition of the vegetative capacity in *De anima* II 4 that situates this definition within the general framework of Aristotle's science of living things.

2.1 What Is the Soul for Aristotle?

I begin with some well-known features of Aristotle's conception of the soul, as he presents it in the famous chapters 1–3 of the second book of his *De anima*. There, he says that the soul is the *formal essence of the living body*. One way to understand this is to conceive of the soul as the principle of the *being* of living things. As a principle of being, the possession of soul makes it that each living thing has the being that it has, i.e. it makes each living thing being what it is, namely a living thing of a certain sort. As Aristotle sometimes (rather infelicitously) says, the living being is "composed" of its matter (*hulê*) – its body – and its essential form (*morphê*) – its soul. As a result, it is possible to address the living thing as the *hylomorphic compound*, as Aristotle does sometimes (*ex amphoin*, *synholon*, *synamphoteron*). But we should take the meaning of "composition" with a grain of salt here. Aristotle does not want to say or imply that the soul and the body are composite parts of the living being. He does not think that the soul and the body each have an independent existence on their own or that they can be separated from one another as composite parts of the "hylomorphic compound". Aristotle does not even think that they are *parts* of the living thing. The reason for this is the fact that the soul is supposed to be the essence of the living thing. The essence of *X* is that which makes it that *X* is the kind of thing it is, it fixes *X*'s identity. A part, on the other hand, for Aristotle, is typically a component of a larger whole. And while we can conceive of a whole as loosing one of its parts without ceasing to be what it is,[2] we cannot conceive of anything loosing its essence without thereby ceasing to be what it is. The essence bestows the identity on whatever thing it is the essence of. It thus makes both the thing and the parts of that thing being what they are. To speak of a thing's identity as a part of that thing is for that reason misleading. And the same goes for the soul with relation to its body. Soul and body are inseparable from each other and there is *no* way for Aristotle in which soul and body can be separated from each other, save hylomorphic analysis. The philosopher can of course *distinguish,* in her thought, soul and body as different ontological aspects of what in reality and nature is just one thing. In this analytical sense, and only in this sense, the essential form is separable from its matter, namely "in account" (*chôriston logôi*, *Met.* H 1, 1042a29).[3] And it seems that this is exactly what happens in Aristotle's own definition of the soul in the *De anima* II 1- III 8 as well, where he conceptually *isolates* the soul as

[2]At least not for Aristotle's way of thinking about parts. See *Met.* Δ 25, 1023b12-25.

[3]Cf. *Phys.* II 2, 194b12-15.

the essence of living things from their bodily aspects. However, he leaves no doubt about the analytic character of that separation of the soul from the body in account when he says that the matter of the soul is the *living* body (*DA* II 1, 412b25-26), i.e. the ensouled body and not the soulless dead body. On that conception of the relation between body and soul, to inquire whether soul and body are one is not necessary, for *obviously* they are:

> That is also why there is no need to inquire whether the soul and the body are one, just as there is also no need to ask whether the wax and its shape are one, nor more generally whether the matter of each thing and that to which the matter belongs are one. For, while one and being are spoken of in several ways, what it is spoken of in the primary way is the actuality. (*DA* II 1, 412b6-9, trans. Miller, modified)

Body and soul are one like wax and shape of a candle. Indeed, the soul, as the actuality of the body, *is* its very being and unity. However, although they are merely metaphysical aspects of one and the same living thing, soul and living body relate to one another also in further interesting and non-trivial ways. As a biologist Aristotle is deeply interested in understanding what living things are and how and why they do what they do insofar as they are alive. And his conception of the body soul –relation will allow him to get a handle on these questions. Most importantly, he conceives of the relation between soul and body as one of *teleological subordination*: the living body, he says, is the tool or instrument (*organon*) of its soul (412a29-b6) comparable to the way an axe is a tool for cutting (b11-17).[4] That means that the body is best understood as executing the life-functions of the living thing as they are defined by its soul; the job of the living body's activities is to bring about the physical reality and actuality of the soul as its goal and final cause (De an. II 4, 416b10-12, 15-21).

Having said that the soul is the essence of the living body, and that the *De anima* is dedicated to defining the soul, I now wish to mention another important aspect of Aristotle's conception of the soul. This is the fact that souls are essences of natural things. Souls are *natures* in the two-place meaning of the word as the "natures of living things". Living things – animals, plants, and human beings – are natural things and as such they are subject to change: they actively do things in the world and they passively undergo processes and affections. And like all natural things, they do this according to certain patterns, which are internal to them and determine in which ways they are susceptible to active and passive change. These patterns of change (of "motion and rest") are their *natures*. Natures are the essences of natural things. Hence, since living things are natural things and souls are their essences, souls are natures.

To illustrate some of the implications of the fact that souls are natures, let us start with a simple example. Take a stone. A stone is a natural thing for Aristotle. Its nature is earthen, i.e. stones have an internal principle of motion and rest that makes

[4]The translation of *organikon* with "organic" strongly suggests an anachronistic concept, namely the concept of an *organism*, which is the concept of different parts of an organism being both means for the other parts of the organism and also ends of the other parts. This idea of a reciprocal teleological relation is entirely absent in Aristotle. See Menn 2002, pp. 108–114.

it that they, like earth, move downwards whenever they are not prevented from doing so by an external hindrance. Now the *nature* of the stone is precisely this tendency to fall. It determines the stone's "behavior" in a primary way, i.e. in such a way that there is no further reason why stones fall downwards if not prevented from doing so. Imagine someone who asks, "why does the stone fall?" One perfectly acceptable answer would be "because somebody threw it up, so that it fell down." However, if the questioner should then again ask "but why did the stone fall down after it moved in the upward direction?" we would suspect either that this person is trying to make a point or that she is insufficiently acquainted with stones. Stones fall downwards because they are *stones* (earthen). To be a stone is to be a certain kind of body that has the capacity and the tendency to fall in a primary way, solely by virtue of being what it is. So there is no further question as to why the stone falls. Falling is just what stones do in such circumstances because that is their nature. In Aristotle's natural philosophy natures are such primary patterns of change and rest. They are the *capacities* of natural things to manifest these primary patterns of change.

In the case of living things, these patterns of change and rest are of course incomparably more complex than those of the stones'; but they still are exactly such patterns. Souls are the capacities for motions and states of rest according to the patterns that are natural for living things. Now, unlike those of the stone, the motions and states of rest that souls enable their possessors to engage in are the ones that, taken together, make up a living thing's *life*. The soul, then, is the essential form in the sense of the nature of the living body. It enables the body to engage in its life-activities and is inseparable from it in every way except in hylomorphic analysis.

But if soul and body are inseparable in this way, why does Aristotle invest so much work in the *De anima* to define the soul and to thus conceptually isolate it from the living body? Is not it better to simply speak of living things as what he anyway thinks they are, namely soul-body compounds?

Although the conceptual isolation of the soul may appear to be an unnecessary complication from the standpoint of Aristotle's ontology of living things, it is fully justified from the standpoint of his general theory of science. Aristotle is an *explanatory essentialist*. This means that he maintains not only that things in the world have essences that allow us to understand what they most fundamentally are, but also that we can render scientific explanations of other, non-essential features of things by appeal to their essences. Indeed, he maintains that essences, if defined properly, can explain *all* non-essential necessary and universal features that hold of things, insofar as they are bearers of their essences. This holds also for living things. Aristotle thinks that the soul, if properly defined, can explain all the necessary and universal attributes that living things have insofar as they are alive. For this reason, he is eager to conceptually isolate the soul from the body in the *De anima*. Because he tries to define the soul as the explanatory essence of the necessary and universal but non-essential features that living things have insofar as they are alive. The "insofar" is very important in this context. Aristotle uses the so-called "*qua*-operator" (*qua, hêi,* "insofar as") to focus on those features of living things that they have *as such* while excluding features that they have insofar as they are something else, like, e.g., insofar as they are physical bodies or insofar as they are identical with themselves. Although such non-essential features hold of living things universally and necessarily as well, they are not what the science of living things is concerned with because they do not hold of them *insofar* as they are living things (but insofar as they are physical bodies in the former case or insofar as they are anything at all in the latter). What are the *per se* features of living things? Aristotle does not tell us much about that on the level of living things generally in the *De anima*. But in the case of *animals*, he offers a detailed list of such *per se* accidents, as he sometimes calls them. He says that the differences (*diaphorai*, which I here take to be the *per se* features) of animals can all be grouped under the following four kinds: their functional body parts, their actions and affections, their character traits, and their ways of life (*HA* I 1, 487a11ff.). These four kinds of features constitute the scientific facts (the "that", *hoti*, or the *phainomena*) of the science of animals. Aristotle's overall zoological program consists in first collecting these facts or phenomena about animals, i.e. the universal and necessary attributes that animals have insofar as they are animals – and these, if I am correct, consist in nothing other than the differences by which animals can be distinguished from each other insofar as they are animals – to then render scientific accounts of the facts by reference to the essences of animals (the essences being the first principles for the "on account of which" the facts hold, the *dihoti*). For Aristotle, to define the soul by determining its "what-it-is" as the essence of living things, therefore, is to define the first principle of an explanatory science of the phenomena (or facts or per se accidents) of living things. And this is what he says explicitly early on in the *De anima*:

> It seems that not only is having discerned the what-it-is is useful for contemplating the causes of the accidents of the substances [= their non-essential, but universal and necessary *per se* features, KC], as for example, in mathematics ascertaining what straight and curved are, or what line and plane are, is useful for seeing how many right angles the angles of a triangle equal, but also, conversely, that ascertaining the accidents plays a great part in knowing the what-it-is. For when we can render an account of all or most of the accidents according to their appearance, we will also then be able to speak best about the substance. For the starting point of every demonstration is the what-it-is, so that those definitions which do not lead us to discern the accidents, or at least to conjecture about them, will clearly and in every case be dialectical and vacuous. (*DA* I 1, 402^{b}16-403^{a}2, transl. Shields, modified)

In the *De anima*, Aristotle defines the soul as the first explanatory principle of the science of living things, i.e. of his biology. He does so in order to explain the phenomena or facts or, as he says here, the (*per se*) "accidents" of living things. This is why he conceptually isolates the soul from the body: because he wants to *explain* the bodily and other non-essential *per se* features of living things with reference to the soul as their first principle and cause (cf. De an. II 4, 415b7-28).

Another commitment of Aristotle's, no less important than his commitment to explanatory essentialism, is to the strictest standards of *methodological economy*. Aristotle is eager to explain each phenomenon (fact or non-essential *per se* feature) at the level of its largest possible extension. Generally, he thinks – and repeats tirelessly in his works – that in scientific explanation, one ought to start with the most general features first and only thereafter add explanations of the more specific features. As a rule, Aristotle holds that one ought to give scientific explanations at what he calls the "commensurately universal" level (*prôton katholou*),[5] meaning that one should give explanations which are as general as possible and as specific as necessary so as to account for of each phenomenon (fact, non-essential *per se* feature) at its largest possible extension. Aristotle adopts this procedure to minimize explanatory work and to avoid repetitive explanations,[6] but also to identify the "right place" for the explanation of every phenomenon or fact in each science. That means each phenomenon should be explained only once in each science. Aristotle insists that only what is known in this commensurably universal manner is scientifically, and therefore genuinely, known.[7]

Taken together, Aristotle's commitments to explanatory essentialism and methodological economy result in a rather demanding conception of science generally and of the definition of the essence of a scientific domain in particular. The definition should contain only the most basic and at the same time most explanatorily powerful propositions from which the scientific explanations of all the non-essential universal and necessary *per se* – features (the facts or phenomena) can be derived. For Aristotle's science of living things this means that the definition of the soul should offer an account at the highest possible level of biological abstraction; it should be capable of grounding the scientific explanations of all universal and necessary facts about living things in nature as a first principle and as the generic essence of all living things. The definition of the soul should contain only the most fundamental and explanatorily powerful propositions. These are the definitions of the most basic life-capacities and not more, while all non-essential universal and necessary *per se* features of living things, including their functional bodily features, are to be scientifically explained by reference to these definitions. This is a crucial, even if often neglected, fact about the scope of Aristotle's project in *De anima*. The treatise is devoted to the definition of the first and most abstract explanatory principle of his science of living beings, and not specifically to the *explanation* of any of the phenomena of living things. The explanation of these phenomena is the task of other treatises, namely the so-called aitiological treatises: the *Parts of animals*, the *Generation of animals*, the *Motion of animals*, the *Progression of animals*, and

[5] Literally, the "first universal", see *Anal. Post.* I 4, 73b25-74a3; a32-b3.

[6] *PA* I 639a15-b5, 644a25-b15, cf. *Phys.* I 1, 189b31-32, *DA* I 1, 402b8-10.

[7] His example for illustrating the importance of commensurately universal explanation in *Posterior Analytics* I 4 is the demonstration of the proposition that triangles have a sum of angles equal two right angles (2R). To know 2R scientifically is to know it in a commensurably universal way, which is to know it as a proposition about triangles *simpliciter*. To the contrary, if 2R were demonstrated on a more general level, say by appealing to the property of figures, then the explanation would've been either redundant (each explanation should only occur once!) or misplaced, since it ought to have been stated as a property of figures. And if, on the other hand, 2R were demonstrated on the level of more specific kinds of triangles, say equilateral triangles, that demonstration, even if it were correct, would only show that 2R holds *qua* equilateral triangles, but not *qua* triangles simpliciter. In fact, it might not even hold for triangles generally, if the demonstration in fact relies on the triangles involved to be equilateral. These contrary examples are unscientific.

the so-called *Parva naturalia*, a collection of short treatises devoted to explain phenomena "common to body and soul", i.e. psychophysical episodes such as perceiving, the corresponding objects of perception, memory, recollection, sleep, dreaming and so on. The *De anima* operates on a level of zoological abstraction higher than the aiotiological treatises, and the aiotological treatises presuppose the framework and definitions the *De anima* provides. We may therefore think of the *De anima* not so much as a part of Aristotle's biology but as his general metaphysics of living things. It investigates and defines what it is to be alive for all living things on an explanatory primitive, and therefore highest, level of biological abstraction.

2.2 What Is a Part of the Soul?

As we've seen, Aristotle conceives of the soul as the fundamental and generic principle in virtue of which all living things are said to possess life and on account of which they have their other, non-essential universal and necessary attributes insofar as they are alive. But "life" is said in many ways, says Aristotle (De an. II 2, 413a20-25): vegetation, perception, intelligent thinking and self-locomotion, though very different from each other, are each sufficient for the attribution of life. Whenever we encounter one of these activities we say that the thing that possesses it has life. And since the soul is the generic principle of life in living things, each of these ways in which "life" is said is going to be part of that principle. And as the soul is furthermore the basic *nature* of living things, and natures are *capacities* to either actively engage in or passively undergo actions and affections, the soul as a whole is going to be the *set* of basic life-capacities by virtue of which living things can engage in their corresponding life-activities. A part of the soul will then be such a basic life-capacity (413b10-13). According to Aristotle's conception, then, a part of the soul is a capacity of living things that determines most basically what living things actively do or passively undergo insofar as they are alive. The notion of parts of the soul plays a fundamental role in Aristotle's science of living things. Their definitions are distinct, non-overlapping accounts of life-capacities that are "separate in account" from each other while jointly covering what it most basically means to be alive for all living things.[8] Despite his at times misleading language ("capacity *of the soul*"), Aristotle does not think that the soul is a bearer of capacities. Such a conception of the soul would be humuncular. The soul, as we have seen, does not *have* capacities; rather, the soul, as the nature and essence of living things, *is* the set of capacities that makes their bearer, the living body, alive. There is no common subject and bearer of these capacities over and beyond the living body. This is why Aristotle says that the best way of defining the soul is to define each of the parts of the soul in turn (De an. II 3, 415a12-13). And defining the parts of the soul in turn

[8] For an extended argument that parts of the soul are parts of the common and generic essence of living things and thus the first principle of the science of living things, see Corcilius and Gregoric 2010, and Corcilius 2017, IX ff.

will also be the basic program for the rest of the treatise after the first commonest account in *De anima* II 1-3. Aristotle discusses each of the parts of the soul in the following sequence: the vegetative part (II 4), the perceptual part (II 5- III 2), the intellectual part (III 4-8), and the capacity of locomotion (III 9-11). Each of the first three discussions ends with a definition of the respective part: the vegetative, the perceptual, and the thinking part. The discussion of locomotion deviates from the discussion of the first three parts. As it turns out the account of that capacity is not independent from those of other capacities, as it includes reference to perception and, in the case of human beings, also to the achievements of the thinking part of the soul. The capacity of locomotion, therefore, while being a capacity of the living body, is not a part of the soul for Aristotle, because it is neither explanatorily basic nor definitionally independent from other basic life-capacities.

The discussions of the three canonical capacities of the soul, the vegetative, the perceptual and the thinking part of the soul, by contrast, do show that they are both explanatorily basic and independent (or separate) in definition from each other. It seems that this is the reason why Aristotle calls them "parts" of the soul in the first place: they are separate in definition (and thus also in being), which is to say that their definitions do not overlap, while they jointly exhaust the definition of the basic life-capacities that together constitute the first principle of the science of living things. If that is correct, then Aristotle's locution of "parts of the soul" is chiefly motivated by conceptual questions that regard the soul as the first principle of the science of living things, and not (or at least not primarily) by the *operational* question that seems to have motivated Plato in introducing parts of the soul in *Republic*. The question that drove Plato to introduce that distinction there was whether the activities corresponding to the different parts of the soul can co-occur, i.e. simultaneously operate, in the same subject or not (Rep. IV 434d–441c).[9] Aristotle's conception of parts of the soul, by contrast, does not reflect operational independence.[10]

However, this picture of parts of the soul as conceptually separate parts of the definition of the soul raises a pressing question. If the parts of the soul are conceptually separate, how can they form a unitary soul? Aristotle cannot seriously hold that living things with a plurality of soul-parts such as animals and humans have multiple souls, because the parts of the soul are the parts of the definition of the natures and essences of the things whose soul-parts they are. As natures and essences, they are the very *principles* of the unity of these living things. This strongly suggests that they should fulfill more demanding criteria of unity than mere aggregation (see De an. I 5, 409b9-11, cf. II 1, 412b5-11 and I 5, 410b10-13).

Aristotle addresses the problem in De an. II 3, even if in a highly abstract way. He says that the parts of the soul of an individual living being form a unitary soul

[9] Aristotle does not think of the parts of the soul as operationally independent powers like Plato does, but first and foremost as parts of the definition of the soul in the sense specified above. Cf. Aristotle's criticisms of the Platonic way of dividing the soul in De an. III 10, 432a22–b7, in which he argues against the very *criteria* by which Plato divides the soul.

[10] On the operational aspects of the parts of the soul in Aristotle, see below p. ***

analogous to the way in which complex geometrical figures "contain" simpler figures. A square, for example, is complex in that it is analyzable into triangles. It "contains" triangles not as actual parts – a square is a square and not an aggregate of triangles – but it contains them in potentiality, namely insofar as it can be analyzed into triangles. Aristotle thinks of the unity of complex souls in a serial and top down way: the being of each complex soul is determined by its most specific and sophisticated, and in that sense "highest", part. Each of these will "contain" the lower parts in the sense that they will be implied in it, albeit not as actual and functionally independent parts. The vegetative part of, say, a dog's soul will be included in the dog's soul but it will be included in it as the vegetative functions of the *dog*, while a human being's vegetative and perceptual parts will be contained in the specifically human soul as a *human's* vegetative and perceptual functions. What that means is that the vegetative parts of a plant, of a brute animal, and of a human being will each differ from each other *in essence*, because in each of these cases what it is to be vegetation and what it is to be perception receive their determinate character from the highest and most sophisticated part. This determination of the lower capacities by the higher ones modifies their very being. A dog's vegetative part will be different in essence from a human's vegetative part because dogs are different from humans. Lower and higher parts of the soul thus do not relate as actual and proper physical parts (each of which retain their identity while being part of a composite whole) but as determinable and determinant, where the being and essence of the former receives its specific character by the latter. A human being's perceptual part is essentially a rational being's perceptual part and is therefore going to be different in essence from a dog's perceptual part, even though *qua* perception there is going to be one common definitional account for both.

This conception of parts of the soul as relating as determinable and determinant is but an application of Aristotle's general views about the parts of the definitional account (*merê tou logou*). According to these views the items that jointly constitute the essence of a given thing along the steps of a definitional tree generally relate in such a way that the last differentia at the very bottom of the tree is the most important one:

> If then a differentia of a differentia be taken at each step, one differentia—the last—will be the form and the substance; (…) Therefore it is plain that the definition is the formula which contains the differentiae, or, according to the right method, the last of these. (Met. VII 12, 1038a25-30)

And the generic distinctions further up the definitional tree will be modified in accordance with the last and most specific determination:

> For I give the name of 'difference in the genus' to an otherness which makes the genus itself other. (Met. X 8, 1057b38-1058a4)[11]

By receiving their determination by their specific differences, the generic features in each case acquire essences that are "other", even though *as generic features* they are

[11] In the Greek: λέγω γὰρ γένους διαφορὰν ἑτερότητα ἣ ἕτερον ποιεῖ τοῦτο αὐτό.

the same. And this kind of determinant/determinable relation is supposed to hold also in the case of the ascending series of the parts of the soul: vegetation, perception, and the thinking part. The last and most specific and sophisticated, and in this sense "highest", part of each living thing determines the very being of the previous and more generic ones in the series.[12]

With this Aristotle is able to offer a good answer to the above question: the soul of each kind of living being, irrespectively of whether it is a complex soul or not, is but one single soul because the essence of the lower and more general parts in the serial order of souls receive their determination from most specific and in that sense "highest" part. The dog's vegetative part will essentially be a dog's vegetative part, and a human being's vegetative and perceptual parts will essentially be a human being's vegetation and perception, with the result that all the parts of any given complex soul will constitute one single nature and essence, despite the fact that each of the definitional accounts of the parts as such are separate from each other. In this way Aristotle can reconcile the ontological unity of complex souls with their conceptual complexity: the natural unity of the lower parts in the serial order of soul-parts with the higher ones trumps, as it were, their definitional separability. As a result, complex souls on each level are homogeneous and without internal differentiation (cf. De an. II 2, 413b16-24). So in regard of the *ontology* of complex souls, we may say that Aristotle holds that each soul is a partless whole, while he at the same time maintains as well that the definition of each soul *conceptually* divides into separate parts. This, at any rate, is what he says about the relation between the rational and the non-rational parts of the human soul in the *Nicomachean Ethics*, and it seems to me that this is what he thinks about parts of the soul generally:

> It does not matter for the moment whether these (parts) are separate like the parts of the body or anything else that can be physically divided, or whether they are two in account while inseparable in nature, like the convex and concave in a curved figure. (*NE* I 13, 1102a28-32, trans. Crisp, modified.)

As convincing as this solution to the problem of the unity of the parts of the soul may appear to those interested in conceptual analysis, it seems exceedingly abstract to satisfy the student of living things. For all it offers is the assurance that there really is no problem with a plurality of soul-parts on the grounds that they, while separate in definition ("two in account"), form a natural unity. But on what grounds? What is it that makes the parts or the soul a natural unity? Aristotle's answer to this question is, again, very much the answer of a biologist. It is his thesis of the *teleological subordination* of the lower parts of complex souls under their highest parts. While the lower and more generally shared capacities of the soul are *necessary conditions* for the existence of the higher parts in the serial order, the latter are the former's natural *goals*. The vegetative capacity of a horse, for instance, exists for the sake of its perceptual part, whereas both the vegetative and the perceptual part of a human being exist for the sake of his or her rational part (cf. De an. II 3, 414a29

[12] Or, to put in differently, it is not the case for Aristotle that parts of the definition are predicated of the other parts as "of something else." (Anal. Post. II 3, 90b34-38).

sqq., De gen. an. II 3, 736b1-8). In this way, the parts of a complex soul in each living thing form a natural unity: they are geared towards a common goal, namely the existence and exercise of the highest capacity in the serial order. The parts of the soul of complex souls, though separate in definition, are *operationally fused*, i.e. they each are dedicated to the existence and well-functioning of their highest part. The parts of the soul, therefore, are not operationally independent modules; they are all of one seamless piece.

2.3 The Vegetative Part of the Soul

Vegetation is the part of the soul at the most general and "lowest" level in the serial order of soul-parts. It is responsible for the basic life activities in all living things. However, given what we've just said, what vegetation is will differ in each kind of living thing because it makes a difference to its very being whether its operations are put to the service of this or that specific soul. But here we are not concerned with what vegetation is *qua* some higher capacity of the soul, and only with what vegetation is *qua* vegetation itself. We are not asking what vegetation is *qua* a dog's or a human's vegetation – but what vegetation is *qua* such and in isolation from any superordinate teleological context. We are seeking the most abstract and purest definition of the vegetative part of the soul.

Aristotle expounds his methodology for defining parts of the soul in De an. II 4. He says that parts of the soul are capacities which as such have to be defined with reference to their manifestations. But to be able to do this, one has to take it into account that capacities of the soul are capacities of a special kind:

> But if one must state what each of them is – for example, the capacities of thought, perception, and nutrition – one must state before still that what it is to think or perceive; for the actualities and the actions are prior in account to the potentialities. And, if this is so, and if their corresponding objects should have been studied beforehand, one would first have to make a determination about those objects and for the same reason, for example, concerning nourishment (*trophê*) and the object of perception (*aisthêton*) and the object of thought (*noêton*). (De an. II 4, 415a16-22, trans. Miller, modified)

Like all capacities, the capacities of the soul have corresponding manifestations. These are the activities of the soul (*hai energeiai kai hai praxeis*). They are prior in account (*logôi*) to the corresponding capacities, and therefore must be determined before the capacities are determined. However, it turns out that the activities of the soul have corresponding objects. Activities of the soul are essentially object-related. Hence, understanding the capacities of the soul, and thus to be able to define them, requires understanding not only the activities and actions that are their manifestations – such as in the case of the definition of brittleness, which involves only the understanding of its corresponding manifestation "breaking" – but it also involves the understanding of the objects that are said to correspond to their manifestations.

These are the objects of the activities of the soul.[13] They are to be determined prior to the definition of the activities and actions "for the same reason", i.e. because they are prior in account to their corresponding activities. To define the capacity of perception, for example, we need to know beforehand what perceiving is, and before we can know that we need to know what the objects of perception are. Perceiving is essentially an object-related activity, and that activity cannot adequately be grasped without previous reference to its corresponding objects. I will refer to this definitional procedure as the *retrograde method*. In the above text Aristotle justifies his method by appeal to the relation of priority in account. Priority in account is priority in definition, which seems to correspond to formal priority. So, by the transitivity of the priority relation, the definition of the object of the manifestation of a capacity of the soul has formal priority over both, the definition of the corresponding manifestation and the definition of its corresponding capacity.

The objects of the basic activities of the soul, the object of vegetation, the object of perception, and the object of intellection thus all have some kind of formal priority over the parts of the soul.[14] However, that formal priority of the objects should not mislead us into thinking that the retrograde method gives us an easy and straightforward way of defining the parts of the soul. For, as we will see presently, the determination of the external object of a given activity of the soul does not render a full determination of the formal object of the corresponding capacity. To see what I have in mind, consider, for instance, the capacity of perception. The capacity of perception no doubt is *of* perceptual objects. One might conclude, on that basis alone, that the definition of the perceptual part of the soul would have to be something like "the capacity to receive and take on perceptual objects." But that would be wrong. Aristotle, as is well known, defines the perceptual part of the soul in a way that is crucially different. He says:

> the capacity to receive perceptible forms without the matter, just as the wax receives the seal of a signet-ring without the iron or gold, and it gets the golden or brazen seal but not in so far as it is gold or bronze. Likewise also the sense (perception) is affected by each thing which possesses colour or taste or sound, but not in so far as each of these things is so called [i.e. a thing that possesses colour and taste and sound] but in so far as it is this sort of thing and according to the proportion. (De an. II 12, 424a18-24, trans. Miller, modified)

So the idea behind the application of the retrograde method in the *De anima* cannot be the simple one according to which the determination of the external object of the activity of perception as "perceptible object" is *sufficient* to specify the full definitional (formal) object of perception; rather, as the passage just quoted shows, the definitional object of the capacity of perception is the perceptual *form* of the external object without its matter and according to the proportion (of, as turns out later, perceptible values on qualitative scales). Aristotle even says that the external objects of modally specific perception are fully actual only while they are being perceived

[13] Since brittleness constricts the range of objects to those that may bring about the breaking, one could think of that constricted range as a kind of correlated object of breaking as well.

[14] Johansen even offers a defense of the thesis that the correlate object of the capacities of the soul is their formal cause; see Johansen 2012, 98 ff.

(De an. III 2, 426a10ff.). So something very important must occur in-between the affection of the perceiver by the external perceptible objects and the matterless presence of the fully-fledged formal objects of the perceptual capacity in the perceiving organism. And whatever this important something is, it is no part of the external object. It rather is something that is done by the perceiving animal, namely the intake of the matterless form of the external object according to qualitative proportions by the perceiving organism. This, and the resulting presence of the matterless form in the organism is the function of the perceptual soul; it is the result of the work that the living body does *qua* possessing such a soul. The definitional object of the perceptual capacity, therefore, can by no means be fully identified without reference to the soul's own work. And what that means is that the formal object of the activity of perception, though prior in definition to the corresponding capacity, is at least in part the *outcome* of the soul's work and actuality. But this is at it should be. We are defining the capacities of the soul, after all, and the capacities of the soul are capacities of the living body to *do* certain things, namely to engage in the living body's life-activities; and it is clear that none of the things that the living body does in virtue of having a soul can be found in external correlate objects.[15] So applying the retrograde method cannot mean that the external correlate object of the activity of the soul allows us to give a full determination of the formal object of perception.

Once this is clear, however, the danger of circularity looms large over the retrograde method. For if the formal object of a given capacity of the soul is at least in part the outcome of the soul's own activity, then defining that capacity by reference to that object seems circular. The *definiendum* would be part of the *definiens*.

Aristotle is, I think, aware of the problem and he systematically tries to avoid it. To illustrate how, take again perception. We have seen that perception is defined as the capacity to receive perceptual forms without the matter (*aneu hylês*) and according to the proportion (*kata ton logon*). However, previously to the definition in De an. II 12, Aristotle has not spoken of form and matter of the perceptual object, and neither has he spoken of the fact that perceptual forms consist in proportions of values on qualitative scales. And while he *does* speak of form and matter and qualitative proportions in the definition of the perceptual capacity, he does so in a conspicuously roundabout way. He says that perception is the capacity to receive forms of perceptual objects "without their matter" and "according to the proportion", yet he does not tell us what these forms and proportions are. This, I take it, is no coincidence, but a deliberate attempt to avoid circularly in the definition of the perceptual capacity. Aristotle applies the retrograde method by defining the capacities of the soul by way of their objects, but he does it in a way that painstakingly avoids positive reference to the achievements of the soul's own contribution in the constitution of its corresponding formal objects. He individuates the senses by reference to their outside objects and the causal effect that they have on the respective sense

[15] This holds even in the case (counterfactual for Aristotle) that the perceptual soul should be a capacity for undergoing an *entirely* passive process. Still in that case the actuality of the reception of perceptual forms would have no counterpart in the outside world, and could for that reason not be defined solely with reference to an outside and, as it were, unprocessed correlate object.

organs, without saying much about their phenomenal features. And this is Aristotle's mode of procedure throughout the discussion of perception in chapters 5–12 of the second book of the *De anima.* After having stated that and how the perceptual capacity is potentially like its objects (II 5), he individuates the different sense modalities by their external objects – color, sound etc., and describes how each of these objects have a causal impact on the bodies in their surroundings – the so called media that have the capacity of mediating the affection they suffer from the perceptual objects –, to then further describe how the media produce perceptual stimuli in the respective peripheral sense organs of the animal. They can have this effect on the sense organs because the sense organs consist of the same kind of bodies as the media so that the sensory affection by the external object can be passed on to them. Importantly, chapters 7–11 mention no further processing of the perceptual stimulus inside the perceiving animal's body, nor do they mention the qualitative features of the perceptual objects, i.e. what Aristotle calls the "differences" of each of the modally specific sense objects, color, sound, smell, taste, and touch, of all of which he thinks as proportions (*logoi*, see De an. II 8, 420a26f.).[16] Having discussed each of the sense modalities, Aristotle, in De an. II 12, then finally turns to what is commonly taken to be the definition of the perceptual capacity as a whole, i.e. jointly for all sense modalities, and defines it as the capacity to take in the forms of external perceptible objects without their matter.[17] This definition is notable because it does seem to be an attempt of Aristotle's to capture by way of definition what the capacity of perception does with relation to its objects – albeit without saying anything specific about what the act of perceiving actually consists in. Up to this point the discussion of sense perception in De an. II described only the *causal ancestry* of perceptual episodes as they occur before the proper act of perception, i.e. the causal history of sensory affection from the causal powers of the external sensory objects up to their arrival in the perceiver's perceptual apparatus (the peripheral sense organs). The definition in II 12 then goes a little further than that by offering a description of the proper act of perception, yet only in the form of a *negative characterization*. To define perception as the capacity of taking on perceptual objects without their matter clearly avoids any positive description of the act of perceiving with its qualitative and phenomenal features. What the actual qualitative and phenomenal features of perception are as they constitute the act of perception is practically not talked about in positive terms in the *De anima*. That this is in fact Aristotle's procedure in the *De anima* is confirmed by his own remarks in his treatise on sense perception and sense objects, the *De sensu et sensibilibus*:

---

[16] Exception are the passage in De an. II 11, 424a5-16, where he speaks about sensory discrimination, and the brief and summary remarks on the object of the sense of hearing in De an. II 8, 420a26-420b4 to which, by the way, Aristotle refers in his *De sensu* (440b27-28). The reason why he anticipates the qualitative description of the sense object there is presumably that Aristotle wants to establish causal correlations between certain of the incoming perceptual stimuli (fast, slow etc.) and the sound experience that they produce in the perceiver.

[17] A number of important manuscripts of the De an. (SUXPE) omit "forms" and write instead: "perception is what is capable of receiving the perceptible objects without their matter".

> Regarding the objects of each of the sense-organs – I mean colour, sound, smell, flavour, and touch – has been stated universally in *On the Soul*, namely what their function (*ergon*) is, i.e. what their activity is in relation to each of the sense-organs. (Sens. 3, 439a6-9).

Aristotle here summarizes his discussions of perceptual objects in the *De anima*. He takes himself to have dealt there only with the effects that perceptual objects have on the *sense-organs*, i.e. with what I have called the causal ancestry of perceptual acts; he does not take himself to have given a description of the intrinsic character of the sense-objects as they are perceived, i.e. as qualities according to determinate proportions of values on qualitative scales. This, if I am correct, is so because the description of that intrinsic character of perceptual acts would have to involve reference to the activity of the perceptual soul, which, if he did include it in the definition of the perceptual capacity, would render the definition circular. Next in the *De sensu*, he announces a treatment of the objects of perception as what they are, i.e. insofar as they cause the actuality of perception:

> We must next consider what account we are to give of any one of them; what, for example, we should say colour is, or sound, or odour, or savour; and so also respecting [the object of] touch. We begin with colour. (….) Let us state now what each of these (objects of perception) are so as to bring about the perception, i.e. the actuality." (*Sens.* 439a6-12 transl. Beare, modified)

This contrasts with the previous characterization of the treatment of the perceptual objects in the *De anima*. Whereas Aristotle took himself to have given an account of the objects of perception *qua* their effect on the sense organs in the *De anima*, he now announces a treatment of these objects *qua* the actuality of perception. And this treatment in chapters 3–5 of the *De sensu* will involve talk of determinate proportions and offer an account of the qualities of the objects of the different sense modalities. What I think follows from this is that for Aristotle the treatment of the perceptual capacity is fully completed only once he has led us through the entire account of perception not only in the *De anima* but also in the *Parva naturalia*. In Aristotle's way of presenting things, we can only fully know what the soul does when we have studied its operations in the living body. At any rate, Aristotle seems to apply the retrograde method with great caution so as to avoid circularity in the definition of the perceptual capacity of the soul. This is why he begins with a characterization of the external object of perception and avoids reference to the soul's own achievements in the process. In this way, the external object, to be sure, will be part of the formal object of the corresponding capacity, but only as a *first and relatively indeterminate part* of it: it will neither exhaust the determination of its formal object, nor will it involve reference to the soul's work.[18] Thus, the full determination of the formal object does not stand at the beginning of the process of defining the parts of the soul – as Aristotle's initial statement of the retrograde method in De an. II 4 might suggest to the reader–, but is its final result. A full statement of the formal objects of the parts of the soul will have to involve a description of the soul's own

[18] On Aristotle's method of procedure in defining the parts of the soul in the De an. see Corcilius 2008, 21–44, and, on the perceptual part, see Corcilius 2014, 33–36.

activities, since these activities are what the capacities of the soul are capacities *of*. Therefore, in the case of the definition of the vegetative capacity in the *De anima* the retrograde method should make us expect to hear only of a first and relatively undetermined part of the formal object of that capacity as well, and we should not expect to learn much about the actual work that the vegetative soul does in the living body.

What is the object of the vegetative capacity with which Aristotle starts off the definition of that part of the soul in the *De anima*? The passage in De an. II 4, 415a16-22, quoted above, says that it is nourishment (*trophê*). However, in the immediate sequel, Aristotle seems to allow for more than one correlate object of the vegetative part of the soul:

> Hence, it is necessary to speak first about nourishment (*trophê*) and the generation of offspring. For the nutritive soul (*treptikê psychê*) belongs to the other creatures too, and it is the first and most common capacity of the soul, in virtue of which all of them have life. Its works/products (*erga*) are to generate and make use of nourishment (*trophêi chrêsthai*). (De an. II 4, 415a22-25, trans. Miller, modified)

Nourishment (*trophê*) and the generation of offspring (*gennêsis*, "begetting") seem to be two different things. While the first is either the corresponding external object of nutrition or the product of the activity of nourishing (Aristotle here uses the term "*trophê*" in both of these meanings, see below), the second is the object or product that corresponds to the activity of begetting. To further complicate matters, in the ensuing discussion Aristotle will add growth as a third activity. How do these different objects and activities relate to each other? This is the *problem of the unity* of the vegetative part of the soul: Aristotle associates three different activities, nutrition, growth, and reproduction, with the vegetative capacity and he speaks of two different corresponding products/objects of that capacity, namely nourishment (*trophê*) and the generation of offspring (*gennêsis*); how can these different products and activities amount to a single object of a single vegetative activity?

Aristotle does not acknowledge the problem. He simply asserts both of the following claims, without offering us any explanation: (i) the vegetative capacity is but one capacity of the soul which is responsible for nourishing and begetting offspring, (ii) that capacity is to be distinguished from the other capacities of the soul by nourishment as its object:

> Since the same capacity of the soul is both capable of nourishing and capable of begetting offspring, it is necessary also to give a definition of nourishment (*trophê*) first; for it is distinguished from the other capacities by this work/product. (De an. II 4, 416a18-21- trans. Miller, modified)

How can he say this? What makes Aristotle so sure that the vegetative capacity is but one capacity, despite apparently having a plurality of correlated objects? And what makes him so certain that nourishment is the object by reference to which the vegetative part of the soul is to be distinguished from the other parts of the soul?

I do not believe that Aristotle would have overlooked the problem of unity, nor do I believe that he unwittingly associates a plurality of activities and objects with the vegetative capacity of the soul. If I am correct, he does not acknowledge the problem of unity because, in his way of thinking about the vegetative capacity and the diverse objects and activities associated with it, the problem does not arise. In order to see how this can be the case, I now turn to the passage in the *De anima* where Aristotle offers his definition of the vegetative capacity. For convenience, I divide the text into four sections. Section (i) identifies nourishment as the correlate object of the vegetative capacity; section (ii) distinguishes growth from nourishment by way of a minimal characterization of nourishment. As a result, growth is ruled out as an essential feature of vegetation; section (iii) clarifies the relation between nutrition and the production of offspring; section (iv) states the definition of the vegetative capacity:

> **(i)** Since nothing is nourished unless it partakes of life, that which is nourished is an animate body in so far as it is animate. Hence, nourishment is related to what is animate and it is not related to it in a co-incidental way. **(ii)** But being nourishment is different from being able to cause growth. For, in so far as that which is animate is of a certain quantity, nourishment is capable of making it grow; while, in so far as that which is animate is a certain this (*tode ti*) and a substance, nourishment is nourishment. For it preserves the substance (*sôzei gar tên ousian* [i.e. of the animate living thing]), which exists as long as it is nourished; **(iii)** and it [i.e. the nourishment] is able to bring about the generation (*genesis*) not of the thing that is nourished but of something which is like the thing that is nourished. For its substance already exists, and nothing brings itself into existence, but it preserves itself; **(iv)** hence, this sort of principle of the soul is a capacity which preserves that which possesses it in so far as it is the sort of thing it is. (De an. II 4, 416b9-19, trans. Miller, modified)

Section (i) identifies nourishment as the object of the vegetative capacity. It also says that there is a *per se* – relation between nourishment (*trophê*) and the vegetative capacity. This is important because the existence of such a *per se* – related object is a requirement for the application of the retrograde method. If the definition is going to be an expression of the essence of the capacity, or of a part of it, which is certainly Aristotle's ambition here, the object referred to in starting to individuate that capacity better be an object that is not accidentally related to it, i.e. it better be a *per se* – related object and thus, in the aforementioned way, a first and relatively indeterminate part of the capacity's formal account. Aristotle identifies this *per se* - related object of the vegetative capacity as nourishment. The term "nourishment" is ambiguous. It can refer either to external foodstuff prior to any digestive processing by the nourishing organism, or to the foodstuff as it is already concocted by the living thing's vegetative capacities and ready to be integrated in, and allocated to, the parts of its body. It is the first sense which is relevant here; the latter is what Aristotle calls the "ultimate nourishment" (416b4-7), which is the result and outcome of the vegetative capacity's work, as it is ready to being integrated into the parts of the animal body (in larger animals this ultimate nourishment is their blood). Aristotle here applies the same method as in the case of perception. He picks out nourishment as the external and yet unprocessed object of the vegetative capacity, presumably in order to provide himself with a first, non-circular, and relatively indeterminate formal description of it. That object, even though it is external to the digestive process, relates to the vegetative capacity in a non-accidental way: it is the external food that enters the process of nutrition – nourishment in the sense of "what gets used" in the process of nutrition.[19] The external nourishment thus is a *per se* - part of that process both (efficient-) causally and formally, even if only a first and relatively indeterminate part. This allows Aristotle to refer to the correlate object of the vegetative capacity without mentioning that capacity's own activity, as he has done in case of the definition of the perceptual capacity. That same argument also allows us to explain why Aristotle does not define the vegetative capacity by reference to offspring as its correlative object. Offspring is the result of the vegetative soul's activity and, for that reason, cannot be described without reference to the actuality and the work done by the vegetative soul. The offspring of a living being is itself a living being and cannot, according to Aristotle's own standards, be defined properly without reference to its soul. Section (ii) takes the next step by distinguishing nourishment from growth. As the analysis of the essence of nourishment and growth ("being nourishment and being able to cause growth") shows, growth is not a separate activity but merely a quantitative aspect of the process of nourishing. Nourishing is prior and more fundamental than growth in the same way in which the category of substance is prior to, and dependent upon, the category of quantity. Nourishment, the corresponding object of the vegetative capacity, says Aristotle, has the capacity to preserve substances (i.e. individual living things, hylomorphic compounds), while the capacity to cause growth is a quantitative aspect of that capacity, namely the capacity to preserve the preserved substance's appropriate size (cf. Gen. et. corr. I 5, 322a20-28). The priority of nourishment over growth is also apparent in the fact that they can come apart: while all living things continue to nourish themselves for as long as they are alive, they very often do not continue to grow in size for their entire life-span (Aristotle, it seems to me, alludes to this by saying that the animate thing is alive for as long as it nourishes itself).[20] Note that he does not detail here how the corresponding activities of growing and self-preservation take place; note also that he talks of the activities of growth and nutrition as qualifications of what the *object* of the vegetative capacity, the external nourishment, is capable of, thus avoiding direct reference to the corresponding activities of the soul. Section (ii) introduces self-preservation as a job description of the vegetative capacity. That description, however, is a neutral one. It does not say how and by which kind of activity the vegetative capacity accomplishes its task. Section (iii) then settles the relation between the activity of nourishing and the activity of begetting offspring. Again, Aristotle speaks of these activities not directly as capacities of the living thing but as capacities of the *object* of vegetation. His argument is this: while the external nourishment is capable of bringing about the self-preservation of substances (living things), it is also capable of bringing about the generation of new individuals of that same species (of something "that is like the thing that is nourished"). This strongly suggests that Aristotle does not regard the generation of offspring as an effect of the vegetative process that is fundamentally different from individual self-preservation,

[19] See 416b19: "nourishment makes it [i.e. the vegetative soul] ready to be active (*paraskeuazei energein*)." "That by which it [the nourishing ensouled body] is nourished is the nourishment.", 416b22-23.

[20] Cf. 415b26-28.

as he seems to be thinking that there is but one common external correlate object of the corresponding capacity, namely the external nourishment. Also, in both cases it is one and the same substance that is preserved, only that in the former case the self-preservation is of it as an individual, and in the latter case of it as species. This way of thinking about the relation between nourishing and begetting offspring as being fundamentally the same is, I think, implied by the definition of the vegetative capacity in section (iv) in the immediate sequel. In that definition Aristotle moves from begetting and nourishing to a higher level of abstraction that covers both cases of self-preservation:

> this sort of principle of the soul is a capacity which preserves that which possesses it *in so far as it is the sort of thing it is* (*hêi toiouton*). (De an. II 4, 416b17-19)

This formulation holds for both kinds of self-preservation, individual and specific. The vegetative capacity is the capacity of preserving a living thing's own substance; and while this may occur in different ways, they still will all be ways of self-preservation. On that account, Aristotle is perfectly justified to say, as he does a little later in the text, that "since it is right to call everything after its end (*telos*), and its end in this case is to beget something like itself, the primary soul will be a capacity to beget something like itself (*hoion auto*)." (416b23-25) Begetting something like oneself is a way to preserve one's own substance. But since its product, the offspring, outlasts the existence and self-preservation of the individual, it makes good sense to think of it as the "end" of the vegetative capacity. But that seems just a matter of perspective. Aristotle seems to think that both kinds of self-preservation are ultimately due to, or originate in, one and the same kind of process (in section iii he said that nourishment as the corresponding object of the vegetative capacity is capable of bringing about offspring as well, which suggests that nutrition is in one way or the other part of the process of generation).[21] In any case, individual and specific self-preservation seem inseparable in the following asymmetric way; we cannot conceive of a living thing that is capable of begetting offspring without simultaneously being able to preserve its own individual substance but we can conceive of a living thing that, though self-preserving, does not have the capacity to generate offspring). So nutrition seems somehow to be causally basic with relation to the other activities of the vegetative soul, while these other activities remain conceptionally distinct from it. Aristotle's definition of the vegetative capacity, I think, is meant to capture this. He defines vegetation as a cluster of capacities which are adequately subsumed under self-preservation "in so far as it is the sort of thing it is". The capacities of growth, nutrition, and begetting all, as it were, "flow" from that abstract capacity. They are, to use Aristotle's locution, "one in number but different in being", i.e. they are one and the same capacity but differ from each other in that they have different relational properties, the one being self-preservation in relation to the individual, the other in relation to its species, and the third in relation to

[21] This will be confirmed by Aristotle's theory of sexual generation in the *De generatione animalium*.

its size.[22] The different sub-capacities of the vegetative capacity are different, but not separate, from each other in account.[23] They all converge on a common conceptual core, which the preservation of one's own substance.

Aristotle's definition of the vegetative part of the soul as the capacity to preserve "that which possesses it in so far as it is the sort of thing it is" is, from a methodological point of view, similar to his definition of perception as the capacity to take on perceptual objects without their matter in De an. II 12. Both definitions work with descriptions of the soul's activity that are, as it were, *from the outside*, i.e. they do not say in virtue of which *specific* activity the soul is responsible for these achievements. This, as I have emphasized, is very likely to be a deliberate move on Aristotle's part. Saying what the soul does in making these achievements will not be possible without a detailed account of the living body's actions and affections. Giving such an account, however, is not part of the program of the *De anima*, which, as we have seen, is a treatise devoted to the definition of the soul as the first and most abstract first common principle of the scientific explanation of the phenomena of living things. This, if I am correct, is as it should be, given the retrograde method and Aristotle's concern with the avoidance of circularity.

To investigate the actuality of the nutritive soul and to give an account of its achievements is part of the investigation of the actions and affections common to body and soul (see *De sens*. 436a1 sqq., and, for the case of vegetation, De an. II 4, 416b30-31). Aristotle's *Generation of animals* is, I think, certainly the most important among these writings. But it is not the topic of this paper.[24] However, the

---

[22] He uses this expression with respect to growth and nutrition in Gen. et corr. I 5, 322a23-28. For other examples see De an. III 7, 431a12-14, Phys. III 3, 202b5-22, Met. XI 9, 1066a29-34. In these passages (and many other passages like them) Aristotle uses the formula "the same but different in being" to extensionally identify things whose definition is different from each other. This can, for example, be the relationally different forms of desire (the identification of the capacities of pursuit and avoidance and their identification with the capacity of perception in De an. III 7) or it can be the identification of the different relata of the causal relation in physical change (the agent and the patient relata of the causal relation in Phys. III 3, and Met. XI 9). The case of desire is particularly interesting because in De an. III 7, 431a12–14, Aristotle identifies the capacities of pursuit and avoidance as one and the same capacity, namely the capacity of non-rational desire. With this he makes a move which, if I am correct, is parallel to the move he makes in the definition of the vegetative capacity in De an. II 4, namely moving upwards from two relationally (and therefore also definitionally) different capacities to a third and basic capacity, see 431a12–14: "... and avoidance and pursuit when actual are the same. And what is capable of pursuing and what is capable of avoiding are not different, either from one another or from what is capable of sense-perception; but their being is different."

[23] For a similar account of the unity of the vegetative part, see Johansen 2012, 106–111, especially, 107: "So, the food is the object of nutrition *qua* preserving the *ousia* of the living being; it [i.e. food] is the object of generation *qua* causing another living being; and it is the object of growth *qua* causing an increase in the size of the living being."

[24] This may also include the allegedly lost treatise "On Nourishment"Aristotle refers to in De somn. 3, 456b6, *Meteor.* IV 3, 381b13, and in De an. II 4, 416b30-31. For further references to the probably lost treatise in Aristotle's oeuvre see Louis 1952. On the workings of the vegetative capacity in the formation of the embryo, see Gen. an. see Gelber 2010, and Connell 2016. See also King 2001, 21 sqq.

investigation of the vegetative soul's achievements will, among other things, also show that there is a strong *causal* connection between the three aspects of self-preservation (over and above their numerical sameness and difference in definition). In the case of nutrition and growth the connection is obvious. Growth, as we have seen, is a quantitative aspect of nutrition. But there seems to be also a direct and strong causal connection between nutrition and sexual reproduction. In Aristotle's theory, both processes crucially involve the *assimilation* of external foodstuff by the living being's own causal agency to its own substance (which he implied above in section iii). That assimilative agency consists in a special, organic kind of "cooking" or "concoction" of foodstuff by the living thing's internal heat (*pepsis*).[25] The overall goal of that process is the living thing's nature (*physis*) in the sense of its form (*eidos*) and substance (*ousia*, Meteor. IV 2, 379b18-380a6). Again, the formulation of the goal of that process in Aristotle's *Meteorology* seems deliberately abstract so as to cover both, individual and specific self-preservation.[26]

References

Aristotle. 1991. De Sensu et sensibilia. Trans. J.I. Beare. In *Aristotle, Complete Works*, ed. Jonathan Barnes. Princeton: Princeton University Press.

———. 2000. *Nicomachean Ethics*. Translated and Edited by Roger Crisp. Cambridge: Cambridge University Press.

———. 2016. *De Anima*. Translated with an Introduction and Commentary by Christopher Shields. Oxford: Oxford University Press.

———. 2018. *On the Soul and Other Psychological Works*. Translated by Fred D. Miller Jr. Oxford: Oxford University Press.

Althoff, Jochen. 1997. Aristoteles' Vorstellung von der Ernährung der Lebewesen. In *Aristotelische Biologie. Intentionen, Methoden, Ergebnisse* (Philosophie der Antike 6), ed. W. Kullmann and S. Föllinger, 351–364. Stuttgart, Franz Steiner.

Connell, Sophia M. 2016. *Aristotle on Female Animals*. Cambridge: Cambridge University Press.

Corcilius, Klaus. 2008. *Streben und Bewegen. Aristoteles' Theorie der animalischen Ortsbewegung*. Berlin: De Gruyter.

———. 2017. *Aristoteles. De anima. Über die Seele. Übersetzt, mit einer Einleitung und Anmerkungen*. Hamburg: Meiner.

Corcilius, Klaus, and Pavel Gregoric. 2010. Separability vs. Difference: Parts and Capacities of the Soul in Aristotle. *Oxford Studies in Ancient Philosophy* XXXIX: 81–119.

[25] The external foodstuff needs to be assimilated, i.e. transformed into flesh and other matter of the living thing, which requires its chemical transformation. That, in turn, requires heat. This is why all living things qua their vegetative activities possess some amount of heat (De an. II 4, 416b27-29). For a description of the similarities see Althoff in Kullmann/Föllinger 1997.

[26] This may also include the allegedly lost treatise "On Nourishment"Aristotle refers to in De somn. 3, 456b6, *Meteor.* IV 3, 381b13, and in De an. II 4, 416b30-31. For further references to the probably lost treatise in Aristotle's oeuvre see Louis 1952, for a recent discussion see Wöhrle in Kullmann/Föllinger 1997. On the workings of the vegetative capacity in the formation of the embryo, see Gen. an. see Gelber 2010, and Connell 2016. See also King 2001, 21 sqq.,

Corcilius, Klaus. 2014. Activity, Passivity, and Perceptual Discrimination in Aristotle. In *Active Perception in the History of Philosophy* (Studies in the History of Philosophy of Mind 14), ed. J. F. Silva and M.Yrjönsuuri, 31–53. Springer.

Gelber, Jessica. 2010. Form and Inheritance in Aristotle's Biology. *Oxford Studies in Ancient Philosophy* XXXIX: 183–212.

Johansen, Thomas Kjeller. 2012. *The Powers of Aristotle's Soul*. Oxford: Oxford University Press.

King, Richard. 2001. *Aristotle on Life and Death*. London: Duckworth.

Louis, Pierre. 1952. Le Traité d'Aristotle sur La Nutrition. *Revue de Philologie, de Littérature et d'Histoire Anciennes*: 29–35.

Menn, Stephen. 2002. Aristotle's Definition of the Soul and the Programme of the *De Anima*. *Oxford Studies in Ancient Philosophy* XXII: 83–139.

Platonis. 1905. *Opera IV*, ed. breviq. adn. cr. instr. Ioannes Burnet. Oxford: Oxford University Press.

Wöhrle, Georg. 1997. Aristoteles als Botaniker, in: W. Kullmann, S. Föllinger (edd.) *Aristotelische Biologie. Intentionen, Methoden, Ergebnisse* (Philosophie der Antike 6), Stuttgart Franz Steiner, 387–396.

Chapter 3
Embodied Intelligent (?) Souls: Plants in Plato's *Timaeus*

Amber D. Carpenter

Abstract In the *Timaeus*, plants are granted soul, and specifically the sort of soul capable of perception and desire. But perception, according to the *Timaeus*, requires the involvement of *to phronimon*. It seems to follow that plants must be intelligent. I argue that we can neither avoid granting plants sensation in just this sense, nor can we suppose that the *phronimon* is something devoid of intelligence. Indeed, plants must be related to intelligence, if they are to be both orderly and good – for intelligence is the cause of normativity and teleology. And yet, plants are not intelligent creatures. While plants must have individual souls if they are to be distinct from each other, each having its own orderly life, the intelligence they require for their individual sensations is not similarly individuated, but is rather that of the cosmos itself. Plants' peculiar partial individuation arises from the fact that their ultimate good is only derivative: unlike animals, it is only by completing the body of the cosmos that a plant's good a Good Thing.

3.1 The Limits of Degenerative Creation

In the *Timaeus*, Plato has the natural philosopher, Timaeus, tell a tale of degenerative creation. The Demiurge creates the most perfect beings: subordinate, created gods, the soul of the cosmos, and – with the leftovers – countless rational souls which will become embodied in order to complete the body of the cosmos. Only when the lesser gods provide these rational souls with bodies, and suitable psychic functions to go with them, will the cosmos embody (as nearly as possible) the perfect Living Thing.

Reprinted, with minor changes, with permission of Brill Academic Publishers from 'Intelligent(?) Souls: Plants in Plato's *Timaeus*', *Phronesis* 55 (2010): 281–303; permission conveyed through Copyright Clearance Center, Inc.

A. D. Carpenter (✉)
Yale-NUS College, Singapore, Singapore
e-mail: amber.carpenter@yale-nus.edu.sg

F. Baldassarri, A. Blank (eds.), *Vegetative Powers*, International Archives of the History of Ideas Archives internationales d'histoire des idées 234,
https://doi.org/10.1007/978-3-030-69709-9_3

Human beings ‘create’ in turn all the other animals – those that fly, swim, and go by land – through becoming morally corrupt: the less rational a human being manages to make his soul, the more morally corrupted it is; such a soul then requires, for its next incarnation, the sort of body which suits the corruption of his soul: greedy, lustful men, for example, become bulls or tigers; flighty, superficial men become birds, and so on.

In a related paper (Carpenter 2008), I investigated the difference between human beings and other animals on such a view, and argued that – in principle – there was none: animals, like humans, are engaged in the very same project of attempting (though usually failing) to restore the natural motion of their rational souls, so that they can become better, inheriting after death bodies corresponding to such rational activity, and eventually no bodies at all, as they return to their native stars.[1]

As animals just are degenerate forms of human beings, it would (I argued) be no more appropriate to eat animals than it is to eat humans. But if we cannot decently eat animals, we can at least eat plants. This is why plants have been provided expressly for this purpose:

> Of necessity, however, it came about that he [the mortal animal] lived his life surrounded by fire and air, which caused him to waste away and be depleted, and so to perish. The gods, therefore, devised something to protect him... These are now cultivated trees, plants and seeds. (*Timaeus* 77a1-3, 6).[2]

This seems straightforward enough: The purpose of plants is to replenish the body of the cosmos which regularly deteriorates due to its environment. With such a physical, even mechanical role, it may seem that the very possibility of intelligence in plants does not arise. And yet, matters are not so simple, and the *Timaeus*’ universe, they cannot be. For ‘soul’ has a dual role: it hosts intelligence, as we see in the degenerative creation tale; and it imports life. And these two features are not so easily prised apart.[3] In order to see this, we may start with what appears to be a technicality.

Timaeus says we may count plants among the ζῷα because, after all, they ζῆν – and “everything that partakes of ζῆν ought justly to be most correctly called a ζῷον” (*Timaeus* 77b1-2). The oddity of this move is hard to capture in English. It is something like Timaeus saying that plants get to be called ‘living beings’ because they ‘partake of living’ – this at least captures the etymological connection on which the argument rests. But it does so without capturing why it is something that need be argued at all, or even mentioned. We could instead translate “Everything animate is rightly called an animal”, if it were generally agreed as obvious that plants are animate. The sort of force with which the claim should strike us ought to be something

---

[1] See also Carpenter 2018 for another view that sees no deep categorical distinction between human and non-human animals, taking them as all having the same end.

[2] Translations follow Zeyl 1997, with modifications.

[3] See *Sophist* 249a, for example, where these are drawn closely together. It might be that in this Plato is following, while modifying, the views of his predecessors; on this, see interesting discussion Menn 1995, Ch. 5, esp. 29–30.

more (though not exactly) like a philosopher stipulating that all plants are rightly called 'animals' because they 'are alive'. True, ζῷον does not exactly mean 'animal'; but its ordinary extensions from that central meaning commonly include things like representations of animals, hence statues, and then depictions of all kinds. Thus the etymological argument gets its bite: since plants indisputably ζῆν (live), there can be no objection to counting them ζῷα ('animals'/living things).

But since plants are ζῷα, they must have souls. These they are immediately granted, although with substantial qualifications.

> What we are talking about now [plants] partakes of the third type of soul...the type totally devoid of judgement, reasoning and intelligence [δόξης μέν λογισμοῦ τε καὶ νοῦ], though it does share in sensation, pleasant and painful, with desire [αἰσθήσεως δὲ ἡδείας καὶ ἀλγεινῆς μετὰ ἐπιθυμιῶν]. (77b3-6)

3.2 The Puzzle Arising: Must Sensation Imply Intelligence?

Allowing plants only the lowest sort of soul may seem both obvious and innocuous, until we recall the description of sensation earlier in the dialogue:

> When even a minor disturbance affects that which is easily moved by nature, the disturbance is passed on in a chain reaction...until it reaches the *phronimon* and reports the property that produced the reaction. (64b3-6)

According to this model of sensation, there is no sensation (αἴσθησις) without the involvement of τὸ φρόνιμον, the intelligent part. Plants are granted sensation and even desire. Are they, then, intelligent? We seem to have four options:

> Option 1: Plants have a *phronimon*, and so are intelligent.
>
> Option 2: Plants have a *phronimon*, but *phronimon* here does not indicate intelligence, but rather something else, rather more minimal – such as consciousness.
>
> Option 3: Plants do not have a *phronimon*, and so sensation in plants is different from sensation in all other living beings.
>
> Option 4: Plato simply overlooked the implication, and so did not notice the problem.

Options 1 to 3 all look unpalatable in one way or another. So I begin with what may seem the least unpalatable, Option 4.

Now it might be that Plato simply did not see the problem. At 64b, Timaeus is well into his detailed physical account of the various modes of perception; by 77b, the discussion has moved on, and he has a different concern altogether in view – namely, to provide for the nourishment of animals. It seems a short space of text for Plato not to have put the two together; but perhaps with the emphatic turning of attention at 77a, Plato is no longer writing with his preceding psychological claims in view. Let it be, then, that the apparent tension here was inadvertent. Nevertheless, it is worth asking how it is that Plato ended up with this muddle (if it is that), and whether it might easily be resolved.

It is not difficult to imagine how the situation might have arisen. Plato, setting out the necessary conditions of perception, has primarily human desires and sensations in his sights; and these arguably *do* arise in an intelligently inflected way – or at least, it seems likely Plato thought they did. When he comes to speak of plants, however, his main concern is to acknowledge that they are alive. He duly casts about for some sort of soul to grant them, and chooses the most primitive sort available to him – the sensing-desiring kind. He would have gone for a still lower form of soul, if it had been available to him, if he had only thought of it. Plato, on this view, is crying out for Aristotle's conception of 'nutritive soul', but just misses it.

This seems to me to suppose a gross failure of imagination on Plato's part. Surely the author of the *Phaedrus* and the *Republic* could have thought up a different sort of soul if he had felt the need – unless we are to suppose he was exhausted by the sheer inventiveness of the *Timaeus*. Such imaginative failure becomes perhaps plausible, and relevant, if regarded as stemming from a conceptual gap: Plato lacks not just Aristotle's 'nutritive soul' but also Aristotle's concept of the purely biological for which 'nutritive soul' stands.[4]

But if Plato does indeed lack a notion of the 'purely biological', this seems to be saying no more than that Plato's understanding of soul, life, order, and purpose were not Aristotle's. For in fact, the notion of 'purely biological' is just the contemporary successor to Aristotle's way of dividing faculties – and there is no need to presume that it is therefore inevitable, correct, or even clear outside of the whole Aristotelian natural philosophy.[5] This does not, however, imply that Plato recognizes no distinction between the way plants are alive and the way animals are alive, for he has Timaeus mark the distinct character or place of plants in several ways.

First, plants are not needed to complete the perfect animal. When the Demiurge looks to the Living Thing Itself, in order to create as good an instance as possible, he lists the various kinds of living things proper to the Idea; plants are not among the kinds enumerated.

> He determined that the living thing he was making should possess the same kinds and numbers of living things as those which, according to the discernment of the intellect, are contained within the real Living Thing. Now there are four of these kinds: first, the heavenly race of gods; next, the kind that has wings and travels through the air; third, the kind that lives in water; and fourth, the kind that has feet and lives on land. (39e7-40a2)

[4] See Johansen 2020 for a recent illuminating dissection of the very real differences between Plato and Aristotle on nutrition and the function of the soul; many points in that account echo those made here.

[5] That self-replication alone should be the sign of life would imply – as the Mīmāṃsā noted – that crystals were alive (see Freschi 2015 for illuminating discussion). It is not at all apparent that there is a coherent notion of 'mere life', separated out from sentience (minimally, the capacity to feel pleasure and pain). Indian Buddhists therefore wrestled with the difficulty whether plants were sentient (therefore, capable of suffering, as the Jaina claimed, so that consuming them *harmed* them), or whether instead plants were not living beings at all. For excellent treatment of the topic, see Schmithausen 1991.

Indeed, it is just *before* the introduction of plants that Timaeus concludes with satisfaction, "So all the parts, all the limbs of the mortal living thing came to constitute a natural whole" (*Timaeus* 76e7-77a1). After this, he makes a fresh start in his discussion, introducing the creation of plants.

Timaeus not only makes a fresh start in the narrative with the introduction of plants, but the narrative depicts the gods themselves engaging in a new activity, creating a new kind of thing.

> They caused to grow a nature, akin to our human nature, though endowed with different forms and sensations [ἄλλαις ἰδέαις καὶ αἰσθωήσεσιν] so that it is a different living thing [ὥσθ' ἕτερον ζῷον εἶναι]. (77a3-5)

Plants must be made out of similar materials; they must be suited to humans and animals, and so be congenial to them. But they are freshly created for that express purpose, and not degeneratively created by the reincarnation of an especially stupid worm. However far into mindless viciousness an intelligent soul descends, it does not stoop so far as to take on the life of a plant. Plato observes a real discontinuity here, between the animal world and plants.[6]

In short, Plato recognizes in the *Timaeus* that fine balance: plants belong with (have kinship with) animals rather than rocks; and yet they are not a sort of animal – their kinship is of a different sort than the kinship that all animals have with each other.

If then the recognition of the 'purely biological' means a recognition that plants differ from other living beings, while yet being living things, then Plato does indeed recognize this and places plants here; it remains an open question, however, what we should say this difference in 'ways' of being alive consists in. The purely biological might, for instance, be conceived from the first as distinguishing plants from animals precisely with respect to agency (including desires) and sentience – 'purely biological' in that case would *just mean* 'life without desire or perception', or 'non-conscious life'. But why should we believe in advance that just this is the correct way of marking the difference between animal and plant life? In fact, if purely biological is taken to mean 'non-conscious life', this in turn can suggest a sort of living consisting in a series of purely mechanical processes.[7] If so, it may be just

[6] As Brisson 2004 notes, "Within the psychic scale mentioned above, we note two discontinuities: (1) A discontinuity between the souls of gods and of demons, which never fall into a body subject to destruction; and the souls of human beings and of animals, which inhabit destructible bodies with diverse appearances. (2) A discontinuity between the souls of human beings and of animals, which are endowed with a rational part, and the souls of plants, which are reduced to the desiring part, or *epithymia*".

[7] It need not suggest this, but one can see how it might: if a life is non-conscious, and yet there is growth and other change to account for, then these changes should be explained by the physical principles of motion governing the smaller parts constituting the plant. Consider again the Mīmāṃsā's crystals: this is what non-sentient self-increase of an organised structure looks like. So if that is the 'living' attributed to plants, by that measure crystals are also alive in the very same way. Menn 1995 (Ch. 6, esp. 40–42) discusses the importance of non-reductive explanation to Plato's motivations for citing *nous* as a source of change.

such a notion of non-conscious, non-agentive, non-experiencing 'life' that Plato is carefully guarding *against* in assigning plants the very same sort of soul shared by animals, and hence assigning them real sensations and desires of their own.[8]

3.3 Option 3: Equivocation

But need we think this is indeed the very same sort of soul? I have already called attention to the ways in which Plato carefully distinguishes plants from other living things in the *Timaeus*. This might incline us by now to find Option 3 more attractive. Timaeus does introduce plant-kind as "a different living thing", with "different forms and sensations". Perhaps this means that we should suppose one of those differences to be a radical difference in the necessary conditions for sensation.[9] If plant perception happens differently, then it may not need any intelligent part, as animal perception does. On this view, not just forms *and* sensations, but forms *of* sensation would be different for plants.

The first problem for this view is that Plato does not say this, although it would not have been difficult for him to have done so. Even if Plato did not wish suddenly to introduce an entirely different sort of soul, never described before, he has ready enough ways of indicating mere similarity rather than sameness. Yet he has Timaeus say that plants have the third sort of soul, not a soul 'in some ways like the third' sort. He specifically grants plants perceptions of pleasure and pain, and desires – not things 'that are in a way akin to' perception and desire.

Moreover, if 'sensation' in plants took different forms from that taken by sensation in *all* other creatures, then we should be entitled to wonder with what right it should be called sensation at all. This is not idle Socratic mimicry. If the mechanisms of sensation varied widely from animal to animal, particularly in respect of the relationship to intelligence, then there would presumably be some other reason that all these were rightly considered the same sort of thing. And plants would share in that. As it is, however, the one thing that all sensing has in common – through all modalities, and across all variations in material realization – is this process of communicating a physical event through the body to the soul via the *phronimon*. So if plants lack this, then they have sensation only metaphorically, or by analogy; they have 'as if' sensation, and not the real thing. But if plants are meant to be a different

[8] *Phaedrus* 245c-e is relevant here: All soul must be self-moving, "this is the very essence and principle of a soul, for every bodily object that is moved from outside has no soul, while a body whose motion comes from within, from itself, does have a soul, that being the nature of a soul" (*Phdr.* 245e4-7). My thanks to Anja Burghardt for helpful discussion of what gets lost if we deny plants desires.

[9] Indeed, it looks as if pseudo-Porphyry, in the *ad Gaurum*, took it in just this way (IV.6). I thank Luc Brisson for drawing my attention to the relevance of the discussion in the *ad Gaurum*, and for providing me with the materials pertaining to this text on which he and colleagues are working.

sort of living being *in just this respect*, why explicitly give them the third sort of soul – the sort of soul that is all about sensations and desires – rather than some different sort of soul altogether?[10]

There is a more crucial issue at stake, part of which will arise again when we consider Option 2. Suppose that some kind of sensation can occur in a quite different way. We then have to concede that in the generic description no work is being done by specifying the terminus of the communicated motion as 'the intelligent part'. Rather, more specifically, no work is being done by *the phronimon* in perception *as such*. Perhaps it is relevant to the sort of perceptions distinctive of humans, or – as Plato thinks – distinctive of humans and animals together. But its involvement in the process does not make the difference between a motion's being a perception rather than not. Instead, it only makes the difference between a perception's being 'our sort', rather than some other sort – some other, more 'bare' perception. But this latter does not look like the contrast that Plato is drawing. As will be discussed further below, he seems here at *Timaeus* 64b to be concerned with the difference between a mere unperceived (non-perceptual) change, and the kind of change that constitutes perception.

It is worth considering, then, what we think the *phronimon* is meant to be contributing in the first place. Is it supposed to be granting something like 'conceptual content' to otherwise content less bare sensations? The *Theaetetus*, for instance, can be read as a clarification of perception through distinguishing it from any sort of cognition whatsoever[11]; and so it might be thought that Plato endorses a notion of 'bare' perception. If so, this might in turn support the hypothesis that here, in the *Timaeus*, plants are supposed to perceive via a different mechanism from that spelled out for animals at 64b (Option 3) – though even here, it must be remembered that the 'bare' perception perhaps carved out in the *Theaetetus* was presumably meant to be a description of perception in all percipient beings, so that we still lack support for the hypothesized equivocation.

Or should we rather suppose that the *phronimon* is responsible in some inarticulate way for the fact that some motions are *sensations* at all? And if so, how? The exact meaning of '*to phronimon*' here is not clear, and is disputed – it may imply no suggestion of intelligence or reason at all. This will be a feature which Option 2 appeals to, so I will discuss it in detail below. But it is worth keeping in mind that, even if we do accept the translation I've offered ('the intelligent part'), 'intelligence' need not mean anything grand or 'Platonic'. Even when Plato speaks of

[10] Again, even if merely artistic considerations prevented him introducing here a never-mentioned fourth sort of soul, Plato has resources enough for indicating that plants have only quasi-perception, and pale imitations of the desiring-sensing soul.

[11] This is perhaps a dramatic way of putting the point made by Frede 1987; Burnyeat 1976 also shows Plato making perception less cognitive, from the *Republic* to the *Theaetetus*. But as his emphasis is on the work contributed by the sense/sense-organs as opposed to the *psyche* (presumably this, and not *noũs* is what he refers to as 'mind'), the question of whether there might be, for Plato, some utterly unintelligent soul with which one *merely* perceives does not arise.

phronēsis, of things being *emphrona*, or of *to phronimon*, it may yet be an open question what this intelligence consists in – whether, for example, it requires any necessary connection to eternal Forms, and even if so then to what sort.[12] If that is so, then what precisely the involvement of the *phronimon* is meant to supply to the process initiated by bodily movements is uncertain.

And whichever way we go, we should remember that we will have to say the same thing about 'desire', also integral to the 'third part of the soul'. This too speaks in favour of regarding intelligence in the *Timaeus* – in general, and in the case of perception – neither as the giver of conceptual content (as Option 3 would have it), nor as the giver of awareness (as Option 2, discussed below). Rather, intelligence primarily introduces normativity and teleology. But since this will be equally relevant to considering the next two options, I will leave this aside for the moment, and leave open the question of Option 3. If granting plants perceptions and desires utterly dissociated from the *phronimon* could do the work meant to be accomplished in assigning them perception at all, then it is still possible (though unlikely, it seems to me, and although he says nothing to indicate it) that Plato meant to be assigning plants merely an *ersatz* perception and desire.

3.4 Option 2: Minimisation

Instead of *aisthēsis* being something different in plants, might it rather be that the *phronimon* involved in *all* perception is different from what we might expect (Option 2)? Perhaps plants have genuine sensations and desires, just as animals do, but in neither case is the *phronimon* involved meant to be especially intellectual. Perhaps the *phronimon* integral to sensation does not indicate intelligence at all, but only bare 'awareness'.[13] It is at least occasionally appropriate to translate φρονέω and cognate words as 'in one's senses', and this nearly means simply 'to be alive' or perhaps 'to be conscious'.[14]

But this option turn out on consideration to be even less promising than the previous one. First, even if φρονέω is occasionally used to indicate whether someone lives, it does this *through* its primary meaning of 'being wise, thinking, being minded, understanding', not by dropping this meaning. For example, consider ἐμφνεῶ, which has a literal meaning of 'having the breath in one': since breath is one of the distinctive features of some (but not all) living beings, its presence or

[12] But see Brisson 1999, 159–160. By contrast, I think even if we do conclude that *to phronimon* is 'the intelligent part' this need not automatically entail that it has to do with recollecting eternal Forms in particular acts of perception; whether it does so will depend upon further views, and arguments, about what thinking for Plato is like.

[13] O'Brien 1997 argues for this interpretation.

[14] LSJ offer as the IVth entry under φρονέω: "*to be in possession of one's senses*, sts. almost = ζῆν".

absence will coincide with the aliveness or otherwise of such beings.[15] When φρονέω or ψυχοῶ are used to mean 'alive' as opposed to 'deceased', this is through a reference to the being in question having or lacking some distinctive mark of living things of the relevant kind – *viz.*, whether they are minded, or ensouled, respectively. In neither case need we suppose that the root meaning of the word has been lost; and so φρονέω should not be supposed to mean '*merely* conscious' as opposed to intelligent. When φρον*-words are used to mean alive, they do so through the primary meaning of the word, namely 'having one's wits about one', 'being smart' – that is, where one's mode of living is distinctively characterized by intelligence. Thus, LSJ cite several instances of the phrase "ζῶν καὶ φρενῶν", which they appropriately render not as a redundancy, but as "alive and in his right mind".[16]

In the specifically Platonic context, reading *to phronimon* in a way that does not indicate some sort of intelligence goes even more obviously against the grain.[17] Plato classes *phronēsis* together with cognitive success-terms too consistently to suppose he means us completely to dismiss those now, even if it is Timaeus, a natural philosopher, speaking.[18] More importantly, he does this in the *Timaeus*. For example, Timaeus describes the world-soul as providing "a divine source of unending and rational [ἔμφρονος] life for all time" (*Timaeus* 36e4-5). As a description of the world-soul, and the motions in which its sole activity consists, it is hardly credible to translate ἔμφρονος βίου with something so weak as '(merely) conscious life'. When these motions become embodied in the heavens, the motion of the Same

[15] Similarly with θαλπῶ: those living things which generate their own heat cease to do so when they die; thus, to say of such creatures that they have or lack heat can refer to their being, or not being, alive. Or again, the dead – at least the long dead, or the dead and buried – are no longer visible or manifest to us, while the living are always visible or manifest (at least in principle). So, one might use ἐπιφάντος to refer to the mark of 'visibility' as a euphemism for indicating whether one is still among the living.

[16] Not, as O'Brien 1997 would have it, "alive and conscious". He points out that "even a chorus of Aeschylus will not simply say of someone dead that he is not alive" ("Mais même un chœur d'Eschyle n'ira pas jusqu'à dire, de quelqu'un qui est mort, qu'il n'est donc pas vivant. Ce dont il est ici question, c'est la conscience."; 301); but it is hardly less vacuous to say, "he is dead and no longer conscious", which is O'Brien's preferred translation of οὐ φρονοῦντι (at 517 of Aeschylus' *Choephori*). Taking φονοῦντι to mean 'have one's wits' at least tells us what is relevant, and lamentable, in being dead.

[17] LSJ cite only one dubious instance from Plato, *Sophist* 249a.

[18] Dixsaut 2000, esp. 106–109, demonstrates how consistently Plato associates *phronēsis* with value, virtue, and intelligence. She argues that *phronēsis* is invoked particularly when the point is that intelligence is a quality of living things, that it is souls that have intelligence. One might, perhaps, dispute whether *to phronimon* should necessarily be read as relating to *phronēsis*; perhaps only the latter form of the root '*phron**' is imbued with value. But since, apart from various other forms of φρονεῶ marking intelligence-words, *to phronimon* is used by Plato in the *Republic* to mean 'the intelligent part' (at 530b6-c1 and 586d4-e3), the burden is to show that it does not mean the same thing here at *Timaeus* 64b. O'Brien attempts to show it cannot mean the same thing (1997, 299–300); but his argument seems to be that since the mortal part of the soul feels the pleasures, the *phronimon* cannot be the immortal and rational part of the soul. But the claim was only ever that the soul has perceptions *in virtue of* the *phronimon* (the rational part), not *in* it.

is described as “the single and most intelligent [φρονιμωτάτης] of revolutions” (*Timaeus* 39c2). Again, φρονιμωτάτης here can hardly be taken for mere consciousness.

Or consider the first introduction of soul into body:

> And they went on to invest this body – into and out of which things were to flow – with the orbits of the immortal soul. These orbits, now bound within a mighty river, neither mastered that river nor were mastered by it, but tossed it violently and were violently tossed by it... All these disturbances [παθήματα] are no doubt the reason why even today and not only at the beginning, whenever a soul is bound within a mortal body, it at first lacks intelligence [ἄνους]. But as the stream that brings growth and nourishment diminishes and the soul's orbits regain their composure...their revolutions are set straight, to conform to the configuration each of the circles takes in its natural course. They then correctly identify what is the same and what is different, and render intelligent [ἔμφρονα] the person who possesses them. (43a4-7, 44a7-b7)

When first embodied, Timaeus tells us, soul is ἄνους, unreasoning, due to the disturbances created by physical change; however, when the circles of Same and Different are re-established in their proper courses, this renders a person ἔμφρονα. Here again, ἔμφρονα must mean *intelligent*, not merely conscious; for, in the first place, we are surely not supposed to think that we are positively unconscious of the strong influx and efflux of youth, and only become *conscious* once the revolutions are set straight; and in the second place, this same regular turning of the circles of the Same and the Different which render one *emphrona* is what *noũs*, and even *alēthēs doxa*, consist in (see especially *Timaeus* 36c-37c). This very form derived from φρονεῶ will appear again, in another disputed passage (*Timaeus* 74e10-75a6, discussed below). So there is some reason to think that not only being *phronimos*, but also being *emphronos*, is a matter of intelligence, not some bare awareness or mere consciousness.

Now consider again more closely the role that the *phronimon* plays in the process of an event or affection (πάθος, 64b4, c2, c3) becoming a sensation. This will also be relevant to the consideration of Option 3, insofar as that option has be left open. First, both here at *Timaeus* 64b *ff.* and at the parallel passage in the *Philebus* (33d *ff.*), Plato is careful to specify that there is no αἴσθησις without a soul's being affected by some bodily change:

> Instead of saying that the soul is oblivious when it remains unaffected by the disturbances of the body, now change the name of what you so far called obliviousness to that of non-perception [ἀναισθησίαν]...But when the soul and body are jointly affected and moved by one and the same affection, if you call this motion perception [αἴσθησιν], you would say nothing out of place. (*Philebus* 33e10-4a5)

There is no ‘unconscious’ perception (αἴσθησις), because having a sensation *is* being aware of a bodily event. If *aisthēsis* just is the awareness or consciousness associated with bodily disturbance, there is no need to suppose an extra element, *to phronimon*, to play that role, to *be* the awareness. The *phronimon* must rather be part of the way in which some disturbances come to cause moments of awareness in us. The question is, *Which* part? What is the *phronimon* doing in the *Timaeus*, that it must be involved in the process of bodily changes amounting to moments of perception or awareness?

Let us go back to *Timaeus* 64b, where Timaeus makes the claim that *aisthēsis* requires the *phronimon*.[19] The passage continues by contrasting these moments of perception with unperceived disturbances:

> On the other hand, something that is hard to move remains fixed and merely experiences the disturbance without passing it on in any chain reaction. It does not disturb any of its neighbouring parts, so that in the absence of some parts passing on the disturbance to others, the initial disturbance affecting them fails to move on into the living thing as a whole and renders the disturbance unperceived. (64b6-c3)

The *pathos* must reach the whole creature (τὸ πᾶν ζῷον, c3) if there is to be perception. If there is a role for the *phronimon* in perception – if there is some reason for it to be included in the process – it could very well be responsible for explaining how it is that the creature *as a whole* is affected by a specifically located bodily change – how it is, for example, that a pain in the finger is not just the concern of the finger, and not just felt *by* the finger, but is rather something felt by the person in the finger, which that person can recognize as painful and decide to do something about. No one part of the soul, inasmuch as it is just a part, will be able to play this role. But intelligence in its soul-unifying role is just what is needed. A recollection of the *Republic* has opened the *Timaeus*, and so we might recall here that, in the *Republic*, this distinctive feature of intelligence as concerned for the whole is quite explicit: "Is not it appropriate for the rational part to rule, since it is really wise and exercises foresight on behalf of the whole soul…?" (*Republic* IV.441e4-5).[20] The rational part of the soul, I suggest, is distinguished by its ability to take account of the whole, and so to communicate what is of concern to the whole creature. This suggests that the *phronimon* ought to retain its connection with intelligence.

Since the *Timaeus* and *Philebus* offer strikingly similar accounts of perception, we might do well to consider the consequences of this starting-point in the moral psychology of the *Philebus*. Pleasures, Socrates goes on to argue there, can be true and false. How? The argument Socrates pursues appeals to the inter-connected nature of mental life, which interconnectedness is governed by cognitive, truth-seeking capacities. Pleasures, it is first observed, never arise apart from the soul being moved as in perception; and when the soul is moved, as in perception, the opportunity arises for deliberation, decision, judgement. The example Socrates offers is of perceiving a shape in a distance, and trying to discern whether it is a statue or a man (*Philebus* 38c-e). This, "if memory and perceptions concur with

---

[19] *Timaeus* 64b3-6, quoted above: "When even a minor disturbance affects that which is easily moved by nature, the disturbance is passed on in a chain reaction...until it reaches *to phronimon* and reports the property that produced the reaction."

[20] The connection intelligence and unity of the whole is also a running *motif* of the *Philebus*, another closely related dialogue – see especially *Philebus* 28d-30c, where intelligence is the *cause* of unity of body and soul, responsible for the fact that the universe is well-ordered, rather than chaotic.

other impressions at a particular occasion" (*Philebus* 39a1-2a), he likens (with some embellishment) to conceiving a hope (*Philebus* 39e4-6), which is a kind of pleasure.

Pleasures can become true or false in this way: as something necessarily perceived, they take shape through taking a place within the whole of our psychological economy.[21] And when a mental event (a *pathos* of the *psuchē*) does that, it becomes related to and answerable to judgement and understanding. These are then able to become experiences relevant to the whole person – thus, the conclusion Socrates draws here is that good persons generally have pleasures of a similar kind (*Philebus* 40a-c).[22] The movement of the soul which is pleasure, then, becomes assessable as true and false, and becomes a characteristic of the whole person because, in moving the soul, judgement and memory – that is to say, intelligence in some form – also get involved. In short, here too, a *pathos* 'reaches the soul', becoming a matter for the whole soul, in virtue of intelligent activity – which is just to say that sensation must involve the *phronimon*. One might object that this should count only for human souls, or only for those souls that have intelligence. But in the *Timaeus*, animal souls are in exactly the same condition – and the very question at issue is whether plants are, too.

We should consider a further passage, a curious discussion of flesh and joints – for it looks at first as though this may support the 'mere awareness' reading of *phronimon*, and its cognates. In fact it supports the notion that being alive and intelligence are for Plato inextricably interwoven.

> All those bones that were more ensouled [ἐμψυχότατα] than others he proceeded to wrap in a very thin layer of flesh, while those that were less ensouled [ἀψυχότατα] he wrapped in a very thick layer of dense flesh. And indeed at the joints of the bones...he introduced only a thin layer of flesh, so that the ability of the joints to flex would not be impeded, a condition that would have made it very difficult for the bodies to move. Another reason was this: if there were a thick layer of flesh there packed extremely densely together, its hardness would cause a kind of insensibility [ἀναισθησίαν], which would make thinking [διάνοιαν] less retentive and more obscure [δυσμνημονευτότερα καὶ κωφότερα]. (*Timaeus* 74e1-10)

Now it might be odd to suppose that more flesh equals less sensitivity, and less flesh equals more sensitivity; and it might be bizarre to speak of joints – of all things – as being 'sensitive'. If he had just left it there, we might suppose that the very fact that Timaeus speaks of joints as 'sensitive' demonstrates conclusively that sensation be meant to be something involving intelligence. Instead, however, Plato here has Timaeus immediately link less sensitivity to decreased cognitive capacity; and he uses for this the undeniably cognitive word, *dianoia*.

It is clear then that here at least Plato is explicitly linking sensitivity to intelligence. And this, it seems to me, counts as much against Option 3 (equivocation on sensation-desire) as against Option 2 (*phronimon* as unintelligent mere consciousness). In the construction of the body to suit ensoulment, intelligence and sensitivity

[21] The dependency of pleasure on the cognitive environment in which it arises is discussed in Carpenter 2011.

[22] The Good Man Argument, and its warrant for the conclusion that good persons have true pleasures, is discussed in detail in Carpenter 2006.

run in tandem, coming together in greater and lesser degrees.[23] It remains only to see that intelligence here is not something different from *phronēsis* or the *phronimon* – and that the passage give us almost immediately.

> This explains why thighs and calves...and all other bodily parts where there are no joints... are all fully provided with flesh. It is because they have only small amounts of soul [ὀλιγότητα ψυχῆς] in their marrow, and so are devoid of intelligence [κενά ἐστιν φρονήσεως]. On the other hand, all those bodily parts that do possess intelligence [ὅσα δὲ ἔμφρονα] are less fleshy, except perhaps for a fleshy thing – the tongue, for example – that was created to be itself an organ of sensation [αἰσθήσεων]. (74e10-75a6)

Here *phronēsis* (75a4) is used as if interchangeable with *dianoia* (74e10); and the tongue being sensitive means it is among the bodily parts that is intelligent. Thus what it means to 'be more sensitive' is not simply to be more conscious, but to be more intelligent.

This is why Timaeus points next to the unfleshiness of the skull. The skull, we learn, has much less flesh than needed to protect itself, so that the head – house of immortal soul – can be the most sensitive, *emphroneined* bit of us. Since

> a shorter lifespan was in every way preferable for everyone to the longer but inferior one... the head [lacks flesh, and instead] has turned out to be more sensitive and intelligent [εὐαισθητοτέρα μὲν καὶ φρονιμωτέρα]... (75c1-2, 5-7)

The head was designed to house the intelligent part of the soul (69c-d, where it is called 'divine'). So the point of the head being φρονιμωτέρα ought to be that it is the more intelligent part of the body, not that it is the "more conscious" part, where "conscious" has the sense of bare, or unintelligent consciousness.

It does not look, then, as if we can make granting plants a *phronimon* palatable by making the attribution something so weak as mere 'consciousness'. And so we are left to consider Option 1, and the unpalatable possibility that plants are intelligent.

3.5 Option 1: Biting the Bullet

If *phronimon* retains its connection to *phronēsis*, so that its proper place is among the 'thinking'-terms, with *noũs*, *episteme*, *dianoia* and so on; and if the process of a bodily change causing a sensation to arise goes via this *phronimon*; and if plants have sensation – then plants have a *phronimon*, and so are intelligent, in some sense.

Let us think again about why intelligence and perception run in tandem, and about why Plato should want to assign such a thing to plants. Recall that the third part of the soul is responsible for desire as well as for sensation, and plants are being granted *both*. If it seems absurd to suppose plants perceive, surely it is even more absurd to suppose they have *desires*. But desire and sensation are deeply embedded in one another, not accidentally bundled together. To have the capacity for sensation

[23] 'more ensouled' tracks 'more *emphroneined*' tracks 'more sensitive'.

is to be able to sense things *as pleasant* and *as painful*, and *therefore* to want the one and want to avoid the other; this in turn is related to the phenomenon mentioned above, that a pain in one part of the organism belongs to the whole, and elicits appropriate responsiveness from the organism.

If Plato gave the matter any thought at all, in assigning the third sort of soul to plants, what sort of perceptions and desires could he have thought they had – and why? Presumably he had in mind the way that plants are responsive to their environments as being either threatening or nourishing, and the fact that they grow accordingly. Leaves turn towards the sun, roots seek water. Whole plants bend in one direction, towards a congenial environment, and away from an uncongenial one. One might explain these phenomena otherwise, but it is not far-fetched to suppose that for Plato these were manifestations of a plant's minimal ability to respond to some things as 'good', and others as 'bad'.[24]

But responding to anything *as good*, therefore wanting anything, is the most minimal and generic teleological behaviour there is, especially as this responsiveness explains the organized, regular and unified nature of plants – in contrast, for example, to mountains, which are not living things. If such teleological behaviour is supposed to happen entirely dissociated from any intelligence whatsoever, it is unclear that Plato could accept it as genuinely teleological and as genuinely explanatory. Among the two sorts of causes recognized in the *Timaeus*, reason and necessity, the latter would never be explanatory all on its own.[25] If intelligence alone introduces genuine explicability and normativity, it seems unlikely that sensation – bound up as it is with desire – could do any of the work it is supposed to do in explaining the organized growth and responsiveness of plants, if it were some special sort of perception utterly dissociated from all intelligence, or if perception and desire *as such* did not involve intelligence, in some way. The point of granting plants percipient soul at all would be undermined if that soul had no association with intelligence – whether because its perception-desiring is mere *ersatz* perceiving-desiring, or whether because *all* perception (and desire?) as such had connection only to bare unintelligent consciousness.

There is, however, a large and obvious objection to this conclusion. The passage dealing explicitly with plants – the same passage that credits them with sensation and desire – explicitly denies them "judgement, reasoning and understanding" (δόξης μὲν λογισμοὺ τε καὶ νοῦ, 77b5); plants have only "the third type of soul" (*Timaeus* 77b3-4).

What exactly is it that plants lack, having only 'the third kind' of soul? Might there be some far inferior sort of intelligence that they have, while lacking

[24] Even so resolute a Darwinian as Daniel Dennett feels the need to build "some brute mechanical capacity to stop a random walk when a Good Thing comes along, a minimal capacity to 'recognize' a tiny bit of progress" (Dennett 1995, 79) to explain successful responsive adaptations.

[25] Lennox 1985, 210–12, makes this point, in the discussion of Plato's wider commitment to intelligence as reasoning 'for the best', and the contrast between this and the mere materials which are so arranged, across several dialogues, including the *Timaeus*.

"judgement, reasoning, and understanding"? Timaeus' account of plants goes on to specify exactly what plants lack, and it seems a fairly comprehensive exclusion:

> For throughout its existence, it is completely passive; and its formation has not entrusted it with a natural ability to discern and reflect upon any of its own characteristics, by revolving within and about itself, repelling movement from without and exercising its own inherent movement. Hence it lives [ζῇ], and is not other than a ζῷου, but it stays put, standing fixed and rooted, since it lacks self-motion [ἑαυτοῦ κινήσεως]. (77b6-c5)

The text is dense here; but lacking intelligence seems to bring together lacking (a) self-awareness, (b) self-reflection, and (c) activity, particularly the ability to originate motion. This pretty well denies plants any version of that activity definitive of intelligence, as described for example at *Timaeus* 89a1-3:

> The best movement is the one that occurs within oneself caused by oneself [ἐν ἑαυτῷ ὑφ' αὑτοῦ], for this is most akin [συγγενές] to the thinking [διανοητικῇ] and to the motion of the universe.

In denying plants *noũs*, *logismos* and even *doxa*, it seems clear that Timaeus intends to deny them all forms of intelligence.

3.6 A Suggested Solution

The problem remains so far insurmountable: Plants lack reason; yet sensation, which plants have, requires reason. To resolve the riddle we must think seriously – and creatively – about ways in which plants might both have and lack intelligence.

This will not be accomplished by granting plants one kind of intelligence, but not another. We must rather find some distinctive *way* that plants might have or 'partake of' intelligence – the relation plants have to intelligence will be different from (1) the relation animals have to intelligence; and so from (2) the way in which animals manifest desire and all the functions of the third kind of soul.

Plants occupy a strange half-way place in the *Timaean* universe. Rocks and rivers are a part of the body of the cosmos by being merely functional, constitutive elements. Their 'living' just is the living of the cosmos – they have no lives of their own. Plants differ in precisely this respect: they do have individual, distinct lives, and so have their own goals and souls. Because they have aims internal to themselves and not merely prescribed by the necessities of the world-soul's body, they have each a soul of their own, responsible for perceiving this or that *as* 'good and attractive' or the opposite.

Plants are thus responsive to their environments, and not in some mechanistic way, reducible to the motions of their parts. To attribute desires to plants is to recognize that they are responsive to their circumstances in their (still unexplained) ability to recognize something as a Good Thing, or a Bad Thing, and respond accordingly. And these goods are recognizably good *for this individual plant*.

But plants differ after all from animals in their 'aiming'. For animals, however confused, are *trying* to aim at *the* good. They try to achieve *for themselves* the same good that the world-soul – and perhaps even the Demiurge's soul, if that is different – aim at. Their good consists in achieving a clear and true understanding of this same good and gaining that. Being 'good-*qua*-horse' (if it even makes any sense in the Timaean universe) does not count for much, for horses like every other animal are ultimately trying to restore the revolutions of their souls to that appropriate to the perfect beings which they once were.[26] All substituends for this true good are failures. Plants' highest and only aim, by contrast, is their own immediate good. This is the force of denying them higher forms of soul, and of breaking the degenerative lineage that binds animals together. To be responsive to what is immediately good and bad for its staying alive in the state it is in counts as *success*, for a plant. Unlike horses and jackdaws, plants can be 'good of their kind', and this *counts*. It counts as success for a plant because plants have, of themselves, no higher aim.

But precisely because, unlike animals, a plant's good is not attainment of The Good, it is not obvious that it should be a good thing *simpliciter* for plants in general to achieve their specific ends. If it is good, it cannot be because they have attained The Good. It can only be good *simpliciter* for a plant to attain its particular good because the universe as a whole is organized in such a way that plants doing what they want contributes to the good-ordering of the cosmos as a whole. Local teleology contributes to the goodness of the universe in virtue of the nature and structure of the cosmos – and not in virtue of the individual organism's end itself being good. In this respect, plants resemble rivers and rocks. Their pursuit of their local ends is used to constitute the healthy cosmic body, and it is *only* a good thing because of the place it holds in that rational structuring.[27]

This dual nature of plants regarding their place in the universe suggests a solution to our riddle of plant intelligence. The explanation of plant functioning, we saw, both requires reason and denies them reason; this tension can be resolved by recognizing plants as simultaneously distinct living things, *and* ultimately only a mere constitutive part of the body of the cosmos. Each plant has its own soul, and so is a ζῷον – a living individual, distinct from all others; but this is only a vestigial soul, abbreviated into the functions that are posterior to, and attendant upon intelligence – an intelligence plants do not have, at least not of their own. The suggestion is that while plants have individual sensate-desiring souls, which move them towards health and flourishing, these souls are only operative in virtue of their relation to the intelligence of the world soul. Plants partake of immortal soul in the same way that stones do: merely in virtue of their functional role within a living being with an immortal soul.

On this view, plants still differ categorically from rock and water and fire by having some soul or another, some life or another, to call their own – plants have the requisite physical and psychic machinery for individuation, with a distinct basis for

[26] This deep contrast with Aristotelian teleology is highlighted by Johansen 2020; the ethical implications of this different are explored in Carpenter 2017, and Carpenter 2021.

[27] This is a thought that Stoicism expands, extending the principle to every organism within the universe. Plato, by contrast, retains the robust individuation of animals.

sensations of pleasure and pain, and for corresponding desires.[28] Their physical and psychological capacities are enabled, however, not via some intelligence which is their own, but in virtue of their connection to the intelligent whole they help to constitute. So their sensation does indeed occur, like all sensation, in virtue of *some* intelligence – but it is the intelligence of the world-soul, and not their own. This is the concrete correlate to the teleological point: Plants have ends of their own; but this is only a genuine *telos* and source of order because of its relation to the cosmos and *its* intelligent ordering. This might, then, count as a version of the discarded Option 3, inasmuch as perception and desire happen differently in plants and in animals. But it will differ from Option 3 by insisting on preserving the vital link between perception, desire and intelligence.

Such a solution is pure invention: Plato simply has not said, and it is tempting to think that Plato did not care much about plants – or about animals, for that matter. His project is a moral one, and his physics and cosmology should be an improving one: we should understand nature in the way that is most edifying for us.[29]

But the problem, even regarded as ethically motivated, will not go away that simply: Plato needs plants to be sufficiently different from human beings that eating them is appropriate.[30] He needs them to be sufficiently the same, however, that he can provide a unified, non-mechanistic account not only of soul, but of the cosmos.

To put this in more familiar Platonic terms, some causes are teleological – that is, directed towards the good, however remotely or confusedly; other causes are what I have been calling 'merely mechanistic' – "those things which, being moved by others are compelled to move others" (*Timaeus* 46e1-2). Unlike the former, these latter "only produce chance effects without order or design" (46e5-6).[31] This distinction between primary and secondary causes is described at 48b as the distinction between intelligence and necessity. As this second way of marking the distinction indicates, the only non-mechanistic source of cause and explanation that Plato recognizes is intelligence.

The problem, then, is not so much that Plato cannot imagine Aristotle's 'nutritive soul'; it is rather that such a so-called 'soul', if conceived as utterly disconnected from rational causation, cannot be the principle of the *organized* life found even in plants. If divorced entirely from reason, it cannot be distinguished from the sort of accidental unity that rocks and rivers have. On the other hand, if the intelligent principle informing plant life is internal to each individual plant, then plants cannot be categorically distinct from animals. In being responsive to the environment, orderly in its changes, and evaluable in terms of success/thriving or otherwise, plants must have recourse to some governing principle – that is, to intelligence. But this governing intelligence is not distinctive of, or personal to, each plant. Their desires and

[28] Being limited to the third part of the soul, all plants could take to be good is pleasure, and pain is the only thing that could from their perspective be bad.

[29] See, e.g., Johansen 2000, and Carone 1997 and 2005.

[30] *Timaeus* 77a1-6, quoted above.

[31] Or, their effects may have order and design, but only insofar as their motion is guided by the former sort of cause; see Lennox 1985, 210.

perceptions – their responsiveness to their specific environments as good and bad in specific ways – are organized and evaluable as good only by reference to their role in the orderly organization of the cosmos as a whole. The goodness of a plant lies only in being food for animals capable of becoming good in their own right.[32] Thus desiderative-perceptive soul individuates each plant as having its own life-trajectory; but this non-mechanistic responsiveness is enabled by the participation of the plant in the wider intelligibly ordered world.

The neo-Platonists would later wrestle with essentially the same problem, in the context of trying to determine the status of a fœtus, which is in some sense alive, but lacks the sort of independence required for having a life (as in, 'a life of one's own').[33] The conclusion they draw is similar – the fœtus has its own soul of the lower sort, which lives vicariously off the higher soul of its mother. We might be inclined to think that this problem, no less than its strange split-soul solution, is just Plato's problem – so much more reason not to be a Platonist. Yet it raises an interesting question: How are we to think of something that has to be sure a life of its own, but *only* in virtue of the fact that this life is organized by its functional role within an organized whole? More generally, how are we to do justice to the individuality of each living organism, with its orderly changes and internal unity, and yet mark the radical discontinuity between the sort of living plants enjoy and the sort shared by animals – the sort of difference that makes it at least potentially a moral issue whether we eat animals, as it just cannot become with regard to eating plants?

References

Brisson, Luc. 1999. Plato's theory of sense perception in the *Timaeus*. *Boston Area Colloquium in Ancient Philosophy* 13: 147–176.

———. 2004. Myths in Plato's ethics. In *Plato Ethicus: Ethics is life*, ed. Maurizio Migliori and Linda M. Napolitano Valditar, 63–76. Sankt Augustin: Academia Verlag.

Burnyeat, Myles. 1976. Plato on the grammar of perceiving. *Classical Quarterly* 26: 29–51.

Carone, Gabriela Roxana. 1997. The ethical function of astronomy in Plato's *Timaeus*. In *Interpreting the Timaeus-Critias: Proceedings of the IV Symposium Platonicum*, ed. Tomás Calvo and Luc Brisson, 341–350. Sankt Augustin: Academia Verlag.

———. 2005. *Plato's cosmology and its ethical dimensions*. Cambridge: Cambridge University Press.

Carpenter, Amber D. 2006. Hedonistic persons: The good man argument in Plato's *Philebus*. *British Journal of the History of Philosophy* 14 (1): 5–26.

———. 2008. Embodying intelligence: Animals and us in Plato's *Timaeus*. In *Platonism and forms of intelligence*, ed. Marie-Elise Zovko, 39–56. Berlin: Akademie Verlag.

———. 2011. Pleasure as *Genesis*. *Ancient Philosophy* 31: 73–94.

[32] Perhaps also in maintaining the temperate climate in which animals thrive; but, unlike food, this is not mentioned by Plato, and may be an anachronistic thought.

[33] Again I am indebted to Luc Brisson, for drawing my attention to this, and providing me with the edited text and French translations of the *ad Gaurum* (see especially section VI).

———. 2017. The unhappiness of the great king. In *Rereading ancient philosophy: Old chestnuts and sacred cows*, ed. Verity Harte and Raphael Woolf, 60–79. Cambridge: Cambridge University Press.

———. 2018. Illuminating community: How to learn from India's lack of a category for non-human animals. In *Oxford philosophical concepts: Animals*, ed. Peter Adamson and Fay Edwards, 63–86. Oxford: Oxford University Press.

———. 2021. Ideals and ethical formation: Confessions of a buddhist platonist. In *Reasons and empty persons: mind, metaphysics, and morality*, ed. Christian Coseru. Cham: Springer. forthcoming.

Dennett, Daniel. 1995. *Darwin's dangerous idea: Evolution and the meanings of life*. New York, NY: Simon & Schuster.

Dixsaut, Monique. 2000. *Platon et la question de la pensée*. Paris: Vrin.

Frede, Michael. 1987. Perception in Plato's later dialogues. In *Essays in ancient philosophy*, 3–8. Minneapolis: University of Minnesota Press.

Freschi, Elisa. 2015. Systematising an absent category: Discourses on nature in Prābhākara Mīmāṃsā. In *The human person and nature in classical and modern India, supplementary volume 2 to Revista delgi Studi Orientali, nuova serie*, ed. Giorgio Milanetti and Raffaele Torella, 45–54. Pisa and Rome: Fabrizio Serra Editore.

Johansen, Thomas K. 2000. Body, soul, and tripartition in Plato's *Timaeus*. *Oxford Studies in Ancient Philosophy* 16: 87–111.

———. 2020. Soul, life, and nutrition in the *Timaeus*. In *Heat, pneuma, and soul in ancient philosophy and science*, ed. Hynek Bartoš and Colin G. King, 121–139. Cambridge: Cambridge University Press.

Lennox, James. 1985. Plato's unnatural teleology. In *Platonic investigations*, ed. D. O'Meara, 195–218. Washington, DC: Catholic University of America Press.

Menn, Stephen. 1995. *Plato on God as Nous*. *Journal of the history of philosophy monograph series*. Carbondale: Southern Illinois University Press.

O'Brien, Denis. 1997. Perception et intelligence dans le *Timée* de Plato. In *Interpreting the Timaeus-Critias: Proceedings of the IV Symposium Platonicum*, ed. Tomás Calvo and Luc Brisson, 291–305. Sankt Augustin: Academia Verlag.

Schmithausen, Lambert. 1991. The problem of the sentience of plants in earliest Buddhism. In *Studio Philologica Buddhica, monograph series VI*. Tokyo: The International Institute for Buddhist Studies.

Zeyl, Donald J., tr. 1997. Timaeus. In *Plato: Complete works*, ed. John M. Cooper. Indianapolis, IN: Hackett.

Chapter 4
The Vegetative Soul in Galen

Robert Vinkesteijn

Abstract Galen of Pergamum developed a new notion of the vegetative soul as seated in the liver, in a synthetic appropriation of Platonic tripartition, Aristotelian hylomorphism, Hippocratic elemental theory and Hellenistic science. The traditional analogy between plant and human being receives a firmer grounding in Galen, making the model of the plant more prominent than ever in the discussion of the vegetative soul. While most of the particular functions of the vegetative soul in human beings are well-defined by Galen, its generative and formative powers remain unexplained, since the stuff to which we attribute these powers seems to lack intelligence.

4.1 Introduction

The notion of the desiderative soul has its roots in Plato's tripartition of the soul, particularly in the *Republic* and *Timaeus*, where it is distinguished from the spirited and rational souls and localized somewhere in or around the belly. It is characterized as irrational and passive – though capable of sensation – and related to pleasure and pain as well as desire in general (*Rep.* IV 435C f.; *Tim.* 44D-45B f., 69C-72D, 73D; Brown 2012).[1] In Aristotle, as often, this Platonic notion acquires more clarity and specificity. It becomes the 'vegetative', 'generative' or 'nutritive' soul, characterized by its specific powers, which are now reduced to basic biological functions. It does not share in sensation or desire in general (*Nic. Eth.* I 13, 1102a32-b31; *De An.* II 3 414a33 f., II 4, 416a19 f.). Later, Galen of Pergamum (129-210/217) created an original synthesis and systematization of this philosophical and biological concept, including besides Plato and Aristotle also Stoic and medical influences. In this chapter, we shall look into Galen's notion of the desiderative soul and the way it differs from that of his predecessors.

[1] See Carpenter in this volume for a more complete view of the *Timaeus*.

R. Vinkesteijn (✉)
Philosophy Department, Leiden University, Leiden, The Netherlands
e-mail: r.w.vinkesteijn@uu.nl

F. Baldassarri, A. Blank (eds.), *Vegetative Powers*, International Archives of the History of Ideas Archives internationales d'histoire des idées 234,
https://doi.org/10.1007/978-3-030-69709-9_4

We shall pay particular attention to the role of plants. Comparisons and analogies with plants have been an important aspect in the tradition of the vegetative soul from its beginnings in Plato, but in Galen's work the role of plants becomes more prominent, and the analogy between human beings and plants becomes more than a mere analogy.

We shall start with a brief introduction to Galen's tripartition of the soul (I). His notion of the human soul and its relation to the body is complex, consisting of a synthesis of Aristotelian hylo-morphism and Platonic tripartition, with a strong explanatory role for the elementary qualities of the most basic bodily substances. After this brief preamble, we shall be equipped to have a look at Galen's version of the desiderative soul in the context of his interaction with Platonic and Aristotelian views (II). It will turn out that the model of the plant becomes much more prominent in Galen's discussions of the subject. Therefore, we shall have a look at the way Galen primarily understands the desiderative soul as the soul of plants (III). Galen's strong analogy between the human embryo and plants, explained in terms of similarity in substance and capacity, suggests that we can understand (new-born) human beings as plants (IV). Finally, the generative and formative power of the desiderative soul, as well as its role as the guardian of life, requires a further analysis of its relation to nature as creator (V).

4.2 Galen's Tripartition of the Soul

As is well-known, the tripartition of the soul is fundamental for Galen throughout his entire oeuvre. His own version of this psychological theory is strongly based on Plato, but incorporates much Hippocratic and Aristotelian doctrine as well (De Lacy 1988; Vegetti 2000; Tieleman 2003; Hankinson 1991, 2006; Donini 2008; Singer 2013; Trompeter 2018).

Galen's main point of departure is Plato's *Timaeus*, and there is a clear reason for this. In the *Republic*, the other main source for the Platonic tripartition, Socrates argues that the soul must be divided into three different parts, considering the empirical fact of psychic conflict and the maxim that opposite activities or affections cannot be predicated of some one thing in the same respect, at the same time and in relation to the same thing (*Rep.* IV 435C f.). For Galen, however, this was not strong enough, since, on the basis of the argument, he takes Plato to have proven merely that the soul needs to be somehow divided into three. This, however, could still mean that the soul has three different powers (δυνάμεις), rather than three different parts (μόρια) or forms (εἴδη) with their own location in the body, as Galen would have it (*PHP*[2] V 7, 336,16 De Lacy).[3] Indeed, according to Galen, this is

[2]All abbreviations used for Galen's works are the standard ones, which can be found in Singer (2013).

[3]Socrates does actually speak of a division in forms (εἴδη) and parts (μέρη), but Galen does not consider this kind of division scientifically proven on the basis of Socrates' argument. See Brown

where Aristotle and Posidonius got it wrong, since, even though they did recognise that the soul had different powers, as opposed to Chrysippus, they wrongly attributed those powers to one single substance (οὐσία) (*PHP* VI 2, 368 De Lacy; Tieleman 2018). Therefore, the tripartition of the *Timaeus* is a step closer to Galen than the *Republic* and is also preferred by him over the horse and chariot metaphor found in Plato's *Phaedrus*. In Galen's reading of the *Timaeus*, the soul does not only consist of three different powers, but of three separate forms and parts (εἴδη τε καὶ μέρη) belonging to specific bodily organs (*PHP* VI 2, 370,13 f. De Lacy).[4]

Each part of the soul is now designated a specific location in the body: the rational in the brain, the spirited in the heart, and the desiderative in the belly (*Tim.* 44D-45B, 69C-72D, 73D). This is crucial for Galen: with some adjustments, this theory can be aligned with the anatomical findings of Hellenistic science, and allows for a further integration of the soul into the body, while still doing justice to Plato's original arguments of psychic conflict (Vegetti 2000; Tieleman 2003; Schiefsky 2012).[5] In *On the Doctrines of Hippocrates and Plato* (*PHP*) and other works, Galen systematizes the tripartition from the *Timaeus*, updating it, as it were, to the science of his day, and attributing to the three souls separate seats in brain, heart and liver (rather than belly), with each having its own specific system of connections with the rest of the body: nerves, arteries and veins respectively. In this manner, the functions of all three souls can be located in concrete bodily organs, that each become a centre of a particular group of psychic functions with its own channels of distribution (*MM* IX 10, X 635 K).[6]

In works such as *The Elements According to Hippocrates* (*Hipp. Elem.*) and his commentary on the Hippocratic *On the Nature of Man* (*HNH*), Galen further elaborates on the constitution of these organs themselves, this time basing himself on a distinctive mix of Aristotelian and Hippocratic doctrine. The organs, according to Galen, consist of a specific configuration of so-called 'homoeomerous' bodies.[7]

(2012) 53 f. and Schiefsky (2012) 331 f. for further discussion of the tripartition in Plato's *Republic* and in Galen.

[4] In the *Timaeus* too, Galen does not consider this point proven scientifically, considering the status of its arguments merely 'plausible' or 'persuasive' (see Chiaradonna (2014) for Galen's use of this term), which is why one of the major aims of *PHP* is to prove, scientifically, that the soul has three separate parts located in brain, heart and liver respectively (see also Tieleman (2018)); cf. *Prop. Plac.* 8, 180,15 f. Boudon and Pietrobelli.

[5] Galen has a remarkably strong aversion to understanding man as a unity that comes to the fore in several ways and that perhaps lies at the basis of his consistent adherence to Plato's tripartition. In his commentary on the Hippocratic *On the Nature of Man*, he disqualifies the notion that 'man is one' as 'bizarre' (*HNH* 36 Mewaldt, XV 67 K)

[6] These systems of distribution are important, especially because one of the major problems with the Platonic notion of soul, according to Galen, is the question of its extension through the body, see, e.g. *QAM* IV 776 K: '... nor, indeed, do I discern how, not being any part of the body, it is able to extend through the whole of it.' (tr. Singer 2013, 382).

[7] The distinction between homoeomerous bodies and organs in this form goes back at least to Aristotle, for Galen it is basic doctrine, see, e.g., *Opt. Med.* I 60 K; *QAM* IV 773-74 K; *HNH* 6,11 f. Mewaldt (XV 7-8 K); *Part. Hom. Diff.* 51 Strohmaier; *Hipp Elem* 126 De Lacy (I 479-81K).

These homoeomerous bodies, in turn, cannot be divided further into actual parts that differ from each other in form (hence the name), but can be conceptually analyzed in matter and form, the matter being without quality, and the form being a mixture of the four elemental qualities of hotness, coldness, dryness and wetness (*QAM* IV 773 K; *Hipp. Elem.* 114, 16 f. De Lacy; *UP* 18,22 f. Helmreich; *Nat. Fac.* I 6, II 12 K). In some of his later work, particularly *That the Powers of the Soul Follow the Mixtures of the Body* (*QAM*), Galen seems to work out the final consequences of this combination of Platonic tripartition and Aristotelian hylomorphism. Combined with his general assumption that what is primarily active is not the organ as a whole but rather its smallest constitutive parts, he experiments with the idea that the three parts of the soul (being *sources of movement*, after all) are identical with the mixtures that constitute their respective organs. Thus, it has been noted that Galen, starting with *PHP*, seems to work towards an increasingly corporeal reinterpretation of the Platonic tripartite soul (Vegetti 2000; Tieleman 2003; Vinkesteijn, 2020). In this process of systematization and somatisation, the soul that was first described by Plato as 'desiderative' (ἐπιθυμητικόν), and that is the subject of this volume under a more Aristotelian name, undergoes some fundamental changes, which we will discuss in the following section.[8]

4.3 Galen's Adaptation of the Desiderative Soul

First of all, Galen locates the desiderative soul in the liver, whereas in Plato its location seemed to have been the belly (*Tim.* 70D-E). His arguments for this specific location are less compelling than those for the location of the other two souls, in Galen's own perception. That the rational part resides in the brain and the spirited part in the heart, can be proven scientifically, after all, on the basis of anatomical demonstration. With the liver as source of the veins and seat of the desiderative soul this is more difficult (Tieleman 2003, 153 f.). Galen does provide several arguments for its relocation, however: the liver connects with the stomach and the rest of the body as the source of the veins; its bloodlike substance indicates that it produces our blood; the liver and veins show a structural similarity to the root system of plants, which would imply that its function is also that of nutrition (*PHP* book VI; Schiefsky 2012, 340 f.). In short, the liver is understood by Galen as the source of nourishment, since it alters the foodstuff that it receives from the stomach into blood, and distributes it through the veins in order to provide each part of the body with its proper nourishment.

From this perspective, it might make sense to locate the desiderative soul here, since it was concerned, among other things, with the desire for food and drink. However, the relocation of the desiderative soul to the liver also involves a strong

[8] 'Vegetative' translating φυτικὸν, distinguished by Aristotle from the ἐπιθυμητικὸν in *Nic. Eth.* I, 13, 1102a32-b31, and said to be the cause of nutrition and growth, present in both embryos and full-grown animals.

restriction of its activities to the functions of nourishment and growth (De Lacy 1988; Vegetti 2000, 77–8; Tieleman 2003, esp. 158–9). The function of sexual desire, prominent in Plato, is muffled away somewhat, and Galen seems to integrate desire as such more into the rational soul, seated in the brain.[9] The upshot of his relocation of the desiderative soul is that the *desiring* that it does, is explained by Galen merely in terms of attraction of nourishment. This interpretation, although it comes with problems of its own, conveniently takes care of the Platonic problem with the attribution of desire in a broader sense to plants. After all, since Plato considered plants to have the desiderative soul as well, he would have to explain how plants have desire. With the notion of desire being restricted to the power for attraction of nourishment through a root-like organic system, this is no longer a problem (*Tim.* 77B; *Prop. Plac.* 15, 189,23 f. Boudon-Pietrobelli; Carpenter 2010).

In this way, the functions of the desiderative soul (indeed, now it becomes somewhat more appropriate to call it 'vegetative' or 'nutritive') become much more similar to the functions actually exercised by plants. One could say that Galen, seen from the perspective of the *Timaeus* and its problematic attribution of desire to plants, invites us to reason the other way around: the question is not how plants are somehow able to have desire and are thus like us to some extent, but rather how part of us is actually like a plant. Another way of putting this, is that Galen's notion of the vegetative soul is less anthropocentric than the one we find in the *Timaeus*. Indeed, the functions of our vegetative soul are sometimes described by Galen as being 'natural', and he remarks on several occasions that it does not matter whether we call this part 'soul' or 'nature' (*UP* IV 7, I 201,19–202,2 Helmreich; *Nat. Fac.* I 1, II 1–2 K; *MM* IX 10, X 635 K; *PHP* VI 3, 374,18 De Lacy; De Lacy 1988, 53 f.; Tieleman 2003, 158–9). In this manner, Galen transforms the third soul as the seat of the psychological function of desire more or less metaphorically located in the belly, into that of the physiological function of nutrition shared by all life-forms and concretely located – in our case – in the organ of the liver (Vegetti 2000).

Plato's thesis that plants take part in the desiderative soul was important to Galen for another reason. It underlined and proved that this soul could exist autonomously, i.e. apart from the other two souls. This served to prove that the three souls are truly separate, and not merely three different functions of one underlying substance, as Aristotle and Posidonius would have it. Thus, whereas in Plato the ascription of the desiderative soul to plants was not much more than a passing remark without further elaboration, in Galen, the model of the plant now comes to play a strong role in the analysis of our tripartite soul.

Galen doubtlessly found precedents for these views in Aristotelian philosophy. In the *Nicomachean Ethics*, when discussing virtue of the soul, Aristotle distinguishes between the vegetative (τὸ φυτικὸν) and the desiderative powers (τὸ ἐπιθυμητικὸν), together making up the irrational part of the soul (*Nic. Eth.* I 13,

[9] In some works, though not in *PHP*, Galen introduces the testicles as a fourth principle responsible for the function of reproduction (cf. Tieleman (2003) 159 f., including references in note 70); See Vegetti (2000), Schiefsky (2012) and Trompeter (2018) for discussion of the complex issue of the relation between desire and rationality in Galen.

1102a32-b31). The general difference between these two kinds of powers is their respective share in rationality: the desiderative power takes some part in rationality, while the vegetative does not take any part in it. Furthermore, the vegetative power is said to be the cause of nutrition and growth, operative from the embryonic state onwards. Aristotle also mentions a nutritive (θρεπτική) and generative power (γεννητική) in *De Anima*, calling them the same power (*De An.* II 4, 416a19 f.). It seems as if Aristotle uses these different names to emphasize the various aspects of one and the same power - namely the one he termed vegetative in the *Nicomachean Ethics* - separating it from the powers that have at least some share in rationality. Thus, in Aristotle's view, the vegetative power seems to be concerned with everything that is needed to generate and sustain a living being and that is carried out without any rational or even voluntary interference. This, indeed, is very close to how Galen came to view the functions of the vegetative soul later.

The problem with the attribution of desire to plants, raised in Plato's *Timaeus*, is also taken up by Aristotle. Since plants do not possess sensation, they cannot, according to Aristotle, have desire. A position that is affirmed by Galen in *On the Use of Parts* (*UP*), although he sometimes also attributes a limited form of sensation to plants (*De An.* II 3, 414a33 f.; De Lacy comm. *PHP* 662; *UP* IV 7, 201–2 Helmreich; *Prop. Plac.* 15, 189,17 Boudon-Pietrobelli). This form of sensation, however, needs to be understood in the same sense as Galen understands the 'desire' of the desiderative soul merely in terms of the attraction of foodstuff: the vegetative soul naturally has the capacity for determining what needs to be attracted, how its nourishing properties need to be retained, how it needs to be altered into the proper humoural substance and to which organs and body-parts it needs to be distributed, what needs to be expelled, etc. These operations, Galen reasons, assume some kind of capacity for rudimentary sensation, if only to determine whether or not a specific substance needs to be attracted and retained or not.

An important difference between Galen and Aristotle is that Galen emphatically considers the vegetative soul as a separate substance having powers, and not itself as a power of a substance. This allows Galen to take the identification of our desiderative soul and the soul of plants a step further: it concerns not merely an analogy or even identity in kinds of powers or functions, but it is actually the same soul, simply existing by itself in the case of plants. In the next section we'll delve further into the role of plants in Galen's discussion of the desiderative soul.

4.4 The Desiderative Soul as the Soul of Plants

When Galen begins his discussion of the vegetative soul and its location in *PHP* book VI, he quickly relates the discussion to plants:

> 'τοιαύτης γὰρ δυνάμεως ἀρχὴ τὸ ἧπαρ οἵα καὶ τοῖς φυτοῖς ὑπάρχει. καλείσθω γὰρ ἔν γε τῷ παρόντι δύναμις, ὕστερον ἐπιδειξόντων ἡμῶν ἀκριβέστερον ὡς πολλῶν ἐστι δυνάμεων ἀρχὴ τὸ ἧπαρ καὶ κάλλιον οὐσίαν ψυχῆς ὀνομάζειν, οὐ δύναμιν, ἐν ἑκάστῳ τῶν τριῶν σπλάγχνων περιεχομένην, ἐν ἐγκεφάλῳ μὲν λογιστικήν, ἐν καρδίᾳ δὲ θυμοειδῆ, κατὰ δὲ τὸ ἧπαρ ἐπιθυμητικὴν...' (*PHP* VI 3, 374,9-14 De Lacy)

'Indeed the liver is the source of the same kind of power that exists also in plants. For the present, let it be called a power, although we shall demonstrate later with more precision that the liver is the source of many powers, and it would be better to speak of a substance of the soul, rather than a power, enclosed in each of the three internal organs: in the brain, the rational substance, in the heart, the spirited, and in the region of the liver, the desiderative…' (tr. De Lacy 1980, 375, slightly modified)

In this passage we recognize Galen's emphasis on the threefold division of the soul being a division in separate substances (οὐσίαι) located in specific organs, rather than a division of powers of a single substance. The liver is to be understood, according to this passage and many others, as the source (ἀρχὴ) of the same kind of power that exists in plants. Since the power itself presupposes a substance as its source and since we can see plants exercising this kind of power, the study of plants may teach us something about the substance of our own desiderative soul. In fact, since plants lack the other two souls and this is the only kind of soul they have, they are most suitable for studying the desiderative soul, according to Galen:

'κάλλιστον οὖν εἰς τοῦτο περὶ τῆς κατὰ τὰ φυτὰ γεννήσεώς τε ἅμα καὶ διοικήσεως ἐπισκέψασθαι πρότερον· εἰκὸς γάρ που μόνην ἐκείνοις ὑπάρχουσαν τὴν ζητουμένην δύναμιν ἐναργέστερα τεκμήρια τοῦ μέρους ὅθεν ὁρμᾶται παρασχέσθαι.' (*PHP* VI 3, 374,29-32 De Lacy)

'For the investigation of this matter it is best to examine first the generation and government of plants; for it is reasonable to expect that since the power we are investigating is the only (power) they have, they will provide clearer indications of the part that is its source.' (tr. De Lacy 1980, 375)

It is important to note here that for Galen it is *actually* the same power that we find both in plants and in the human liver: he simply refers to 'the power under investigation' (τὴν ζητουμένην δύναμιν). Since this power is present in plants by itself, it is easier to analyse.[10] Galen's basic assumption seems to be that the next grade on the 'scala naturae' (whether one moves from plants to animals or from animals to human beings) consists in principle of an addition with respect to the previous grade, rather than an alteration, and that therefore the component parts of an advanced grade can be understood through analysis of a less advanced grade.

As we already noticed, Galen infers a similarity in substance from the similarity in power (the latter being observable). In the context of Galen's theoretical framework, this makes sense. For, as he explains in *QAM* and elsewhere, we should understand the cause or principle of an activity always to be a substance (or a 'nature', in this sense the two are synonymous for Galen), while the word 'power' or 'capacity' (δύναμις) is merely a name that we apply to describe the relation of causality between that substance and something else (*QAM* IV 769-71 K). In Galen's view, we predicate of a certain thing that it has a capacity for acting upon something or being acted upon by something in a certain way, when we do not know

[10] Cf. *Foet. Form.* 68,21 Nickel (V 665-6 K): 'As their soul is a single thing, of one type – for they have neither a spirited nor a rational part – there is some hope that we shall find the management of plants to be a pure unadulterated thing too.' (tr. Singer 1997, 183)

the actual cause for this action or affection, that is, when we do not fully understand the substance of the thing (*Prop. Plac.* 14, 187,14 Boudon-Pietrobelli; *Nat. Fac.* I 4, II 9–10 K; Frede 2003, 94; Hankinson 2003, 51; Tieleman 2003, 144–51). Thus, the fact that plants exhibit the same powers as our liver, suggests that they have a substance or nature that is, to some extent at least, similar to that of our liver. This is exactly how Galen interprets Plato's remark on plants taking part in the desiderative soul as well, as we can see from one of the fragments of his (mostly lost) commentary on the *Timaeus*:

'... ὁ Πλάτων εἶδος ἔφη ψυχῆς εἶναι τὸ ἐπιθυμητικόν, οὐχ ἑτέραν μὲν ἔχον ἐν φυτοῖς, ἑτέραν δὲ ἐν ἡμῖν φύσιν, ἀλλὰ τὴν αὐτήν, ᾠκίσθαι τε κατὰ τὸ ἧπαρ...' (Schröder 11,25)

'... Plato says that the desiderative is a form of soul, that does not have a different nature in plants than it has in us, but the same nature [in both], and it dwells in the liver...'

Book VI of *PHP*, about the desiderative soul and its location in the liver, is indeed full of analogies with plants, but as we can see here the similarity is taken further than that of mere metaphor or analogy by Galen: *the nature (or substance) of the soul of plants is the same as the nature (or substance) of our desiderative soul.* This explains why the similarity between the powers of the liver and those of plants is so strong for Galen. He sometimes even seems to get lost in the metaphor, as we can see in this passage from *PHP* VI:

'οἰκειότερον δ᾽ ἦν ἄρα μήτε τὴν ἐκ τοῦ ἥπατος ὁρμηθεῖσαν φλέβα πρέμνῳ προσεικάζειν εἰς πολλοὺς κλάδους σχιζομένῳ μήτε τὰς ἐσχάτας τελευτὰς βλαστήμασιν, ἀλλὰ ταύτας μὲν αὐτοῖς τοῖς πέρασι τῶν ῥιζῶν τοῖς εἰς τὴν γῆν ἐμπεφυκόσι, τὴν μεγίστην δὲ φλέβα τῇ συμπάσης τῆς ῥιζώσεως ἀρχῇ. φαίνονται γάρ πως εἰς τὴν μήτραν αἱ τελευταὶ τῶν φλεβῶν ἐρριζωμέναι τροφὴν ἐξ αὐτῆς ἀναφέρειν ἐπὶ τὸ ἧπαρ ὡς ἀρχὴν ἅπαντος τοῦ φυτοῦ.' (*PHP* VI 6, 402,11 De Lacy)

'Actually it would have been more appropriate not to compare the vein issuing from the liver to a trunk being split into many branches, or the furthest ends to shoots, but to compare these (latter) to the very tips of roots growing into the earth and the largest vein to the source of the whole root-growth. For in a way the ends of the veins, rooted in the uterus, appear to bring nutriment from it to the liver as the source of the whole plant.' (tr. De Lacy 1980, 403)[11]

Oddly, while starting with a comparison between liver and plant, at the end of this passage Galen seems to call the liver the source of a plant. This does not make sense at first sight, since plants do not have livers. Perhaps Galen writes sloppily in this passage and simply gets lost in his own analogy, accidentally mixing up humans and plants? Again, I think we should reason the other way around to make sense of this passage: if Galen seems to call the liver source of a plant, we should not look for livers in plants, but we should see the extent to which those beings endowed with

[11] Cf. *PHP* VI, 3, 382, 17 f. De Lacy, where Galen considers the vein connecting liver and stomach similar to the roots of plants, the stomach being like the earth, and the vein from the liver to the rest of the body as similar to the stem of plants. Thus, the liver itself is compared to that part of the plant from which both roots and stalk sprout, which Galen calls ῥίζωσις and which he considers to be the seat of the plant's soul; cf. *Sem.* 90,16 f. De Lacy (IV 539-40 K)

a liver could be considered plants. And indeed, for the latter perspective we can find excellent parallels in Galen, as we shall see below. Moreover, for the comparison, and even identification of the human being with a plant, Galen can draw on an extensive tradition. In the previous tradition, however, the comparison seems to exclusively concern the human being in the earliest stages of its development, as an embryo. As we shall see, Galen takes this idea a step further, comparing the human being to a plant not merely in its embryonic stage, but even beyond. In what follows, we shall delve deeper into Galen's radicalization of the metaphor of the plant for the human liver and vegetative soul, after some brief allusions to a few precedents for regarding the human embryo as a plant.

4.5 The Human Being as a Plant

The Hippocratic treatise *On the Nature of the Child* (*Nat. Puer.*) must be one of the oldest Greek texts that we have in which the human embryo is extensively compared to plants (Lonie 1988, esp. 231–9). The familiar analogies used in Aristotle and Galen can already be found there: the mother is to the embryo like the earth is to a plant and the various channels and the veins and nerves growing from the organs are compared to branch- and root-structures (*Nat. Puer.* 22,1-27,1, 13–18 Lonie). After a remarkably lengthy description of the functioning of plants and trees, the author concludes that 'from beginning to end the process of growth in plants and in humans is exactly the same'. This must have been to Galen's liking, considering his own identification of the powers of plants and our powers of growth and nutrition, and in fact Galen refers to this work (*Nat. Fac.* II 3, II 86 K).

Presumably inspired by this Hippocratic text or later medical tradition, Aristotle in his *On the Generation of Animals* compares the life of the embryo to that of a plant, by virtue of it 'being fastened on to something':

> 'διὰ μὲν οὖν θατέρου λαμβάνει τὴν ἐκ τοῦ ὠχροῦ τροφήν, τὸ δ᾽ ὠχρὸν γίνεται πλέον· ὑγρότερον γὰρ γίνεται θερμαινόμενον, δεῖ γὰρ τὴν τροφὴν σωματώδη οὖσαν ὑγρὰν εἶναι καθάπερ τοῖς φυτοῖς, ζῇ δὲ τὸ πρῶτον καὶ τὰ ἐν τοῖς ᾠοῖς γιγνόμενα καὶ τὰ ἐν τοῖς ζῴοις φυτοῦ βίον· τῷ πεφυκέναι γὰρ ἔκ τινος λαμβάνει τὴν πρώτην αὔξησιν καὶ τροφήν.' (*GA* III 2, 753b25 f.)[12]
>
> 'Through one of these cords the embryo receives the nourishment from the yolk; and the yolk increases in bulk, becoming more fluid as it is heated, since the nourishment, being corporeal, must be available in fluid form, just as it must for plants, and the embryos that are in process of formation, either within the egg or within the uterus, are to begin with living the life of a plant, since their first growth and nourishment they obtain through being fastened on to something.' (tr. Peck 1942, 295)

[12] cf. also V, 1, 779a1 f.: 'But on the other hand, if it is necessary that the animal should have sensation and if it is first an animal when it has acquired sensation, we ought to consider the original condition to be not sleep but only something resembling sleep, such a condition as we find also in plants, for indeed at this time animals do actually live the life of a plant.' (tr. Barnes 1984, 1204)

The human embryo lives the life of a plant. Aristotle here justifies this comparison in terms of the dependency of the embryo upon something else for its growth and nourishment, and not in terms of the identity of capacities or substance. For plants, it is the earth which they are dependent upon in this way and we find several analogies in Aristotle's biological works, where the mother (and interestingly also, after birth, the stomach) is compared with the earth, and the roots of plants with the mouths of animals (*GA* IV 6, 774b25 f.; *Iuv.* I, 468a9-12; *IA* 4, 705b6-8; *PA* II 3, 650a20-27; *PA* IV 4, 678a12-14). Aristotle must have derived these analogies from the medical tradition and they can be found extensively in Galen as well (Kovacic 2001, 78 f., 125 f.).

For Aristotle, the human being temporarily lives like a plant, while it is still directly dependent upon its mother for nourishment, and becomes animal after it acquires sensation and becomes (potentially) locomotive, so that it can acquire its own food.[13] Thus, in Aristotle's analogy there is a strong emphasis on the dependency for nourishment upon something else, an emphasis that we also find in the Hippocratic *On the Nature of the Child* (22,1, 13 Lonie).

We find the analogy of the human foetus with plants also in the Stoics, although they do not ascribe 'soul' to plants (Long 1982, 38 f.; Tieleman 1991; Gourinat 2008). But according to Galen, as we have seen, this is a matter of mere terminology. Indeed, the Stoic terminology, where the plant's powers of growth, nourishment and reproduction are explained in terms of 'nature' rather than 'soul', seems more suitable to Galen's view on the desiderative soul in general.

In any case, we can safely assume that Galen was thoroughly familiar with these Hippocratic, Peripatetic and Stoic precedents, and that he continues the idea of the human foetus as a plant on their basis. Given his strong emphasis on the tripartition of the soul into three separate substances, however, for him it must be the case that the human foetus has one soul, like plants, but does not have the other two (operative) yet. Also, for Galen, the reason why the human foetus can be considered a plant, is not so much its dependency upon something else or the lack of soul or certain functions, but much rather the similarity in powers, that in turn must be dependent on a similarity in substance. This is actually the reason why Galen rebukes Aristotle for separating the 'material' and 'formative' principle in animals (the material principle being contained in the female, the formative in the male): Aristotle did not separate those in plants and thus 'explains the workings of nature differently in plants and in animals' (*Sem.* I 9, 94,24 f., translation De Lacy). The key issue for Galen here is that we understand the generative processes of plants and animals as essentially the same, since the soul that guides this process is the same in both (Kovacic 2001, 79). In fact, in the passage right after rebuking Aristotle, Galen proceeds to give a lengthy comparison between the generation of plants and animals, that serves to prove that with regard to their generation they are functionally

[13] Cf. also *GA* II 4, 740a24-b2: 'Since the embryo is already potentially an animal but an imperfect one, it must obtain its nourishment from elsewhere; accordingly it makes use of the uterus and the mother, as a plant does of the earth, to get nourishment, until it is perfected to the point of being now an animal potentially locomotive.' (tr. Barnes 1984, 1148)

identical. The comparison does not stop at mere analogy for Galen: 'For if the nature of the seed does not produce plants with one set of powers and animals with a different set, then you must make a transfer to animals from what you see in plants.' (*Sem.* I 9, 96,3 translation De Lacy). Galen's point is, again, that the powers of plants are actually the same as those of the vegetative soul in animals and that we therefore must infer a similarity in nature or substance as well. Again, this implies that we can learn a lot about the human embryo through the study of plants, as Galen also remarks in his *On the Formation of the Embryo* (*Foet. Form.*):

> '...τὸ κυούμενον οὔτ᾽ ἀρτηριῶν ἔχον ἀναγκαίαν χρεῖαν ἐν ἀρχῇ τῆς γενέσεως οὔτε σφυγμῶ οὔτε καρδίας ὥσπερ οὐδὲ τὰ φυτὰ. καὶ μέντοι καὶ περὶ τῆς τῶν φυτῶν γενέσεως ἐσκέφθαι τι χρὴ πρότερον. ἐκ γὰρ τῶν εἰς ταῦτα ἀναγκαίων ἔνεστι καὶ γιγνώσκειν, ὁποίων τε καὶ ὁπόσων δεῖται τὸ κύημα μέχρις ἂν ὑπὸ μιᾶς διοικῆται ψυχῆς ὡς τὰ φυτά.' (*Foet. Form.* 68,7 Nickel, V 665 K)

> '... the embryo has no need of the function of the arteries in the first stages of its formation, nor of pulses, nor of the heart – no more than do plants. We should perhaps have made some enquiry into the formation of plants earlier. From a consideration of the conditions necessary for plants, we shall be able to learn exactly what needs the embryo has during the period in which it is still managed by one soul in the same way as plants are.' (tr. Singer 1997, 182-3)

Here, too, Galen puts emphasis on a comparison in terms of function. The difference between plants and animals is simply that plants are managed by only one of the three souls, whereas animals and human beings have more souls added to their first one, with other powers. The crucial point is that the first stage is the same in all, and that the substantial difference between plant, animal and human being, consists of an addition much rather than an alteration, in Galen's view. This comes to the fore clearly in another passage from Galen's *On Semen* (*Sem.*), again shortly after rebuking Aristotle for explaining the workings of nature differently in plants and animals:

> 'εἰ δὲ τριπλοῦν ὁρᾷς τὸ φυτόν, ἢ ἔναιμον, ἢ σαρκοειδές, οὐ χρὴ τούτων ἕνεκα νομίζειν ἄλλο τι καὶ οὐ φυτὸν ὑπάρχειν αὐτό. τὸν γοῦν δημιουργὸν τὸν αὐτὸν ἔχει ἀμφότερα, τὴν φυτικὴν ψυχήν. ἀλλ᾽ ἐκεῖνο σκόπει, ὅτι τὸ φυτὸν τοῦτο μέλει γενήσεσθαι ζῷον, οὐκ ἀποβαλὸν ἥν εἶχεν ἐξ ἀρχῆς δύναμιν, ἀλλ᾽ ἑτέραν ἐπικτησάμενον.' (*Sem.* I 9, 96,16-21 De Lacy)

> 'If you see that the plant is three-fold, or has blood in it, or is fleshy, you ought not for those reasons suppose that it is something else and not a plant. Both have the same craftsman, the vegetative soul. Look at it this way: this plant is going to become an animal not by losing the power that it had from the beginning, but by acquiring another one.' (tr. De Lacy 1992, 97)

It is not entirely clear, at first, what Galen means here by 'three-fold'. In the previous passage, he said that plants 'have a two-fold growth' from their seeds, namely the downward root-growth and the upward growth of stalk and branches, and that 'so also the embryos have much-divided outgrowths consisting in arteries and veins that extend as stalks to the whole foetus and as roots to the uterus.' (*Sem.* I 9, 94,18 f. translation De Lacy) Thus, it seems that with three-fold, the addition of the third system of connections besides arteries and veins, the nerve-system, is included. Likewise, after the passage cited above, Galen clarifies in what way three-fold is meant: the embryo needs to be capable of locomotion after being released from the uterus, and therefore 'acquired two additional sources' that keep it warm (a main function of the heart) and enable voluntary movement (a main function of the brain). Again, 'three-fold' seems to refer to the three souls or their respective organs with their systems of connections throughout the body. After all, the organs can already be found in the embryo even when not all three souls are active yet. Thus, what Galen seems to be saying here, is that even though something possesses not only a liver but a heart and brain as well, and even though it has blood and is fleshy, it can still be called a plant, as long as its primary functions are the same as those of a plant. From this perspective, what happens if the human embryo develops further, it seems, is not a negation or perhaps even change of its plant-nature, but rather just an addition of extra functions. The condition for being a plant, it seems, is just to be crafted by the vegetative soul. In that sense, we have the same craftsman as any plant, and we are not only a kind of ex-plants, as seemed to be the case with the Hippocratics, Peripatetics and Stoics, but rather, we could say with Galen that in as far as we are alive, *we are still primarily plants.*[14] This is why, for Galen, explicitly opposing himself here to Aristotle, the first stage of life is the same for all living beings: all are crafted by the same vegetative soul. For Galen, the plant is therefore not only the basic model of life as such – as it was for much of the previous tradition as well – but all life must also remain plant as long as it lives. In this sense, an animal is simply a plant with the powers of locomotion and sensation added to it, just as a human being is an animal with rational powers. Therefore, just as some evolutionary neuro-psychologists of today hold that the study of animals could teach us about human emotions because they assume humans to be animals with further capacities, Galen could say that the study of plants could teach us about the human capacities for nourishment and growth, because both animals and humans are plants with further capacities. That is the case, for Galen, because all three share a common cause, the substance of the vegetative soul as the cause for life as such.

But, we might ask at this point, what is this mysterious substance? In the passage from *Sem.* cited above, the vegetative soul was called 'δημιουργός' by Galen, translated as 'craftsman' by De Lacy. This is remarkable, since usually Galen refers to 'nature' (φύσις) as the creator of beings, and not to the vegetative soul particularly. The question of what this nature is and how exactly it relates to the beings it creates, as well as the question of its relative immanence or transcendence, is not an easy one in Galen (Kovacic 2001; van der Eijk 2014). Besides, Galen is also notorious for taking apparently contradictory views on speculative subjects such as these, or for eventually refraining from commitment to any clear position. In this passage,

[14] Cf. Kovacic (2001) 195: 'Der Embryo ist gerade noch Pflanze, die im Begriffe ist, Tier zu warden. Die Pflanze wirft nicht die Fähigkeit weg, die sie von Anfang an besitzt, sie bekommt vielmehr – im Übergang zum Tier – noch andere Fähigkeiten dazu: die Loslösung von der Gebärmutter und die örtliche Bewegung. Daher hat das Tier etwas Pflanzenhaftes in seiner Physiologie; die Pflanze ist wiederum nicht grundverschieden vom Tier, sie ist sogar sein Modell.'

however, Galen clearly considers the vegetative soul to be the creator, in some sense at least, of both plants and animals, including human beings. This raises the question of the creative power of the vegetative soul. If the vegetative soul is the first of the souls to come to be, and if the other two souls presuppose its existence, can it somehow be considered as their maker? But, then again, would it not be odd if the soul that is 'lowest in hierarchy' turns out to be the creator of the other two? And what is it that possesses this creative power, what is the substance of the vegetative soul? In the final section of this chapter, we shall address these questions.

4.6 The Creative Power of the Vegetative Soul?

In his *On the Natural Faculties* (*Nat. Fac.*), Galen considers the artistry of nature in discussion with Erasistratus. He uses the standard metaphor of artists making statues and poses the question: what is the cause of animal generation, besides seed and blood, that governs the proportionate combination of these two substances? According to Galen, Erasistratus would have answered: 'Obviously, the semen itself' (*Nat. Fac.* II 3, II 84 K). Interestingly, in Galen's metaphor, the seed is compared to the artificer, and the blood to the matter the artificer uses to construct his statue.[15] All the powers that Galen usually attributes to nature as the creator and guardian of living beings, are one by one attributed to the semen itself in this passage: the power of attraction of appropriate substance (more specifically: the right amount of blood), the alternative power (for properly altering the underlying substance, considered to be operating at the level of the creation of homoeomerous bodies, particularly), and the shaping power (for giving it the proper place, outward form and configuration, considered to be operating at the level of the organs, particularly).[16] Especially the latter power is consistently considered artistic in Galen's work. At the same time, though, Galen is careful to not attribute intelligence to the semen itself (*Nat. Fac.* II 3, II 85 K; *Foet. Form.* 104,15 De Lacy). This is somewhat confusing, since Galen generally assumes that nature must be intelligent, because it is the cause of the generation of such well-designed creatures – this could be said to be the main point of his *On the Use of Parts* (*UP*). But here, the powers through which nature generates living beings, are attributed to the seed, while

---

[15] In *Nat. Fac.* Galen seems to adopt more of an Aristotelian perspective than in other works (esp. *Sem.*): he seems to simply accept the hylomorphic division into male seed as form and female blood as matter here. Partly, this may be due to the polemic with Erasistratus.

[16] *Nat. Fac.* II, 3, 85–6 K; the specific terms are: ἑλκτικήν τινα δύναμιν, ἡ ἀλλοιωττική δύναμις, ἡ διαπλαστική [δύναμις]; cf. in particular I, 5, 10 K for attribution of the latter two terms to nature as creator: 'Genesis, however, is not a simple activity of Nature, but is compounded of alteration and of shaping.', where the same hylomorphic metaphors are employed as well. The power of attraction is related to nature as a guardian of her creatures, that does not abandon them after creation, see, e.g., *Nat. Fac.* II, 3, 80 K; in *Foet. Form.* 62,14 f. Nickel (V 659 K f.), we find the same notion of the seed constructing the embryo, there also compared with the construction of trees; cf. also *Sem.* 90, 7–92,12 f. De Lacy (IV 539–41 K)

intelligence is denied of it. Does that mean that we have to consider nature as something intelligent that employs its powers *through* the seed? But then, why would Galen liken the seed itself to the artificer, and not rather to a mere material used by the true artificer? As is often the case with Galen (and perhaps for good reasons), it is difficult to extract from his works a clear and definitive answer on speculative questions such as these. However, the crucial thing to note from the passage in *Nat. Fac.*, is that seed and nature are actually equated, and it is said of the seed that it 'becomes a nature' when it is in touch with its proper material (the blood):

> 'ὃ γὰρ ἦν πρότερον σπέρμα, τοῦθ', ὅταν ἄρξηται φύειν τε καὶ διαπλάττειν τὸ ζῷον, φύσις τις γίγνεται.' (*Nat. Fac.* II 3, II 83 K)
>
> 'For that which was previously semen, when it begins to bring forth and shape the animal, becomes a specific nature.'

Again, here Galen says of the semen itself that it 'brings forth' and 'shapes' the living being. Likewise, in *Foet. Form.*, Galen says that the semen bears the 'formula of the creator' (*Foet. Form.* 86,17-8 Nickel). This seems to imply that the substance of the vegetative soul, at least during the stage of generation, must be either the semen itself or contained in the semen somehow, since its powers are attributed to it. Indeed, while it now seems to become difficult to uphold the difference between the generative powers of nature and those of the seed, the same difficulty applies between the generative powers of the seed and those of the vegetative soul. We already noted, in the passage from *Sem.* quoted above, that Galen called the vegetative soul the creator (δημιουργός) of both plants and animals. The same creative role that he ascribes to the powers of the semen, are also attributed by him to the power of the vegetative soul, as becomes clear from the following passage:

> 'τὴν φυτικὴν δ' ἀρχὴν ἁπάντων πρώτην ἔχει δημιουργοῦσαν οὐκ ἐξ αἵματος, ἀλλ' ἐξ αὐτοῦ τοῦ σπέρματος ἀρτηρίαν καὶ φλέβα καὶ νεῦρον, ὀστοῦν τε καὶ ὑμένα...' (*Sem.* 98,1-3 De Lacy; cf. *Sem.* 78,24 and 82,12-3 De Lacy)
>
> 'But (the foetus) has first of all the vegetative source, which creates not from blood but from the semen itself artery and vein and nerve, bone and membrane...' (tr. De Lacy 1992, 99, modified)

Thus, we may conclude from this that Galen considers the vegetative soul to be contained in the seed and that the powers of the seed are none other than the powers of the vegetative soul. This must be what he means when he says that the semen contains the 'formula of the creator'. Another passage from *Sem.* affirms this reading, where the powers that are attributed to the semen, are exactly the natural powers that are attributed to the vegetative soul in *Nat. Fac.*[17] This must imply, however, that the vegetative soul exists, in some form or other, before the liver does and that it

[17] *Sem.* 84,16 f. De Lacy (IV 534 K): 'As much of the semen, then, as touched the uterus immediately became membrane, as was shown a little earlier. But as for all the rest, it too, of course, had inborn powers, the power to attract what was congenial, the power by which it would retain and alter it and turn it into food for itself, and the power to expel what was alien and superfluous.' (tr. De Lacy 1992, 85); *Nat. Fac.* I, 12, II 28 K.

forms its own organ. Perhaps this is not so odd, considering the fact that it also exists in beings that do not have a liver. And indeed, as we have seen, this formative power in the seed, the vegetative soul, continually constructs all organs of both plants and animals, as one and the same power, and continues to work in them after creation, to ensure subsistence. Thus, to at least some extent, Galen identifies the vegetative soul with nature as such and locates its primary substance in the seed. This come to the fore nicely in the following passage from *Foet. Form.*, where Galen uses 'the power in the seed' and 'nature' synonymously, and makes it clear that it is the sole artificer of both animals and plants:

> 'ἐν ᾧ δὲ χρόνῳ διαπλάττει ταῦτα ἡ ἐν τῷ σπέρματι δύναμις, εὔλογον δήπου καὶ ἄλλα τινὰ διαπλάττεσθαι παρακείμενά τε τοῖσδε καὶ μεταξὺ κείμενα τούτων τε καὶ τῆς μήτρας. οὐδέποτε γὰρ εἰκὸς ἵστασθαι τὴν περὶ τὰ φυτὰ καὶ τὰ ζῷα διαπλάττουσαν δύναμιν, ἀλλὰ πᾶσαν δύναμιν ἅπασιν ἅμα τοῖς μέρεσιν ἐπιφύειν τι καὶ προσαύξειν. οὔκουν ἀποστήσεται τῆς τῶν ἄλλων δημιουργίας ἡ διαπλάττουσα τὰ ζῷα φύσις, ἀλλὰ τὰς φλέβας ὥσπερ γε καὶ τὰς ἀρτηρίας ἀεὶ προάξει κατασχίζουσα καὶ περφύσει ταύταις τἆλλα σπλάγνα, καθάπερ ἧπαρ τε καὶ καρδίαν ἐλέχθη περιφύειν, ἅμα καὶ τὸ σχῆμα τὸ πρέπον καὶ θέσιν ὅσα τ'ἄλλα τοιαῦτα προσήκει τοῖς μορίοις ἔχειν ἀπεργαζομένη προσηκόντως.' (*Foet. Form.* 86,21-88,2 Nickel)

> 'But at the same time that the power in the seed is constructing them, we may reasonably suppose that certain other parts are being constructed too: those adjacent to these and those between them and the womb. For the power that constructs plants and animals will never stand still; all the power will attach itself to all parts simultaneously, and augment them. So the nature which constructs animals will not refrain from the artificing of the other parts, but will develop the veins and arteries continuously by a process of subdivision, and the other organs in connection with these (it has been explained how liver and heart grow out from the vessels), and will in the appropriate manner bring about the correct shape and position and all the other qualities which the parts should have.' (tr. Singer 1997, 191-2)[18]

Interestingly, Galen here is talking about a power (singular) that continuously constructs plants and animals (plural). Since the power concerned is that of the vegetative soul and any vegetative soul itself is mortal for Galen, we must infer that he distinguishes – at least conceptually – between these vegetative powers as such and the same powers as the powers of specific substances. As such, apparently, this soul does not belong to any individual being and is rather common to all life, as its continuous prerequisite, while remaining anonymous itself. Again, the same goes for Galen's notion of nature: he speaks of nature in a general sense, as the artificer and guardian of all life, and in a more specific sense, as the nature of individual beings, or as semen can turn into a specific nature, forming a specific being (Kovacic 2001). Therefore, a particular vegetative soul could be said to be the specific instantiation

[18] Cf. Hipp. Elem. 128,11 De Lacy (I 482 K), where Galen says that the four elementary qualities 'alone, by altering the underlying substance, cause the elements to change into each other, and they are the artisans [δημιουργοί] of plants and animals.' (tr. De Lacy 1996, 129); there is, however, often a tension in Galen, between explaining things in terms of the elemental qualities or the mixtures they form on the one hand, and the intelligence of nature implied by the observed design of its individual beings on the other – see also the final quotation of this chapter (and see van der Eijk (2014) on this subject).

of the powers of nature that form and sustain life, operating primarily through the specific mixture of elemental qualities principally contained in the seed and later in the liver.

However, as is often the case with Galen, his reasoned scientific scepticism in the end does not allow us to interpret his writings in too dogmatic a fashion. Just as we might think we have rightly interpreted the eternal artificer of life as the vegetative soul contained in and transmitted through the seed and have identified it with nature as it operates in the formation and sustenance of particular beings, he casts his doubts: seed does not appear to be intelligent, while the elaborate design of living beings most certainly assumes an intelligent cause. Given the importance of these kinds of doubts throughout Galen's work, his strong ability to recognize the limits of his knowledge and the tension in his work between the conviction of intelligent design and the explanatory power of mixture (van der Eijk 2014), it seems apt to let him have the last word on this matter, for now, and leave matters somewhat undecided:

> 'ἐγὼ μὲν οὖν ἀπορεῖν ὁμολογῶ περὶ τοῦ διαπλάσαντος αἰτίου τὸ ἔμβρυον. ἄκραν γὰρ ὁρῶν ἐν τῇ διαπλάσει σοφίαν τε ἅμα καὶ δύναμιν οὔτε τὴν ἐν τῷ σπέρματι ψυχήν, τὴν φυτικὴν μὲν ὑπὸ τῶν περὶ τὸν Ἀριστοτέλη καλουμένην, ἐπιθυμητικὴν δ' ὑπὸ Πλάτωνος, ὑπὸ δὲ τῶν Στωϊκῶν οὐδὲ ψυχὴν ὅλως, ἀλλὰ φύσιν, ἡγοῦμαι διαπλάττειν τὸ ἔμβρυον οὐ μόνον οὐκ οὖσαν σοφήν, ἀλλὰ καὶ παντάπασιν ἄλογον, οὔτ' αὖ πάλιν ἀποστῆναι τελέως αὐτῆς δύναμαι διὰ τὴν πρὸς τὰ γεννήσαντα τῶν ἐγγόνων ὁμοιότητα.' (*Foet. Form.* 104,15 Nickel)

> 'And so I confess that I do not know the cause of construction of the foetus. For I observe in this construction the utmost intelligence and power, and I cannot allow that the soul in the seed, which Aristotle calls vegetative and Plato desiderative, and which the Stoics consider not to be soul at all, but nature, constructs the foetus, since this kind of soul is not only not intelligent, but entirely devoid of reason; nor, however, can I entirely distance myself from that opinion, in view of the similarity of the offspring to the parents…' (tr. Singer 1997, 200)

Acknowledgments The research for this chapter has been made possible by the Nederlands Organisation for Scientific Research (NWO) and falls within the project "Human Nature: Medical and Philosophical Perspectives in the Work of Galen of Pergamum" directed by prof. Teun Tieleman.

References

Barnes, Jonathan. 1984. *The Complete Works of Aristotle*. Princeton, NJ: Princeton University Press.

Bazou, Athina. 2011. *Γαληνοῦ Ὅτι ταῖς τοῦ σώματος κράσεσιν αἱ τῆς ψυχῆς δυνάμεις ἕπονται*. Athens: Ακαδημία Αθηνῶν.

Boudon-Millot, Veronique, and Antoine Pietrobelli. 2005. Galien ressuscité: edition princeps du texte grec du De Propriis Placitis. *Revue des Études Grecques* 118: 168–213.

Brock, Arthur John. 2006/1916. *Galen. On the Natural Faculties*. Cambridge, MA/London: Harvard University Press.

Brown, Eric. 2012. The unity of the soul in Plato's Republic. In *Plato and the divided self*, ed. Barney, Brennan, and Brittain. Cambridge: Cambridge University Press.
Burnet, John. 1958–62; 1900–3. *Platonis Opera*. Oxford: Clarendon.
Bywater, Ingram. 1957/1894. *Aristotelis Ethica Nicomachea*. Oxford: Clarendon.
Carpenter, Amber. 2010. Embodied intelligent (?) souls: Plants in Plato's Timaeus. *Phronesis* 55: 281–303.
Chiaradonna, Riccardo. 2014. Galen on what is persuasive (pithanon) and what approximates truth. In *Philosophical themes in Galen*, ed. Adamson, Hansberger, and J. Wilberding. London: Institute of Classical Studies, University of London.
Cooper, John, ed. 1997. *Plato. Complete works*. Indianapolis: Hackett.
De Lacy, Phillip. 1978–1980 *Galeni de placitis Hippocratis et Platonis*. CMG V 4, 1, 2, Berlin: Akademie Verlag.
———. 1988. The Third Part of the Soul. In *Le opere psicologiche di Galeno*, ed. Manulo and Vegetti. Napoli: Bibliopolis.
———. 1992. *Galeni de semine*. CMG V 3, 1, Berlin: Akademie Verlag.
———. 1996. *Galeni de elementis ex hippocratis sententia*. CMG V 1, 2, Berlin: Akademie Verlag.
Donini, Pierliuigi. 2008. Psychology. In *The Cambridge companion to Galen*, ed. Hankinson. Cambridge: Cambridge University Press.
Eijk, van der, Philipp. 2014. Galen on the nature of human beings. In *Philosophical themes in Galen*, ed. Adamson, Hansberger, and Wilberding. London: Institute of Classical Studies, University of London.
Frede, Michael. 2003. Galen's Theology. In *Galien et la philosophie: huit exposés suivis de discussions*. Entretiens sur l'antiquité classique tome XLIX. Fondation Hardt.
Gourinat, Jean-Baptiste. 2008. L'embryon végétatif et la formation de l'âme selon les Stoïciens. In *L'embryon. Formation et Animation*, ed. Brisson, Congourdeau, and Solère. Paris: Vrin.
Hankinson, Jim. 1991. Galen's Anatomy of the Soul. *Phronesis* 36 (2): 197–233.
———. 2003. Causation in Galen. *Galien et la philosophie*. Barnes and Jouanna. Entretiens sur l'antiquité classique tome XLIX. Genève: Fondation Hardt.
———. 2006. Body and soul in Galen. In *Common to body and soul*, ed. King. Berlin: de Gruyter.
———. 2017. Teleology and necessity in Greek embryology. In *Teleology in the ancient world. philosophical and medical approaches*, ed. Rocca. New York: Cambridge University Press.
Helmreich, Georgius. 1968. *Galeni de usu partium libri XVII*. Amsterdam: Hakkert.
Johnston, Ian, and G.H.R. Horsley. 2011. *Galen. Method of medicine*. Cambridge, MA/London: Harvard University Press.
Kovacic, Franjo. 2001. *Der Begriff der Physis bei Galen vor dem Hintergrund seiner Vorgänger*. Stuttgart: Franz Steiner Verlag.
Kühn, Karl Gottlob. 1964–1965; 1821–1833. *Claudii Galeni Opera Omnia*. Hildesheim: George Olms.
Long, Anthony. 1982. Soul and body in Stoicism. *Phronesis* 27: 34–57.
Lonie, Iain M. 1988. *The hippocratic treatises "On Generation", "On the Nature of the Child", "Diseases IV"*. Berlin/New York: Walter de Gruyter.
Mewaldt, Johannes. 1914. *Galeni in Hippocratis De natura hominis*. CMG V 9, 1, Leipzig, Berolini: Teubner.
Nickel, Diethard. 2001. *Galeni de foetuum formatione*. CMG V 3, 3, Berlin: Akademie Verlag.
Peck, Arthur Leslie. 1942. *Aristotle. Generation of animals*. Loeb Classical Library 366, Cambridge: Harvard University Press.
Schiefsky, Mark. 2012. Galen and the tripartite soul. In *Plato and the divided self*, ed. Barney, Brennan, and Brittain. Cambridge: Cambridge University Press.
Schröder, Heinrich Otto, and Paul Kahle. 1934. *Galeni in Platonis Timaeum commentarii Fragmenta*. Lipsiae; Beroline: Teubner.
Singer, Peter. 1997. *Galen. Selected works*. Oxford: Oxford University Press.
———, ed. 2013. *Galen: Psychological writings*. New York: Cambridge University Press.

Strohmaier, Gotthard. 1970. *Galeni De partium homoeomerium differentia*. CMG Supplementum Oriëntale, Berlin: Akademie Verlag.

Tieleman, Teun. 1991. Diogenes of Babylon and Stoic embryology. Ps. Plutarch, "Plac". V 15.4 Reconsidered. *Mnemosyne* 44: 106–125.

———. 2003 Galen's Psychology. in ed. Barnes and Jouanna, *Galien et la philosophie*, Entretiens sur l'antiquité classique tome XLIX. Genève: Fondation Hardt.

———. 2018. Galen and doxography. In *Aëtiana IV: papers of the Melbourne colloquium on ancient doxography*, ed. Mansfeld and Runia. Leiden: Brill.

Trompeter, Julia. 2018. The actions of spirit and appetite: Voluntary motion in Galen. *Phronesis* 63: 176–207.

Vegetti, Mario. 2000. De caelo in terram. Il Timeo in Galeno (De placitis, quod animi). In *La filosofia in eta imperiale*, ed. Brancacci. Napoli: Bibliopolis.

Vinkesteijn, Robert. 2020. Mixing body and soul: Galen on the substance of soul in *QAM* and *De Propriis Placitis*. *Phronesis* 65 (2): 224–246.

Chapter 5
Avicenna on Vegetative Faculties and the Life of Plants

Michael Fatigati

Abstract This essay presents an overview of what Avicenna (Ibn Sīnā, d. 1037) has to say across his corpus about vegetative faculties and plant life. It begins with a look at more general concerns, including Avicenna's efforts to enumerate various vegetative faculties according to sound principles, and to not only distinguish them from animal and human faculties, but also explain their integration with those higher faculties. The second half of the essay examines Avicenna's contributions to more specific issues related to the vegetative life, including zoophytes, reproduction and embryology in his *Book of Animals*, and the distinction between the concepts of "life" and "soul" in his *Book of Plants*.

Avicenna (Ibn Sīnā, d. 1037) is well known for his innovative discussions of the internal senses and the rational soul, so it is not surprising that he has much to say about the vegetative soul, as well. However, since little of Avicenna's work on the vegetative soul has previously been examined, my goal here is just to give a sense of what he discusses, and where. I will begin by looking at some very general concerns Avicenna brings up about enumerating vegetative faculties, and distinguishing them from the faculties of the animal soul. Avicenna's conclusions in this area are not particularly unique, but he does have interesting ways of reaching them. We will then see that Avicenna builds on these general concerns, highlighting the shared goals and interrelations between vegetative faculties, and higher faculties. That is, vegetative faculties are not only conditions for higher faculties but, when they occur together, the latter ennoble the former.

In the second half of this essay I will present some of the more specific issues Avicenna treats related to the vegetative soul: first looking at zoophytes (plant-like animals), and then some concrete issues concerning reproduction and embryology, in which Avicenna as a medical doctor took a keen interest. This will pave the way, finally, for a look at the concept of "life" in his work dedicated to plant biology.

M. Fatigati (✉)
Centre for Medieval Studies, University of Toronto, Toronto, Canada
e-mail: michael.fatigati@mail.utoronto.ca

F. Baldassarri, A. Blank (eds.), *Vegetative Powers*, International Archives of the History of Ideas Archives internationales d'histoire des idées 234,
https://doi.org/10.1007/978-3-030-69709-9_5

The relevant Avicennian works and his sources will become evident as we proceed, but to say a little up front, Avicenna is primarily influenced in his discussions of the vegetative faculties by Aristotle's *De anima*, to which he had access in multiple Arabic translations (Elamrani-Jamal 2003, 346–58).[1] There were those in his tradition, such as A.B. Al-Rāzī, who were more Platonic in their psychology (Plato being understood via Galen, especially the Arabic translation of Galen's summary of the *Timaeus*; see D'Ancona 2013), but Avicenna explicitly distinguishes himself from them when it comes to psychology (Avicenna 1959, 5.7, p. 252).[2] The broadest relevant difference between the two approaches was whether one identified the concupiscible appetite with the vegetative soul, locating both in the liver (the Platonic/Galenic approach), or whether the vegetative faculties were considered distinct from the animal faculties (Aristotelian).

When it comes to the biological particulars of vegetative life, as distinct from more general faculty psychology, Avicenna incorporates Aristotelian and Galenic streams of thought. For example, his *Book of Animals* and *Book of Plants* (which we will discuss) are broadly Aristotelian in terms of their topics and structure. Yet—as will be discussed towards the end of this essay—the biological details are updated via Galen, whose influence is especially evident in Avicenna's *Canon of Medicine*.

5.1 Arguments for Enumeration and Distinction of the Vegetative Faculties

While Avicenna does not distinguish new vegetative faculties in the way that he distinguishes new types of internal senses (e.g., estimation), he takes pains to clarify the nature of the vegetative faculties, and justify the divisions that he does make. There are two notable points of development in this area of Avicenna's thought from his earliest summative work on psychology, part of his *Compendium*, to his mature *Healing: Psychology*. Firstly, in the *Compendium*, vegetative faculties are described as subordinated to one end (namely, reproduction of the species), while in the *Healing* Avicenna gives more attention to the individual ends of each faculty. This may simply be a difference in emphasis, but it reflects his trademark effort to distinguish faculties based on sound principles. And secondly, there is a development in terms of his argumentative focus: in the *Compendium* his arguments center around how we know vegetative faculties are distinct from one another, while in the *Healing* he assumes those distinctions, and focuses on how we know that vegetative and sensitive faculties are distinct, when they occur together in animals.

In the *Compendium*, Avicenna distinguishes the faculties of growth, reproduction, and nutrition, by noticing that living bodies exist at various points with some but not others of them. The assumption is that separability in terms of existence

[1] On Aristotle's vegetative soul, see Corcilius's contribution in this volume.

[2] On Plato and Galen, see respectively Carpenter's and Vinkesteijn's contribution to this volume.

entails being a distinct faculty. For example, we see living bodies that take nourishment, but don't reproduce, such as plants that are not yet ripe, and animals that have not reached puberty (Avicenna 1875, section 4, p. 349). Similarly, we see living bodies that take nourishment, but do not grow, such as the elderly who continue to be nourished and, rather than growing, in fact decline. On the basis of these observations, Avicenna says that these three faculties must be distinct.

There is nonetheless, he says in the *Compendium*, an ordered relation between the vegetative faculties. Nutrition is the starting point, and present throughout. Reproduction is the end of the vegetative soul, and has teleological priority over the other two faculties. Growth, on the other hand, is the means which binds (*al-wasiṭa al-rabiṭa*) the starting point to the end (Avicenna 1875, section 4, p. 349). He goes on to explain this later point by saying that although nature's goal is continuation of a species, new bodies cannot be made all at once (in the sense that a given body is not ready to reproduce at the moment of its creation). So there is a need for the potential parent to be sustained, which is why we have nutrition, and a need to mature the potential parent to the point of reproduction, which is why we grow. So growth binds nutrition to reproduction in the sense that nutrition is a prerequisite for growth, and growth orients the nourished body towards its goal.

Turning to the later *Healing: Psychology*, as mentioned, Avicenna is less interested in arguing for the distinctions between each vegetative faculty, and instead dedicates more time to explaining the nature of the vegetative faculties in more detail, and arguing for the distinction between vegetative and animal faculties more generally. In section 1.4 of *Psychology*, Avicenna raises some aporias that lead to his famous principles of faculty enumeration. For example, he wonders why we should posit a distinct concupiscible and irascible appetite, instead of just being content with one appetitive faculty. Similarly, we typically enumerate the various vegetative faculties, but at some point growth stops, and degeneration (*dhubūl*) begins. He wonders why, then, we wouldn't posit a degenerative faculty, in addition to the usual three vegetative faculties (Avicenna 1959, 1.4, p. 35). The reason for not needing to posit a degenerative faculty becomes clear in Avicenna's principles of faculty enumeration, as follows.

When Avicenna sees different objects that cannot be reduced to one another, or that exist separately, he is willing to posit distinct faculties. He also distinguishes sensitive faculties based on whether they are active of passive, and whether they retain or do not retain information (Black 2000, 59–60). But we do not need to posit a distinct faculty just to account for strong or weak acts in relation to one object (e.g., certainty and belief can be accounted for by the same faculty) or even superficially contrary acts (e.g., doubt and belief can likewise be accounted for by the same faculty). The later point explains why we would not need a faculty of degeneration, in addition to nutrition. Degeneration is not a new process, but rather a gradual failure of nutrition.

With his principles of faculty enumeration in the background, Avicenna proceeds in *Psychology* 2.1 to describe the object of each vegetative faculty, ending with the following schema:

> In brief, the nutritive faculty (*al-ghādhiya*) is intended for the preservation of the substance of the individual, while the faculty of growth (*al-nāmiya*) is intended for the completion of the substance, and the generative faculty (*al-muwallid*) is for the persistence of the species, since love of perpetuity emanates from God onto everything. [...] For the nutritive brings about a substitute for what is dissolved from the individual, whereas the generative brings about a substitute for what is dissolved from the species (Avicenna 1959, 2.1, 54–55).[3]

Avicenna elaborates on these three ends as follows: the faculty of nutrition (*al-ghādhiya*) transforms food into a likeness of various organs throughout the body. That is, it is a fact of nature that organs do not remain intact, but dissipate (*yataḥallil*), and need to have their substance gradually replaced in order to stay alive. On its own, nutrition is distributed evenly to all parts of the body, and merely replaces what has been lost (e.g., restoring muscle tissue after a run). However, from infancy to adulthood, people typically grow more in terms of height than width. The faculty of growth (*al-nāmiya*, whose act Avicenna sometimes refers to as *al-tarbiya*, or "nurturing") is responsible for directing nutrition so that it supplies more to certain parts of the body, to lengthen what is required according to the norms of a particular species (Avicenna 1959, 2.1, 54–55).

The growth process continues soaking up resources until puberty is reached, at which point growth slows down, and some resources are freed up, to be made use of by the generative, reproductive faculty (*al-muwallid*). The generative faculty is responsible for two related acts, according to Avicenna: first, creation of seed and the semen which carries it, and secondly, governance of the initial process whereby seed comes together with menstrual fluid, and a new, structured living thing is made. This new living thing thereafter has its own faculties of nutrition and growth (but no active reproductive faculty yet), and these eventually take over to continue development in the womb (Avicenna 1959, 2.1, p. 54–55). But the parents' generative faculty is necessary to sustain the new being, until it has its own living momentum.

Having given detailed accounts of vegetative ends, Avicenna also takes pains in *Psychology*, as mentioned, to distinguish vegetative from animal faculties more generally. The standard way of showing that vegetative faculties are distinct from animal faculties is to show that they are separable in terms of existence. Avicenna initially offers this standard take, quickly articulating a distinction between three levels of soul based on the faculties shared by all living beings (vegetative faculties), those shared by animals and humans (sensitive and locomotive faculties), and those distinctive to humans (theoretical and practical reason) (Avicenna 1959, 1.4, p. 37).

But while this is suggestive, Avicenna thinks we need to go further in showing vegetative faculties are distinct from the higher faculties *in the same animal*. It could be, perhaps, that these faculties are essentially one, and yet for some reason only some of them are manifest in different organs and species. What he may have in mind here is the idea— associated with the Stoics, whose psychological thought he had access to through Galen and Alexander of Aphrodisias—that all psychological faculties are essentially one throughout the body, but differences show up

[3]All translations of the Arabic are my own. For *Healing: Psychology*, I have at times consulted a draft translation by Michael Marmura and Deborah Black.

because of the dispositions of organs (see Sharples 2004, section 118.5, Sharples 2012, section 27.1). On this view, animals and plants would have the same faculties from the point of view of their souls, but more of those faculties would be manifest in animals, due to their more complex biology.

To this line of argument, which would erase any essential distinction between vegetative and animal faculties, Avicenna responds with the following tortuous but unique argument:

> For were the psychological powers [reducible to] one, and were the vegetative acts to issue from the [same] power from which animal acts primarily issue, then the lack had by plant-bodies and animal-organs that are nourished and [nevertheless] do not sense what is hard or soft to sensation, is either because of:
>
> [1] the lack of the power [of sensation in those bodies/organs]
> [2] or because the matter [of those bodies/organs] is not affected by [the causes of sensation].
>
> But it is impossible to say [2], i.e., that matter is not affected by hot and cold, nor influenced by them, nor by strong tastes or smells—for matter is [obviously] affected by these things.
>
> So it remains that [1], i.e., that this [lack of sensation in organs that are nourished] is because of the lack of the [sensitive] power affecting it, even though there does exist there the nutritive power. Hence, the two [types of] powers [vegetative and sensitive] are different (Avicenna 1959, 1.4, 37–38).

Avicenna begins this argument with the observation that there are some animal organs that have at least one vegetative faculty, nourishment, but which do not sense hard and soft. That is, there are many internal organs that simply do not feel, despite the fact that they are in contact with other parts of the body. Were your kidneys to have sensation, we would be bombarded by all kinds of jostling feelings throughout the day. Yet the kidneys are obviously nourished, since otherwise they would degenerate.

So the fact that kidneys (in our example) have the nutritive faculty, but not sensation, is due to one of two things, in the above argument: either (1) the fact that the matter kidneys are made of is impermeable, or unaffected, by the sorts of things our external senses recognize as hard and soft, or hot and cold, or (2) it is due to the kidneys' not having sensation. Avicenna takes it as obvious that (1) is not the case, since the matter of kidneys seems just as permeable by that which causes us to sense hard and soft, cold and hot, as, say, the matter of the tongue. So the root difference between the two organs must not be some difference in matter, but a difference in faculty. It is not as though the sensitive faculty is there, latent in the kidneys, just not being used (what was understood as the Stoic idea), but rather, this must be an instance where the nutritive faculty exists apart from any sensitive faculty *in the same subject.*

Underlying this conclusion is the that the notions of "matter" and "faculty" for Avicenna both describe something physical, but at different levels of complexity. To put the conclusion in other words: physically, the kidneys and the tongue obviously have *some* properties in common (e.g., they are both soft and permeable). But some of their physical make-up is more uniquely structured, and that physical difference is what grounds the difference in faculties (these different levels of physical complexity are discussed at Avicenna 1959, section 4.4, 197–198, and Avicenna 1947, 75–76).

5.2 Plants and Souls

Healing: Psychology 2.1 continues with one final set of arguments related to vegetative faculties, making use of the above-mentioned distinction between different levels of physical complexity. Avicenna deals with the question of whether plants are *alive* in *Healing: Book of Plants,* which we will look at in the final section of this essay, but in *Psychology* he primarily addresses the question of whether plants have *souls*. Having shown that nutritive organs and sensitive organs have distinct faculties, despite low-level material similarities, Avicenna also wants to show that the vegetative faculties are not simply functions of these low-level material properties. This is a concern which Aristotle previously raised in *De Anima* 2.4 (Aristotle 1961), targeting Empedocles, and Avicenna's response is similar to Aristotle's.

The objectionable stance Aristotle and Avicenna both respond to is the idea that the growth of plants seems to be natural, as opposed to psychological: roots spread downward, supposedly, due to being primarily composed of earth, which tends downwards, while branches spread upward due to their elemental nature being mostly fire, which tends upwards. For Empedocles's view to seem plausible, we have to keep in mind that by "fire" he did not just mean the sort of thing we might see in a burning building, but rather an airy, dry, intensely active substance.

Aristotle, and with him Avicenna, objects to Empedocles's view, because it reduces plant activity to the level of natural, non-psychological movement. Granted, plants do not have voluntary motion, like animals and humans (which depends on imagination and appetite), but neither do they simply move according to the basic principles of matter. Rather, they have a kind of inner form that directs their growth in a way that a simple clump of earth, or mass of fire, would never proceed.

According to Avicenna, there are two pieces of evidence that plants have a *psychological* principle, and not just a material, elemental one. The first is that fire does not stop consuming and increasing, so long as there is sufficient matter to consume: "Were fire [said] to be nourished, and were its rules the same as the rules that govern the nutrition of [living] bodies, then it would not be necessary that bodies would stop growing; for fire, as long as it has matter, does not stop, but increases without ceasing (Avicenna 1959, 2.1, 55)." In other words, if tree branches' growth were due to fire, there does not seem to be any reason why it would ever stop, since the fuel supply for the supposed fire remains unchanged as the plant gets older. But plants obviously *do* deteriorate, pointing to a life span dictated by something other than sheer fuel supply.

Secondly, psychological behavior is evident even in the objector's own admission that plants, beginning from seed, grow in two separate directions. The objector attributes this merely to the relative lightness and heaviness of fire and earth, but if there were merely matter here, with nothing psychological involved, we would not expect one thing to grow in these contrary ways. But plants *do* grow in both of these ways, which results in a puzzle: "Why would there not be some sort of specific difference to explain this movement, that is, the way the light [moves] differently than the heavy? Were that due to the governance of the soul, then let the rooting and branching also be due to the soul (Avicenna 1959, 2.1, 55)." There is plainly some additional factor governing movement in plants other than elemental heaviness and lightness, to account for its contrary growth patterns. If the objector does not want to say this more complex behavior is due to a soul, that is one thing, but then they are left with no way of explaining the difference between a tree, on the one hand, and two distinct masses of fiery and earthly matter, on the other.

5.3 The Ennobling of the Vegetative by Higher Faculties

We have so far mostly discussed how Avicenna argues for the distinctiveness of the vegetative faculties, but Avicenna also emphasizes how vegetative faculties are *integrated* with higher faculties. This integration can be understood chiefly in the sense that lower faculties are *ennobled* by higher faculties, discussed in Avicenna's early *Treatise on Love*. The treatise is so called because it focuses on how everything—from inanimate objects to celestial bodies—moves on account of a love for that which is suitable for it. Avicenna's basic point here is teleological: everything flows from God, the source of existence, and naturally desires to return to God. That is, everything, owing to the sort of thing that it is, tends towards the Good to the extent that it is able. It is suitable for rocks to tend downwards, and to this extent they partake of the Good. Humans naturally desire knowledge, which is much closer to the nature of the Good. Plants' characteristic acts—nutrition, growth, reproduction—also reflect the Good, insofar as they aim at preservation of the species, which is an imitation of divine eternality (Avicenna 1983, 267; also discussed in Avicenna 1875, section 4).

The integration of the faculties arises in *Treatise on Love* as Avicenna explains how humans' love for their suitable objects manifests itself. The issue he appears to be concerned with is that—having just described the objects of love for the vegetative and animal faculties—it seems as though humans, with still yet a third set of faculties in their nature, will be pulled in too many directions. He gets around this by arguing that, when combined in the same subject, lower faculties are ennobled, and thereby subordinated to the highest faculty's goals. The following passage is his most explicit statement to this effect[4]:

[4]Avicenna was possibly influenced in this by Alexander of Aphrodisias: "For this reason, even though there are many soul powers in [living] things in which the rational power is present, the soul constituted from them all is one, because no subsequent power can occur without the power that comes before it. Rather, they all belong to it as parts, where successive [powers] are joined to the

> Every one of the psychological faculties, whenever it is joined to a higher, more noble faculty, there occurs in it [lit: passes over into it] an increase in refinement and nobility, on account of this connection and the height of [the higher faculty's] beauty, such that actions proceeding from the lower faculty experience an increase, either in terms of number, or excellence of mastery, or elegance of their method and quest towards their end. For every higher faculty has an ability to support the lower and strengthen it, and to prevent the lower from harm in such a way that it causes the lower to receive an increase of beauty and nobility [...] like the support of the concupiscible animal faculty for the vegetative faculties, and how the irascible faculty protects [the vegetative faculties] from something that could harm their matter (though they cannot prevent the vegetative faculties' natural tendency towards degeneration) [...] and likewise the rational faculty adapts the animal to its goals (Avicenna 1983, 254).

In this text, we see how the various faculties work together in several ways. First, higher faculties help the lower faculties by *defending* them from harms. For example, the animal soul helps the nutritive faculty avoid harm in animals (the nutritive faculty in plants get no such help). Second, higher faculties provide *better methods* to achieve lower goals, as with how humans acquire nourishment much more efficiently than most animals, freeing them up for other tasks. Finally, lower faculties can be given *more objects* than they would otherwise have. For example, animals only know to eat what instinct tells them to, but humans have invented all manner of new dishes.

All these modes of integration of the higher and lower can be summed up by the phrase in the last line of the text, where Avicenna says that higher faculties "adapt (*tawqīf*)" lower faculties to higher goals. Avicenna elsewhere calls this a "partnership (*ishtirāk*; Avicenna 1983, 256)." This adaptation or partnership does not essentially change the lower faculties, but it does integrate them with the higher. Of course, Avicenna acknowledges that sometimes lower faculties are disordered and need reigning in (Avicenna 1983, 259), so this integration is not always a given.

Avicenna describes the partnership between vegetative faculties and a higher faculty most at length when it comes to the vegetative faculties and concupiscible appetite (Avicenna 1983, 256). We have already seen that the generative faculty is oriented at reproduction of the species. The concupiscible appetite, likewise, while it is *ultimately* oriented towards sensory pleasure, often seeks to find sensory pleasure in acts related to reproduction. A major difference between the generative faculty and the concupiscible appetite, however, is that the concupiscible appetite is a faculty for pursuing goals in a voluntary manner (*bil-ikhtiyār*), whereas the vegetative faculties act in a natural, albeit psychological, way (*bi-nawʿ ṭabīʿī*). When in partnership, however, Avicenna says the vegetative faculties can concur (*tuwāfiq*) with the concupiscible appetite, such that it *seems* vegetative faculties reach their goal in a voluntary manner. In other words, when partnering with a voluntary faculty pursuing the same generic sort of goal, vegetative faculties experience the benefits of voluntary action, which means seeking a goal on one's own terms, in one's

preceding ones and the preceding ones are because of this expanded and developed (Avicenna 2012, section 30)."

own way. A plant cannot choose when to blossom and scatter seed, but animals and humans can choose mates.

Outside of *Treatise on Love*, Avicenna discusses the integration of the higher faculties on vegetative faculties in one other place, albeit indirectly. The example comes early on in *Healing: Psychology*, in the midst of Avicenna's initial effort to define the soul as a substance and first actuality of the body. There, he explains the soul's relationship to the body in a way that highlights the effects from distinctly human emotions on the vegetative faculties. The point of this in context is just to show that the human soul's being potentially separable does not prevent it from having important connections to the body in this life. The passage is somewhat tortuous in its prose, so it has been numbered throughout for the sake of reference below:

> And the governance of the soul over the body causes the strengthening of the faculty of *growth* and its weakening, in view of [1] the soul's awareness of propositions [*qaḍāyā*] that it [2] hates or loves with a hate that is not at all bodily. And that is when there is [3] occurring to the soul [4] a certain assent [*tasḍīqan mā*], and that is not at all that which affects the body insofar as it is a belief, but rather there follows from that belief [5] the affect [*infiʿāl*] of joy [*surūr*] or pain [*gham*], and that also is one of the psychological apprehensions, and does not occur to the body insofar as it is body, and [that second joy or pain] affects the *growing, nutritive* power such that there occurs in it, on account of [6] the occurrence that first occurs to the soul—let that be intellectual joy [*al-faraḥ al-naṭqī*]—a strength and effectiveness in its act, and on account of the contrary occurrence—let that be intellectual pain [*al-ghamm al-naṭqī*], in which there is no physical pain—a weakness and an impotence that corrupts its effectiveness. And perhaps the temperamental disposition will be broken down through that [weakness and impotence] (Avicenna 1959, 1.3, 31–32, emphasis mine).

To clarify, humans sometimes experience intellectual joy. We know that by this he is referring to an act of the theoretical intellect, because assenting (see [4]) to propositions (see [1]) is one of the intellect's primary purposes (Avicenna 1959, 1.5, 42 and Avicenna 1959, 5.3). Avicenna elsewhere defines joy or pleasure as what happens when any faculty reaches something that counts as its goal, and is aware of doing so (Avicenna 1960, 8.3, 11 and 15). So presumably intellectual joy means attaining some sort of awareness of universal truth. At [5], we see that there follows from intellectual assent or intellectual denial the affect of joy or pain, and other effects in the vegetative faculties.

So, in addition to, or perhaps as a part of, the emotion of intellectual joy, the vegetative faculties are strengthened, and a kind of weakening occurs from intellectual pain. Avicenna does not describe what he means here any further, but it is not difficult to fill in the blanks: as a medical doctor, Avicenna's basic point is probably just the observation that positive emotions effect the whole person, whereas negative emotions do the opposite. Examples are most easily identifiable in what we know today about the negative physical effects of mental distress: stress can affect fertility, digestion, and more.

Though Avicenna does not say so, it is possible to see how this would fit in with the idea of lower faculties being caught up in the goals of higher faculties: emotions direct us in various ways, and the positive or negative activation of the vegetative faculties just described could add to this. Distress primarily inclines us to stop doing whatever it is we are doing, and having unpleasant signals come from a variety of sources makes the message more difficult to ignore.

5.4 Metaphysical and Temporal Priority Among Faculties

The transformative power of rational and animal faculties over vegetative faculties is supported by the general sense in which each type of living organism's species defining faculties are *prior* to any others it may have. Avicenna explains the metaphysical relation of the vegetative soul to animal and rational souls early on in his *Healing: Psychology*. In 1.3, where he is explaining the notion of the soul as a substance, a worry arises as to how animal and human souls can actually be substances—by which Avicenna means "that which is not in anything else as in a subject (Avicenna 2005, 2.1)"—if the vegetative soul is prior to them. One might think that the vegetative soul first makes its matter subsistent, and is followed by the animal and human faculties, from which it would follow that the later inhere in a preexisting vegetative substance (Avicenna 1959, 1.3, 29). In other words, if vegetative faculties are necessary for the existence of animal and human faculties, it seems as though the animal and human faculties inhere in the vegetative, making the vegetative soul the substance, and animal and human faculties dependent (Avicenna 1959, 1.3, 30).

To this, Avicenna responds that what the objector is actually pointing to, but misdescribing, is the fact that vegetative faculties are a *condition* for the existence of animal and rational souls, which is not the same as being prior in terms of substance. It is true that animal and human faculties only exist *alongside* the vegetative (with the possible exception of the theoretical intellect), but the lower faculties are not the species-defining, substance-creating aspect of the coalition of faculties that occur in animals and humans. Rather, Avicenna says, confusion arises when we speak of anything like a vegetative "soul" existing in plants and animals. As Avicenna clarifies shortly thereafter, soul talk, properly speaking, is reserved for the species-defining aspect of some substance:

> Were it not for custom, it would be better to say that that which is first [i.e., the vegetative] is [rather] a condition mentioned in the description of the second [i.e., animal or rational], if we were wanting to describe the soul, and not some psychological faculty that belongs to the soul on account of its act. For actuality [*al-kamāl*] has to do with the definition of soul, not the definition of psychological faculty (Avicenna 1959, 1.5, 40).

The most basic sort of soul is the vegetative soul. As Avicenna says later in 5.7, in the palm tree, vegetative faculties are identical to what we would call the soul. But in humans there is no such thing as a vegetative soul. Rather, there is a rational soul, which enacts a substance that has rational, animal and vegetative faculties. And those vegetative faculties in a human, while generically similar to those in a plant, are specifically different. As he says again in 5.7, highlighting these different uses of soul and faculty, "The plant soul in the palm does not share a species with the faculty of growth in humans at all, for [the palm's] faculty of growth is not able to be joined to an animal at all (Avicenna 1959, 5.7, 259–260)." Although we can abstractly group together some aspects of plants, animals and humans according to their function, the actual manifestation of the generative faculty in animals is otherwise different than the generative faculty in plants. This is similar to the way that a

Mac and a PC both have the ability to browse the internet, though the hardware and software differs for both, leading the some different ways of browsing the internet, and different integrations with the overall operating system. Vegetative faculties in humans, in virtue of having been caused by a rational soul, and part of a rational substance, are different than vegetative faculties in something with *only* a vegetative soul.

One might wonder, however, whether this picture of metaphysical priority has missed the common-sense concern underlying the worry described above: how can we assert the priority of the rational faculties, when animals and humans appear to develop their vegetative life prior to being able to sense, move or reason? Such a view is proposed in prior authors like Al-Farabi, who explicitly says that the vegetative faculty is the first thing to arise in humans (Al-Farabi, 1985, 4.10)—but this is hard to square with the metaphysical priority of the rational soul just described. Here is one place Avicenna's work in the field of medicine and biology comes to the forefront.

Perhaps metaphysical priority could be preserved even if vegetative faculties were temporally prior in animals and humans, but Avicenna argues that there is no such temporal priority: *all* the faculties—in animals and humans—are virtually present in the vital spirit (*rūḥ*), prior to the full-fledged development of any organs for nutrition, sensation, motion or reason. Avicenna has much to say about the nature of the vital spirit, but what is important for our purposes is that the vital spirit is a lightweight, fine physical substance produced by the heart (Avicenna 1984, 221). The first two organs to emerge in humans, supported by the generative powers of the parents (Avicenna 1959, 2.1, 54–55), are the heart and the liver. The liver produces the coarse matter that gets formed into other organs, while from the heart vital spirit flows to those not-yet-animate organs, giving them life, and enabling them to receive their respective vegetative, sensitive or motive faculties [...] (for a discussion of *pneuma* in Galen, Avicenna's main influence for this theory, see the introduction to Galen 1916; Galen himself is of course influenced by Aristotle's theory of pneuma, on which, see Peck 1943).

Avicenna makes the point that psychological faculties are virtually contained in the vital spirit in three separate places: *Canon of Medicine*, *Cardiac Drugs*, and *Healing: Psychology*. As he states baldly in the later text, "We say that the first thing physical psychological powers are carried about by is [...] the vital spirit (Avicenna 1959, 5.8, 263)." In other words, despite not yet having appropriate organs, all psychological faculties (which includes vegetative faculties) are in some way present in the vital spirit from the very inception of life. Similarly, in the *Canon* and *Cardiac Drugs*, Avicenna discusses there being a single vital spirit, which gets transformed in various ways as it is distributed throughout the body, so as to underly the vegetative and sensitive functions of various organs (Avicenna 1982, 1.6.4, 127).

It seems we should think of the vital spirit as holding a kind of code, or DNA, wherein all psychological faculties are virtually present. Thus, not only is the vegetative soul not essentially prior to sensitive and rational faculties, it is also not even temporally prior. Of course, these issues do not arise for plants, despite their lack of vital spirit, because plant vegetative faculties simply are the species-defining faculties of their soul.

5.5 The Dividing Line: Plant-Like Animals and Voluntary Motion

Avicenna clarifies the distinction between vegetative and animal faculties by discussing what are known as zoophytes, or plant-like animals. The classic examples of this sort of creature are invertebrates like sea-sponges or anemones. Aristotle discusses zoophytes in *De Anima* 2.3, as he is reviewing the distinction between the three types of souls. He notes that within these groupings some faculties appear to be more fundamental than others. For example, in most animals we see the full range of sensitive faculties, alongside locomotion. In some animals, however—such as zoophytes—we find the sense of touch without any of the other senses, or even locomotion. Themistius, commenting on that passage in *De Anima*, says that if the tradition is correct in classifying zoophytes as animals, they must be capable of voluntary motion, using concupiscible appetite (Gr: *epithumia*, Ar: *shahwa*) to voluntarily shrink away from painful things and towards beneficial objects (Themistius 1973, 62.12–14). They lack, however, the cognitive apparatus for irascible appetite. So zoophytes not only have some of the senses without others, they also have some of the appetites without others.

Avicenna's own discussion of zoophytes hits upon many of the same themes, but his focus is on the sense in which such living beings can properly be said to be capable of voluntary motion. Zoophytes arise in *Psychology* 2.3, his chapter on the sense of touch (his examples are of oysters and sponges). Zoophytes, Avicenna says, have touch without any other senses, and also appear to be limited in terms of motion.

Yet voluntary movement is the most important distinguishing mark between plants and animals, so it is important to understand in what sense these ambiguous creatures have it, if we are to consider them animals. So he draws a distinction between two types of voluntary motion: motion in place, and motion from place to place (i.e., locomotion; Avicenna 1959, 2.3, 68). He says there is an ordered relation between these types of motion that parallels the ordered relation between the senses. Having touch without having any other senses goes hand in hand with being capable of motion in place without having locomotion. So despite the apparent similarity between zoophytes and plants, zoophytes have the bare minimum to be considered animals, since they have the sense of touch, some rudimentary cognitive apparatus, and motion in place.

5.6 Reproduction in Avicenna's General Biological Works

With the psychological framework established, we can now turn to biology. The key work here is Avicenna's *Book of Plants*, which we will look at in the next section, but we begin by looking at Avicenna's more general biological works, with an eye to what he has to say there about vegetative life. A detailed overview of Avicenna's biological work has been written by Musallam, in which he focuses on Avicenna's

Book of Animals (*Kitāb al-Hayawān*) and its relation to his *Canon of Medicine*.[5] In the *Book of Animals* Avicenna delves into animal and human biology in detail, which of course includes a discussion of our vegetative organs. It is structured according to the flow of Aristotle's works on animals, and in that regard is reliant on Ebn al-Beṭrīq's translation of the Aristotle source material into Arabic. Books 1–8 summarize *History of Animals*, and Books 15–18 *Generation of Animals*, with little significant variation having been noted thus far.

Books 9–14, according to Musallam, is where things get interesting. 11–14 proceeds along the lines of Aristotle's *Parts of Animals*, discussing the organs for nutrition, growth and reproduction. But rather than being a summary of Aristotle, Avicenna has largely substituted ideas from his own more Galenic *Canon of Medicine*. As Musallam (1989) attests, "Over seventy-five percent of the text of books 11–14 [...] and fully forty percent of the text of the whole *Ḥayawān* [...] came from the *Canon*." The effect of this, aside from changing and updating some of the outmoded Aristotelian biology related to vegetative life, was to focus far more on human biology (and less on non-human animals) than Aristotle.

Books 9–10 do not neatly fit into the threefold structure of Aristotle's *Book of Animals*, but rather contain Avicenna's own discussions of controversial issues related to the reproduction. In Book 10, which is much shorter, Avicenna deals with conception and pregnancy loss in humans. In book 9, however, as Musallam details, Avicenna lays out various positions on male and female contributions to reproduction: first Hippocrates's argument, based on parental resemblance in the child, that both mother and father must contribute some sort of "semen." Aristotle rejected this idea, famously claiming that only the father contributes semen, the active element, while the female passively receives this in menstrual blood. After the discovery of ovaries, Aristotle's position became harder to take wholesale, and Galen revived the notion that both sexes make an active, formal contribution to reproduction. Into this fray comes Avicenna, who seems to split the difference between Galen and Aristotle: he argues that male and female *both* contribute seed, but that the male seed is active while the female is passive. As Musallam points out, however, this is still perhaps closer to the Aristotelian view in spirit.

5.7 Avicenna's *Book of Plants* and the Concept of Life

Avicenna's work dedicated exclusively to the specifics of plant biology is his *Book of Plants* (*Kitāb al-Nabāt*). It is the seventh part of the physical sciences portion of his magnus opus, *The Healing*, nestled after *Psychology* and just before his *Book of Animals* (both discussed above). His primary source here appears to be the Pseudo-Aristotelian *De Plantis*. Translated into Arabic by Isḥāq ibn Ḥunayn in the 800 s, it

[5] Most of what follows in this section is a summary of the information relevant to vegetative faculties gleaned from Basim Musallam (1989), as little other work has been done in this area.

was eventually recognized in later Arabic manuscripts as a reworking by Nicholas of Damascus of the lost Aristotelian original (Lulofs 1989, 123–127), but it is unclear whether Avicenna was aware of that distinction. He would also have had access to the Pseudo-Galenic *De Plantis*, which exists only in its Arabic translation (Gutas 2010), but the structure and topics Avicenna deals with clearly follow Pseudo-Aristotle. A fuller study would be necessary to determine exactly to what extent Avicenna is innovative in this text, but what follows is a brief overview to show his general approach and concerns.

Avicenna begins his *Book of Plants* with a consideration of whether plants should be said to have appetite and/or to be alive. These issues also arise at length at the beginning of the Pseudo-Aristotle text, but Avicenna seems to think that they are fairly straightforward, and primarily depend on the way in which we define terms. Plants, he says, do not have appetite, but rather are attracted to their objects in the way that everything, whether ensouled or not, tends towards its end. He begins the *Book of Plants* by making this point, as follows:

> Plant's attraction [*jadhb*] to nourishment is via the attraction of organs different than ours, for they are attracted through a natural faculty, not from a sensitive appetite [*shahwa ḥissiya*] particular to each organ. That is, there is a [natural] attraction, distinct according to each organ. On the other hand, [sensitive] appetite only occurs with some imaginative act, and such appetite necessarily is found in that which moves on account of desire for obtaining nourishment, as in a human and [their] prey, or even simple expanding and contracting movements, like the oyster and its shell. As for that which has no way of obtaining nutrition through movement, it rather only gets the nutrition that it is [already] in contact with, as is the case with plants. [Plants] are attracted to something not through volition but through [the natural attraction of] organs, such that they have no appetite, and have no need for the benefit of [appetite]. So it is appropriate that a plant is not given sensation. For, were it given sensation, it would be for nothing, since there is no way for a plant to escape from harm, or seek benefits (Avicenna 1965, 3).

There is a sense, Avicenna says, in which plants are "attracted" to nutrition as an object, and their behavior is guided by this (e.g., roots grow towards water), but this does not mean we should think they have animal-like appetites. "Appetite [*shahwa/shawq*]" is a term reserved for attraction in response to an imagined object, which object serves as the goal for voluntary motion. This is different than the sense in which *every* nature has a telos, and naturally tends towards, or is "attracted" towards [*jadhb*], that end.

Immediately following the issue of appetite, Avicenna considers the definition of "life [*al-ḥayā*]," and whether it should apply to plants. Discussing similar Greek authorities as discussed in Pseudo-Aristotle, Avicenna argues that this is primarily a verbal dispute:

> One strays from truth if they argue that plants have sensation for the sake of [i.e., because they also have] thought and understanding, like Anaxagoras and Empedocles and Democritus. Now if activity in the realm of nutrition were called life—so that a body, if it persists through nutrition, is [called] living, and then, if it is unable to maintain its individuality through nutrition, outside corruption has power over it so that its mixture changes, and its power dissipates, then it [is said to] die—then [if that is what we mean by life] it is appropriate to call a plant living. And if [calling a plant living] is on the condition that [life means] perception and non-voluntary movement, then it is not permitted to call plants living in any respect. And this dispute is mostly verbal (*lafẓī*; Avicenna 1965, 3).

Because plants are capable of nutrition, growth and reproduction, some call them alive. As we saw in the previous section, Avicenna technically thinks that having the vital spirit is what makes something living, but he can also see why we might in some contexts say that having vegetative faculties constitutes life. But others, conflating the idea of life with thought, understanding, and sensation, assume that plants must have the later faculties in some measure as well, if they are to be considered alive. This is a mistake, Avicenna thinks, because we know those later faculties are only given to living beings capable of voluntary motion and, as mentioned in the prior paragraph (above), plants cannot act on the basis of information concerning harms and benefits. So it is appropriate to call plants living, provided we understand that there is life capable of voluntary motion (animals and humans), and life which is not (plants).

This exposition runs contrary to a recent article, titled "Avicenna's Denial of Life in Plants (Tawara 2014)." In this article, Tawara argues that while Avicenna early on in the Compendium attributed life to plants (because they have souls), he later revised his view in the *Book of Plants*. The key text for Tawara's claim is the one just presented, where Avicenna says towards the end that "it is not permitted to call plants living in any respect (Tawara 2014, at footnote 10)." But as we have seen, this statement is part of Avicenna's permissive stance that we can either call plants alive or not alive, *depending on what we mean by life*. So Avicenna does not decisively deny life to plants in this text.

Tawara claims to find further support for his view in Avicenna's *Canon of Medicine*. There, according to Tawara, Avicenna says life is not present in plants, because plants lack the "animal faculty." The animal faculty—which could also be translated "vital faculty," equivalent with the vital spirit discussed in the prior section—is "the faculty that makes the organs prepare for accepting the faculty of sensation and movement, and the functions of life (quoted at Tawara 2014, 135)." Based on this, it may seem Avicenna is denying life to plants in the *Canon*: life is associated with sensation and movement, and if anything is certain, plants do not have sensation and movement.

But context suggests that Avicenna is, again, not going so far as to deny plants life. In the *Canon*, Avicenna is dealing primarily with humans from a medical point of view. As we saw in section 4, the vital spirit is indeed a condition for life in animals and humans. But just because plants lack the same conditions for life as humans, does not mean that we have to say that plants lack life.

Again, in a work written in the same later period of his life as *Healing: Book of Plants*, *Healing: Psychology* insists that plants have a soul, and says that he is perfectly happy with calling something alive, if by alive we mean "has a soul (Avicenna 1959, 1.1, 15; also see Sect 5.2, above)." So viewed, Avicenna's later statements concerning of life of plants are ambivalent and less dramatic.

The rest of Avicenna's *Book of Plants* touches on many of the same specific issues as Pseudo-Aristotle, as can be seen from the following table of contents (the table of contents is not necessarily Avicenna's, but it is in some of the manuscripts):

1. On the generation and nourishment of plants, their male and female [organs], and the origin of their temperamental mixture.
2. On the organs of plants at the beginning of growth, and after that.
3. On nutritive principles, and on procreation and generation in plants.
4. On the state of generation of the parts of plants and the state of their difference, and the difference between plants according to countries.
5. On defining the distinct states of trunks, and branches, and leaves.
6. On that which is produced by plants in terms of fruit, seed, thorn, resin, and suchlike.
7. In which there is general talk about types of plants followed by talk about temperamental mixtures of things which have a nutritive soul.

Most of these chapters contain specific lists and examples, though Avicenna gets more theoretical again in Chap. 7. There he discusses the temperamental mixtures of plants, explicitly for the sake of clarifying principles of medicine. Plants have primary temperamental properties, and knowing how these interact with other plants allows one to develop secondary mixtures, which are of use to doctors. He offers some examples, but ultimately does not want to get into too many specifics.

He justifies this fly-by approach in the final sentence of the treatise, with which we can conclude, as follows: "And this account suffices for our giving a foundation, so let us end our discussion of plants. For were we to delve into [more] distinctions concerning their concrete properties and actions, we would descend into a particular science (Avicenna 1965, 38)." His *Canon of Medicine* and *Cardiac Drugs* contain far more information about specific mixtures that can be used to treat illness, and it is likely to these works that he is deferring.

References

Alexander of Aphrodisias. 2004. *Supplement to* On the Soul (*Includes* De Intellectu). Trans. R.W. Sharples. London: Duckworth.

———. 2012. On the soul: Part I*: Soul as form of the body, parts of the soul, nourishment, and perception.* Trans. Victor Caston. New York: Bloomsbury.

Al-Fārābī. 1985. Al-Farabi on the perfect state [virtuous city]. Edited and translated by Richard Walzer. Oxford: Clarendon Press.

Aristotle. 1961. In *De Anima*, ed. W.D. Ross. Oxford: Clarendon Press.

Avicenna. 1875. Fī N-Nafs ʿalā Sunnat al-Ikhtiṣār (compendium on the soul). In "Die Psychologie des Ibn Sīnā," *in ZDMG (Zeitschrift der Deutschen Morgenländischen Gesellschaft)*, ed. Samuel Landauer, 29:335–418.

———. 1947. "Al-Taʿalīqāt ʿalā Ḥawāsh Kitāb al-Nafs Li-Arisṭāṭālīs [Marginal comments on Aristotle's *De Anima*]." In *Aristu ʿinda Al-ʿArab*, ed. A. Badawī, 75–116. Cairo: al-Qāhirah maktabat al-nahḍah al-miṣrīyah.

———. 1959. *Al-Shifāʾ, Kitāb al-Nafs [Healing: Psychology]*, ed. Fazlur Rahman. London: Oxford University Press.

———. 1960. *Al-Ishārāt Wa-al-Tanbīhāt [Remarks and Admonitions]*, ed. Sulaymān Dunyā. Vol. 4, al-Taṣawwuf [Mysticism]. Cairo: Dār al-Maʿārif bi-Miṣr.

———. 1965. *Al-Shifāʾ, Kitāb al-Nabāt [Healing: Book of Plants]*, ed. ʿAbd al-Halīm Muntasir. Al-Hayʾah al-Miṣrīyah al-ʿĀmmah lil-Kitāb. Cairo.

———. 1982. *Al-Qānūn Fī al-Ṭibb [Canon of Medicine]*, ed. Institute of history of medicine and medical research. Vol. 1. New Delhi: Vikas.

———. 1983. Risāla Fi-al-ʿishq [Treatise on love]. In *Al-Tafsīr Al-Qurānī*, ed. Ḥasan ʿĀṣī. Beirut: al-Muʾassasa al-jāmiʿiyyah.

———. 1984. Al-Adwiyya al-Qalbiyya [Cardiac drugs]. In *Min Muʾallafāt Ibn Sīnā al-Ṭibbiyyah*, ed. Muḥammad Zuhayr al-Bābā, 221–94. Tunisia: al-Munaẓẓamah al-ʿArabīyah lil-Tarbiyah wa-al-Thaqāfah wa-al-ʿUlūm.

———. 2005. *Metaphysics of the healing [Al-Ilahīyyat Min al-Shifā', with Facing English-Arabic]*, ed. and trans. Michael E. Marmura. Provo: Brigham Young University Press.

Black, Deborah. 2000. Imagination and estimation: Arabic paradigms and Western transformations. *Topoi* 19 (1): 59–75.

D'Ancona, Cristina. 2013. Greek sources in Arabic and Islamic philosophy. *The Stanford Encyclopedia of Philosophy*. http://plato.stanford.edu/archives/win2013/entries/arabic-islamic-greek/. Consulted November 25, 2019.

Elamrani-Jamal, Abdelali. 2003. De Anima: Tradition Arabe. In *Dictionnaire des philosophes antiques: supplement*, ed. Richard Goulet, 346–358. Paris: Editions du Centre national de la recherche scientifique.

Galen. 1916. *On the natural faculties*. Trans. A. Brock. Vol. 71. The Loeb classical library. Cambridge: Harvard University Press.

Gutas, Dimitri. 2010. Appendix B: Medieval translations: Greek philosophical works translated into Arabic. In *The Cambridge history of medieval philosophy*, ed. Robert Pasnau and Christina van Dyke, 802–814. Cambridge: Cambridge University Press.

Lulofs, Drossaart, and E.L.J. Poortman. 1989. *Nicolaus Damascenus De Plantis: Five translations*. Amsterdam: Brill.

Musallam, Basim. 1989. Avicenna: §X, Biology and Medicine. In *Encyclopaedia Iranica*, ed. E. Yarshater, 3: 94–99. Boston: Routledge & Kegan Paul. http://www.iranicaonline.org/articles/avicenna-x. Consulted on 25 Nov 2019.

Peck, A.L. 1943. Appendix B: On the function of sumphuton pneuma [...]. In *Generation of animals*, ed. A.L. Peck, 576–593. Loeb Classical Library. Cambridge: Cambridge University Press.

Tawara, Akihiro. 2014. Avicenna's denial of life in plants. *Arabic Sciences and Philosophy* 24 (1): 127–138.

Themistius. 1973. In *An Arabic translation of Themistius's commentary on Aristotle's De Anima*, ed. M.C. Lyons, vol. 2. Oriental Studies. Columbia, SC: University of South Carolina Press.

Chapter 6
Can Plants Desire? Aspects of the Debate on *desiderium naturale*

Marilena Panarelli

Abstract During Antiquity and the Middle Ages, two opposite traditions – namely the Aristotelian and the Platonic – expressed antithetical positions regarding the question whether plants were able to desire. The issue in question is closely related to the question of the sense of plants and their possibility to have sense perception. Over time, the two philosophical issues developed differently. In the Aristotelian tradition, the question of sense in plants has continuously been treated in a way that denies plants sensation, while the issue of desire in plants developed differently, opening a more heated debate. Some of the positions expressed by the most authoritative exponents of the two traditions will be taken into account in this chapter, starting with Plato and Aristotle, examining in particular the pseudo-Aristotelian *De plantis*. This will be followed by an examination of the Neoplatonic positions, which represent a moment when Aristotelian and Platonic doctrines somehow intersect; thus, Plotinus and Isaac Israeli's position will be considered. Finally, some of the most important medieval exponents will be considered, precisely Averroes, Roger Bacon and Albert the Great, who show that the question of desire in plants has metaphysical implications.

One of the most complex questions concerning plants that has stimulated philosophical debate is whether or not they are capable of feeling and, moreover, of experiencing desire. This issue has recently been reconsidered by studies in the field of Plant Neurobiology (PNB), revealing that plants can be considered sensory and communicative organisms (Baluška et al. 2009). This contradicts the position that has long been held within the Aristotelian tradition, namely that plants are merely passive beings, unable to feel. Indeed, the Aristotelian tradition ascribes the vital

---

M. Panarelli (✉)
Università del Salento, Lecce, Italy
e-mail: mpanarel@hotmail.it

F. Baldassarri, A. Blank (eds.), *Vegetative Powers*, International Archives of the History of Ideas Archives internationales d'histoire des idées 234,
https://doi.org/10.1007/978-3-030-69709-9_6

functions of plants to the vegetative soul, which is only able to generate, grow and feed. Yet, another tradition—which can, regarding certain aspects, be defined as Platonic and Neoplatonic—assigns a wider range of functions to plants. In light of recent discoveries, the purpose of this paper is to reconstruct the often heated debate on the *desiderium naturale*, as it developed from Antiquity to the Middle Ages, scrutinizing to what extent exponents of the Aristotelian tradition admitted the possibility that plants might desire.

6.1 The Pseudo-Aristotelian *De plantis* on Desire in Plants and the Perspective of Ancient Sources

On several occasions in his research on plants,[1] Aristotle refers to a work, which also Diogenes Laertius attributes to him, entitled *peri phyton*.[2] Nonetheless, a purely Aristotelian work on plants has not yet been found. While during the Middle Ages, the apocryphal *De plantis* was ascribed to Aristotle, it is nowadays almost unanimously ascribed to Nicolaus Damascenus,[3] a peripatetic of the first century BC, who wrote various commentaries to Aristotle's works. Although this text was not directly written by Aristotle, it bears truly Aristotelian features, transmitting a doctrine on plants which can be defined as *Aristotelian*. The text successfully transmitted a philosophical examination on plants into the medieval Arabic and Latin culture, while the borader works written by Theophrastus were not successful during the Middle Ages and, moreover, did not deal directly with the issue in question (Repici 2009, 84).

The pseudo-Aristotelian treatise opens with a discussion of the powers of the plant's soul, questioning the possibility of ascribing sense and desire to plants. The theoretical positions of Anaxagoras, Abrucalis—who is, in fact, Empedocles—and Plato all admit this possibility and are all examined and reported by Nicolaus, for the sake of refuting them later.[4] Furthermore, the authority of Democritus (I, §10), to whom the opinion that plants possess intelligence is ascribed, is mentioned. One of the most noteworthy aspects of this pseudo-Aristotelian work is that it is a kind of doxographic work, able to transmit the opinions of ancient authors on questions

---

[1] Aristotle mentions his work on plants in *De sensu* IV 4, 442b24-26, *De longitudine et brevitate vitae* 467b, *De generatione animalium* I 1, 716a1, *De partibus animalium* II 10, 656a2-3, *Historia animalium* V 1, 539a20-21. See Repici 2009, 79. See Corcilius, Chap. 2 in this volume.

[2] Laertius 2008 V 25, 323.

[3] On Nicolaus of Damascus, see Moreaux 1973, 445–514, Damascenus 1841, 3. Nevertheless, the attribution to Nicolaus of Damascus has been debated; see, for instance, Fazzo and Zonta 2008, 681–90; Herzoff 2016, 135–87.

[4] Constaret enim utrum habeant necne plantae animam et virtutem desiderii dolorisque et delectationis discretivam. Anaxagoras autem et Abrucalis desiderio eas moveri sentire quoque et tristari delectari quoque asserunt. Quorum Anaxagoras animalia has esse, laetari quoque et tristari dixit, fluxum foliorum argumentum assumens. Abrucalis autem sexum in his permixtum opinatus est. Plato siquidem desiderare tantum eas propter vehementem nutrimenti necessitatem ait. Quod si constet, gaudere quoque et tristari sentirique eas consequens erit. (*De plantis* I, §2–3)

concerning plants. The quoted authors represent the most authoritative voices against the Aristotelian position,[5] who can nevertheless be regarded as belonging to the same tradition with regard to this issue.

Democritus' and Anaxagoras' positions on this matter are transmitted through a documentary source. However, it remains uncertain whether the source, quoted in Diels-Kranz A 116, reports an authentic position or whether it is merely strongly influenced by that of Plato. As has been shown by Luciana Repici, it is not possible to be entirely certain of the validity of these sources and the fact that these opinions are reported by the pseudo-Aristotelian text is not sufficient to establish their authenticity (Repici 2000, 67).

A similar case is that of Empedocles, as no other source assuredly confirms the diverse theses regarding the psychic aspects of plants reported in *De plantis*. Although most of the doxographic fragments attest to a particular interest in the plant kingdom, and to biological issues in general, there is once again no trace of the debate on the psychic capacities of plants; thus, *De plantis* remains the only source reporting such statements.

In contrast, Plato's well-known position on the matter is different.[6] In Plato's *Timaeus* 77b-c, the fact that plants are able to perceive and desire is explicitly stated:

> We may call these plants "living things" on the ground that anything that partakes of life has an incontestable right to be called a "living thing". And in fact, what we are talking about now partakes of the third type of soul, the type that our account has situated between the midriff and the navel. This type is totally devoid of opinion, reasoning or understanding, though it does share in sensation, pleasant and painful, and desires. For throughout its existence it is completely passive, and its formation has not entrusted it with a natural ability to discern and reflect upon any of its own characteristics, by revolving within and about itself, repelling movement from without and exercising its own inherent movement. (*Timaeus* 77b-c)

In this passage, Timaeus states that everything that lives can be adequately called *zoa*, that is, animal. As plants possess the third type of soul, they are also able to desire. The passage quoted above is one example of the Platonic method of dichotomic division.

On the contrary, Aristotle openly denies the fact that plants could be animals in *De anima* 413a1, stating in *De partibus animalium* 61a10 that living beings that are not yet animals are classified between *zoa* and *azoa*. The discussion of the Aristotlian and Platonic position continued throughout the centuries and, consequently, shaped medieval discussions.

The text that medieval scholars most commonly refer to and comment on is the pseudo-Aristotelian *De plantis*. The conclusion of this pseudo-Aristotelian treatise, regarding the issue in questions here, is that plants have no sense organs and, therefore, can neither perceive nor desire, as desire is always accompanied by sensitivity.[7] Later, however, this statement was again questioned.

---

[5] The three are quoted together also by Plutarch at the very beginning of his *Quaestiones physicae*.

[6] Plato 1997, see Skemp 1947.

[7] Dico ergo quod plantae nec sensum habent nec desiderium: desiderium enim non est nisi ex sensu, et nostrae voluntatis finis ad sensum convertitur (*De plantis* I, §11)

6.2 Plotinus and the Contemplation of Plants

In *Enneads* I 4, [1–5], Plotinus questions whether beings different from humans are able to experience happiness,[8] focusing mainly on plants. Although he attributes the perfect state of happiness to humans who exercise rationality only, he also admits that if other beings may achieve the goal of their own nature—their *ergon*—, they are also able to attain a state of well-being.[9]

In this passage, Plotinus discusses the Aristotelian idea of the teleology of nature—which is expressed by Aristotle in *Nicomachean Ethics* I, 7, 1097a-1098a—, coming to the conclusion that happiness is only possible for those human beings who exercise complete rationality. This, however, is not coherent with the Aristotelian teleology itself, which implies that if a natural being can achieve its own *telos*, it can attain a state of happiness. It is impossible, according to Plotinus' objection, not to assign happiness to plants on the basis of this teleological framework; thus, Aristotle's arguments are fallacious.

While revealing this incoherency in Aristotle's approach, Plotinus also offers a less reductionist position than that of the Stagirite regarding the possibility to assign well-being to plants. Plotinus' idea of true well-being as the prerogative of humans does not exclude that plants may also be capable of a certain degree of well-being.

Plotinus' position in this regard becomes clearer in his objection against those—namely, the Epicureans—who state that well-being coincides with pleasure and, as such, might be attained by sensible beings but not by plants. While he denies that plants have sensibility, Plotinus admits that they have the same chances as animals to experience pleasure, as they are living beings (Repici 2011, 172). In fact, they are able to experience a state of pleasure simply because they live and not because they possess any kind of sense. Nevertheless, true well-being does not coincide with pleasure and it is, as already stated, achieved only by humans. In *Enneads* I 4 [2], he claims that sensation is a kind of consciousness of pleasure, which, therefore, does not coincide with pleasure itself:

> Those who deny it to plants because they have no sensation run the risk of denying it to all living things. For if they mean by sensation being aware of one's experiences, the experience must be good before one is aware of it; for example, to be in natural state is good, even if one is not aware of it, and so is to be in one's own proper state, even if one does not yet know that it is one's own proper state, and that it is pleasant (as it must necessarily be). (*Enneads* I 4 [2])

The presupposition of pleasure is thus not the sensation as such, but the mere fact of living. Pleasure as such is irrational and happiness cannot be found in it just because reason exists on a higher level than pleasure. In this objection, by stating that pleasure does not coincide with happiness, Plotinus admits that plants can experience pleasure. Afterwards, he considers the Stoic position, namely that happiness is in

[8] See Linguti 2000.

[9] In his English translation, A.H. Armstrong decides to translate the term *eudaimonía* with 'well-being' and not with 'happiness'.

reason, which is also criticized by Plotinus, as the Stoics are not able to explain why this is supposedly the case.

Thus, in order to find a solution to this intricate problem, Plotinus analyzes the different meanings of the term *life*. Indeed, in *Enneads* I 4 [3], 11–15, Plotinus explains that the term *life* does not mean the same for each being to which it refers, as it is an equivocal term:

> We, however, intend to state what we understand by well-being, beginning at the beginning. Suppose we assume that it is to be found in life; then if we make "life" a term which applies to all living things in exactly the same sense, we allow that all of them are capable of well-being, and that those of them actually live well who possess one and the same thing, something which all living beings are naturally capable of acquiring; we do not on this assumption grant the ability to live well to rational beings, but not to irrational. Life is common to both, and it is life which by the same acquisition [in both cases] tends towards well-being, if well-being is to be found in a kind of life. So I think that those who say that well-being is to be found in rational life are unaware that, since they do not place it in life in general, they are really assuming that it is not life at all. (*Enneads* I 4 [3], 11–15)

Moreover, the difference among living beings is not a specific difference, but the same difference as that between what causes and what is caused.[10] This premise suggests that if a univocal meaning of life does not exist, a univocal meaning of well-being cannot exist either.

The difference of meanings actually consists in the difference of gradations of the clarity or dimness of life itself. Thus, life and well-being do not mean the same thing for humans and for plants; it is a mistake to examine well-being while attempting to make the meaning of well-being equivalent for plants and humans. Although the highest level of well-being pertains to the intellective soul, well-being as an attribute of life manifests itself on different levels and to different degrees. Thus, also plants can experience it.

The principal aspect of Plotinus' argument is that the clearest and most perfect level of well-being is achieved only by humans who live an intellective life; well-being is, in this sense, an attribute of life and is the most perfect kind of life. Nevertheless, animals and plants can somehow also *be well*, but at a lower and more obscure level.

This concept is also connected to the Plotinian doctrine of the productive contemplation of nature, which is at the heart of Plotinus' philosophy (Chiaradonna 2009, 155). The doctrine is introduced by an ironic question that once again mentions plants:

> Suppose we said, playing at first before we set out to be serious, that all things aspire to contemplation, and direct their gaze to this end – not only rational but irrational living things, and the power of growth in plants, and the earth which brings them forth – and that all attain to it as far as possible for them in their natural state, but different things contemplate and attain their end in different ways, some truly, and some only having an imitation and image of this true end – could anyone endure the oddity of this line of thought? (*Enneads* III 8 [1])

[10] In I 3 [15], Plotinus states: "I do not mean "another kind" in the sense of a logical distinction, but in the sense in which we Platonists speak of one thing as prior and another as posterior."

Hence, the fact that plants are able to experience well-being is related to the Plotinian idea that plants, as part of nature, participate in contemplation on the lowest level. Every action in nature is conceived by Plotinus as a form of contemplation. Thus, the simple activities of plants should also be understood in this way. It is significant for the issue in question that Plotinus held that contemplation as such is always moved by a kind of desire to achieve its own form and that this desire is the cause of contemplation. Thus, if every kind of nature participates in a certain form of contemplation, differentiated by degree, being moved by desire, a form of desire is present in plants, which leads them to feed, grow and generate fruit.

In sum, Plotinus conceives of well-being as an attribute of life. This can fully be possessed by humans—who have the highest degree of life, which finds expression in the full state of happiness—, but it can also be possessed by irrational beings when their vital functions are successfully performed. The success concerning the functions of irrational beings like plants is due to their desire to attain their designated form. It is precisely this desire that causes contemplation even in plants.

6.3 Isaac Israeli on the *Sensus Naturalis*

The Platonic approach to the topic was further developed in the tenth century AD.[11] Indeed, one of the sources that strongly influenced this debate is Isaac Israeli's *Liber elementorum*, in which he distinguishes three different types of sense that can be categorized as the *intellectual, animal* and *natural* sense.[12] In order to understand Israeli's approach, it is essential to point out that he defines sense as being related to pleasure and pain; each level of sense correlates to a certain degree of pleasure and pain. This definition is quite different from Aristotle's, and it is precisely on the basis of this definition that one can extend the argument in order to recognize that plants have some sort of sensibility. On the contrary, Aristotle, in *De anima* 423a, relates the perception of sense to the organ of sense. And even in the pseudo-Aristotelian *De plantis*, the argument denying sensibility in plants claims, as has been said, that plants cannot perceive and desire because they do not have any organs. Israeli's definition seems, to some extent, to be an extension of Plotinus' argumentation. Sense is here understood as the natural tendency of living beings to avoid pain and seek pleasure.

[11] See Altmann and Stern 1958.

[12] Dicemus ei, quoniam sensus est secundum tres modos: ipsius enim est naturalis et animalis et intellectualis. Et dolor quidem absolutus est sensus rei absolutae cum dolore exitus a natura et essentia sua ad diversum, secundum quod ipsa est eius contrarium. Et delectatio quidem absoluta est sensus rei absolutae cum delectatione eius, quod convenit naturae et essentiae eius (*Liber elementorum* III, ed. Lyons 1515, f. 10).

Therefore, it is possible to assign a kind of *natural* sense to plants, namely a certain sensibility through which they are able to feed or not. Within this Platonic and Neoplatonic tradition, the possibility to feed is closely related to the capacity to desire. Israeli states that plants can feed and grow precisely because they have this kind of sensibility, which enables them to seek and then desire pleasure.[13] Thus, nourishment is sought for by plants, and when it is appropriate for the *complexio* of the plant itself, the plant can experience a sense of pleasure. Yet, when the *complexio* is too different from that of the plant, it feels a sense of pain. Israeli indeed offers some concrete examples,[14] for instance: sweet water is preferred by plants; this is evident from the fact that when plants feed on sweet water, they grow best and produce fruits. On the contrary, nourishment and water with opposite features lead to the dryness of plants. The different behaviors as results of the different features of nourishment testify to the fact that plants are not indifferent to the kind of nourishment they receive.[15] What is notable here is that Israeli considers fructification as the natural goal of plants, which is, moreover, emotionally qualifiable.

The fact that plants possess the *sensus naturalis* is stated by Israeli to be *manifestum,* that is, evident. Although a certain connection with Plotinus' position becomes apparent, there is still a huge difference between the two positions. Plotinus, as previously stated, avoids admitting that plants have any sensibility, despite their being capable of pleasure. Here, sense has a different semantic value. Contrary to Aristotle's conception of sense as the object of the sense organ, sense can be defined as the instrument to measure pleasure and pain and, as such, is possessed by plants.

[13] Si igitur sentiens fuerit naturale, eius sensus erit naturalis; et si fuerit animale, erit sensus eius animalis; et si fuerit intellectuale, erit sensus eius intellectualis. Et quoniam ipsa sunt naturalia et non animalia: propter hoc sentiunt sensu naturali, quod convenit naturae et complexioni suae ex nutrimento, et delctantur eo, et recipiunt ipsum assidue sibi, et refugiunt ab eo, quod diversum est a natura et complexione ipsorum, et expellunt ipsum a se (*Liber elementorum* III, ed. Lyons 1515, f. 10).

[14] Significatio autem illius est, quod ipsa recipient aquas dulces, et delectantur aqua dulci et pingui terra et aere temperato et caliditate leni, et refugiunt ab aquis nitrosis et aluminosis et sulphureis et terra calefacta et aere superfluae caliditatis et frigiditatis. Et ex eis iterum, quae significant illud, est, quod videmus de augmento et fructificatione ipsarum in terra pingui et aqua dulci et aere commensurato; et de exsiccatione ipsarum, penuria nutrimenti, et extenuatione et infructuositate in terra et aquis et aere diversis ab illis (*Liber elementorum* III, ed. Lyons 1515, f. 10).

[15] Iam igitur manifestum est, quod arbores et plantae, quamvis non sentiant sensu animali, tamen sentiunt sensu naturali, et delectantur eo, quod eis est conveniens ex nutrimento et refugiunt ab eo, quod ab eis diversum est. Et illud quidem significatur ex membris nostris. Invenimus enim ipsa delectari eo, quod convenit complexioni eorum et recipere ipsum ad se, etsi non sentiant illud, et refugere ab eo, quod ab eis diversum est, et expellere ipsum a se, etsi non sentiant illud sensu animali. [...] Sensus inest animalibus tribus, scilicet animae vegetabili et animae animali et animae rationali. Et vegetabili quidem inest sensus naturalis, quo sentit in nutrimento et augmento suo. (*Liber elementorum* III, ed. Lyons 1515, f. 10).

6.4 Averroes' *Long Commentary* on Aristotle's *Physics*

Medieval Latin scholars approached the debate through the reading of many other texts, mostly written by Arabic philosophers who did not confine the *scientia plantarum* to comments on the pseudo-Aristotelian treatise.

One of the texts that further enhanced the understanding of the issue is Averroes' *Long Commentary* on Aristotle's *Physics*. Commenting on *Physics* I 9192a, Averroes introduces a distinction not present in the Aristotelian text. In order to explain the Aristotelian statement that matter desires form, Averroes speaks about the *appetitus naturalis*, perceived as being different from the *appetitus animalis*. The *appetitus naturalis* is defined as *appetitus materiae ad formam* and it is conceived of as the cause of movement in living beings. The *appetitus naturalis* is therefore present in plants, being the kind of desire through which plants feed.[16]

In an Aristotelian framework, Averroes admits a *desiderium naturale* in plants, thus providing future scholars with another significant interpretative key to the debate. Moreover, it has a significantly metaphysical implication for the topic of plants, namely that matter is not completely passive. Plants, and more generally matter, cannot be conceived of as merely passive, precisely because of the causes of their modifications and functions.

In *Physics* 192b8, Aristotle's division between natural and artificial beings positions plants among natural beings, which have the principle of movement within themselves. The most important difference between natural and artificial beings is that, to use Averroes' paraphrase, natural beings possess an *impetus mutationis innatus*. In his interpretation, Averroes conceives of the movement of the growth of plants as caused by the *appetitus naturale.*

However, his argumentation does not focus on plants alone, but deals with desire and movement in a rather general way. While it appears that Averroes wrote a commentary on the *De plantis*, only a few fragments of a Hebrew translation[17] have survived. In these, the question concerning desire in plants is not addressed. Nevertheless, these few lines have shifted the perspective of how to analyze the question, admitting the possibility that plants too could desire in a certain way, even in Aristotelian terms.

---

[16] Et intelligit hic per appetitum illud, quod materia habet de motu ad recipiendum formam. Appetitus enim alius est naturalis sine sensu, ut in plantis ad nutrimentum et alius est cum sensu, ut appetitus animalium ad nutrimentum. In materia igitur est appetitus naturalis ad recipiendum omnes formas recipit igitur eas alternatim, quando forma agens est praesens. (*De Physico auditu* I, 5, 46) Cf. Averroes 1562.

[17] Damascenus 1989, 363–371.

6.5 Plants' Desire by Roger Bacon

The *De plantis* was commented on by several Latin masters,[18] each one of them dealing to a greater or lesser degree with the question of desire in plants.

Roger Bacon, whose commentary *per modum quaestionis* introduced an originality to the arguments dealt with in the *De plantis* that gave a new impulse to the question. Bacon opens the debate on the sensibility of plants with one *obiectum* from *De sensu et sensato* 4, 442a, which states that the cause of nutrition is sweet flavors (Bacon 1932, 177–179 ll. 22–12). The *obiectum* goes on stating that the nutritive faculty is connected with the reception of the sweetness of the water, which is the object of the sense of taste; plants must therefore possess the sense of taste. In response to this objection, Bacon distinguishes between the way plants and animals receive nutrition. While animals receive nutrition *per naturam et voluntatem*, plants receive nutrition *per naturam solum*. Sweetness, which is mentioned in *De sensu et sensato*, must therefore be understood according to a distinction: either in relation to the *subiectum*, that is, everything that feeds on something sweet, or considered as the *dulcis in quantum dulce*, in which case only animals feed on it. At the conclusion of the question, the meaning of *sapor* as *passio nutrimenti* is redefined. Taste is not *passio nutrimenti* in reference to its substance, but only in reference to its *ratio*. Thus, it only belongs to the nutrition of animals, whose discerning faculty makes them able to understand precisely its *ratio*. Similarly, Bacon concludes that plants do not even have the sense of *tactum,* as they are immobile and as such they cannot *fugere nociva*. It is thus clear that Bacon's position opposes that of Israeli, who assigns the sensibility of avoiding pain and attaining pleasure to plants. Although Bacon offers a purely Aristotelian position with regard to sense, the question *Utrum in plantis sit desiderium* (Bacon 1932, 187–189, ll. 28–2) nevertheless represents a different position, as he distances himself from the statement expressed in the *De plantis*. Indeed, in order to find a solution for this intricate question, Bacon appeals to Averroes' argument.[19] As a matter of fact, it was only through the words of the *Commentator* that the Aristotelian position could be partially reconsidered. He thus distinguishes between two different kinds of desire: one related to animals and the other related to natural things more generally, including plants. It is hence possible to state that there is a kind of *desiderium* in plants.

---

[18] Lohr lists eleven commentaries on the *De plantis*; see Lohr 1967. *Ibid.* 1968, *Ibid.* 1970, *Ibid.* 1971, *Ibid.* 1972, *Ibid.* 1973, but the number of them has recently been reconsidered; see Panarelli 2019, 2020.

[19] Quod concede, quia desiderium proprie et maxime proprie reperitur in animalibus, quia additur affectus ibi supra commune desiderium, et additur similiter cognitio supra rationem desiderii communiter sumpti, et hoc intendit hic; tamen loquendo de appetitu vel desiderio communiter loquendo, secundum communem rationem appetitus prout sic extendit ad naturalem et voluntarium, sic potest dici quod in plantis est desiderium, et sic patet solutio rationum ad utrumque (Bacon 1932, 187–189, ll. 28–2).

While Bacon is commenting on a text that he attributes to Aristotle, he accepts a position far from that of the Stagirite, admitting that plants were able to experience a kind of *desiderium naturale*. By doing so, his position is at a fundamental crossroads of traditions.

6.6 Albert the Great's Summary of the Question

The text that constitutes the most salient moment in the history of the reception of the *De plantis* is the *De vegetabilibus* by Albert the Great. The issue in question is approached in detail in *De vegetabilibus* I, 1, 10, where the Dominican master shows his fidelity to the pseudo-Aristotelian treatise, devoting a large part of his discussion to the question in order to validate the statements made in the *De plantis*.

The problem concerning sense and desire in plants first refers to Plato's authority, but is expounded through Isaac Israeli's doctrine of a natural sense. Conversely, the peculiarity of Albert's method is the accurate explanation of a subject that he eventually does not support. First, Albert explains that philosophers like Plato and Isaac attribute sense to plants.[20] However, he emphasizes, and by doing so apprehends his counter-argument, that this does not imply that plants possess the same kind of sense as animals.[21] Albert distinguishes here between external and internal senses, specifying that these philosophers intend to attribute external senses alone to plants. In Albert's explanation of the argument, the difference between the animal sense and the natural sense regards the way in which the *sensibilia*, the sensible perceptions, act on the subject which receives it. For instance, in the case of animals, the sensible perceptions age *immaterialiter*, whereas in the case of plants, the sensible perceptions act *materialiter*, namely through the qualities of matter.

Hence, in this passage, Albert reveals his interest in reporting, in the most detailed way possible, Israeli's opinion. Albert explains that his idea that plants can desire

---

[20] Ipsi enim, sicut testatur Isaac, duplex distinxerunt desiderium et duplicem sensum, unum quidem, quod est cum apprehensione desiderati et sensibilis; et aliud, quod est sine apprehensione omni. Et ideo quando sensum attribuerunt plantae, non dederunt ei sensum et desiderium cum apprehensione sensibilis et desiderati, sed sine his (*De vegetabilibus* I 1 10 n. 67, 36)

[21] Et ideo non dederunt plantis sensum, quo utuntur animalia perfecta, qui est per medium extrinsecum; sed tantum illum, quo utuntur animalia imperfecta, qui est per medium intrinsecum, sicut tactus et gustus, secundum quod est quidam tactus. [...] Animaliter autem inesse dicunt sensum, quando inest secundum solum animae actum vel passionem; et hoc est judicium sensibilium et apprehensio, quam sola facit anima, quando recipit formam sigilli immaterialiter omnino, sicut cera figuram sigilli, nihil omnino recipiens de auro vel alia sigilli materia. Naturaliter autem inesse sensus dixerunt, quando sensibilia insunt per actiones qualitatum materiae et per esse materiale, quodhabent in materia extra, sicut calidum nest calefacto, et dulce ei; quod infunditur dulci substantiae, et sic de aliis; naturam agentis et patientis constituitur sensibile in esse materiali et naturali. Nec fuit alia ratio eorum, quare hos sensus duos naturaliter inesse plantis dicerent, nisi quia per tales modos agentium naturalium et patientium qualitates talium sensuum efficiuntur in plantis, quando auguntur et nutriuntur. (*De vegetabilibus* I 1 10 n. 67–68, 36–37)

takes a kind of middle position between the idea of desire as *sequela sensus communis*—which pertains to animals, as desire is here conceived as a consequence of sense—and the idea of desire as *appetitus animae matarie naturalis primae*, which regards natural beings more in general and coincides with the idea of desire found in Averroes. Isaac considers natural sense an *appetitus animae qui a corpore non separatim agit.* Hence, in order to clearly explain what his mistake consisted of, the Dominican master aims to clarify the exact intention of the Hebrew philosopher. Although Albert recognizes that Isaac's opinion is supported by *multi valde in philosophia excellentes*, he observes that they *peccant in sophisma aequivocationis in nomine sensus et appetitus*, namely that the term *sensus* is used by Isaac Israeli in an equivocal way. Meanwhile, Aristotle establishes the definitions of sense and desire that cannot be referred to plants as such.[22] Albert denies sensibility and desire to plants, explaining his argument as follows: if the plant's soul was able to perceive, it should somehow be separated from the body. Thus, sense always implies a sort of knowledge of the perceived object.[23] Moreover, it is not possible to conceive of sensation, as Isaac does, as totally involved in matter. Sensibility, on the contrary, implies for Albert a sort of separateness from matter. While sensitive and intellective souls are more or less separate from the body, the vegetative soul is, according to Albert, completely dependent on the body and, for this reason, cannot perceive.

Nevertheless, although he saves the Aristotelian position, Albert does not return to the idea of matter as being completely passive, which is proven by his doctrine on the *inchoatio formae*, strongly influenced by Averroes, as Rodolfi shows (Rodolfi 2004). Even in Albert, the Aristotelian doctrine is thus influenced by Arabic sources, specifically Averroes. What is, however, most remarkable is that Albert, in order to avoid misunderstandings, clarifies the exact meaning of the term *desiderium*, establishing that, as such, it cannot refer to plants.

With this detailed argumentation, Albert places the two traditions that emerge from the ideas of Plato and Aristotle in opposition to each other. On the one hand, there are the analyzed arguments from the pseudo-Aristotelian text; one the other, there are those from Isaac Israeli's *Liber elementorum.* Albert takes stock of the debate and aims at shedding light on every aspect of it. Thus, in these few pages, the Dominican master summarizes a very long discussion that had lasted for centuries, establishing the impossibility to refer sense and desire to plants as Aristotelian, on the basis that the vegetative soul performs only material functions—growing, feeding and reproducing—, whereas sense and desire are conceived as exceeding the body.

[22] Sensus enim, secundum suum nomen, animalem quondam dicit esse sensibilium perceptionem, et non naturalem solum. Similiter autem desiderium dicit appetitum excitatum ex delectabili nuntiato. Et neutrum convenit plantis, et ideo neque sensus, neque desiderium insunt eis. (*De vegetabilibus* I 1 10 n. 73, 39)

[23] Anima autem sensibilis, cum sit minus separata, accipit in instrumento, sed iudicat in seipsa de conceptis. (*De vegetabilibus* I 1 10 n. 74, 39)

6.7 Concluding Remarks

The attempt to reconstruct this debate has shown that, although the bipartition of the philosophical tradition origins during Antiquity with Plato and Aristotle, it underwent significant re-elaborations during Late Antiquity and the Middle Ages. This long debate, which has thus far received only little scholarly attention, involved many very important philosophers.

As has been shown, the debate on desire in plants is closely connected to the debate on sense in plants. While the possibility to assign sensibility to plants has always been neglected in the Aristotelian tradition, the case of desire in plants is different. Although the clear statement of the pseudo-Aristotelian *De plantis* and several passages of Aristotle's work deny plants desire, the philosophical tradition reveals a vivacity of perspectives in this regard.

Plotinus' position regarding this issue has shown to be quite intricate, assigning a kind of contemplation to plants that implies that they are, in a certain way, absolved of their state of total passivity. The case of Isaac Israeli then represented the most antithetical position regarding the Aristotelian position, assigning desire and even sense to plants. Nevertheless, these approaches somehow influenced a purely Aristotelian philosopher, Averroes, who, in commenting on an Aristotelian passage, assigned desire to plants, subtly modifying Aristotle's arguments. A metaphysical need is expressed through the idea that matter requires its own form. This reformulation implies that not only plants but even matter loses its state of total passivity, thus being able to experience the *desiderium naturale*. On the basis of this reformulation, an English master, Roger Bacon, also distanced himself from the Aristotelian statement on the issue and agreed with Averroes that there should be a form of *desiderium naturale* in plants. Nevertheless, the purely Aristotelian position is saved by the explanation of Albert the Great, who—through an accurate analysis of the counter-argument—repositioned Aristotle's arguments correctly. Plants cannot desire because they are completely involved in matter and it is not possible to assign to matter any kind of perception and desire, which is conceived of as a consequence of perception, as perception requires a certain separateness from matter. Nevertheless, although plants, and thus matter more generally, appear to return to a more Aristotelian framework in Albert, his concept of matter, and therefore of plants, cannot be reduced to mere passivity if his idea of the *incohatio formae* is taken into account.

Thus, the debate on desire in plants has shown less continuity in comparison to that on sense. This is the case because a more general requalification of the metaphysical relation between matter and form is hidden behind the idea of the desire of plants. Hence, the debate on this issue can be understood as being embedded in a more general reconsideration of Aristotelian philosophy, as absolving plants of their passivity also means absolving matter of the state of total passivity, thus deeply revaluating one of the fundamental concepts of philosophical investigation.

Acknowledgments This contribution was part of the project "PRIN 2017 - Averroism. History, Developments and Implications of a Cross-cultural Tradition 2017H8MWHR".

References

Adam of Buckfield. 2013. In *Glossae super De vegetabilibus et plantis*, ed. J. Long. Leiden-Boston: Brill.

Alfredus of Sareshel. 1985. Super librum De vegetabilibus,. In *Medieval Studies*, ed. R. J. Long, 47.

Altmann, Alexander, and Samuel Milkos Stern. 1958. *Isaac Israeli. A Neoplatonic Philosopher of the Early Tenth Century*. Oxford: Oxford University Press.

Averroes. 1562. *De physico auditu*, ed. Venetiis apud Junctas, Frankfurt am Mein: Minerva.

Bacon, Roger. 1932. Quaestiones supra De plantis. In *Opera Hactenus Inedita Rogeri Baconi*, ed. R. Steele. Oxford: Clarendon.

Baluška, František, Stefano Mancuso, Dieter Volkmann, and Peter W. Barlow. 2009. The 'root brain' hypothesis of Charles and Francis Darwin. Revival after more than 125 years. *Plant Signaling and Behavior* 4 (12): 1121–1127.

Chiaradonna, Riccardo. 2009. *Plotino*. Roma: Carocci.

Damascenus, Nicolaus. 1841. In *De plantis libri duo Aristoteli vulgo adscripti*, ed. E.H.F. Meyer. Leipzig.

———. 1989. *De plantis*, Five translations. Amsterdam/Oxford/New York: North-Holland Publishing Company.

de Alvernia, Petrus. 2003. In *Sententia super librum De vegetabilibus et plantis*, ed. E.L.J. Poortman. Leiden: Brepols.

Diels, Hermann, and Walther Kranz. 1906. *Die Fragmenten der Vorsokratiker*. Berlin: Weidmann.

Fazzo, Siliva, and Mauro Zonta. 2008. Aristotle's Theory of Causes and the Holy Trinity. New Evidence about the Chronology and Religion of Nicolaus 'of Damascus'. *Laval théologique et philosophique* 64 (3): 681–690.

Herzoff, B. 2016. Wer war der Peripatetiker Nikolaos, der Verfasser des Kompendiums der Philosophie des Aristoteles und Bearbeiter seiner Schrift über die Pflanzen? *Antike Naturwissenschaft und ihre Rezeption* 26: 135–187.

Israeli, Isaac. 1515. *Liber elementorum* in *Opera omnia Ysaac in hoc contenta cum quibusdam aliis opusculis, cum tabula et repertorio omnium operum et questionum in commentis contentarum*. Lyon 1515: Iohannes de Platea.

Laertius, Diogenes. 2008. In *Vitae philosophorum*, ed. M. Marcovich, vol. I. Berlin: de Gruyter.

Linguti, Alessandro. 2000. *La felicità e il tempo. Plotino, Enneadi*, I 4–I 5. Milano: LED.

Lohr, Charles. 1967. Medieval Latin Aristotle Commentaries. *Traditio* 23: 313–413.

———. 1968. Medieval Latin Aristotle Commentaries. *Traditio* 24: 149–245.

———. 1970. Medieval Latin Aristotle Commentaries. *Traditio* 26: 135–216.

———. 1971. Medieval Latin Aristotle Commentaries. *Traditio* 27: 251–351.

———. 1972. Medieval Latin Aristotle Commentaries. *Traditio* 28: 281–396.

———. 1973. Medieval Latin Aristotle Commentaries. *Traditio* 29: 93–197.

Long, James. 1985. Alfred of Sareshel's Commentary on the Pseudo Arisotelian De plantis: A Critical Edition. *Mediaeval Studies* 47: 125–167.

Magnus, Albertus. 1867. *De vegetabilibus libri VII*, ed. E. Meyer and K. Jessen. Berlin: Reimer.

Moreaux, Paul. 1973. *Der Aristotelisumus bei den Griechen. Von Andronikos bis Alexander von Aphrodisias. Erster Band: Die Renaissance des Aristotelismus im I. Jh. V. Chr.* Berlin/New York: de Gruyter.

Panarelli, Marilena. 2019. John Krosbein's Commentary on Pseudo-Arisotetelian *De* plantis: A Critical Edition. *Bulletin de philosophie médiévale* 61: 125–152.

Panarelli, M. 2020. Albert the Great's De vegetabilibus and its unique position among the medieval commentaries on De plants. In *Περὶ φυτῶν. Trattati greci di botanica in Occidente e Oriente*, eds. M. F. Ferrini and G. Giglioni, 137–162. Macerata: Eum.

Plato. 1997. Timaeus, *Complete Works*, ed. J.M. Cooper. Indianopolis: Hackett Publishing Company.

Plotinus. 1989. *Enneads I 1-9*, ed. A.H. Armstrong. Cambridge, MA: Harvard University Press.

Repici, Luciana. 2000. Uomini capovolti. In *Le piante nel pensiero dei Greci*. Rome/Bari: Laterza.
———. 2009. Il *De plantis* pseudo-aristotelico. In *Le monde végétal. Médicne, botanique, symbolique*, ed. A. Paravicini Bagliani. Firenze: Sismel.
———. 2011. Piante, animali e uomini nel mondo antico: analogie, discontinuità, gerarchie. In *Passaggi. Pianta, animale, uomo*, ed. L. Repici, B. Cavarra, and V. Rasini, 19–44. Milano-Udine: Mimesis-Edizioni.
Rodolfi, Anna. 2004. *Il concetto di materia nell'opera di Alberto Magno*. Firenze: Sismel.
Skemp, Joseph Bright. 1947. Plants in Plato's 'Timaeus. *The Classical Quarterly* 41: 53–60.

Chapter 7
Disclosing the Hidden Life of Plants. Theories of the Vegetative Soul in Albert the Great's *De vegetabilibus et plantis*

Amalia Cerrito

Abstract In his *De vegetabilibus et plantis* (ca. 1256), Albert the Great reorganizes the entire botanical knowledge of his time, creatively resorting to Aristotelian epistemic principles. Plant complex physiology is theoretically relevant for the theological-philosophical discourse. Nevertheless, plant life is mostly hidden, preventing direct investigation. This is borne out by unclear and fragmentary arguments ascribed to the 'philosophers' in the pseudo-Aristotelian *De plantis*. To rebut 'those who treated the life of plants confusedly', Albert explains the vegetative soul in a more rigorous way, exploring innovative themes such as sensitivity and the levels of touch. Going deeper into the reasons for the vegetative power, Albert also highlights the relationship between plant and matter (i.e., the elemental properties) to more general issues (e.g., sleep in plants; the night closing of flowers; sexual differentiations; the place of plants in the *scala naturae*).

In 1248, the Dominican order entrusted Albert the Great (1200–1280) with the foundation of the *studium generale* of Cologne, i.e., a university for Dominican friars. In order to provide his fellow Dominicans with an updated competence in several disciplines—and receptive of the philosophical debates of the time — he wrote his commentaries on Aristotle (Weisheipl 1980, 28–32). Within this cultural program, in 1256 ca., Albert also commented on the *De plantis*, a short treatise mistakenly attributed to Aristotle and translated into Latin by Alfred of Sareshel (twelfth century). This work bears only scant traces of the rigor and the epistemic principles one would expect in an Aristotelian treatise. In fact, this is apocryphal, and scholars today ascribe it to Nicolaus of Damascus (Hugonnard-Roche 2003). In his commentary, though, Albert never explicitly questions the attribution to Aristotle.

A. Cerrito (✉)
University of Pisa, Pisa, Italy

University of Florence, Florence, Italy
e-mail: amalia.cerrito@unifi.it

F. Baldassarri, A. Blank (eds.), *Vegetative Powers*, International Archives of the History of Ideas Archives internationales d'histoire des idées 234,
https://doi.org/10.1007/978-3-030-69709-9_7

Nevertheless, he literally rewrites the *De plantis* in a more consistent way: rather than just explaining the text, he reorganizes it and adds supplementary sections, so that the original two books turn into seven in Albert's own *De vegetabilibus et plantis*.[1] In doing so, Albert patterns his *De vegetabilibus* on the solid epistemic structure of Aristotle's zoological treatises (i.e., *History of Animals, Parts of Animals, Generation of Animals*).

Albert's main strategies can be summarized as follows:

(i) Integration and upgrading of contents via well-structured digressions (*digressiones*) which also include extra-Aristotelian sources (as in the alphabetical 'files' of books VI–VII) (Stannard 1980; Reeds and Kinukawa 2013, 575–576).
(ii) Textual reorganization for a clearer argumentative flow (especially in books I and IV).
(iii) Inclusion of original sections on the life of plants in the central books (II, III and V) based not only on the brief mentions of plants scattered in Aristotle's works, but also on Peripatetic and Arabic sources.

Applying such strategies, Albert goes well beyond the plain task of commenting on the *De plantis*, and gathers the wealth of information available in the thirteenth century, including medical theories. Moreover, through analogies (*secundum proportionem*) with animal physiology, Albert builds up an extremely consistent morphology and physiology of plants.

In this paper, two examples of such strategies will be examined. The first part is centered on Albert's integrative digressions (an original discussion on the powers of the vegetative soul). The second part is devoted to the central books (II–III and V) of the *De vegetabilibus*, where Albert deals with the powers of nutrition, growth, and reproduction.

7.1 A Simple Soul in a Simple Body

According to the author of the *De plantis*, plant life is almost hidden when compared to the more evident and visible animal life. Therefore, a direct and rigorous investigation of plants and their functions is not easy. This finds confirmation in the numerous philosophical positions discussed in the *De plantis*, which diverge significantly from Aristotle's doctrines (Nicolaus of Damascus 1989, 517). Yet, in his digressions, Albert rebuts all of these positions. He is well aware that, in order to disclose the hidden life of plants, it is necessary to answer a preliminary question: what sort of life do they live? Consequently, in the first book of the *De vegetabilibus et plantis*, he devotes fourteen chapters to this issue (Albert the Great 1867, 33).[2]

[1] For an overview of the content and structure of Albert's *De vegetabilibus*, see Wöllmer 2013. On the medieval readings of pseudo-Aristotle's *De plantis*, see also Panarelli, Chap. 6 in this volume.

[2] Ernst Meyer, the critical editor of Albert's *De vegetabilibus*, also authored a monumental *Geschichte der Botanik*, where a chapter is devoted to Albert's digressions on plant animation. See Meyer 1854–1857, 40–42.

The first philosophical position presented in the *De plantis*—and attributed to Democritus and Anaxagoras—ascribes to plants a sort of intellective soul, which would be made ineffective by the matter of the body.[3] Such a theory sounds unreasonable to Albert: the powers of the intellective soul do not require the body as their instrument; therefore, no intellective soul can be oppressed by matter, as its functions are immaterial. In turn, the vegetative faculty with its material substrate is more clearly separable from sensible and intellective powers. Thus, the vegetative soul of plants is the simplest and it is designed only to feed, grow, and reproduce: 'Its activities begin and end in the plant body' (Albert the Great 1867, 33).

Additionally, how does the soul-body relationship work in plants? Generally, in living beings the organic disposition and composition go hand in hand with the complexity of the soul that animates them. A simple soul with simple powers requires a simple organic body. Thus, in plants a particularly complex differentiation in organs it is not necessary, since plants can only grow, feed, and reproduce. Unlike animals, plants are not diversified in their parts. The parts of plants are as simple as their nourishment (or humor) is. Animals are provided with a sensitive soul that clearly has a wider range of powers, thus their soul requires more organic differentiation. Hence, their nourishment is more complex than that required by plants. For example, plants lack a liver, nerves, and veins because the vegetative soul does not have any kind of powers that would use these as instruments. The organic body of plants is composed of homogeneous parts: they are almost evenly (*aequaliter*) soft or hard and all bodies present the same organic arrangement. As Albert exemplifies: 'several plants exhibit almost the same disposition of roots and branches, as one can see when comparing the pear and the apple tree; but the difference between their species is evident in the diversity of fruits' (Albert the Great 1867, 35).

7.1.1 Theories of Plant Perception

Still dealing with the issue *an planta vivat/whether plants live*, Albert ponders over the Aristotelian idea of a certain continuity (*synécheia*) between natural beings.[4] This does not exclude intermediate stages or gradations. The principle of a physiological continuity allows Aristotle to place ambiguous living beings, such as

[3] For a discussion of Anaxagoras' and Democritus' theory that plants have intellect (as witnessed in ps.-Aristotle's *De plantis*) see Curd 2008, 237–238.

[4] See, e.g., Aristotle 1991, 588b 5–12: 'Nature proceeds from the inanimate to the animals by such small steps that, because of the continuity, we fail to see to which side the boundaries and the middle between them belongs. For, first after the inanimate kind of things is the plant kind, and among these one differs from another in seeming to have more share of life; but the whole kind in comparison with the other bodies appears more or less as animate, while in comparison with the animal it appears inanimate. The change from them to the animals is continuous, as we said before.'

the zoophytes, halfway between animals and plants. Thus, when Albert discusses the philosophical positions of those who deny life in plants due to their lack of motion, he refers to this philosophical principle. According to Albert, the boundary between life and non-life cannot be represented by perception or sensation, which is the origin of motion. This is evident in the animation of the heavens: they possess a regulating power (*regitiva virtus*) and govern the sublunary world, which is subject to the corruption of matter, unlike the incorruptible heavens (Albert the Great 1867, 17). Although the heavenly spheres live a higher life than animals, as they have the faculty to manage the entire sublunary world, they do not have the power of sensation. This is evidence that not all those who are alive have the power of perception.

'Being alive'—and not the power of motion—is the trait that distinguishes a living being from a stone. There is no intermediate stage between life and not-life: 'being alive' is equivalent to 'being not dead'. One can also consider ageing or illness as something that brings us closer to death. However, these two stages also only pertain to a living being. In turn, death is privation of life *tout court*. Plants are endowed with at least one part of the soul and they proceed from youth to death through ageing and even undergoing sickness as any animated being (Albert the Great 1867, 17).

Albert claims that perception is not the dividing line between living and non-living beings. But even if it was, he explains that also plants manifest a certain type of sensory activity. In fact, there are those who hold the theory of natural sensation in plants (Albert the Great 1867, 18). If compared to that found in animals, which entails sensible or rational judgment, natural sensation in plants is imperfect. Therefore, one cannot consider something that feeds as an inanimate (or non-living) being.

Here, Albert condenses a theoretical issue already raised in his *De somno et vigilia* (1256 ca.). In this work, Albert deals with the relationship between natural sensation and nourishment. He explicitly refers to Isaac Israeli's *De elementis* (tenth century) where sensation is divided into natural, animal, and intellectual (Isaac Israeli 1515, fol. 10r).[5] The natural sense belongs to plants and acts via the elementary qualities of the body. It consists in perceiving the quality of the elements: the body somehow grasps the advantage or disadvantage of intaking a particular nourishment according to its organic complexion. This kind of perception is only typical of plants because in the intellectual or animal sensation it involves the judgment of sensitive forms.

Albert slightly modifies Isaac Israeli's theory by claiming that this kind of sensation does not involve the sense of touch or taste as feeding senses ('Isaac Israeli claims that plants have the sense of touch and the sense of taste' Albert the Great

[5]The *Book of Elements* was Latinized in the eleventh century by Gerard of Cremona. On Isaac's philosophical activity, see Altmann and Stern 2009.

1890, 25).[6] Although Isaac never talks specifically about touch and taste in plants, Albert considers them as the senses involved in the natural perception. This is probably linked to the question of the localization of the inner senses. Although Aristotle addresses the question only marginally, the organic localization of the five senses became central to the later Peripatetics (Steneck 1974, 1980; Di Martino 2008).

Albert too attempts to locate the internal senses within the body. In animals the sense of taste is placed in the tongue and the judgment of flavors follows the perception of taste. The sense of taste is instrumental for preservation, as animals perceive what is pleasant or harmful to eat. According to Aristotle, the organ of touch is the flesh as the medium all tactile perceptions go through. In Albert's opinion, instead, not all flesh is sensitive and he seems to reduce the medium of touch uniquely to the presence of nerves (Steneck 1980, 276–277).

This is the theoretical framework underlying the reference to the natural sensation in Albert's *De vegetabilibus*. So, when Albert claims that plants cannot perceive because they are not endowed with any organs of perception, he is well aware of which kind of organs are involved in perception. In fact, in this instance as in many others, Albert is concerned with the disambiguation of the functional identity of nerves and roots, specifying that actually the nerves of plants (*nervi plantarum*) are insensitive nerves (Albert the Great 1867, 62).

Such a view is also the theoretical basis of Albert's analysis of a position that the *De plantis* ascribes to Plato, according to whom plants manifest a sort of desire for food. Two passions stem from such desire: pain (*tristitia*), caused by the lack of food, and joy (*gaudium*), caused by the intake of food. Such a theory, though, is untenable both for the author of the *De plantis* and for Albert, since plants lack perception organs, therefore they cannot experience any desire-driven passion.

Yet this 'Platonic' theory linking plants' desire for food and the resulting pair of passions (a sort of pain and a sort of joy) is not entirely useless. Albert subtly reshapes it to explain some vegetal phenomena such as the intake of food for survival (Albert the Great 1867, 7–10). Such appetites in plants would be inexplicable without involving a sort of perception. Animals are endowed with the power of perception and the power of estimation of what is pleasant or harmful to eat.

What about plants? How can they discriminate what is pleasant or harmful for them to eat without resorting to a sensory power? In order to explain this natural phenomenon, Albert links (i) Isaac's theory of natural sensation, (ii) Aristotle's well-known theory according to which 'everything that feeds is alive' (Aristotle 1984, 424a 22–26), with (iii) the theory of the organic localization of sensation.

The theory of the natural sensation of plants is clearly at variance with Aristotle who claims that only animals can perceive, since sensation involves apprehension and motion. Albert is well aware of such contrast and gives an alternative account. He explains that the plant—concerning the powers of its soul and organic body—is a *res imperfecta*. In fact, plants exercise only the faculties strictly connected to the

[6] Here Albert is concerned with answering the question 'whether plants sleep', an issue that will be further discussed in his *De vegetabilibus*, as we shall see in the next section.

material forms, without involving either apprehension or movement. Furthermore, the organic body is imperfect because the plant has no organs devoted to the faculty of motion or apprehension. However, the body of plants undergoes a quantitative and qualitative change. Indeed, the desiderative power is twofold: it acts either (i) in the presence of the object of desire; (ii) or without its presence. Moreover, Albert adds that while in perfect animals the desiderative power involves an extrinsic medium (*desiderum cum apprehensione sensibilis et desiderati*), in the imperfect ones only an intrinsic medium is required (i.e. the organ of touch and taste). For this reason, quite often the roots of plants are compared to animal mouths (*ori similes*).[7]

Even though the vegetative soul neither requires apprehension, nor does it involve the senses of touch and taste, in every animated being there are both agents (*agentia*) and patients (*patientia*) as instruments of the soul. So, how do these instruments work in plants? Albert proceeds to comment on Isaac's theory of natural sensation. While animal sensation always entails a movement of the soul (namely, the apprehension of the sensitive forms), natural sensation consists in the perception of the quality of matter (e.g. the sweetness in what is sweet; the warmth in what is warm and so on). If natural sensation acts through the body (namely, via the complexion and the composition of the matter), there should be no reason why minerals and stones would not be able to perceive. But Albert explains that the natural perception depends on the soul which informs, as the first agent (*primum agens*), the active quality of the bodies. Even though the qualities act *within* and *via* the body (*corporaliter*), their actions are always informed by the soul. On the contrary, minerals are inanimate things, so they cannot perceive, not even in a corporeal way.

Once again, using an analogy, Albert makes it clear that the perception in plants is entirely embedded in their bodies, as opposed to animal perception. Whereas in animals the powers of sensation, memory or imagination are neither hot, nor cold, in plants the digestion and the indigestion of nourishment are, respectively, hot and cold. So, it is evident that plants undergo alterations of the body instead of movements of the soul. In the same vein, Albert reduces the passions of *tristitia* and *gaudium* to a mere phenomenon explainable in terms of the physiological alteration of digestion and indigestion.

7.1.2 Do Plants Sleep?

The fact that flowers close at night is one of the favorite pieces of evidence for those who claim that plants do sleep. Albert ascribes that opinion to Socrates and Plato, reshaping it into a three-stage argument:

(i) The 'night closing of flowers' is evidence of the capacity of plants to sleep. Indeed, the changing of light between night and day seems to stimulate the

[7] Cf. for example Aristotle 1984, 416a 2–5. For more on the analogy of roots of plants to the mouth of animals, and the reversal of up and down, see Repici 2000.

plants. During the night the flowers are closed (*deperire*), but when the sun rises, the flowers are stimulated to open up (*aperire*), just as if they were awakening (Albert the Great 1867, 40).

(ii) According to the Aristotelian treatise *On Sleep* (Aristotle 1984, 456b 18–30), sleep is also an effect of digestion, for the digestive activity produces evaporation that flows to the head causing sleepiness. Since plants have nutritive powers and feed, there is no reason why they should not be able to sleep.[8]

(iii) Moreover, feeding in plants seems to differ between winter and summer. During summer, the heat of the sun weakens the plants, so they quickly consume their nourishment. This is why, during summertime, plants mostly feed at night. In turn, the cold temperature of winter appears to compress the plants on the outside, even though they contain more nourishment internally than in summertime.

Following Aristotle, Albert argues that the nutritive powers always involve intake and emission of heat, which is released when the nourishment is distributed to all the parts of the organism. In animals, this process causes sleep and wakefulness.

Albert addresses these issues by recalling the Peripatetics' position (*sententia Peripateticorum*). Generally speaking, there are two types of cold (*frigiditas*): (i) one that numbs or leads to death (*stupefaciens sive mortificans*); (ii) another that is complexional, or physiological, whose role is to retain the humoral balance in limbs. The first type, the accidental paralyzing cold from outside, far from causing sleepiness, if anything, wakes animals from their sleep. On the contrary, the natural complexional cold (that balances the earthy and watery elements) compresses and closes the organic parts producing sleep (Albert the Great 1867, 40–41). Several animal activities involve the production of heat. Natural sleep in animals re-establishes the right body temperature. Albert adds that men cover themselves during sleep in order to prevent the external cold from interfering with the restoration of temperature.

In plants, the humoral cold closes the external parts to protect them from the cold and to re-establish their initial temperature, but it does not cause sleepiness. The reason is that sleep is the suspension of sensory activity, an activity not to be found in plants, meaning that a plant does not require sleep (Albert the Great 1867, 42). Thus, the phenomenon of the closing and opening of flowers is not evidence of plant sleep. Rather it is the effect of humoral cold, which is to be found in all living beings. Without involving any sleepiness, the humoral cold closes the flowers, as well as the rest of the external organs of the plants, at night or in winter. Conversely,

---

[8] Cf. Albert the Great 1867, 40: 'Dicunt enim quod omne quod cibatur, cibo distributo per omnes partes proportionaliter, aliquando intrahit calorem et spiritum ad locum, unde hauritur cibus distributus, et aliquando emittit calorem et spiritum in membra et partes, quibus cibus distribuitur. Planta autem praedicto modo cibatur: oportet igitur quod dicto modo calorem et spiritum intrahat et emittat. Talis autem intractus caloris et spiritus et emissio eurundem causant somnum et vigiliam, sicut patet per dicta in primo de somno et vigilia.'

the heat of the sun spreads to all the plant parts, arousing the opening up of the flowers.[9]

Eventually, Albert enhances the third argument concerning the difference between feeding in winter and summer. It is true that in summer plants consume their nourishment quickly, while, in winter, this is stored and consumed slowly. However, he adds, the difference lies in the nourishment itself (*succus*): due to the summer heat, it is less condensed, thus easier to consume and assimilate (Albert the Great 1867, 43).

7.1.3 Male and Female Plants?

The connection between the two sexes and the four causes is a core tenet of the Aristotelian theory of animal generation. Within the process of generation, the four causes are divided into active and passive principles. While the male role (formal, efficient and final cause) is active, the female is passive (material cause). Strictly speaking, the generative process is activated by the male seed, which informs the menstrual blood of the female. In the case of plants, the active and passive roles are preserved, but they are combined within the same individual plant. In dealing with this topic, Albert stages a virtual discussion between the ancient philosophers (*antiqui sapientes*), Aristotle, and himself.

According to the *antiqui sapientes*, the distinction between male and female in plants is manifested in a morphological differentiation within the same species: the male is characterized by very narrow leaves and its seeds are smaller than the female's; this holds true for the peony, the olive, and all other plants. A comparison with animals is very telling: 'in animals, the female and the male genitalia are attributed to one of the two sexes by accident (the accident, they said, is not having the seed within); if they had it in themselves, like the plant, they would not mate to transmit the seed, and nature would provide them with such [*viz.* differentiated] genitalia.' (Albert the Great 1867, 45).

Additionally, any generated being has a generating agent, and its generation entails the passage from potentiality to actuality. Thus, in the generative process, the generating factor and the generated being differ in their substance: 'In the same substance [...] this agent cannot be identical to the patient (otherwise the same thing would act and undergo, would be act and potentiality, would exist and not exist). The generating agent, in fact, already exists, while what is about to be generated does not exist yet, but will in the future' (Albert the Great 1867, 44–45). Since every generation always involves both active and passive principles, these must differ in their substance while being identical in form and species. So, this is also true of plants, where the substantial distinction between male and female occurs.

[9] Cf. Albert the Great 1867, 42: 'Et quia frigiditas <complexionalis> comprimit, ideo contrahuntur flores in nocte, et de die, laxante et extendente calore partes exteriors et subtiliante humorem et tumefaciente eundem, exenduntur flores et dilatantur.'

Aristotle, instead, holds that the sexual differentiation between male and female individuals only belongs to animals. Of course, the generation of plants also involves active and passive principles, but these are already combined within the same individual plant (Albert the Great 1867, 45–46). In order to generate a being endowed with a sensitive soul, both active and passive principles are necessary: one cannot generate without the other. This is evident in the so-called *ovum venti* or 'wind egg', i.e. the unfertilized egg, produced by the female alone, without a male.[10] As one finds in Aristotle, the female of certain species of birds and fish is able to produce such eggs, but these, in turn, are unable to produce proper offspring. Thus, Albert glosses, 'the wind egg lives just the potential life of the plant and not the animal life'; to produce offspring, 'it needs the active generating factor, more perfect and more formed, able to generate from its own substance; and this is the male' (Albert the Great 1867, 46). However, since plants are less perfect than animals, they do not require such a substantial distinction between male and female.

In Albert's opinion, proper generation happens only via sexual intercourse, where the male emits the semen that the female receives. Through mating, the male seed spreads the soul of its own species, while the female seed provides the matter for formation and growth.[11] Plants, instead, do not reproduce by mating. Thus, the proper animal generation must be distinguished from its imitation (*imitatio virtutis sexus*) in plants (Albert the Great 1867, 47). What activates the process of plant reproduction is already in the seed. It only needs an external stimulus to start growing: for some plants this is the heat of the sun, while for others it is wet ground, or cold temperatures.

The very fact that not all plants need the same external stimulus or condition shows that their generative process is only an imperfect imitation. Thus, it is no surprise that plants exhibit only a weak sexual differentiation and that male and female are combined within the same individual plant. The external agents seem to perform the paternal and maternal role in the generative process, as Protagoras claims by saying that 'the sun is the father of every plant, while the ground is the mother'. However, the sun does not cover all of the male and paternal functions in

---

[10] The syntagm *ova venti* renders the Greek *hypēnémia*, that occurs often in Aristotle's *History of Animals* to designate those eggs produced by the female but not fertilized by the male. Cf. Mayhew 2004, 45–46. Albert discusses this many times in his *De animalibus*: see, e.g., Albert the Great 1999, 489; 530–532; 534; 838; 257.

[11] Here Albert hints at the female role in the generative process, a theme he will deal with at length in his later *De animalibus*. While Aristotle reduced the female to a mere provider of nourishment for the embryo, in the 2nd c., thanks to his anatomical observation, Galen claimed the presence of female seminal vassels (or ovaries) producing the female seminal fluid. Through Avicenna's *Canon of Medicine*, Albert was aware of the Galenic embryological theories and of their contrast with the Aristotelian position. Albert points out that the 'female sperm' has three components, namely the passive seed, the menstrual matter, and the nourishment. Cf. also Albert the Great 1920, 1049: 'Sed sexus qui distinctus est in dictis animalibus, omnino commixtus est in arboribus. In his enim et generaliter in plantis virtutes sexuum commixtae sunt nec distinguitur in eis mas a femina per actum generationis [...].' Cf. Demaitre and Travill 1980, and Miteva 2018. On Albert's medical knowledge, see Siraisi 1980.

the generative process. Indeed, Albert points out that the proper father not only provides the seed, but also bequeaths the active generative power to his offspring. This does not happen with sun and plants. Thus, Albert concludes, one can call the heaven 'father of plants' only metaphorically (*metaphorice*) (Albert the Great 1867, 105–106).

This is not the case of the maternal role of the ground. The female functions one sees in animals match entirely the power of the ground: just like the female does in the animals (*sicut matrix in animalibus*), so the ground provides all the matter necessary for feeding and growing (Albert the Great 1867, 106).

7.2 Disclosing the Physiology of Plants

7.2.1 Plant Development

One of the less visible processes is the formation of plants. In the fifth book of his *De vegetabilibus,* Albert lists seven aspects involved in plant development. These are gathered in two groups. The first one comprises the efficient factors: (i) the heat of the heavenly spheres (*calor caelesti circuli*); (ii) the heat from the environment (*calor locis*) where the generation takes place; (iii) finally, the heat of the seed matter (*calor materiae seminalis*), which corresponds to the vital heat (*calor vitalis*). The heavenly heat is the first factor that gives life to the seed. Yet, a sort of equilibrium between the seed and the environment is required: this needs to be neither too hot, nor too cold. When the two are well-balanced, the heat of the seed matter can form the seminal moisture (*humidum seminale*), which fosters the plants' feeding capacities (Albert the Great 1867, 290–292).

The second group consists of three material aspects which regulate all the material substance of plants: (i) the natural moisture (*humor naturalis*); (ii) the ground moisture (*humor loci*); (iii) and the pluvial moisture (*humor pluviarum*). Albert points out that these material substances balance the same four basic elements (i.e. air, earth, fire, and water).

Natural moisture stimulates the seed to sprout in the ground. It contains all the plant's matter (and the *genus* information as well), from which the roots result. When the natural moisture dries, it gives way to the ground moisture; in analogy with the animal womb, its role is to provide the nourishment for the sprouting of plants. At this point, the pluvial moisture (i.e., rain, snow) humidifies the nutritional moisture (*humidum nutrimentale*). Finally, the last element needed is the air (*aër*), which surrounds plants during the formation process. If the air is healthy (*bonus*), this helps the formative process; if it is unhealthy (*malus*), it corrupts the plants (Albert the Great 1867, 291).

7.2.2 *Roots and Marrow*

Roots are the first organs that are formed in the generative process and the main organs involved in life activities. As already outlined, for plants 'being alive' means having the power of feeding, and the roots fulfill this task (in analogy with what the mouth does in animals). Additionally, Albert claims that a functional identity of heart and root can also be found since both organs distribute the vital heat to the body (*calor vitalis*), which is essential to the nutritive process (King 2001; Studtmann 2004).

To explain the function of the roots, Albert draws an analogy between them and the *paterfamilias*:

> A plant is composed of parts that refer to the root as if it were the father who takes care of them (*pater et procurator*), from whom they receive the substance of food and powers. Through the peculiar task of each one, these parts convert such substance in wealth and continuity of the species. The same happens in the government of the house (*in oeconomicis*), where the plurality of sons and servants [...] use the substance received from the *paterfamilias* for the overall wealth of the house (*domus*). (Albert the Great 1867, 127).[12]

This passage highlights the central role of the roots, whose function is to accumulate nourishment and distribute it to all the organs according to their needs. As fathers do with their families, so roots promote the survival of the species. Fittingly, like family members (sons and servants) perform their tasks for the sake of their *domus*, so plants' complementary organization is aimed at the prosperity of the species. The sons and servants stand for the marrow, which preserves the nourishment taken from the ground. According to Albert, one can consider the marrow as vicarious of the root in reason of its power to contain and distribute the 'plant spirit' (*spiritus plantae,* i.e. the vital spirit) within the body. The marrow supplies nourishment to the body parts that are most distant from the roots.

7.2.3 *Nutrition and Death: The Role of Pores*

Plants do not have digestive organs. It is the ground that provides their nourishment, after elaborating it via a peptic process (*concoctio*) (Boylan 1982). The moisture from food (*humidum nutrimentale*) is assimilated by plants, to form their parts, grow and, replace the radical moisture (*humidum radicale*).[13] However, each plant species cannot grow more than allowed by its nature. As Albert remarks, every

[12] Already in his earlier *Commentary on Peter Lombard's Sentences,* Albert had sketched the analogy between root and *paterfamilias*. Cf. Albert the Great 1893, 325: 'Semen dicit vivere sicut ovum venti dicit vivere Philosophus in eodem libro: et hoc est ab actu nutrimenti et potentia: quia semen assimilatur filio egredienti de domo patris, qui partem substantiae secum portat, et illam negotiando multiplicat [...].'

[13] On the role of the *humidum nutrimentale* and its connection with the radical moisture, see Reynolds 1999, McVaugh 1974; Crisciani 2005. Perfetti 2009.

natural species has a maximum and minimum width, length, and thickness. Even if certain parts sometimes grow more than others, such irregularity never affects the primary organs (Albert the Great 1867, 109).

It has already been mentioned that plants lack those organs that effect the conversion of food in animals, i.e. stomach and liver. Even if plants receive food that is already concocted, how do they take it in? Albert explains that this process involves hidden pores (*pori occulti*), a spongy organ that filters nourishment (Albert the Great 1867, 108).

In the *De plantis* there is no mention of the pores as spongy parts. The passage of nourishment from the root to the different parts of the plant takes place through ducts (*meatus*), which can open and close. Quite often, Albert juxtaposes the term 'duct' (*meatus*) and the term 'pore' (*porus*), suggesting that they are two different parts of the root.[14] By describing 'the very narrow duct of the pores', he distinguishes the structure of the pore (the entry point of nourishment) from that of the conduit, through which the nourishment spreads.

The distinction between pores and ducts is already attested in Bartholomeus Anglicus' *De proprietatibus rerum*, composed between 1234 and 1245.[15] Yet, Albert is much more accurate both on the material composition of the pores and on their functions. He specifies that the pores are both ducts allowing access to nourishment and spongy organs. Thus, where the *De plantis* simply states that the root is able to absorb and emit moisture (*fluxibilitas*),[16] Albert explains the *fluxibilitas* of the roots in terms of their porosity (*porositas radicum*).[17]

Besides absorbing nourishment, the pores perform other functions too. The pores (hidden or visible) exude (*exudare*) the residue (*superfluitas*) of plants (Albert the Great 1867, 215). The physiology of the hidden pores is based on the functional identity of the pores of plants and those of animals, namely, to expel the impurities through sweating. If plants draw their nourishment already elaborated by the ground and ready to be assimilated, what sort of residue can they exude? Their residue is the surplus of the nutrimental moisture (Aristotle would have labelled it 'useful residue': *chrēsimon períttōma*)[18] which, through sweating (*per sudorem*), flows towards the external parts, firming them up (*pinguedinem facit corporum*) (Albert the Great 1867, 218).

[14] Nicolaus of Damascus 1989, 555: 'Quod deorsum tendit, meatus coangustantur, cumque digesta fuerit materia […]'. Albert the Great 1867, 270: 'Quarum autem nutrimenta deorsum tendunt, habent meatus pororum valde strictos, et forte transversos.'

[15] Cf. Bartholomeus Anglicus 2005, III, XII, 159, 51–57; XVII, II, 23, 327–330; XVII, II, 26, 447–451; XVII, II, 30, 577–582.

[16] Cf. Nicolaus of Damascus 1989, 557; and Albert the Great 1867, 273; 277; 287.

[17] See Albert the Great 1867, 275. Albert's distinction between the pore as a spongy organ and the pore as duct is perhaps modeled on Aristotle's twofold notion of *póros*, used to describe both the anatomical cavities, and the permeability of a spongy tissue. See, e.g., Aristotle 1984, 659b 5; 549a 1–4.

[18] E.g., Aristotle 1984, 724b 21; 725a 13. Cf. Mayhew 2004, 33.

The pores are functional to the life of plants, but they determine their death as well, as Albert clearly outlines in his commentary on the *Book of Job* (1272 ca.). In *Job* 14:7–10 the biblical verses draw a bitter comparison: human life is ephemeral when compared to the life of plants; thanks to the vegetal potential for regeneration, there is hope (*spes*) for a tree to sprout out again, even when it seems to be dead. In commenting on these biblical verses, Albert explains that the death of plants is due to the obstruction of the pores. When this happens, the roots stop taking the moisture from the ground, and the plant stops producing its fruit. But, Albert adds, thanks to the agricultural practice of grafting, roots and pores can be regenerated. The practice of grafting had already been dealt with at length in his earlier *De vegetabilibus.*[19] However, in the later biblical commentary on *Job*, Albert explicitly refers to his source, the Roman agricultural author Palladius' *Opus agriculturae,* providing a consistent description of grafting, step by step: one must free the largest roots from the bark, cut slots in them, wedge stones into the slots to prevent them from closing: then, after filling the slots with manure, they can be covered with soil (Albert the Great 1904, 183). The artificial cutting of roots allows the plant to feed again, and so to return to a new youth.[20]

This panoply of botanical details is designed to stress the difference between plant life and the human condition. While the possibility of regenerating themselves—even by an artificial method of grafting — belongs to the plants' nature, humans—according to Job's bitter outlook—have no means whatsoever to escape death.

7.3 Before and After Albert's *De vegetabilibus*: Alfred of Sareshel and Peter of Auvergne

Albert's systematic botanic endeavor was without parallel in his times. This can be ascertained by comparing Albert's discourse with the few other medieval commentaries on the *De plantis,* written between the twelfth and thirteenth centuries.

The translator Alfred of Sareshel composed his own commentary probably between 1200–1210 (Long 1985, 135). Of course, considerably more than the other commentators, Alfred was aware of the problematic nature of the *De plantis*. Despite this, quite often he refrains from commenting on those problematic passages, which, on the contrary, Albert will deal with at length.[21] It is not clear whether Albert was

---

[19] Compare, e.g., Albert the Great 1867, 92.

[20] In his biblical commentaries, Albert is constantly concerned with providing his Dominican fellows with an up-to-date competence in several philosophical and scientific disciplines. In the *De vegetabilibus* Albert does not specifically devote a paragraph or a chapter to such a subject, while in his commentary on *Job* he gathers all the botanical information scattered in his work and provides a more consistent treatment of this subject. See Cerrito 2018; Perfetti 2018a.

[21] The editor points out that Alfred's commentary leaves twenty-two chapters of the *De plantis* unglossed. Cf. Long 1985, 134.

aware of Alfred's commentary, but some of their passages seem to share similar theoretical bases and, sometimes, argumentative strategies too.

For example, even Alfred comments on Plato's opinion on plants' desire for food, concluding that plants do not experience hunger, since this is felt only in the stomach, which plants are not provided with (Alfred of Sareshel 1985, 148–149). Of course, Albert agrees with that. Both Alfred and Albert observe that desire for food can only proceed from sensory activities, but plants do not have any specific organ designed for sensation. Nevertheless, according to Alfred, the immobility of plants is evidence of their insensibility (Alfred of Sareshel 1985, 153), and they not even possess what Albert labels the nerves of plants (*nervi plantarum*).

According to Albert, instead, the fact that plants cannot move does not mean that they do not go through bodily changes. Thus, while, in Alfred's view, plants fulfil their needs without involving sensory activities, Albert appropriates Isaac Israeli's theory of the natural perception involved in plant feeding and in their physiological alteration of digestion and indigestion.

Naturally, Alfred too compares animals and plants, taking his cue from some passages of the *De plantis*. However, whereas Albert employs the analogy to explain or to scale down the functional identities of animals and plants, Alfred uses it to establish a dividing line between vegetative and sensitive life. Alfred and Albert agree on the centrality of the root and marrow too. Both consider it involved in the nutritive process. Nevertheless, while Albert clearly sets the relationship between marrow and root on their own terms, Alfred deals with these body parts by referring to the animal digestion. He explains that in plants there is a twofold digestion: the first one takes place when 'the plant starts feeding' via the root, while the second digestion occurs in the marrow. Finally, Alfred reports that there is one more step in animal digestion: the third decoction (*decoctio*), to feed the plurality of body organs (Alfred of Sareshel 1985, 160).

The theoretical issue of the generative factors combined in an individual plant is overlooked by Alfred.[22] Once again, he explains plants reproduction in contrast with animals'. Since plants have no sensation, they lack desire for mating (*coitus*): accordingly, plants reproduce by dissemination (Long 1985, 133). While according to Albert the plant's seed contains the active and passive generative factors already combined, Alfred points out that any generation cannot occur before the seed starts to receive nourishment (Alfred of Sareshel 1985, 161). On the contrary, Albert clearly specifies the timeline of plant formation: the root is the first organ to be formed from the seed.

Before being replaced by Albert's *De vegetabilibus*, Alfred's commentary was widely used in the twelfth century, as witnessed by manifold quotations scattered within Bartholomeus Anglicus' and Robert Grosseteste's works (Long 1985, 141).

Almost twenty years after Albert's *De vegetabilibus,* Peter of Auvergne (*magister artium* at the University of Paris between the 1270 and 1290) composed the *Sententiae super de vegetabilibus et plantis,* for didactic purposes (Poortman 2003,

[22] Nevertheless, Alfred planned to deal with such a topic: 'Qui autem circa sexu plantarum dicendum in sequentibus determinabitur'. See Alfred of Sareshel 1985, 147.

ix-x). The didactic nature of this work is mirrored in its size: Peter's commentary is much shorter than Albert's. Peter has his merits: he was the first to question the authenticity of the *De plantis*, and eventually ascribed this work to Theophrastus. Quite often Peter quotes Alfred's commentary on *De plantis* and only occasionally he explicitly mentions Albert (six times, according to the editors), nevertheless, he owes a great deal to Albert's *De vegetabilibus*: [i] he appropriates several examples of what Albert himself had seen and experienced;[23] [ii] he tacitly borrows from Albert the references to other Aristotelian works; [iii] he even reproduces Albert's mistakes, like taking the name Abrucalis occurring in the *De plantis,* as applying to 'Protagoras', rather than to 'Empedocles, as Albert had done; [iv] as the editors point out, his *divisio texus* of the *De plantis* is based on Albert's *De vegetabilibus* I and IV. Further, quite often Peter seems unable to distinguish between Albert's Aristotelian quotes and Albert's self-quotations (Poortman 2003, xviii).

Unlike Albert, Peter is neither concerned with clarifying the argumentative flow of *De plantis,* nor he is willing to parallel Albert's attempt at textual reorganization. Peter simply follows the textual flow of the *De plantis.* Only occasionally does he emulate Albert's purpose to integrate and upgrade its botanical contents. For instance, when commenting on plants' desire for food, Peter fully includes Isaac Israeli's theory of natural sensation. As Albert had done, Peter too embraces the distinction between natural and animal sensation. However, glossing over Albert's explanation of plant natural sensation, he merely concludes that plants, only metaphorically have *sensation* (Peter of Auvergne 2003, 27).

7.4 A Botany Not for Science's Sake

Although Albert's botany may be considered as a major achievement in the history of science, it is to be stressed that his endeavor was not driven by purely scientific aims. Indeed, all of his botanical, zoological, natural-philosophical works—as well as all of his other philosophical commentaries—were conceived to educate his Dominican fellows and prepare them to face the challenging tasks of contemporary preaching. Albert always envisages a link between theological-philosophical training and concrete ecclesial activities (homiletics, preaching against heretics etc.). Therefore, it is hardly surprising that, even when commenting on the Bible, he takes the opportunity to verify, enhance, or modify some natural philosophical issues, such as the death of plants.

Such crossing of boundaries may seem absurd today, if compared to current epistemologies. In Albert's mind, though, the Bible is a book written by God, that also refers to the natural world, which He designed and put into being. A thorough knowledge of all fields of natural philosophy allows the friars to read between the lines of the powerful language of the Bible, discovering the wealth of its literal and allegorical meanings (Cerrito 2018; Perfetti 2018a, b).

---

[23] Cf. Peter of Auvergne 2003, 146: 'Quaedam autem faciun <hoc> interposito tempore, ut quaedam piri et pomi versus Rhenum.' Cf. Albert the Great 1867, 274.

7.5 Conclusion

The vegetative soul is entirely embedded in the body of plants, which is designed only to grow, feed, and reproduce. As we have seen in the first section of this paper, quantitative and qualitative changes are ruled by vegetative powers exclusively within and through the body.

Other powers of plants too depend on the vegetative soul (Sect. 7.1.1). Plants are resistant to the environmental stress: the humoral cold (*frigiditas complexionalis*) closes the external parts in order to protect the plant from the external cold and to restore a balanced body temperature (Sect. 7.1.2). Furthermore, several vital operations are triggered by environmental stimuli. The seed, Albert claims, does not start growing without external stimulation (Sect. 7.1.3); the development of feeding capacities depends on a dynamic interaction between the seed and its ground (Sect. 7.2.1). Finally, environmental factors assist plants in fulfilling their nutritive function: in summer, the ground provides a nourishment that is easier to consume (Sect. 7.1.2); ground and pluvial moisture stimulates the roots to sprout from the seed (Sect. 7.2.1); and – in functional, though not morphological, analogy with the animal digestive organs – the ground provides plants with their nourishment already elaborated through a specific peptic process (Sect. 7.2.3).

The synergic relationship between plants and their environment is guaranteed by the roots. Albert devotes an original treatment to the material composition and the function of the pores which allow the roots to take in food from the ground (Sect. 7.2.3). Resorting to an analogy with the *paterfamilias*, Albert establishes a connection between the ends of roots and fathers, namely 'wealth and continuity of the species'. The household organization is somehow mirrored in plants, whose body parts are arranged and designed for the sake of their species (Sect. 7.2.2).

The hidden life of plants is disclosed by Albert, who reshapes the pseudo-Aristotelian *De plantis* into a voluminous theoretical treatment of plants, in seven books, resorting to the Aristotelian epistemic principles in a critical and dynamic relationship with the thirteenth-century developments in natural philosophy.

Acknowledgments I would like to express my gratitude to professor Stefano Perfetti, who gave me numerous suggestions and helpful advices for the elaboration of this chapter.

References

Albert the Great. 1867. In *De Vegetabilibus Libri VII: Historiae Naturalis Pars XVIII*, ed. E. Meyer and C. Jessen. Berlin: Reimer.

———. 1890. De somno et vigilia. In *Alberti Magni Opera Omnia IX*, ed. A. Borgnet. Paris: Ludovicus Vivès.

———. 1893. *II Sententiarum*, Alberti Magni Opera Omnia VII, ed. Borgnet A. Paris: Ludovicus Vivès.

———. 1904. *Commentarii in Iob*, ed. Weiss M. Freiburg: Herder.

———. 1920. *De animalibus libri XXVI,* ed. Stadler H. Münster: Aschendorff.

———. 1999. *On Animals. A Medieval* Summa Zoologica, translated and annotated by Kitchell K. F. Jr and Resnick I. M. Baltimore-London: The Johns Hopkins University Press.

Alfred of Sareshel. 1985. *Commentary on* De plantis. In: Long, James R. 1985. Alfred of Sareshel's Commentary on the Pseudo Aristotelian *De plantis*: A Critical Edition. *Medieval Studies* 47: 125–167.

Altmann, Alexander, and Samuel M. Stern. 2009. *Isaac Israeli. A Neoplatonic Philosopher on the early tenth century*. London/Chicago: The University of Chicago Press.

Bartholomeus Anglicus. 2005. *De proprietatibus rerum*, De proprietatibus rerum. In: *Texte latin et réception vernaculaire. Lateinischer Text un volkssprachige Rezeption*, eds. Van den Abeele B. – Meyer H. Turnhout: Brepols.

Aristotle. 1984. *The complete Works of Aristotle,* ed. Barnes J., 2vols. Princeton: Princeton University Press.

———. 1991. *History of Animals, Books VII–X*, ed. and trans. D.M. Balme. Cambridge, MA/ London: Harvard University Press.

Boylan, Michael. 1982. The Digestive and Circulatory System in Aristotle's Biology. *Journal of the History of Biology* 15 (1): 89–118.

Cerrito, Amalia. 2018. Botany as Science and Exegetical Tool in Albert the Great. *Aisthesis* 11 (1): 97–107.

Crisciani, Chiara. 2005. Aspetti del dibattito sull'umido radicale nella cultura del tardo medioevo (secoli XIII-XV). *Arxiu de Textos Catalans Antics* 23 (24): 333–380.

Curd, Patricia. 2008. Anaxagoras and the Theory of Everything. In *The Oxford Handbook of Presocratic Philosophy*, ed. P. Curd and D.W. Graham, 230–249. New York: Oxford University Press.

Demaitre, Luke E., and Anthony A. Travill. 1980. Human Embryology and the Development in the Works of Albertus Magnus. In *Albertus Magnus and the Sciences. Commemorative Essays 1980*, ed. J. Weisheipl, 405–440. Toronto: Pontifical Institute of Medieval Studies.

Di Martino, Carla. 2008. *Ratio particularis. Doctrines des sens internes d'Avicenne à Thomas d'Aquin*. Paris: Vrin.

Hugonnard-Roche, Henri. 2003. Pseudo-Aristote, *De plantis*. In *Dictionnaire des philosophes antiques. Supplement*, ed. R. Goulet, 499–505. Paris: CNRS-Éditions.

Isaac Israeli. 1515. *De elementis*. In: *Opera Omnia Ysaac*, ed. Bartholomeus Trot, Lyon.

King, Richard H.A. 2001. *Aristotle on Life and Death*. London: Duckworth.

Long, James R. 1985. Alfred of Sareshel's Commentary on the Pseudo Aristotelian *De plantis*: A Critical Edition. *Medieval Studies* 47: 125–167.

Mayhew, Robert. 2004. *The Female in Aristotle's Biology: Reason or Rationalization*. Chicago/ London: The University of Chicago Press.

McVaugh, Michael. 1974. The *humidum radicale* in Thirteenth Century Medicine. *Traditio* 30: 259–283.

Meyer Ernst, H.F. 1854–1857. *Geschichte der Botanik*, 4 vols. Königsberg: Verlag der Grebrüder Bornträger.

Meyer Reeds, Karen, and Tomomi Kinukawa. 2013. Medieval Natural History. In *The Cambridge History of the Science*, Vol. 2. Medieval Science, ed. D.C. Lindberg and M.H. Shank, 569–583. Cambridge: Cambridge University Press.

Miteva, Evelina. 2018. "Iam patet igitur veritas eius quae dixit Aristoteles et causa deceptionis Galieni." Philosophers vs. Medics in Albertus Magnus' Account on Conception. In *Irrtum-Error- Erreur*, ed. Andreas Speer and M. Mauriège, 107–122. Berlin: De Gruyter.

Nicolaus of Damascus. 1989. De plantis. In *Nicolaus Damascenus*, De plantis, Five translation, ed. Drossaart Lulofs and Poortman, 465–561. Amsterdam/Oxford/New York: North-Holland Publishing Company.

Perfetti, Stefano. 2009. Rigenerazione degli animali? Alberto Magno tra *Parva naturalia* e *De animalibus*. In *Vita longa, vecchiaia e durata della vita nella tradizione medica e aristotelica antica e medievale*, ed. C. Crisciani, L. Repici, and B. Rossi Pietro, 149–167. Firenze: SISMEL-Edizioni del Galluzzo.

———. 2018a. Biblical Exegesis and Aristotelian Naturalism: Albert the Great, Thomas Aquinas, and the Animals of the Book of Job. *Aishtesis* 11 (1): 81–96.

———. 2018b. Filosofia naturale e trasformazione morale. Alberto Magno interprete del *Cantico della vigna* (Isaia, 5, 1-7). In *Edizioni, traduzioni e tradizioni filosofiche (secoli XII-XVI). Studi per Pietro B. Rossi*, ed. L. Bianchi, O. Grassi, and C. Panti, 341–352. Canterano: Aracne.

Peter of Auvergne. 2003. *Sententia super librum 'De vegetabilibus et plantis'*, ed. Poortman Leiden/Boston: Brill.

Poortman, E.L.J. 2003. Introduction. In *Sententia super librum 'De vegetabilibus et plantis'*, ed. E.L.J. Portman, ix–xxxix. Leiden/Boston: Brill.

Repici, Luciana. 2000. *Uomini capovolti, Le piante nel pensiero dei Greci*. Roma-Bari: Laterza.

Reynolds, Philip L. 1999. *Food and the Body. Some Peculiar Questions in High Medieval Theology*. Leiden/Boston/Köln: Brill.

Siraisi, Nancy. 1980. The Medical Learning of Albertus Magnus. In *Albertus Magnus and the Sciences. Commemorative essays 1980*, ed. James Weisheipl, 379–404. Toronto: Pontifical Institute of Medieval Studies.

Stannard, James. 1980. Albertus Magnus and Medieval Herbalism. In *Albertus Magnus and the Science. Commemorative essays 1980*, ed. James Weisheipl, 355-378. Toronto: Pontifical Institute of Medieval Studies.

Steneck, Nicholas H. 1974. Albert the Great on the Classification and Localization of the Internal Senses. *Isis* 65 (2): 193–211.

———. 1980. Albert on the Psychology of Sense Perception. In *Albertus Magnus and the Sciences. Commemorative Essays 1980*, ed. James Weisheipl, 263–290. Toronto: Pontifical Institute of Medieval Studies.

Studtmann, Paul. 2004. Living Capacities and Vital Heat in Aristotle. *Ancient Philosophy* 24 (2): 365–379.

Weisheipl, James. 1980. The Life and Works of St. Albert the Great. In *Albertus Magnus and the Sciences. Commemorative Essays*, ed. James Weisheipl, 13–51. Toronto: Pontifical Institute of Medieval Studies.

Wöllmer, Gilla. 2013. Albert the Great and His Botany. In *A Companion to Albert the Great. Theology, Philosophy and Sciences*, ed. I.M. Resnick, 221–268. Leiden/Boston: Brill.

Chapter 8
On the Natural Generation of Human Beings: The Vegetative Power in a Thought Experiment by Some Masters of Arts (1250-c. 1268)

Paola Bernardini

Abstract This paper focusses on a thought experiment devised by some Masters of Arts (John de Sècheville, *De principiis naturae*, 1956, 128–131; Anonymi Magistri Artium *Quaestiones super librum de anima*, ed. P. Bernardini, 2009, q. 60; Ps.-Petrus Guentin de Ortemberg, *Quaestiones super Physicam*, II, q. 1, mss. London, Wellcome Hist. Med. Libr., 333, fols. 26rb-27va; Firenze, Bibl. Naz. Centr., Conv. Soppr. A.V.563, fols. 29rb-31ra) in three works devoted to naturalistic issues and composed around 1250–1268. The experiment concerns the possibility that the human species be transmitted just by natural generation, that is thanks to the vegetative power rooted in the human being and without any need for God's intervention. The experiment belongs to the set of arguments employed by the Masters in support of the thesis of the natural specific difference of man. Such a thesis was judged heretical and included in the Parisian Syllabus of 1277 (see art. 113: *a human being is a human being independently of the rational soul*). The three texts examined here show a strikingly naturalistic approach, which is deeply in contrast with both the common view of contemporary Masters of Arts and standard thirteenth-century theological theories of human generation.

8.1 Historical and Theoretical Context

The aim of the present paper is to study the functions of the vegetative soul, in particular its generative power, as far as the transmission of the human species is concerned. The focus will be on three works (Anonymus 2009; Sècheville 1956; ps.-Ortemberg, mss. L and F) composed in the Faculties of Arts in Paris (1250-c. 1268) and characterized by a strongly naturalistic approach, which is therefore in

P. Bernardini (✉)
Dipartimento di Scienze Storiche e dei Beni Culturali, Siena, Italy
e-mail: bernardinip@unisi.it

F. Baldassarri, A. Blank (eds.), *Vegetative Powers*, International Archives of the History of Ideas Archives internationales d'histoire des idées 234,
https://doi.org/10.1007/978-3-030-69709-9_8

contrast not only with the most common attitude among contemporary theologians but also with that of the three authors' colleagues, who were usually aware of the constitutive and crucial role played by God's direct intervention in the generation of any human being.

What is at issue is the possibility that the human species be produced just by nature, i.e. through the generative act of the parents (especially of the father, who transmits the substantial form to the offspring), without the intervention of God, the Creator of the intellect.[1] The authors of the three works we are focussing on thus shared the major thesis according to which the intellect is not the specific difference of the human being. Such a thesis, clearly in contrast with the Christian faith, is included in the Parisian Syllabus of 1277:

> art. 113: a human being is a human being independently of the rational soul.[2]

This amounts to suggesting that the specific difference of the human being is not the intellect, but a substantial form, produced entirely by nature, transmitted by generation and rooted in the sensitive soul, yet nobler than the sensitive soul, which is common to all animals.

Different names are given to this 'generated' substantial form, which is usually described as *educta de potencia materie*, i.e., drawn from the potency of matter, on the grounds that, in this theoretical context, the production of the human species is said to be due to a form that derives integrally from matter.[3] In the texts under scrutiny, among the arguments in support of the thesis of a natural specific difference, we find an intriguing thought experiment concerning the generation of living creatures. The experiment is about the 'production' of an odd individual, named *A* (or *Progenitum A*), who, as far as its body and the lower faculties of the soul are concerned, is identical to all other human beings, but is not endowed with intellect.[4]

8.2 The Thought Experiment of the *Progenitum A*. Examination of the Texts

I will now set out to explore the different facets and variants of the experiment, in order to show the prominent role it assigns to the vegetative soul and its generative power. All three versions offer interesting insights, but, as we will see, the general philosophical assumption that best justifies the pronounced naturalistic approach of

---

[1] On this subject, see Bernardini 2017; Bernardini 2019, 2022 (forthcoming).

[2] Hissette 1977, 184 and Piché 1999, 82: 'Homo est homo praeter animam rationalem'. The same formulation can be found in Ps.-Petrus Guentin de Ortemberg, L, 27va; F, 31ra: 'Et ita ultra sensitivam est alia forma ulterior et nobilior sensitiva qua homo est homo preter intellectivam. Et hoc concedo'. Obviously, what is meant here by *anima rationalis* is the intellectual soul.

[3] In the textual tradition, the formal principle whose existence is being discussed here is referred to via different expressions. See Bernardini 2017, 51–72 and Bernardini 2022 (forthcoming), n. 20.

[4] On this topic, see Bernardini 2017, 53–54; 57–58; 62; 66; and Bernardini 2022 (forthcoming).

the three authors is to be found in an anonymous commentary on the *Physics*, previously attributed to Petrus Guentin de Ortemberg.

8.2.1 *The Shorter Version of the Experiment: John de Sècheville,* De Principiis Naturae

John de Sècheville is a remarkable figure at the Faculty of Arts in Paris (he was *rector* around the year 1250) and the only author we know by name among the ones who supported the heretical thesis mentioned above and featured in the *Progenitum A* experiment.[5] John avails himself of the thought experiment in the section of his *De principiis naturae* entitled D*e participatione materiae*, where he discusses the thesis of the unicity of the intellect proposed by Averroes.[6] According to John, two equally absurd consequences follow from the unicity of the intellect for all human beings:

either 1) all human beings are only one human being;
or 2) human beings differ from one another by species (Sècheville 1956, 127, ll. 18–20).

In order to avoid these senseless conclusions, John introduces the thesis of a natural specific difference of human beings: the human being does not belong to a species by means of an intellective form that represents his perfection, but, as in the case of irrational animals, by means of a natural form. John admits that this thesis might seem unlikely, yet he is convinced that it is true beyond any doubt.[7]

John employs the thought experiment quoted above simply in order to prove that the constitution of the human species is due to a natural form and not to the intellective one. He starts by suggesting that we imagine the generative process of a man which is not brought to completion, because God has not infused the intellect into the newly animated body in due time. By the end of the process, a new living entity will actually exists, even though it is devoid of intellect; let us call this living being produced only by nature *progenitum A* (Sècheville 1956, 128, ll. 17–25).

[5] A Master of Arts around the year 1245, John of Sècheville was active in Oxford and Paris for roughly 20 years. In all likelihood, it was during his second sojourn in Paris (after 1263) that he composed the *De principiis naturae* (1265). For further biographical information on John of Sècheville, see Weijers 2003, 165–167.

[6] Here John de Sècheville discusses the problem of the multiplicability of forms. See Sècheville 1956, 126, ll. 21–25. It must be supposed that the production of *intelligentiae* and *intellectus* is not a generation, and that they are not material beings: according to John, they could be multiplied only by God, thanks to His omnipotence: Sècheville 1956, 127, ll. 4–9.

[7] See Sècheville 1956, 127, ll. 20–23: 'Non enim collocatur homo in specie per formam intellectivam quae est ejus perfectio, sed per formam naturalem, sicut leo et asinus; et nos ostendimus hoc esse indubitabile verum, licet videatur valde improbabile (…)'.

A finds itself in a paradoxical condition: insofar as it derives from a *human* generative process, and nature has shaped and arranged its limbs according to the aims and functions proper to the human species, it is undoubtedly similar to a man; on the other hand, it does not possess a rational soul, that is the specific difference which usually identifies the human species itself. A crucial question arises, then: What is *A*? To answer this question, what is required is a long and subtle chain of reasoning, in the course of which John fruitfully and systematically uses the method of comparison. In order to establish what species *A* belongs to, John first of all contends that *A* cannot differ from irrational animals only in virtue of an accidental difference: for in this case, *A* should be identical by species, for example, to a deer or a lion. Its body, however, is evidently predisposed to receive the intellective soul (Sècheville 1956, 128, ll. 29–30); we must therefore suppose an essential difference between *A* and brutes, responsible – as Averroes explicitly states[8] – for the peculiar configuration of its limbs (Sècheville 1956, 129, ll. 14–26).

John also specifies that not even a material difference could be responsible for distinguishing *A* from brutes, because a material difference is just what distinguishes males and females *within* a given species, and all members of the same species – as everybody knows – possess the same substantial form.[9]

In short, John admits that *A* differs from all brutes by a *differentia essentialis et formalis* (Sècheville 1956, ll. 35–36). A few lines above, this difference had been described in the following terms:

> (…) that difference, drawn from the potency of matter and added upon the form of the genus, necessarily constitutes a new species within the genus itself (Sècheville 1956, 128-9, ll. 36-1).

Consequently, even though *A* lacks an intellect, it belongs to the genus *animal* and it is different by species from all brutes. By virtue of the similarity of its life with human life (Sècheville 1956, 129, ll. 3–4), John can therefore invalidate the two embarrassing consequences deriving from the thesis of the unicity of the intellect for all human beings: for there is something else, different from the intellective soul, which establishes the individual's belonging to the human species.[10] That something else is the soul of *A*, which constitutes its specific difference.[11] This soul of *A* is produced by nature through the *virtus formativa* (which operates in the seed), whose final end is the production of an animal capable of receiving the last created form, i.e. the intellective soul. John thus infers that the human species, just like the brute species, is transmitted by natural generation, through a *natural* substantial form that

[8] The source of this argument is Averroes 1953, I, 53, 75: 'Membra leonis non differunt a membris cervi, nisi propter diversitatem animae leonis ab anima cervi'. It is one of the major passages explicitly quoted by the Masters of Arts engaged in the debate on the natural specific difference. See Bernardini 2017, 52.

[9] See also Sècheville 1956, 128, ll. 32–35.

[10] See Sècheville 1956, 129, ll. 4–5: 'in re [i.e. in homine (et in A)] est aliquid aliud ab intellectiva, unde specie differens a cervo, leone et aliis'.

[11] See Sècheville 1956, 129, ll. 24–26: 'Ergo anima ipsius A erit differentia constitutiva speciei sub hoc genere animal, sicut anima leonis vel anima cervi, vel multo forcius'.

predisposes man, and man alone among all living creatures, to the reception of the intellect. Indeed, according to a famous sentence, nature shows greater *solicitudo* towards human generation than it does towards the generation of irrational animals.[12]

2. After having compared *A* to brutes, John expands his argument by comparing *A* to a human being endowed with the intellective soul. Whereas, compared to brutes, *A* is different by species, and is *verius ens in specie* (Sècheville 1956, 131, l. 1), compared to a man *A* is not complete and perfect, for it stands only in potency to this superior form. Therefore, John concludes that *A* is in a separate species; a species, however, which is not modified by the addition of the intellect and must therefore be conceived of as the proper human species.[13]

The status of the odd individual *A* now becomes clear: whereas the intellect represents the most noble act of matter, the specific difference of the human being – we can infer – is the most noble natural form of *A*, that is, the human specific difference, sufficient to distinguish it from brutes. This amounts to saying that a human being, insofar he is a human being, admittedly needs the intellect to reach perfection and completeness, but nonetheless belongs to the human species thanks to a form drawn from the potency of matter.

Actually, we may observe, the intellect is here understood as the ultimate perfection of man considered in himself, that is in close relation to his own nature and his higher rank in the *scala naturae*, but, *in quantum animal* – i.e. compared to all other species – he is identified by generation through an essential, formal difference which is drawn from matter, exactly as in the case of the differences of irrational animals.

---

[12] See Sècheville 1956, 130, ll. 6–13: '(...) sed virtus formativa membrorum operatur in semine opus simile artificio, et organizat ipsum et figurat modis convenientibus suae speciei, sicut manus figuli operatur in olla, et illa eadem virtus in fine operationis fit forma rei. Dicere enim quod virtutes formatrices in fine operationum suarum corrumpuntur non est bonum. Ergo, qua necessitate virtus formativa membrorum cervi fit in fine operationis forma ejus substantialis, dans ei esse specificum sub hoc genere animal et distinguens ipsum secundum speciem ab aliis animalibus, eadem necessitate forma ipsius A consimiliter se habebit ad ipsam, immo, multo fortius, cum longe major sit sollicitudo naturae in perficiendo materiam humanam, ex qua factura est materiam, quae est quasi necessitas respectu intelligentiae, quae est nobilissima perfectio materiae'.

[13] See Sècheville 1956, 130–131, ll. 31–2: 'Apparet ergo quod A (...) comparatum ad hominem habentem intelligentiam actum et perfectionem sui, debeat dici incompletum seu inperfectum, tamen ipsum comparatum ad quodcumque brutum, est longe dignius et perfectius eo, et verius ens in specie, quia longe plus abundat a genere quam aliquod brutum; ipsum autem non potest reponi in aliqua specie animalis irrationalis; erit ergo speciei separatae'. See also Sècheville 1956, 131, ll. 9–13: '(...) Quando huic formae quae, sicut visum est, est nobilissima forma quam natura potest educere de potentia materiae ad actum, advenit ab extra intellectus, qui est per se perfectio hominis secundum quod homo et nobilissimus actus materiae, non constituitur nova species per ejus adventum (...)'.

8.2.2 *The Earliest Version of the Experiment:* Quaestiones super Librum de Anima, *q. 60*

The *Quaestiones super librum de anima*, preserved in ms. Siena, Biblioteca Comunale degli Intronati, L.III.21, fols. 134ra-174va, are, to my knowledge, the earliest testimony to our thought experiment.[14] The text shows here an earlier, well developed argument, whose articulated structure attests to a keen interest in naturalistic issues.

The experiment of *A* is the core of q. 60, but some traces of the doctrine of the natural specific difference are detectable in other questions on book II as well.[15] First of all, it must be remarked that, in the theoretical context in which the experiment is developed, there is no reference at all to the doctrine of the unicity of the intellect: the anonymous Master simply aims to explain the relations between the different faculties of the human soul. He upholds the idea of the plurality of essentially distinct faculties in the human soul, a theory based on an interpretation of the Aristotelian analogy between souls and rectilinear figures (*De anima*, II.3, 414b28-30) that was widely known and discussed by the Masters of Arts. According to this theory, the various potencies of the soul differ from one another *per essentiam*, and the soul as a whole is conceived of as an *aggregatum*, *compositum*.[16] The doubtful issue under discussion in q. 60 concerns the possibility that a human being be generated via a nobler sensitive form, just before the infusion of the intellect by God.[17] The experiment of *A* is presented in the two series of *Sed contra* and, compared to John de Sècheville's version, deals with the issue in a much fuller and more detailed way. The experiment is divided into two sections:

---

[14] This anonymous collection of *quaestiones* on books I and II of the *De anima* dates from 1250 to 1260 and was probably composed by an English Master of Arts just before or after Albert the Great's commentary on the *De anima*, since certain sections of the two works are in close agreement. For further information on the text, see Bernardini 2009, IX–XVIII.

[15] Anonymus 2009, q. 60 (a, b,c), 195–203: 'Queritur utrum sit dare aliquam formam via nature eductam aliam quam intellectivam, que reponat hominem in specie animalis, an ipsa sit intellectiva'. See also q. 54, 175–180: 'Queritur utrum vegetativa et sensitiva in homine sint una substantia'; q. 59, 194–195: 'Queritur utrum intellectiva exit in esse per creationem'.

[16] In Bernardini 2017, 42–51, esp. 42–43, I have tried to demonstrate that this theory is the doctrinal background of the thesis of the natural specific difference of the human being. Even if this theory can be connected to the theory of the plurality of substantial forms, the two are not identical, because this textual tradition attests to several expressions for the potencies of the soul (not only *formae substanciales*, but also *differencie*, *substancie*, *essencie*, and *animae*). Nevertheless, all these texts share the view that the vegetative-sensitive faculties and the intellectual faculties are ontologically irreducible, on account of their different origins: the first two faculties are drawn from the potency of matter, whereas the third has a divine origin. On the thesis of the plurality of forms, see at least the classic works Callus 1939, Zavalloni 1951. As far as the Masters of Arts are concerned, see Bazán 1969 and Bernardini 2014; for a more extensive discussion of this topic, see also de Boer 2013, 33–44.

[17] See above, n. 15.

1. In the first section, the author provides the premises of the argument: let us suppose that *A* is a subject naturally produced by a human seed, but not in possession of the intellect.[18] This gives rise to several problems, which stem from the following question: is there a natural specific difference form (*sensitiva contracta*, not *commune*) capable of placing *A* in its species? And which species does *A* belong to?[19]
2. In the second section, the author expands his argument: even though *A* is devoid of intellect, it nonetheless possesses the vegetative soul and the generative faculty related to it. Let us suppose, then, that *A* generates another subject, *B*: unlike *A*, *B* receives an intellect by divine infusion.[20] Given these premises, two further doubts arise:

1') Do *A* and *B*, father and son, belong to the same species? That is: Does *A* belong to the human species, even though it lacks an intellect?

2') Does *A* differ by species from an irrational animal, e.g. a donkey, or not? And, if so: What is the difference between them?[21]

Despite its apparent importance, the first question is not subjected to detailed examination[22]; by contrast, the second one is the object of a lengthy discussion, in the

[18] Anonymus 2009, *Contra I*, 196, ll. 37–41: 'Accipiatur semen hominis; in hoc semine operatur vegetativa virtus nutrendo et augendo, et qualitates active et passive operantur alterando; operatur etiam virtus sensitiva formando sibi membra et organa. Demus ergo quod illud progenitum in summo dispositum ad recepcionem intellective, et ponamus quod Creator non infundit intellectivam'.

[19] Anonymus 2009, *Contra II*, 197, ll. 49–72: 'Set ista differencia sic contrahens constituet hunc fetum in specie animalis, non nisi in specie hominis, ergo et cetera; quod patet quia membra in hoc semine non sunt membra bruti, set membra racionalis creature. (...) Igitur, si ultra formam sensitive in semine bruti educatur forma specifica, igitur multo forcius in semine hominis; ultra sensitivam simpliciter educetur forma aliqua specifica consituens speciem. Et dicens quod racio pocius concludit oppositum, quia ex hoc quod semen hominis in tanta equalitate consistit, ideo est in potencia ad tam nobilem formam quam naturam non potest inducere, et ideo sua perfectio ab extra est'.

[20] Anonymus 2009, *Contra II*, 198, ll. 96–101: 'Item accipiatur istud progenitum A et [non] interdicatur ei intellectiva. In isto est virtus vegetativa et sensitiva: poterit ergo exercere operaciones suas que sunt nutrire, augere, generare, sentire. Poterit ergo anima generare: generet ergo illud progenitum aliud, sit B; tunc sicut A est in summa disposicione ad intellectivam fit B; eodem modo, summe dispositus ad intellectivam detur, ergo ei intellectiva constat, quod tunc est homo'.

[21] On the comparison between *A* and the donkey, see also Bacon 1911, IV, 4, 2, 284–285, ll. 35–8. Bacon harshly criticizes the thesis of the natural specific difference of the human being: see Bacon 1911, 286, ll. 7–16: 'Item si hec differencia perficeret hominem in specie animalis, tunc si non adveniret anima racionalis, esset sicut species alia animalis que caret intellectu, ut asinus vel aliud habens solam sensitivam. Et possit vivere et moveri et opera animalis exercere secundum proprietatem sue speciei, sicut asinus secundum suam. Set asinus quia sic facit non est in potencia ad ulteriorem formam specificam (...) quia non poterit anima racionalis recipi in hac specie animalis, sicut nec in asino vel leone'. See Bernardini 2017, 55–56. On the relationship between the human species and irrational animals, extensive documentation is available in Köhler 2008, 341–443.

[22] By assuming that we are dealing with a form produced by nature, the author simply supposes that *A* and *B* belong to the same species. See Anonymus 2009, 198, ll. 103–107.

course of which we again find – with some linguistic differences – the same arguments used by John de Sècheville in his comparison between *A* and brutes.[23]

Finally, the solution to q. 60 sets out the Master's opinion about the true nature of man, on the grounds of a decisive distinction:

> (…) it is one thing to say that something belongs completely to a species (*esse in specie completa*) and another thing to say that something has a complete being (*habere esse completum*) (Anonymus 2009, 202, ll. 208–209).

On one hand, then, the absence of the intellective soul does not prevent *A* from belonging to the human species; on the other hand, it prevents it from having a "complete being". The point of the Master is clear: *homo* can be said in two distinct ways, that is: (1) as a living entity having the sensitive soul (namely, the *sensitiva contracta*) as its substantial, specific difference, and (2) as a perfect whole including the intellect created and infused *ab extra*.[24] One last remark is in order: the structure of q. 60 is anomalous, and this may be a sign of its incompleteness. We cannot therefore be sure that the Master's last words on the subject are the ones just reported. Nevertheless, we may presume that the notion of *esse completum* plays an important role here,[25] as a way to preserve – or, at least, to try to preserve – the orthodoxy of the Master's view.

8.2.3 The Last Version of the Experiment: Ps.-Petrus Guentin de Ortemberg, Quaestiones super Physicam *II, 1*

The last version of the *A* experiment is found in the *Quaestiones super Physicam* II, q. 1, erroneously attributed to Petrus Guentin de Ortemberg.[26] Here the discussion is focussed on the famous Aristotelian sentence *homo generat hominem et sol* (*Physica* II, 2, 194b13). Accordingly, the problem at issue is again the possibility of a natural generation of the human being, the relevant *quaestio* being entitled

---

[23] See Anonymus 2009, 199–200, ll. 108–141. In particular, the arguments concern the absurd consequences deriving from the assumption of the identity between *A* and the donkey, namely: 1. the identity between *A* and any other brute; 2. the potency of the donkey to the intellective soul. The following reasoning rules out the possibility of an accidental or material difference between *A* and the donkey, so we must accept that *A* and the donkey differ by a substantial, formal difference, and hence by species. Ultimately – also considering argument 1′ – *A* must have a natural specific difference by which it belongs to the same species as human beings.

[24] See Anonymus 2009, 202, l. 215–203, l. 1: 'Unde homo uno modo nominat sensitivam cum forma illa ulteriori inducta per naturam, et sic dicitur quod homo generat hominem et sol. Alio modo nominat illud totum cum anima intellectiva superaddita (…)'.

[25] On the *esse* as what constitutes the difference between *A* and a perfect man, see also the critical discussion in Sècheville 1256, 129, ll. 7–12.

[26] For a preliminary description of the composite nature of the *Quaestiones*, based on the London codex, see Donati 1993, 58–60 (with a full list of the *quaestiones* on books I and II, still based on the London codex: see 78–84). The attribution of this work has been disputed with good reason: Petrus Guentin de Ortemberg is a much later author (d. ca. 1450). See Schwarz 2012, 703.

> utrum homo constituatur in specie animalis per animam intellectivam aut per aliquid aliud quod sit in potencia materie eductum de ipsa potencia per actum agentis (Ps.-Ortemberg, F, 29va, ll. 10-13).

After the presentation of three arguments against the natural constitution of the human being, the thought experiment[27] is introduced, as in the case of the anonymous *Questiones de anima* of Siena, in the *Oppositum*.[28] This is not the only feature which the two anonymous texts share, as far as the *A* experiment is concerned. At least four further common elements are worth recalling[29]:

---

[27] All these arguments emphasize the crucial role of the intellective soul as the unique specific difference of the human being: see Ps.-Ortemberg, F, 29va, l. 14-29vb, l. 5: 'Set quod istud 2um sit impossibile ostenditur, quia, si sit forma in potencia materie que educta de ea per actum agentis constituat hominem in specie animalis, oportet quod illa sit nobilior quam forma generis per quam constituir genus in suo esse, que videlicet est anima sensitiva: animal enim constituitur in esse per animam sensitivam. Oportet ergo quod sit nobilior quam sensitiva quoniam ipsa est posterior quam sensitiva, et aliquid addit super eam, et ita nobilior. Set vita est nobilior non vita, ut dicit Aristotiles, et vivens non vivente; set sensitiva anima est vita sive vivens, ergo oportet quod illa forma educta de potencia generis, si sit nobilior quam sensitiva, sit vita, et etiam vita nobilior quam sensitiva. Ergo ultra sensitivam est alia vita nobilior, sive forma, quam sensitiva, que tamen non est eadem cum anima intellectiva, quod est impossibile. Ergo homo constituetur in specie animalis non per illam formam, set per animam intellectivam. // Item, si homo constituatur in specie per formam illam eductam de potencia generis sive materie, tunc dicetur completum in esse specificum per illam habitam, non habita tamen anima intellectiva. Est ergo repositum in specie per illam formam, set ulterius advenit anima intellectiva, ergo adhuc constituet speciem inferiorem: ergo una species fuit in potencia ad aliam speciem constitutam per formam additam prime post sui complectionem in esse specifico, quod est impossibile.// Item Porphirius dicit quod animal racionale mortale est diffinicio hominis, et similiter Boetius et Aristotiles quod animal racionale est eius diffinicio /29vb/[hominis, *exp.*]. Set racionale est racionale ab anima intellectiva: set racionale est differencia reponens hominem sub specie animalis, ergo sub specie animalis reponitur per animam intellectivam et non per formam eductam de potencia materie'.

[28] Ps.-Ortemberg, F, 29vb, ll. 9–17: 'Et ad hoc ostendendum ponatur quod natura operata fuerit in semine hominis, et introduxerit de potencia ipsius seminis in ipso fetu animam vegetativam et sensitivam; et ulterius operata fuerit quantum potest ad preparacionem introductionis anime intellective, ita quod istud generatum iam sit in propinqua sive in proxima disposicione dispositum ad recipiendum animam intellectivam. Et ponatur per impossibile quod denegetur ei (...)'.

[29] Several parallel passages may be adduced here: see e.g. the three arguments against the thesis of *sensitiva nobilitata* quoted above, n. 21 and Anonymus 2009, q. 60a, 196, ll. 10–35. Then, in support of the thesis of the *sensitiva nobilitata* as a form able to constitute the human species, an analogy with light is provided: when air receives the form of light (considered as a substantial, not accidental, form), coming from outside and not from the potency of the air, the air in itself does not change its nature: in analogous way, the intellective soul which is created by God and comes *ab extra* does not change the species of the individual constituted by that natural form, the noblest of all. See Anonymus 2009, q. 60a, 203, ll. 228–233; and Ps.-Ortemberg, F, 31va, l. 40 -31vb, l. 11: '(...) non quecumque forma adveniens speciei constituit speciem, set forma substancialis educta de potencia materie; set anima intellectiva, licet adveniat homini post complectionem sui /31vb/in esse specifico, tamen, quia non est educta de potencia materie, set totaliter ab extrinseco, non oportet quod constituat novam speciem. Et datur similiter ad ostendendum quod forma substancialis que est ab extrinseco totaliter non constituit novam speciem, quia aer recipit lucem que est forma substancialis, et tamen non differt secundum speciem ab aere existente sine luce, quia illa forma substancialis non educitur de potencia aeris, set est totaliter ab extrinseco. Et sicut lux non constituit ibi novam speciem, ita nec anima intellectiva in homine'.

1. The premises of the argument: thanks to the *vis formativa* contained in the seed,[30] *A* grows and takes the shape of a human being, even though, at the end of the process, he does not receive a rational soul, that is the intellect, by God.[31]
2. The development of the experiment: as in the Siena *Quaestiones*, the argument is quite extensive, including as it does a comparisons between *A* and brutes, *A* and its son (a perfect man), and brutes and *A*'s son.[32] In order to demonstrate the existence of a natural specific difference present in *A*, the argument then focusses on a comparison between *A* and a donkey[33]: according to the Master, they do not differ in virtue of an accidental, or material difference, but in virtue of a formal one, that is a substantial (or essential) form.[34]
3. Arguing against an unknown *aliquis*, who rejects the thesis of the natural specific difference of the human being on the grounds that the difference between *A* and the donkey does not consist in a definite *esse*, the author partially reproduces an argument already present in the Siena *Questiones*,[35]

[30] For a general reconstruction of the Aristotelian embryological theory and its reception, see Dunstan 1990. On the role of *vis formativa* in generation, see e.g. Van der Lugt, 2008, 233,254. Medieval embryology is a domain that has been far more extensively explored in the last few decades, especially as far as the Thomistic theory and bioethics are concerned; on this topic, see e.g. the discussion on the beginning of human life in Pasnau 2002, 100–130, its criticism in Haldane and Lee 2003, and Pasnau's reply in Pasnau 2003.

[31] See above, n. 28.

[32] See Ps.-Ortemberg, F, 30ra, ll. 15–24: 'Item, ut habitum est, vegetativa nihil ad intellectivam de suis operacionibus: set A habet vegetativam, ergo in ipso erunt virtutes vegetative, scilicet nutritiva, augmentativa, generativa. Habet ergo virtutem generativam: generet, ergo generabit aliquid aptum ad recepcionem anime intellective, quoniam generabit sibi simile. Et ipsum A est tale: tunc illud generatum ab A, cum sit aptum recipere animam intellectivam, recipiat eam, tunc est homo, ut manifestum est cuilibet'.

[33] See above, esp. n. 21.

[34] The similarity between Ps.-Ortemberg and Anonymus 2009 here is striking. See Ps.-Ortemberg, F, 30va, ll. 26–43: 'Item A differt ab asino, ut patet eo quod ponunt in numerum. Aut ergo differt ab ipso differencia substanciali aut accidentali. Si accidentali: set que solum differunt differencia accidentali, conveniunt in numero, idem est ergo A cum asino, et tunc habetur propositum, quoniam sicut asinus reponitur in specie per differenciam sive formam additam priori forme scilicet generis, ita quod A reponeretur in specie per eandem, et tunc addit formam. Si autem differat ab asino differencia substanciali, aut ergo materiali aut formali. Si materiali: set differencia materialis non facit diversitatem in specie, quoniam solus actus dividit, idest sola forma facit diversitatem in specie: ergo erit in specie asini, et tunc habetur propositum ut prius. Si autem differat differencia formali substanciali ab asino, habebit ergo aliam differenciam formalem et substancialem quam asinus; set asinus habet formam substancialem generis, ergo istud progenitum ultra formam generis addit differenciam formalem substancialem, et sic habetur propositum'.

[35] The *aliquis* mentioned by Ps.-Ortemberg also objects that this substantial form is 'less' than the genus's form (*in minus*). See Ps.-Ortemberg, F, 30va, l. 43- 30vb, l. 23: 'Et dicet aliquis quod tale pro/30vb/genitum est differens ab asino secundum esse, set non per essenciam. Set hec responsio adhuc non valet, quia quero: cum unumquodque habeat causam immediatam qua posita ponitur ipsum et qua remota removetur ipsum, que sit causa per quam A differt ab asino secundum esse et non per essenciam in qua sit causa istius esse quod facit differenciam, qua causa posita ponitur hoc esse et hec differencia, utrum scilicet hec causa sit materia aut forma. Si causa istius esse sit materia, cum illa, ut ostensum est, que solum differunt secundum esse causatum a materia, possint esse

4. In addition to that, the author replies to another *aliquis*, who objects that this substantial form is incomplete, insofar as it stands in potency to the intellective soul.[36] According to our Master, such a form cannot be incomplete, for it must have the same ontological status as the specific differences of brutes, which are capable of putting them in their genus.

The point emerges quite clearly when a solution is finally offered to the crucial problem concerning the relations between the natural specific difference and the intellective soul:

> The evidence of the fact that this form is not the intellective one is that: when the nature is moving towards a form, this form must be somehow in what it is introduced by the operation of nature, because the nature draws the form from the potency of the matter, and that form is somehow in the matter. But the intellective form cannot be in any way in the matter, because it is externally, and totally externally, infused. Therefore, the form drawn by the nature from the potency of the matter is different from the intellective one and, as we have seen, the nature draws from the potency of the matter that substantial form beyond the form of the genus through which the human being differs by species from other animals and through which it is placed in the species of an animal. Therefore, beyond the sensitive form there is another, nobler form through which the human being is a human being independently of the intellectual one. And I grant that thesis.[37]

sub eadem specie, A erit in specie asini, et tunc ut prius. Si autem causa istius esse facientis hanc differenciam sit forma, aut ergo accidentalis aut substancialis: si accidentalis, tunc hec non causabit diversitatem in specie, et tunc ut prius. Si substancialis, aut ista forma substancialis est ibi (in F) eque cum forma generis, aut in plus aut in minus. Set hec forma non est ibi (in F) eque cum forma generis, quia sicut in forma generis non habent diversitatem, immo magis ydemptitatem, et ita non diversum esse; ita nec in ista habebunt diversitatem, nec diversum esse. Eadem racione non potest ista forma esse in plus quam forma generis, ergo erit in minus et inferior, per additionem se habens respectu forme generis, et sic habetur propositum'. See Anonymus 2009, q. 60b, 200, ll. 145–154.

[36] Ps.-Ortemberg, F, 31rb, ll. 16–31: 'Sed opponet aliquis: dicis quod ista forma est eadem per essenciam cum sensitiva et non differunt nisi penes magis completum et minus, quia sensitiva iam est sub compleciori, iam sub incompleciori. Dicis ita. Similiter video quod vegetativa et sensitiva non differunt nisi penes completum et incompletum, et non per essenciam, et sicut ibi fuit sensitiva nobilitata, ita similiter hic est vegetativa nobilitata. Set video quod, licet sensitiva non sit alia per essenciam a vegetativa, set vegetativa est nobilitata, attamen sensitiva addit super vegetativam novam potenciam, et hoc quia est sub esse nobiliori; ergo, si sensitiva nobilitetur ulterius, forma adveniens addet adhuc aliam potenciam supra potenciam sensitive, et hoc ideo, quia ista forma est sub esse nobiliori. Set hoc est inauditum, quod sit alia potencia anime post sensitivam nisi sola intellectiva'. See again Anonymus 2009, q. 60a, 202, ll. 198–201.

[37] Ps.-Ortemberg, F, 31ra, ll. 11–24: 'Quod autem ista forma sit alia ab intellectiva ostenditur quia quando natura movetur ad aliquam formam, oportet quod illa forma sit aliquo modo in eo in quo inducitur ab operacione nature, quia natura educit formam de potencia materie et ita illa forma est aliquo modo in materia. Set intellectiva nullo modo est in materia, quoniam infunditur ab extra et etiam totaliter ab extra. Ergo illa quam educit natura de potencia materie est alia ab intellectiva, et ut visum est, natura educit de potencia materie illam formam substancialem ultra formam generis per quam homo differt in specie ab aliis animalibus, et reponitur in specie animalis. Et ita ultra sensitivam est alia forma ulterior et nobilior sensitiva qua homo est homo preter intellectivam. Et hoc concedo'.

8.3 Concluding Remarks

In conclusion, we will consider the thought experiment as a unitary object (while taking account of some significant peculiarities of the different versions); and we will do so mainly on the basis of the text of the ps. Guentin de Ortemberg, which states the thesis condemned in 1277 with utmost clarity.

If we reflect on the very core of the thought experiment, we realize that it points to the autonomy of the operations of the vegetative soul in respect to the higher faculties of man: indeed, the vegetative powers are said to subsist and to be able to perform their operations in a living being such as *A* independently of the presence of the rational soul.[38]

The vegetative soul is at work in the embryo through an active principle, the *vis formativa*, from the beginning of the generative process, before the other faculties develop; and its operations in the individual to be born – nourishment, growth, the organization of the body and of bodily functions – tend towards a form that is *ultra sensitivam*, namely, the final substantial form of the living being (in this case, of *A*). Actually, according to John de Sècheville and Ps.-Ortemberg, this (natural) final substantial form derives from the seed, in which the active principle mentioned above is present: this principle, in addition to causing generation, directs the development of the *progenitum* until the completion of the whole process.[39]

It is a matter of everyday experience that, in order to perform its functions, the vegetative soul does not need the contribution of the rational soul; and it is precisely because of this that our author can hypothesize a natural compound like *A*, totally devoid of intellect, who thanks to its vegetative soul is able not only to feed itself and grow, but also to generate another individual and transmit to it the human species, exactly as a complete and perfect man would.[40]

This soul deriving from the seed is the final substantial form that identifies the species of the living being; and it is because of this form that man, and even *A*, differ

---

[38] On the autonomy of the vegetative soul's power, see above, n. 32; and Ps.-Ortemberg, F, 29vb, ll. 22–33: 'A habet animam vegetativam, ergo vegetativa operabitur suas operaciones et, licet ibi non sit intellectiva, non minus operabitur, quoniam vegetativam quantum ad suas operaciones nihil ad intellectivam, quoniam velit nolit intellectiva, vegetativa operatur suas operaciones. Set operaciones vegetative sunt nutrire, augmentare et generare, ergo ipsa nutriet A et augmentabit donec perduxerit ad quantitatem perfectam et ibi conservare intendit. De quodlibet circa quod operatur, illud productum ad suam quantitatem <perfectam> habet membra alia secundum speciem a membris cuiuslibet bruti, ut patet'.

[39] See above nn. 12, 18, 19. The process of development, set in motion by the *vis formativa*, is directed towards this nobler *sensitiva*, which is the cornerstone of the process itself: see Ps.-Ortemberg, F, 30ra, ll. 8–15: 'Virtus agens ulterius erit ultima forma rei, et non virtus sensitiva, sive anima sensitiva: adhuc si virtus sensitiva formaret membra, tunc equaliter in semine hominis formaret membra asini, sicut et in semine asini [vel sicut] membra hominis, in quantum est sensitiva. Ergo oportet ponere aliam formam ultra sensitivam que primo movet ad formacionem membrorum, et ultimo fit forma rei'.

[40] See Ps.-Ortemberg, F, 30ra, ll. 24–25: 'Ergo A est ita conveniens ad generandum hominem intelligentem sicut et ego vel tu'.

by species from all irrational individuals. Such a soul is not the form of the genus – i.e. an *anima sensitiva communis* – which is common to all animals, but something nobler and peculiar to the human species. The logical approach adopted by the Masters (especially by Ps.-Ortemberg), and which is focussed on the genus/species relations between rational and irrational individuals, aims to confirm the physical conclusions reached through the study of the different phases of the natural process of generation, from the conception to the constitution of the whole individual.[41] As the author emphasizes, through the vegetative power of generation nature draws from the potency of matter what is needed to produce that form which specifies the sensitive soul; what is potentially present in matter are the specific differences according to which different individuals will be placed in rational or irrational species. For this reason, in respect of the other species of its genus, *A* must be placed in the rational species; compared to perfect and complete men, however, it reveals its imperfection and incompleteness, due to the lack of an intellect. In short, the Masters affirm that what belongs to the genus of living beings, whatever its final substantial form may be, can reproduce itself according to its own nature and generate an individual identical to it by species; and this also applies to the human being, which makes God's intervention not strictly necessary.

Generally speaking, it is not easy to place the thought experiment of the *Progenitum A* within a clear-cut philosophical tradition. The experiment appeals to the omnipotence of God, who hypothetically could choose not to infuse the intellect at the due time, thus inhibiting the constitution of the human being in its integrity. On the basis of this hypothesis, one may ask whether the product of natural human generation can be correctly considered "man" by virtue of its bodily and organic structure (due to its naturally generated form), which is similar to the one of a 'complete' human being endowed with intellect.

In the next and final step, our authors reduce their opponents' thesis to absurdity through a comparison between *Progenitum A* and other animal species (without divine intervention *A* and the donkey would belong to the same species!), thus providing intuitive support to their own thesis: man is man by virtue of natural generation.

It seems as though our authors are intent on opposing a tradition of thought, rooted in the Faculty of Theology, which gradually became dominant during the thirteenth century and presented itself as the 'true' Aristotelianism. According to this tradition, a human being can be considered such if and only if both natural and

[41] See Ps.-Ortemberg, F, 30rb, ll. 27–40: 'Unde genus per Porphirium potencia habet utrasque differencias, actu vero nullam. Educuntur igitur differencie opposite de potencia generis ultra formam ipsius, set ex ypotesi brutum et homo sunt species opposite, ergo habent oppositas differencias eductas de potencia generis ultra formam (ipsius *exp.*) suam; set forma generis est anima sensitiva, ergo ultra animam sensitivam advenit tam in bruto quam in homine nova forma sive differencia, educta de potencia generis, constituens utrumque in esse specifico. Ergo adhuc sequitur, si dicatur sic, quod homo reponatur in specie animalis per formam eductam de potencia generis ultra formam ipsius generis'.

supernatural causes contribute to his production; the Christian God, in the strict sense, is the true author of human nature.

In this context, the thought experiment of the *Progenitum A* supports a version of Aristotelianism that is quite different from that mentioned above on account of its strongly naturalistic character and which, as such, is condemned as 'heterodox'.[42] What we have, then, is an Aristotelianism of the Masters of Arts: a "philosophical" rather than "theological" Aristotelianism, we might say, which also makes use of logical tools and counterfactual analysis to lay bare the weaknesses of theories other than its own.

Acknowledgments I am very grateful to Silvia Donati, who let me look at her transcription of some portions of Ps.-Petrus Guentin de Ortemberg, *Quaestiones super Physicam*, II, q. 1, mss. London, Wellcome Hist. Med. Libr., 333, fols. 26rb-27va ("L"); Firenze, Bibl. Naz. Centr., Conv. Soppr. A.V.563, fols. 29rb- 31ra ("F"). Donati is currently preparing a critical edition of Ps.-Petrus Guentin de Ortemberg's *Quaestiones super Physicam* I-II. Unless otherwise specified, quotations of this text are from my transcription of the Florentine manuscript. All the translations of Latin texts are mine.

References

Anonymus Magister Artium. 2009. *Quaestiones super librum de anima* (ms. Siena, Biblioteca Comunale degli Intronati L.III.21, fols. 134ra-174va), ed. Paola Bernardini. Firenze: Sismel–Edizioni del Galluzzo.

Averroes, Cordubensis. 1953. In *Commentarium Magnum in Aristotelis de anima libros*, ed. Frederick Stuart Crawford. Cambridge, MA: Mediaeval Academy of America.

Bacon, Roger. 1911. In *Communia Naturalium, pars IV*, ed. Robert Steele. Oxford: Clarendon.

Bazán, Bernardo Carlos. 1969. Pluralisme de formes ou dualisme de substances? La pensée prétomiste touchant la nature de l'âme. *Revue philosophique de Louvain* 67: 30–73.

Bernardini, Paola. 2009. Introduzione. In *Anonymus Magister Artium, Quaestiones super librum de anima (ms. Siena, Biblioteca Comunale degli Intronati L.III.21, fols. 134ra-174va)*, ed. Paola Bernardini, IX–LXXXV. Firenze: Sismel–Edizioni del Galluzzo.

———. 2017. From First to Second Averroism. The attribution of article 113: *Quod homo est homo praeter animam rationalem* (Paris 1277). *Recherches de Théologie et Philosophie Médiévales* 84 (1): 37–73.

———. 2014. *Temporibus autem meis*. Theologians' Errors with regard to the Human Soul. In *Roger Bacon's Communia Naturalium*, ed. Paola Bernardini–Anna Rodolfi, 139–158. Firenze: Sismel-Il galluzzo.

———. 2019. *Plus quam specie differt*. La formazione dell'essere umano in Alberto Magno e nei commenti aristotelici dei maestri di arti (1240–1270 ca.). *Documenti e studi sulla tradizione filosofica medievale*.

———. 2022. *Ante adventum intellectus*. Matthew of Aquasparta and the Masters of Arts on the Constitution of the Human Species (Paris, 1277). In *Actes du Colloque International* Societas Artistarum *Commentaries on the* De anima, *from Petrus Hispanus to Francisco Suarez*. Forthcoming. Turnhout: Brepols.

de Boer, Sander W. 2013. *The Science of the Soul. The Commentary Tradition on Aristotle's* De anima, *c. 1260-c.* 1360. Leuven: Leuven University Press.

---

[42] On this subject, see Bernardini 2017.

Callus, Daniel Angelo. 1939. Two Early Oxford Masters on the Problem of Plurality of Forms. Adam of Buckfield–Richard Rufus of Cornwall. *Revue néoscolastique de philosophie* 63: 411–445.

Donati, Silvia. 1993. Per lo studio dei commenti alla *Fisica* del XIII secolo. I commenti di probabile origine inglese degli anni 1250-1270. Parte II. *Documenti e studi sulla tradizione filosofica medievale* 4: 25–178.

Dunstan, Gordon Reginald, ed. 1990. *The human embryo. Aristotle and the Arabic and European traditions*. Exeter: University of Exeter Press.

Haldane, John Burton Sanderson, and Patrick Lee. 2003. Aquinas on Human Ensoulment, Abortion and the Value of Life. *Philosophy* 78: 255–278.

Hissette, Roland. 1977. *Enquête sur les 219 articles condamnés à Paris le 7 mars 1277*. Louvain: Peeters.

Köhler, Theodor Wolfram. 2008. *Homo animal nobilissimum. Konturen des spezifisch Menschlichen in der naturphilosophischen Aristoteleskommentierung des dreizehnten Jahrhunderts*. Leiden/Boston: Brill.

Ps.-Ortemberg, Petrus Guentin de. *Quaestiones super Physicam*, II, q. 1, in mss. London, Wellcome Hist. Med. Libr., 333, fols. 26rb-27va; Firenze, Bibl. Naz. Centr., Conv. Soppr. A.V.563, fols. 29rb-31ra.

Pasnau, Robert. 2002. *Thomas Aquinas on Human Nature: A Philosophical Study of* Summa Theologiae *1a 75-89*. New York: Cambridge University Press.

———. 2003. Soul and the Beginning of Life (A Reply to Haldane and Lee). *Philosophy* 78: 521–531.

Piché, David. 1999. *La condamnation parisienne de 1277*. Paris: Vrin.

de Sècheville, Jean. 1956. *De principiis naturae*, (mss. Oxford, Merton Coll. 292, fols. 70-85; Paris, Bibl. Nat., Lat. 6552, ffols. 4-25v) ed. Raymond-Marie Giguère. Montréal-Paris: Vrin.

Schwarz, Brigide. 2012. *Kurienuniversität und stadtrömische Universität von ca. 1300 bis 1471*. Leiden: Brill.

Van der Lugt, Maike. 2008. L'animation de l'embryon humain et le statut de l'enfant à naître dans la pensée médiévale. In *Formation et animation de l'embryon dans l'Antiquité et au Moyen Âge*, ed. Luc Brisson, Marie-Hélène Congourdeau and Jean Luc Solère, 234-254. Paris: Vrin.

Weijers, Olga. 2003. *Le Travail intellectuel à la faculté des arts de Paris: textes et maîtres (ca. 1200–1500)*. Turnhout: Brepols.

Zavalloni, Roberto. 1951. *Richard of Mediavilla et la controverse sur la pluralité des formes. Textes inédits et étude critique*. Louvain: Éditions de l'Institut Supérieur de Philosophie.

Chapter 9
Thomas Aquinas on the Vegetative Soul

Martin Pickavé

Abstract This short chapter explores Aquinas's teaching on the vegetative soul. At first glance, Aquinas does not seem too interested in the vegetative soul, and this type of soul certainly takes last rank compared with the sensory and the intellectual souls, which are of more relevance when it comes to human perfection and morality. However, this does not mean that Aquinas's teaching on the vegetative soul lacks sophistication. The chapter first examines why there is a need to posit a vegetative soul in the first place. It then turns to the three main functions of the vegetative soul – nutrition, growth, and generation – and how they are related. After addressing in what sense, according to Aquinas, human beings possess a vegetative soul, the chapter closes with a reflection on the relative obscurity of the activities of the vegetative soul.

9.1 Introduction

In the works of Thomas Aquinas, the vegetative or nutritive soul – Aquinas seems to use these expressions interchangeably – does not often come up as a subject of philosophical or theological inquiry. This does not mean that Aquinas does not mention it in many places, rather the opposite. Yet his interests are obviously very different from those of his teacher Albert the Great, who spent considerable amounts of ink to write about various vegetative life-forms.[1] The reason for Aquinas's relative disinterest in the vegetative soul can perhaps be extracted from a remark he makes in his *Summa Theologiae* (I, q. 78, a. 1, proem.), where he explains that the theologian – and by extension, one might conclude, also the philosopher – is more interested in the intellective and appetitive powers (*potentiae intellectivae et appetitivae*) of the soul, because it is in them alone that we find the virtues of which human beings are capable. And since it is impossible to understand the functioning

[1] See also Amalia Cerrito's contribution to this volume.

M. Pickavé (✉)
University of Toronto, Toronto, ON, Canada
e-mail: martin.pickave@utoronto.ca

F. Baldassarri, A. Blank (eds.), *Vegetative Powers*, International Archives of the History of Ideas Archives internationales d'histoire des idées 234,
https://doi.org/10.1007/978-3-030-69709-9_9

of the intellective powers and their virtuous dispositions in abstraction from sensory cognitive powers, without which human beings would be incapable of thought and appetite, this comment surely also covers Aquinas's considerable attention to sensory cognition; but not much further. Elsewhere Aquinas adds in the same spirit that human beings are called as such because they possess reason and as a result their acts are called human acts insofar as they participate in reason. Since, however, the activities of the vegetative or nutritive soul do not in any way participate in reason, they are not called human acts (e.g., *Super Sententiarum*, lib. 2, dist. 25, q. 1, a. 3 ad 3). Certainly, unless someone possesses an unhinged curiosity, there are issues more worthy of concern than the vegetative soul and its activities.

But is it indeed the case that the activities of the vegetative soul are, as Aquinas frequently emphasizes, not subject to our control?[2] No doubt some vices concern nutrition and generation, two main functions of the vegetative soul. Just think of gluttony and sexual depravity in human beings. How can these vices be blameworthy if it isn't in our power to avoid them? And does the presence of potential vices not also indicate the existence of corresponding virtues? However, according to Aquinas, this sort of reasoning is a bit rushed. What is vicious or virtuous in the present cases are not the activities of the vegetative soul itself, but the agent's attitude towards such activities, including the attitude towards the pleasures associated with them. For the relevant vices stem from an inordinate desire for the pleasures of food and sex and from their inordinate use. But desire and use are rooted in the sensory, not the vegetative soul.[3] The latter really only acts "through the necessity of nature (*per naturae necessitatem*)" (*Summa contra Gentiles*, lib. 2, cap. 23), which puts it outside the moral order.

Even though the vegetative soul is of little to no relevance when it comes to human perfection and morality, this doesn't mean that it should be dismissed as a topic of philosophical inquiry. On the following pages, I will try to explore some topics in Aquinas's philosophy that taken together add up to an outline of a fuller treatment of the vegetative soul. I shall begin with why there is a need to posit a vegetative soul in the first place (Sect. 2), before I look more closely at the functions of the vegetative soul and how they are related (Sect. 3). Finally, I shall address in what sense, according to Aquinas, human beings possess a vegetative soul (Sect. 4).

[2] See, for instance, *Summa Theologiae* I, q. 82, a. 4; I-II, q. 17, a. 8; and III, q. 19, a. 2.

[3] See *Summa Theologiae* I-II, q. 17, a. 8 ad 3: "Ad tertium dicendum quod virtus et vitium, laus et vituperium, non debentur ipsis actibus nutritivae vel generativae potentiae, qui sunt digestio et formatio corporis humani, sed actibus sensitivae partis ordinatis ad actus generativae vel nutritivae, puta in concupiscendo delectationem cibi et venereorum, et utendo secundum quod oportet vel non secundum quod oportet."

9.2 Why Is There a Need for a Vegetative Soul?

Aquinas famously holds that some material beings are ensouled and others not, and that among the ones that have souls we can find three different kinds of souls: (1) vegetative souls, as in plants, (2) sensory souls, as in non-rational animals, and (3) intellective souls, as in human beings. But given the stark contrast he draws between powers of the vegetative souls on the one hand and those of the sensory and intellective souls on the other, one may wonder why he thinks that there need to be vegetative souls in the first place. Obviously, plants have powers, but even inanimate material beings do as well; fire, for instance, has the power to heat and the power to move upwards; rocks are heavy, which gives them the power to move downwards etc. Moreover, just like the powers of inanimate material beings, the powers of the vegetative soul are not in our control. Last but not least, in contrast to plants and inanimate material beings, beings endowed with sensory or intellective souls can pick up information that allows them to perceive, to understand, and to adapt their behaviour. In a certain sense, even inanimate beings can, literally, be "informed"; a stone can become hot by receiving the form of heat or acquire momentum by receiving the form of a motion. In all these cases of "information" new powers may arise: the moving stone may now have the power to, say, destroy certain objects in its trajectory. Yet the case of the sensory and intellective souls is different, since there the "information" leads to the animated subject being aware of what the form represents and we have perception and cognition, and the activities caused by them, whereas here in the case of inanimate material beings and plants none of this occurs.[4]

To say that Aquinas subscribes to the existence of vegetative souls because of his commitment to Aristotle's teaching would hardly be helpful. The key to understand Aquinas's commitment to vegetative souls comes from his understanding of life. Life is a special kind of existence, since not all things that exist are also alive, but everything that is alive also exists. What is special about being alive? "Those things are alive in the proper sense (*proprie sunt viventia*) that move themselves with respect to some kind of change (*seipsa secundum aliquam speciem motus movent*)" (*Summa Theologiae* I, q. 18, a. 1). Note that this description of life is meant to encompass all sorts of changes, from locomotion and alterations to mental activities such as thinking and perceiving. For Aquinas, therefore, to be alive is to be a self-mover.[5] Plants obviously move themselves. Although they may need environmental

---

[4] See, for instance, *Summa Theologiae* I, q. 84, a. 2. It may look as if plants have perception, just like animals, when they turn themselves towards the sun. But for Aquinas this is not a case of perception. What the plant is doing is merely interacting with the heat caused by the sun and on which it relies for growth.

[5] See also *Quaestiones disputatae de anima*, q. 13; *Summa contra Gentiles*, lib. 1, cap. 97; *Scriptum super libros Sententiarum*, lib. 3, dist. 35, q. 1, a. 1. For discussion see Carreño 2015. Aquinas's identification of life with self-motion makes one wonder whether it commits him to saying that simple material objects like the elements are also alive. For elements have the principle of their movement in themselves and they move through themselves (*seipsis*). However, as Aquinas explains elsewhere, elements or simple material objects such as heavy or light bodies, although

factors such as food, water, and the sun for nutrition and growth, it is they themselves which are the immediate agents of their own activities, which can thus count as vital activities.

Obviously, there are varying degrees of self-motion. For Aquinas, they neatly line up with different kinds of animate beings. As he writes:

> We can find things that move themselves not with respect to a form or an end that pertains to it by nature, but only with respect to the execution of a change; but the form through which they act and the end for the sake of which they act are by nature determined for these things. Of this kind are plants, which move themselves with respect to growth and decay according to a form that belongs to them by nature. But other things move themselves in a further degree not only with respect to the execution of a change, but also with respect to the form that is the principle of change, a form they acquire by themselves. And animals are of this kind, for the principle of their change is not a form that belongs to them by nature, but one that is taken in through the senses (*Summa Theologiae* I, q. 18, a. 3).

As this passage makes clear, the scope of a plant's self-movement is rather limited compared with that of non-rational animals. It amounts to little more than initiating and performing, along certain fixed paths, acts of growth, nutrition, and generation. However, even though animals move themselves to action on the basis of forms they acquire for themselves through perception (and are thus in their self-motion more adaptive to their environment), animals still act for the sake of ends that have been determined for them. In contrast, beings endowed with intellect and reason are capable, to a certain extent, of setting their own ends for themselves. To the extent that the ability of these latter beings – which include humans – to move themselves transcends that of non-rational animals and plants, their life too transcends that of these other life-forms.

That plants are alive and that their life consists in some limited sort of self-motion does not yet show on its own that we need to posit the existence of vegetative souls. To see why there need to be vegetative souls as a distinct class of things in Aquinas's ontology we first have to pay attention to why a vegetative soul, like any other soul, is not just a body. For Aquinas, souls cannot simply be reduced to bodies, for souls are first principles of vital actions in a living thing. It is possible that *some* bodies are principles of *some* vital actions, as the eye is a principle of sight, yet by soul we commonly do not mean just any principle of vital action, but a first such principle. Now, it is fairly obvious that bodies qua bodies are not first principles of vital actions. For if this where so, every body would be a living being and a principle of life, which is obviously wrong. "Therefore it holds of some body that it is living, or that it is even a principle of life, through its being *such* a body"

---

they move through themselves (*seipsis*), do not move themselves (*seipsa*) (*Quaestiones disputatae de veritate*, q. 24, a. 1). In *Summa Theologiae* I, q. 18, a. 1 ad 2, Aquinas returns to the comparison with the motion of simple bodies and he clarifies that they move themselves only when they are not at their natural location. This is very different from living beings that move themselves regardless of their location. In *Sententia de anima*, lib. 2, lect. 3, ll. 173–176, Aquinas points to yet another difference between plant and elemental motion: whereas the latter only move towards one place, the former move in two opposite directions (namely, upwards – stems, leaves etc.; and downwards – roots). For this see also Pasnau 2002, 212–213.

(*Summa Theologiae* I, q. 75, a. 1).[6] What makes a body be *such* a body is its actuality or substantial form. Therefore, Aquinas concludes that the soul, as the first principle of life, is the actuality or substantial form of a (living) body.

The last point could leave open the option that vegetative souls are just simple bodies (say, fire) or mixtures of simple bodies. But this is ruled out by the fact that the self-motion of plants goes beyond the activities and motions of simple bodies or mixtures of them. Yes, to some degree the activities of vegetative life have a close connection with the powers of simple bodies: vegetative life depends on the power of heat, for instance, which plays an important role in the nutritive functions. This shows rather that vegetative life makes use of the powers of elemental qualities than that it is the result of a mixture of elemental bodies.[7]

Now that the existence of vegetative souls has been established the question arises of how many kinds of soul there are and how they differ from each other. In *Summa Theologiae* I, q. 78, a. 1, Aquinas defends the traditional division into three kinds of soul by pointing out different ways in which a soul can be related to the corporeal nature of the body it animates:

> The reason for this discrepancy is that different souls are distinguished in keeping with the different ways a soul's operation surpasses the operation of corporeal nature. For all corporeal nature lies under the soul, and is related to it as its matter and instrument. So there is one operation of the soul that exceeds corporeal nature to such an extent that it is not even exercised through a corporeal organ. This is the operation of the rational soul. There is another operation of the soul, below that one, which is brought about through a corporeal organ, but not through any corporeal quality. This is the operation of the sensory soul. For even if hot and cold, wet and dry, and other such corporeal qualities are required for a sense to operate, still this is not in such a way that the sensory soul's operations gets carried out mediated by the power of such qualities; they are instead required only for the proper disposition of the organ. Finally, the lowest of the soul's operations is that which is brought about both through a corporeal organ and by the power of a corporeal quality. Still, it surpasses the operation of corporeal nature, because the motions of bodies come from an external source, whereas operations of this sort come from an internal source. For this is common to all the soul's operations; for everything with a soul moves itself in some way. The operation of the vegetative soul is of this lowest kind.[8]

This passage again draws a connection between different vital operations and self-motion. Here Aquinas even denies that simple bodies, despite how it may look at first, are moved by themselves, that is, from an internal source. The latter is the case only in living beings.

[6] For translation of passages from *Summa Theologiae* I, qq. 75–89, I follow Pasnau's translation in Aquinas 2002. All other translations are mine.

[7] See, for instance, *Summa contra Gentiles*, lib. 2, cap. 62, n. 8: "Operatio autem animae nutritivae etiam excedit virtutem qualitatum elementarium: probat enim Aristoteles, in II de anima, quod *ignis non est causa augmenti, sed concausa aliquo modo, principalis autem causa est anima,* ad quam comparatur calor sicut instrumentum ad artificem. Non igitur potest anima vegetabilis produci ex commixtione elementorum." See also *Quaestiones disputatae de anima*, q. 1.

[8] See also *Quaestiones disputatae de potentia*, q. 3, a. 9; *Quaestiones disputatae de veritate*, q. 10, a. 1 ad 2.

One obvious upshot of the fact that the vegetative soul is the actuality and form of a living body is that vegetative souls perish with the dissolution of a living body. Unlike intellective souls, which have activities on their own, vegetative souls cannot subsist without their corresponding bodies (see, for instance, *Summa Theologiae* I, q. 75, aa. 2–3) and are thus destroyed with the destruction of their bodies. And as their disappearances are connected, so is the way in which they come about. A new vegetative soul comes into being when a new living body is generated. The vegetative soul does not inform the matter of the living body from the outside, as is the case with the intellective soul, but comes forth "from the potency of matter" in the generation of the soul-matter composite, the living body (see, for instance, *Quaestiones disputatae de potentia*, q. 3, a. 11; *Quaestiones de quolibet* IX, q. 5, a. 1).

Last but not least, since the vegetative soul is a substantial form of a living being, it follows that it is in the whole living body and in each part of it. In other words, the vegetative soul of, say, a tree is in the whole tree that it animates and in each part of it: its roots, its trunk, its leaves etc. Vegetative souls, like sensory and intellective souls, are not just the body's external movers. However, unlike in the case of more complex living beings, such as higher-level animals or human beings, the bodies of plants are not very complex. In particular, the various vital powers of plants do not require such a complicated differentiation of the body as we can find in animals where the sensory powers are each instantiated at different locations in the body. We can notice this when we cut a plant in half and it still continues to live, something plants have in common with very simple animals, such as certain worms. As a result, we can even conclude that the *whole* vegetative soul is in the whole body and in each of its parts, which is true of the other souls only in a very attenuated sense.[9]

9.3 The Functions of the Vegetative Soul

As Aquinas admits, the life of plants is "obscure and hidden (*obscura et latens*)" (*Sententia de anima*, lib. 2, lect. 7, l. 60),[10] especially in comparison with that of animals (see *Summa Theologiae* I, q. 18, a. 1). The self-motion of plants is slow, often too slow to be observed, and plants are incapable of locomotion. At least it is clear what Aquinas thinks the main vital activities of plants are: nutrition, growth, and generation.

To better understand the role the vegetative soul is supposed to play we need to focus on the object of its activities. Again, the comparison with the sensory and intellective souls is instructive:

---

[9] *Summa Theologiae* I, q. 76, a. 8; *Quaestiones disputatae de anima*, q. 10; *Summa contra Gentiles*, lib. 2, cap. 72; *Sententia de anima*, lib. 1, lect. 14 and lib. 2, lect. 4. See Aristotle, *De anima* I.5 411b19–27 for background.

[10] See also *Summa Theologiae* I, q. 69, a. 2 ad 1.

> But there are three levels at which the objects of the soul's operations can be considered. For the object of one capacity of the soul is only the body united with the soul. This kind of capacity of the soul is called the vegetative; for the vegetative capacity acts only on the body to which the soul is united. There is another kind of capacity belonging to the soul, a kind concerned with a more universal object, namely, with every sensible body, and not only with the body united to the soul. There is still another kind of capacity belonging to the soul, one that is concerned with a still more universal object, namely, not only with sensible bodies, but universally with all being (*Summa Theologiae* I, q. 78, a. 1).

Strictly speaking, Aquinas here is talking about the capacities or powers of one soul rather than about different souls. This is due to the context, for he is describing the various sets of capacities of the human soul. Nevertheless, the underlying idea gives us some idea of what the vegetative soul, which comprises the vegetative powers, is about, namely, that it acts only on the living body. But this leads us to a new question: for what purpose does it only act on the body to which it is united? In the article immediately following the one from which the last quotation was taken, Aquinas says that "with respect to the body, three operations of the soul are necessary" (a. 2).

He doesn't say explicitly why they are necessary, but the description of the operations gives us some idea of why they might be so. For there is one operation through which the body acquires existence, another through which the living body acquires an appropriate size, and a third through which the body is preserved, with regard to both existence and size. The generative power (*potentia generativa*), the augmentative power (*potentia augmentativa*), and the nutritive power (*potentia nutritiva*) are each the principle of one of these three operations. But regardless of the differences between them, all of the operations of the vegetative soul have in one way or other to do with supporting the existence of an individual living body. It would thus be natural to conclude that the three operations are necessary for the existence of any living body. This is exactly what Aquinas states elsewhere, where he explains:

> For natural and bodily life three things are *per se* necessary, and a fourth thing only necessary *per accidens*. For it is first required that a thing receive life through generation or birth; second, that it arrive at the appropriate size and strength; and third, for the conservation of the life that has been acquired through generation and for its growth, nutrition is necessary. And these things are *per se* necessary for natural life, because without them, life of a body could not advance. Therefore, three natural powers (*vires naturales*) are assigned to the vegetative soul, the principle of life, namely, the generative, augmentative, and nutritive powers. However, since there can be impediments to the life of a body, on account of which the living thing is weakened, a fourth thing is necessary *per accidens*, namely, the healing of a sick living being (*Summa contra Gentiles*, lib. 4, cap. 58).

That Aquinas calls healing (*sanatio*) necessary for life need not worry us much. When a doctor heals a patient we can say that she – or better the healing brought about through her actions – is in some sense a cause of the patient's life. But obviously healing is not an essential cause of the life of the patient, nor is it an essential cause of life in general.

Note that in the passage just quoted Aquinas calls the three powers of the vegetative soul "natural powers". This is on purpose because these powers perform together exactly the same role that a natural external agent performs in the case of inanimate material objects. The main difference is that these powers perform their functions, as it were, from the inside, and in an "elevated way (*altiori modo*)".[11]

The initial characterization of the three powers of the vegetative soul and their operations indicates an order between them. Obviously, the nutritive power serves the generative and augmentative powers, for without nutrition no growth or reproduction would be possible. Why an augmentative power is necessary is not immediately obvious. Why can't plants simply retain the size they have in the first moment of their existence? For Aquinas, the need for an augmentative power is primarily explained by the fact that plants are naturally generated by means of seeds, which are of small size.[12] So there needs to be some mechanism that accounts for growth. In other words, the need for an augmentative power is rooted in the way in which the generative power operates, an indication that the augmentative power too plays a subordinate role (*Quaestiones disputatae de anima*, q. 13 ad 15). However, the most curious case is that of the generative power, which, although it operates on the living body, is obviously not directed at the living body to which it belongs, but rather at another body of the same kind, which it helps to bring about.[13] In the way it transcends its own body, the generative power thus resembles our sensory powers, and for this reason Aquinas calls it the "more final, more chief, and more complete (*finalior et principalior et perfectior*)" of the three powers of the vegetative soul (*Summa Theologiae* I, q. 78, a. 2). In his commentary on Aristotle's *De anima*, Aquinas even goes so far as to suggest that because of the priority of the generative power the vegetative soul is adequately defined as the soul "that is capable of generating another that is alike in species (*generativa alterius similis secundum speciem*)" (*Sententia de anima*, lib. 2, lect. 9, ll. 245–247).[14] Prioritizing the generative power among the powers of the vegetative soul has some interesting, but perhaps not surprising, implications. For unlike the other vegetative powers, the generative power is not about preserving the individual living body, but about preserving the species to which the body possessing the power belongs. So it seems as

---

[11] *Quaestiones disputatae de anima*, q. 13, ll. 205–212: "Nam ad hoc quod indiuiduum producatur in esse ordinatur potentia generatiua, ad hoc autem quod quantitatem debitam consequatur ordinatur uis augmentatiua, ad hoc autem quod conseruetur in esse ordinatur uis nutritiua. Hec autem consequuntur corpora inanimata ab agente naturali extrinseco. Tamen, et propter hoc, predicte uires anime dicuntur naturales." See also ibid. ad 14: "Dicendum quod potentie anime uegetabilis dicuntur uires naturales quia non operantur nisi quod natura facit in corporibus, set dicuntur uires anime quia altiori modo hoc faciunt"; *Sententia de anima*, lib. 1, lect. 5.

[12] This is the normal way in which plants are generated. Aquinas, however, also allows for the spontaneous generation of plants and lower-level animals. See *Scriptum super libros Sententiarum*, lib. 2, dist. 18, q. 2, a. 3 ad 5; *Sententia de anima*, lib. 2, lect. 7.

[13] The idea that "a generator generates something that is similar to it in kind (*generans generat sibi simile in specie*)" is one of basic principles of Aquinas's natural philosophy. See, for instance, *Quaestiones disputatae de potentia*, q. 3, a. 9 ad 6.

[14] This 'definition' is based on Aristotle, *De anima* II.4416b23–25, but Aristotle most certainly did not intend to present it as a definition. The section in Aquinas's commentary presents a clear order among the three powers of the vegetative soul. The first, and lowest, is the nutritive power, the second and "more perfect (*perfectior*)" is the augmentative power, and the generative power is the "most perfect (*perfectissima*)".

if nature prioritizes the preservation of natural species over the preservation of individuals, although species can obviously be preserved only as long as there are individuals pertaining to them.

Does the appetitive soul consist of only three powers or are there more? And how should we conceive of the just mentioned three powers: are they simple powers or bundles of simple powers? Aquinas doesn't say much about this, but some comments he makes about hunger (*esuries*) are instructive in this respect. One might suspect that every being endowed with a nutritive power is also able to experience hunger. For after all, hunger arises from the absence of food. Do vegetative beings thus have a power for hunger? Aquinas seems to agree with this suggestion to some extent, but not without distinguishing between different appetites for food, of which hunger, strictly speaking, is only one. As he writes:

> There is a twofold appetite for food (*appetitus cibi*): one natural appetite, according to which an appetitive, retentive, digestive, and expulsive power (*vis appetitiva, retentiva, digestiva et expulsiva*) serves the nutritive power, which is a power of the vegetative soul. And hunger is this appetite, which does not result from some apprehension, but follows a need of nature. Another appetite is the sensitive appetite, which follows apprehension (*Quaestiones disputatae de malo*, q. 14, a. 1 ad 4).[15]

As this passage suggests, the nutritive power comes with a set of other, usually unnamed powers. This seems plausible: for nutrition to occur there needs to be an attraction to food, and it needs to be possible to retain food and digest it, and later undigested material needs to be expelled from the organism. This makes it sound as if the nutritive power comes with a whole set of associated powers among which it has some priority insofar as the other powers are directed towards it. Hunger, understood as a basic appetite for food, is one of these basic powers. That we usually understand appetites or desires as something that is triggered by sensory perception – say, my appetite for chocolate triggered by what I saw in my cupboard – does not rule out that there are more basic appetites that are not triggered by perception, but result whenever there is a lack in the organism. Note that in order to be aware of hunger a subject needs to be capable of sense perception. But if the awareness of hunger can be separated from hunger, then there is no problem with there being hunger in vegetative beings, although it would be wrong to say that they can experience hunger.

This brings me to my last point regarding the vegetative powers and their relationship to the vegetative soul itself. As Aquinas will often remind his readers, the vegetative powers are special among the powers of the soul insofar as they are active, "whereas in the sensory part they are all passive, and in the intellective part there is an active and a passive component" (*Summa Theologiae* I, q. 79, a. 3 ad 1).[16] It is not difficult to see why our sensory powers are passive: their activity depends on the presence of sensory objects or their representations. The vegetative powers are, as it were, always on and need not to be triggered by anything outside.

---

[15] See *Quodlibet* IV, q. 11, a. 1 for a similar discussion.

[16] See also *Quaestiones disputatae de anima*, q. 16, a. 1 ad 13; ibid., q. 26, a. 3.

Obviously, their nature as active powers does not mean that vegetative powers are loftier than sensory or intellectual ones. While it is true that active powers are to be considered higher, if active and passive powers have the same object, the objects of the vegetative powers, which are the generation, growth, and nourishment of an individual body, as opposed to a more general object, mean that they rank lowest (*Summa Theologiae* I, q. 79, a. 2 ad 3). But because the vegetative powers are always active, it seems as if they have a closer relationship with the soul itself to which they belong than the other powers do. In the jargon of thirteenth-century philosophy: it seems as if the vegetative powers are identical with the essence of the soul. As Aquinas says: "If the very essence of the soul were the immediate basis of its operation, then anything that has a soul would always actually have the operations associated with life – just as anything that has a soul is always actually living" (*Summa Theologiae* I, q. 77, a. 1). Now, due to the vegetative powers' active nature, everything possessing a vegetative soul seems to have immediately all the activities associated with the nutritive, augmentative, and generative powers. What better argument for the identification of the soul and its vegetative powers?

Nevertheless, Aquinas does not think that the vegetative powers are identical with the soul's essence. The main reason for this is that the vegetative powers are powers that belong to both soul and body, as can already be gathered from the fact that they all operate through bodily organs (see, for instance, *Summa Theologiae* I, q. 77, a. 5).[17] They thus don't just belong to the soul or are identical with it; rather, they are properties (*propria*, *proprietates*) of the living being brought about by the soul, or – to use Aquinas's preferred expression – "flowing from the soul's essence". Technically speaking, "the soul's essence is the cause of all its capacities", including its vegetative powers, "as their end (*finis*) and their active source (*principium activum*)", that is, as their final cause and efficient cause, "and the cause of some of its capacities as their recipient" (*Summa Theologiae* I, q. 77, a. 6 ad 2). It is only of the intellective powers that the soul the sole material cause; for powers whose subject is the soul-body composite, the composite serves as their material cause.

9.4 The Vegetative Powers in Human Beings

Human beings share vital activities with plants, even if the way in which human beings procreate, grow, and have nutrition is implemented very differently from what we find in plants. Similarly, we share vital activities with animals, such as sense perception, sensory desire, and locomotion. Again, our human sensory powers do not always operate in precisely the same way as what we find in non-human

[17] Of course, there are other reasons too. For those see, for instance, *Summa Theologiae* I, q. 77, a. 1; *Quodlibet* X, q. 3, a. 1; *Quaestio disputata de spiritualibus creaturis*, a. 11; *Quaestiones disputatae de anima*, q. 12. For Aquinas's position on this topic as it compares with his predecessors and contemporaries see Künzle 1956).

animals, but this doesn't take too much away from what is common to both. If it is due to their vegetative souls that plants are alive and due to their sensory souls that animals can sense and possess the capacity for locomotion, then it is not implausible to assume that human beings too possess vegetative and sensory souls, on top of which they also have an intellective one that accounts for the specific human features of human animals, such as rationality and free will. This view, also called the plurality-of-forms thesis, was quite popular during Aquinas's lifetime. Notoriously, however, Aquinas rejects it as misguided. For him, there can be only one substantial form in a living being, and so there can be only one soul. In fact, after Aquinas's death, his teaching on the unicity of the substantial form even became the subject of censure and condemnations.[18] In light of all this opposition, many Dominican theologians and philosophers after Aquinas took up the defence of the unicity of the substantial form against its opponents.

Aquinas argues for the unicity of the substantial form in many of his works and in a variety of ways.[19] His main reason for holding that there can be only one substantial form or soul in a living being has to do with worries about substantial unity. In *Summa Theologiae* I, q. 76, a. 3 he explains:

> An animal with several souls would not be one thing unconditionally. For nothing is unconditionally one except through the one form through which that thing has existence, because a thing's being existent and its being one thing comes from the same source. For that reason, things that are characterized by different forms are not one thing unconditionally (e.g., a white human being). Therefore if a human being were to be *living* through one form (the vegetative soul), an *animal* through another (the sensory soul), and *human* through a third (the rational soul), then it would follow that a human being would not be one thing unconditionally.[20]

This is not the place to worry about the soundness of this argument and its presupposition. But it should be clear what the unicity of the substantial form entails for the human soul. For if the human being does not have a vegetative soul, then the vegetative powers must be contained somehow in the one substantial form of a human being, namely, the intellective soul. This is exactly what Aquinas has in mind when he writes: "Therefore, the intellective soul virtually (*virtualiter*) contains whatever is possessed by the sensory soul of brute animals and the nutritive soul plants" (ibid.). The expression *virtualiter* is key here and the English expression "virtually" may not convey precisely what Aquinas has in mind. The point is that the intellective soul contains everything that is possessed by the lower souls by way of comprising all their powers. This is an important point. It is true that what

---

[18] Robert Kilwardby, the archbishop of Canterbury and a Dominican confrere, condemned in March 1277 thirty theses at Oxford. Some of the condemned theses can be traced back to Aquinas's teaching on the unicity of the substantial form. For background see Callus 1955. The unicity of the substantial form was again condemned by John Peckham, a Franciscan and Kilwardby's successor as archbishop, in 1286. For a more recent overview of the debate see Boer 2013, cap. 2.4.

[19] See Wippel 2000, 327–351, for a detailed overview of the texts in which Aquinas addresses the unicity of the substantial form.

[20] See also *Quaestiones disputatae de anima*, q. 11; *Summa contra Gentiles*, lib. 2, cap. 58.

distinguishes us as human beings are our intellectual capacities. But it doesn't follow from this that our human soul is essentially intellective; rather it is "essentially vegetative and sensory as well as intellective".[21] Note also that there is an order among the psychological powers of the intellective soul. We have already seen earlier how Aquinas conceives of the relationship among the vegetative powers, and that he thinks the activities of the vegetative soul are necessary for the existence of living bodies. Yet there is more to be said. Obviously, we can only think with the help of sensory images (*phantasmata*), and the sensory capacities through which we are able to entertain such images use corporeal organs. Through their activities the vegetative powers contribute to the conservation and well-being of the body and its organs. In other words, the vegetative powers support the higher powers, which couldn't operate without them.[22]

Adhering to the unicity of the substantial form commits one to maintaining certain views about the beginning and end of life. If there is only one substantial form, namely, the intellective soul, then there seems to be a problem with, for instance, how there can be continuity between the dead corpse of person x and the living body of person x before her death. For the living body and the corpse have completely different substantial forms. More important for the issue at hand, however, is that even though the intellective soul is not destroyed when the body is destroyed and the intellective soul virtually contains what belongs to the vegetative soul, it is not true that the vegetative powers remain after the destruction of the body. The reason for this is obvious: the vegetative powers exist not in the soul alone but in the soul-body compound. But since the compound ceases to exist, so do the vegetative powers, as do the sensory ones. The most that can be said is that these powers "remain in the soul as in their source or root (*in principio vel radice*)" (*Summa Theologiae* I, q. 77, a. 8).

At the beginning of life, the thesis of the unicity of the substantial form forces Aquinas to hold that the development from the embryo to a full-fledged human being is a series of acquisitions and losses of substantial forms. He agrees with many of his contemporaries that the embryo first possesses only a vegetative soul, and then acquires a sensory soul, and finally an intellective soul. At each stage the previous soul prepares the body for the activities that it receives at the next stage (see, for instance, *Summa Theologiae* I, q. 77, a. 4). But since this cannot occur by "adding" one soul to the next, the previous soul needs to depart for the next one to inform the body. This again raises questions about the continuity between the stages.[23] And it means that the process of generation of a human being or an animal

---

[21] This is emphasized rightly by B. Carlos Bazán in his excellent introduction to Aquinas 2016, 75.

[22] See *Quaestiones disputatae de veritate*, q. 13, a. 4. In this text Aquinas also highlights the close connection between our intellective and nutritive powers by pointing out that heightened intellective activity such as contemplation can sometimes interfere with the activity of the nutritive power.

[23] Aquinas discusses various rival theories of embryonic development in *Quaestiones disputatae de potentia*, q. 3, a. 9 ad 9 and *Summa contra Gentiles*, lib. 2, cap. 89. For more about his embryology see Amerini 2013. For Aquinas's discussion of material identity see now Fitzpatrick 2017.

is strictly speaking not just one generation, but a succession of multiple generations (and corruptions) that follow upon each other. But this is a bullet Aquinas is willing to bite.[24]

9.5 Conclusion

In this short chapter, I have briefly examined the nature of the vegetative soul, the vegetative powers, and those powers insofar as they belong to the human soul. In concluding I want to come back to the relationship between the vegetative soul and life. Obviously, this relationship is very close. Every living being has a vegetative soul or at least a soul that comprises the powers characteristic of the vegetative soul. It is thus true to say that a living being is alive because it has a vegetative principle. However, Aquinas is keen to point out that it does not follow that a living being is alive only on account of a vegetative principle. For Aquinas, life is not just generation, growth, and nutrition; life is, as we have seen, self-motion. And the vegetative principle is only the principle of one sort of self-motion, and a very limited one. But given its ubiquity in living beings and its temporal priority in the generation of higher life-forms Aquinas is happy to call the vegetative principle a "first principle" with respect to life.[25] Yet the life of which the vegetative soul and its powers are a principle is for Aquinas only the "most imperfect and obscure life (*imperfectissima vita et occulta*)". Higher degrees of life can be found among sentient beings, although it varies considerably between the different sorts of animals. Finally, and unsurprisingly, the "most perfect degree of life (*perfectissimus gradus vitae*)" is to be found in human beings (see *Summa Theologiae* I, q. 72 ad 1).[26]

Here Aquinas reminds us again of the obscurity of vegetative life, and not for the first time. This brings me back to the beginning of the chapter and the reasons for Aquinas's relative disinterest in the vegetative soul. Could it also have to do with fact that it is strangely obscure to us? We learn about the nature and essence of

---

[24] See, for instance, *Quaestio disputata de spiritualibus creaturis*, a. 3 ad 13: "Relinquitur dicendum quod in generatione hominis aut animalis sunt multe generationes et corruptiones sibi inuicem succedentes: adueniente enim perfectiori forma deficit imperfectior. Et sic, cum in embrione primo sit anima uegetatiua tantum, cum peruentum fuerit ad maiorem perfectionem, tollitur forma imperfecta et succedit forma perfectior, que est anima uegetatiua et sensitiua simul; et ultimo cedente succedit ultima forma completissima, que est anima rationalis."

[25] *Sententia de anima*, lib. 1, lect. 14, ll. 149–157: "Item, uegetabilis potest esse sine sensibili et intelligibili, set hec non possunt esse sine uegetabili: nullum enim animal habet sensum seu intellectum sine hac, scilicet uegetabili. Sic ergo uiuere attribuitur isti principio, scilicet uegetabili, sicut sentire tactui, non tamen quod animal per solum uegetabile uiuat, set quia est primum principium in quo manifestatur uita." The remark might be intended as a criticism of Averroes, who seems to have identified life with the activities of the vegetative soul. For this interpretation of Averroes see Doig 1974.

[26] Obviously, this classification of degrees of life is meant to be only about living bodies; otherwise God should exemplify the highest form of life.

something by starting our inquiry from its activities and observable properties. This is common to all forms of knowledge. The knowledge of the soul (*scientia de anima*) is no exception in this respect. Aquinas is surprisingly upbeat about knowledge of the soul. First, such knowledge is of a very worthy subject; second, it has a claim to certainty. It is certain, "for this anyone experiences (*experitur*) in himself, namely, that he has a soul and that the soul is what gives life" (*Sententia de anima*, lib. 1, lect. 1, ll. 93–95).[27] But even if we share Aquinas's optimism in this case, which is by no means a given, the way in which we experience that we have a soul is presumably by some sort of introspective awareness of our mental life, which then becomes the basis for an inference to the underlying thing that makes such a mental life possible in the first place. It may be easy for me to be aware of my past and present perceptions, desires, and thoughts and to reflect on them, but it is not clear that our vegetative life can in the same way be the object of our awareness. With respect to our own vegetative powers we seem to stand in the same relationship to them as to the vegetative powers of some other organism. So it is perhaps no surprise that in Aquinas's optimistic psychology, the vegetative soul and its powers are not at the center of attention.

References

Amerini, Fabrizio. 2013. *Aquinas on the Beginning and End of Human Life*. Cambridge, MA: Harvard University Press.

Aristotle. 1956. In *De anima*, ed. W.D. Ross. Oxford: Clarendon Press.

Callus, D.A. 1955. *The Condemnation of St. Thomas Aquinas at Oxford*. London: Blackfriars.

Carreño, Juan Eduardo. 2015. From Self-Movement to *Esse*: The Notion of Life and Living Being in Thomas Aquinas. *Angelicum* 93: 347–376.

Boer, Sander de. 2013. *The Science of the Soul. The Commentary Tradition on Aristotle's De anima, c. 1260-c. 1360*. Leuven: Leuven University Press.

Doig, James C. 1974. Towards Understanding Aquinas' *Com. in De Anima*. A Comparative Study of Aquinas and Averroes on the Definition of Soul (*De Anima* B, 1-2). *Rivista di filosofia neoscolastica* 66: 436–474.

Fitzpatrick, Antonia. 2017. *Thomas Aquinas on Bodily Identity*. Oxford: Oxford University Press.

Künzle, Pius. 1956. *Das Verhältnis der Seele zu ihren Potenzen. Problemgeschichtliche Untersuchungen von Augustin bis und mit Thomas von Aquin*. Freiburg: Universitätsverlag.

Pasnau, Robert. 2002. *Thomas Aquinas on Human Nature*. Oxford: Oxford University Press.

Thomas Aquinas. 1882—. *Opera omnia iussu Leonis XIII P.M. edita*. Rome et al.: Commissio Leonina.

———. 1929–1947. *Scriptum super Sententiis*, edited by Pierre Mandonnet and Maria Fabianus Moos. 4 vols. Paris: Lethielleux.

———. 2002. *The Treatise on Human Nature: Summa Theologiae* 1a 75–89. Translated, with Introduction and Commentary, by Robert Pasnau. Indianapolis, IN: Hackett.

———. 2016. *L'âme e le corps. Somme de théologie. Première partie, questions 75 et 76. Introduction par B. C. Bazán, traduction par J.-B. Brenet*. Paris: Vrin.

Wippel, John F. 2000. *The Metaphysical Thought of Thomas Aquinas. From Finite Being to Uncreated Being*. Washington, DC: The Catholic University of America Press.

[27] For a fascinating survey of the later medieval debate about how knowledge of the soul is possible see Boer 2013, cap. 3.3.

Chapter 10
The Vegetative Powers of Human Beings: Late Medieval Metaphysical Worries

Martin Klein

Abstract In this chapter, I investigate the metaphysical assumptions that medieval thinkers considered necessary in order to integrate the vegetative powers and processes into their conception of human beings as composed of a material body and an immaterial soul. My aim is to show that vegetative powers and processes are central to the late medieval debate on faculty psychology and on the unity or plurality of substantial forms. The chapter has two parts. First, I present three different accounts of the *ontological status* of the vegetative powers in relation to the body and the soul, as found in Thomas Aquinas, William of Ockham, and John Buridan. All three maintain that the vegetative powers belong to the soul but disagree as to whether they are identical with it or somehow distinct from it, especially when the soul in question is immaterial. Second, I investigate how medieval thinkers accounted for the metaphysics behind the *operations* of the vegetative powers in human beings, who are endowed with immaterial souls, in particular when it comes to procreation and digestion. A pressing problem for the vegetative powers of human beings is how they can produce a new substance when this composite is supposed to include an immaterial soul which is not naturally generated. Regarding procreation, one can distinguish two different approaches: *dispositional generation of material conditions* (Aquinas, Buridan) and *substantial generation of the body* (Henry of Ghent, Godfrey of Fontaines). A similar problem for human beings arises in nutrition and growth once this process is taken to be some kind of substantial change, namely, the corruption of food and the production of new parts of the living being. Two accounts can be distinguished: (i) *"transmateriation"* of the matter of food into the nourished body (Aquinas, Buridan) and (ii) *generation* of parts of the body (Francis of Marchia).

M. Klein (✉)
University of Würzburg, Würzburg, Germany
e-mail: martin.klein@uni-wuerzburg.de

F. Baldassarri, A. Blank (eds.), *Vegetative Powers*, International Archives of the History of Ideas Archives internationales d'histoire des idées 234,
https://doi.org/10.1007/978-3-030-69709-9_10

10.1 Introduction

Medieval thinkers took the vegetative capacities of living organisms to be necessarily *powers of the soul*. It is in virtue of the soul that a living being possesses the vegetative powers which make life possible in the first place. Following Aristotle, medieval thinkers distinguished among the following three vegetative powers, which are fundamental for all living beings, from plants to animals to humans: (i) the nutritive power, that is, the capacity to nourish and maintain the body; (ii) the augmentative power, that is, the capacity by means of which a body grows; and (iii) the generative power, that is, the capacity to produce a new body by procreation. Only living beings are properly said to have these vegetative functions, precisely on account of their soul, which is said to be their substantial form. Living beings, like many non-living things, are natural substances which consist metaphysically of matter and substantial form. The substantial form of living beings is their soul, which, by being united with matter, constitutes a body; it is in virtue of the soul that this body is a living organism that exhibits the life functions of nutrition, growth, and procreation.[1]

Once vegetative powers are postulated as powers of the soul, the problem arises of how to account for the precise relationship among the animated body, the soul, and the vital powers in question. Medieval philosophers and theologians debated the question of where to place the vegetative powers in the metaphysical make-up of living beings, human beings in particular. In human beings, the picture of body, soul, and powers gets more complicated, since they are considered both to be biological organisms and to be endowed with an immaterial rational soul. While plants are living beings which have only vegetative powers, and non-rational animals have in addition sensitive powers (sensations and appetitions), human animals also have intellective powers of understanding and willing. For medieval thinkers, this variety of powers raised the question of how the human soul can be the single principle of all these powers; this question is further complicated by the claim that the souls of human beings are incorruptible. Unlike the souls of all other living beings, and substantial forms in general, the human soul does not perish along with the body when the substantial composite of which it is a part is destroyed, but will continue to exist even after the body ceases to exist. In other words, human souls are separable from the body. They owe this separability to their special nature: while the souls of plants and animals are generated in a natural process and originate from a material potency, the human soul is the product of God's creation, and is supernaturally infused into a human body.[2]

Medieval thinkers tried to reconcile the incorruptibility and immateriality of the human soul with the vegetative processes of the human organism, which were

---

[1] On Aristotle's treatment of vegetative powers, most prominently in *De anima* II.4, see Polansky 2007, 200–223.

[2] On the medieval science of the soul and its theological parameters, see Dales 1995; de Boer 2013 and Klein 2019.

clearly seen as material or organic in nature. Insisting that the vegetative powers are powers of the soul in human beings no less than in other living organisms, medieval philosophers and theologians discussed how to integrate those bio-psychic functions into the peculiar body-soul composite of human beings. In what sense are they powers of the soul? Are they distinct from the soul as its principle, or are they identical with it? Moreover, vegetative powers are intrinsically related to the body: the soul does not operate by itself by means of them; rather, it operates in and through the body as its organizing principle. What then is the precise connection between the vegetative powers of the soul and the body? Once accounted for such that they fit the metaphysical make-up of a human being, the vegetative powers of the soul should provide us with some explanation of how living organisms work through them. Thus, vegetative powers in particular should help us to explain how nutrition, growth, and generation can come about in an organism by virtue of its soul, even if this soul is immaterial. How can food be ingested into a body whose substantial form is immaterial, and can such a body grow? And if the human soul is produced by God, in what sense can human beings engage in sexual reproduction?

I shall first present three different accounts of the ontological status of vegetative powers in relation to the body and the soul. I will then present several accounts of how vegetative powers perform their operations in human beings. I will not be concerned in either section with biological considerations, but rather with the metaphysical assumptions medieval thinkers considered necessary in order to integrate vegetative powers and processes into their conception of human living bodies. I will show that, depending on how the theory of vegetative powers is adjusted to the soul as an immaterial substantial form of the body, the metaphysics of vegetative processes makes it hard to conceive of vegetative processes without a material substantial form. As I hope to show, vegetative powers and processes are central to late medieval discussions of faculty psychology and the unity or plurality of substantial forms.

10.2 Where Should the Vegetative Powers Be Placed in Human Beings?

How are the vegetative powers related to the immaterial soul and the body of a human being? I will present the different accounts of Thomas Aquinas (c. 1225–1274), William of Ockham (c. 1288–1347), and John Buridan (c. 1300–1358/60).[3] All three of them maintain that the vegetative powers belong to the soul, but disagree as to whether they are identical with the soul or somehow distinct from it, especially when the soul in question is immaterial. There are further disagreements regarding the relationship of the powers to the body, and whether

[3] For a more detailed discussion of Aquinas's account of the vegetative soul, see Martin Pickavé's contribution in this volume.

there is a sense in which the dispositions of the body can be identified with vegetative powers. Aquinas thinks that although the principle of the vegetative powers is the soul, their subject is the animated body, whereas Ockham considers it impossible for vegetative powers to pertain to an immaterial soul. Buridan even claims that the vegetative powers are identical with an immaterial soul, but seems to argue that the vegetative powers are purely bodily instruments through which the soul acts.[4]

10.2.1 The Soul as Distinct from Its Powers

According to Aquinas, human beings are the substantial composite of matter and *one* substantial form, namely, the soul. The vegetative powers are powers of the soul and are distinct from it. Since the soul is an immaterial form, the vegetative powers are immaterial as well; but since they are powers which the soul actualizes through the body, the subject of the vegetative powers is the animated body. Aquinas gives a concise account of the relationship among the body, the soul, and its powers in his major work, the *Summa theologiae*. Living beings are material substances; that is, they are composites of prime matter and substantial form. Their soul is identified as the substantial form of this composite and, by informing matter, makes up the living body. Living beings can be grouped according to the different kinds of soul they have. While plants have only a vegetative soul, non-human animals also have a sensitive soul, and human beings have additionally a rational soul or intellect. This does not mean that human beings have three different souls; rather, when we speak of the vegetative, sensory, and intellectual "souls" of a human being, what we really mean are different powers of one and the same soul.[5]

The intellective, sensitive, and vegetative powers of a human being are all related in the same way to the soul, but differently to the body. Aquinas's view of the relation between the soul and its powers can be labelled the *Distinction Thesis*, which can be formulated as follows:

(DT) Powers of the soul are not identical to the essence of the soul, but are accidents which are inseparable from the soul.

Thus, there is a difference between the soul and its powers. However, in calling them proper accidents (*propria*) of the soul, Aquinas makes clear that, although they are not identical with the essence of the soul, they are nevertheless inseparable from it.[6]

The relation of a power to the body is different for each set of powers (intellective, sensitive, vegetative) according to how they are instrumentally related to the

[4] On the relationship among the soul, the body, and its powers in medieval philosophy generally, see Wood 2011; de Boer 2013, 227–252; Perler 2015, 114–123 and Bakker 2019, 66–68.

[5] See Aquinas 1889: *ST* Ia q. 76 art. 3 and q. 78 art. 1 (ed. Leon. 5, 220–221 and 251).

[6] See Aquinas 1889: *ST* Ia q. 77 art. 1 ad 5 (ed. Leon. 5, 237b).

body, that is, how the soul acts through the body by means of its powers. Among the various powers of the human soul, the vegetative powers are the most fundamental for a body to be a living being, since every living being is equipped with them, even if, like a plant, it lacks sensory and intellective capacities. They are also the most closely connected to the body, since they operate through bodily organs *and* bodily qualities.[7] By means of the action of heat, the operations of the vegetative powers are carried out. Regardless of what exactly the vegetative powers and their operations are, they all have in common that they "move" the body from within. In every vegetative operation some motion inside the organism is at work in virtue of a principle that is intrinsic to the body, namely, the soul. In this sense, the soul moves the body with respect to vegetative processes by internally orchestrating certain operations. The example Aquinas gives is digestion: digestion is instrumentally performed by means of bodily organs such as the stomach, and in virtue of a bodily quality, namely, heat.[8]

What complicates Aquinas's theory of the powers of the soul is the assumption that the human soul is incorruptible and is separable from the body, unlike the souls of other living beings, which perish with the body. But how is it possible for the human soul to be incorruptible and immaterial yet at the same time have powers which are carried out in a material organism? When a plant or animal dies, its vegetative and sensitive soul perishes, unlike the human intellective soul. It seems therefore that we should assume that there are multiple souls in a human being, one incorruptible (the intellect), the others corruptible (the sensitive and vegetative souls). The intellect and the other souls seem to be essentially different and cannot be subsumed under one and the same soul, for then the human soul would have contradictory features, in being incorruptible on account of its intellect yet corruptible on account of its sensitive and vegetative soul.[9]

But Aquinas is a staunch defender of what we can call the *Unity Thesis*:

(UT) Every material substance is the essential unity of matter and only one substantial form.

This is also true for all living beings: besides the soul as the substantial form of the body there is no other substantial form present in the hylomorphic composite. Aquinas's main reason for denying a plurality of substantial forms in substantial composites is that the unity among multiple substantial forms could not be essential but only accidental.[10] In reply to the argument in favor of a plurality of souls,

[7] On the other hand, the intellective powers are exercised neither through an organ nor through corporeal qualities, whereas the sensitive powers operate *through* an organ which is disposed in a certain way thanks to its corporeal qualities, such as hot and cold, dry and wet. However, they do not operate through those qualities themselves, but merely make use of an organ which is corporeally structured in a certain way; see Aquinas 1889: *ST* Ia q. 78 art. 1 (ed. Leon. 5, 250b).

[8] See Aquinas 1889: *ST* Ia q. 78 art. 1 (ed. Leon. 5, 250b–251a).

[9] See Aquinas 1889: *ST* Ia q. 76 art. 3 arg. 1 (ed. Leon. 5, 220a).

[10] For a detailed discussion of Aquinas's Unity Thesis, see Pasnau 2004, cap. 3 and de Boer 2013, 130–141.

Aquinas forcefully defends the unity of the soul in human beings and states that the human sensory and vegetative powers are incorruptible as well, but only on account of the intellect, which "gives incorruptibility" to them. A vegetative or sensory soul per se is in fact corruptible, as we can see in animals and plants, but if it is connected to an immaterial intellect, it becomes incorruptible.[11]

Aquinas's denial of a plurality of souls in human beings—one intellective and incorruptible, the others sensitive and vegetative, and corruptible—forces him to claim that it is the entire human soul, including all its powers, which is incorruptible. Since powers are proper accidents of the soul, they are still present in the soul even if it is separated from the body. The separated soul still has these powers "virtually", as Aquinas puts it. We might spell this out as saying that they are non-functional powers of the soul without a body. Yet vegetative (and sensitive) powers are distinguished from other powers precisely because of their close relationship to the body by means of which they operate. There seems to be a tension in Aquinas's ontology of vegetative powers. On the one hand, there is one single soul in a human being, which possesses vegetative powers; this soul is incorruptible on account of its intellective power, which also gives incorruptibility to the vegetative powers of the same soul. On the other hand, the vegetative powers are distinct from the intellective powers, since they perform *operations* which are very different in nature.

Aquinas thinks that this tension can be resolved by distinguishing between the *principle* and the *subject* of "operative powers". While the soul is the principle of all of its powers, since it somehow causes them to exist, it is the subject only of those powers whose operations the soul itself is capable of performing. Generally, a power is said to be in something as its subject only when the power's operation can be performed by its subject. Since only intellective operations are performed without the body, the soul is the principle and the subject of all of its intellective powers. The operations of all other powers—that is, the vegetative and sensitive powers—which are performed by means of corporeal organs, have the soul-body composite as their subject. The soul is the principle of those powers in being the first substantial act of the body, that is, in bringing about the existence and functionality of a body. It is through the soul that the soul-body composite is able to perform its operations and can thus be the subject of powers.[12]

Since Aquinas speaks here of "operative powers", his solution seems to be designed not for powers *qua* powers but *qua* being activated.[13] What appears to be a tension between calling the soul-body composite the subject of vegetative operations, and holding at the same time that the body is the instrument or means for the soul to perform vegetative operations, seems to be exactly the crucial point in Aquinas's strategy: the body parts are "body organs"—that is, instruments for vegetative and sensitive functions, the principle of which is the soul—precisely insofar

[11] See Aquinas 1889: *ST* Ia q. 76 art. 3 ad 1 (ed. Leon. 5, 221b), and 1961: *SCG* II cap. 89 n. 1747 (ed. Marietti, 255b).

[12] See Aquinas 1889: *ST* Ia q. 77 art. 5 (ed. Leon. 5, 244b–245b).

[13] This would allow Aquinas to call the soul to be the *subject* of *all* of its powers, as he does in 1889: *ST* Ia q. 77 art. 6 ad 2 (ed. Leon. 5, 246b).

as they are subjects of the vegetative and sensitive soul. Yet the question of how to distinguish the body as the instrument and subject of powers from the soul as the subject and principle of powers is a central concern in the accounts offered by Ockham and Buridan in the fourteenth century.

10.2.2 Soul(s) and Powers: Identical Yet Distinct

William of Ockham distinguishes powers of the soul as dispositions of the body, which serve for the soul to carry out its operations, from powers of the soul itself. In the first sense, powers are bodily dispositions; in the second sense, they are identical with the "material" (i.e., body-bound) vegetative-sensitive soul. In human beings there is another soul in addition to the vegetative-sensitive soul and really distinct from it, namely, the immaterial intellect. Ockham concentrates more on the sensitive powers, but states that in principle the same relation to soul and body also holds for the vegetative powers.

Ockham develops his account by criticizing Aquinas's conception of the composite of body and soul as the proper subject of the corporeal powers. Ockham thinks that there is a tension between saying on the one hand that corporeal powers are *in* the composite of body and soul, and on the other hand that the operations of these powers are performed by means of the body. One might say that it is the soul-body composite that serves as the subject of powers to be performed by differently structured organs; for example, the organ of sight has different properties than the organ of smell. Even then, Ockham thinks, it is more plausible to hold that the proper subject of powers is the soul, which requires certain dispositions of the body in order to perform its operations. This would also fit better with the view that the body serves as an instrument or means for the soul to perform material operations. As we have seen, Aquinas also uses this way of speaking in describing the vegetative soul's operation of digestion by means of bodily heat.[14]

But while Aquinas tries to reinterpret Augustinian-dualistic vocabulary in such a way that it makes good sense in an Aristotelian context, Ockham adopts from the start an Augustinian-dualistic interpretation of the soul-body relationship. Consequently, he thinks that the relation of a power to body and soul is different from what Aquinas describes. In one sense, powers of the soul are dispositions of the body which are needed for the soul to operate; in another sense, they are what belongs to the soul when it is active.[15] Unlike Aquinas, Ockham holds that in the second sense of powers of the soul—that is, powers properly *of* the soul and not in relation to the body—there is no real distinction either between powers and the soul or between powers themselves. We can call this the *Identity Thesis*:

---

[14] See Ockham 1982: *Sent.* III q. 4 (ed. OTh 6, 131–134).

[15] According to de Boer 2013, 248, this distinction is an innovation of Ockham and had a significant impact on later authors in the fourteenth century, among them Buridan; see below.

(IT) The soul is identical with its powers, i.e., there is no real distinction between the soul and its powers.

Ockham thinks that it is superfluous to posit a distinction here, since one can explain how the soul operates without appealing to powers as distinct entities. Powers of the soul are only conceptually distinguished by the different activities of the soul, that is, by the different *acts* it can directly bring about, such as an act of hearing or an act of seeing. But those acts do not indicate that there are really distinct powers in the soul, but only connote the soul in different ways according to what it is capable of doing. In the first sense of powers of the soul, i.e., dispositions of the body, they are distinguished from each other and from the soul: those bodily dispositions in relation to each other, Ockham takes to be really distinct according to the different organs which are necessary for distinct psychic operations. From the fact that these dispositions are qualities of the body and not of the soul, it follows that powers in this sense are also distinct from the soul.[16]

But Ockham also seems to accept a sense in which powers as powers of the soul are really distinct from the soul. Although it is the entire vegetative-sensitive soul which is capable of, for instance, seeing, hearing, or digesting, it can perform those operations in the body only when there are parts of the body with the right material dispositions. These bodily parts divide the soul into parts; for instance, the seeing part of the soul is in the eye and the hearing part in the ear, because the soul is materially generated and extended by quantified parts of matter, and hence is divided according to the different quantitative parts of matter in the body. Thus, powers in the vegetative-sensitive soul are in some sense, as Ockham says, divided into really different parts. As evidence for this, he points to the fact that the soul can lose its power of sight because of damage to the eye while other parts of it remain, for instance the power of hearing. When we take powers of the soul to include the material dispositions and parts of the body, powers of the soul are really distinct from the soul.[17]

At the same time, Ockham, like almost all thinkers of his time, holds that human beings have an immaterial soul which is capable of intellective operations. Those immaterial powers can hardly be identical with a material soul as just described. Ockham's conception of powers of the soul and material operations forces him to introduce a real distinction within the human soul, between the vegetative-sensitive soul, which has powers distributed materially in the body, that is, those which include dispositions of the body, and the intellective soul, which performs immaterial operations. We can call this the *Plurality Thesis*:

(PT) The sensitive-vegetative soul and the intellective soul are two really distinct substantial forms.

---

[16] See Ockham 1982: *Sent.* III q. 4 (ed. OTh 6, 135–136), and 1981: *Sent.* II q. 20 (ed. OTh 5, 435–436).

[17] See Ockham 1982: *Sent.* III q. 4 (ed. OTh 6, 136–137).

Ockham invokes an argument for this thesis similar to the one which Aquinas had tried to reject: because the sensitive-vegetative soul is different in nature from the intellective soul, they must be two really different forms. They are different in nature because the sensitive-vegetative soul is materially extended while the intellect is not. That the sensitive-vegetative soul is extended is due to the fact that it is partitioned according to the quantitative extension of matter, which as we have just seen, serves as the explanation for the variety of its operations that depend on material parts of the body.[18]

It is thus not just the identity of the soul and its powers insofar as they belong to the soul which leads Ockham to the *Plurality Thesis*. Even if there were a distinction between the essence of the soul and its powers, the soul could perform material operations only if it were materially generated and extended. A soul which has the powers of corporeal operations has to be so closely linked to matter that it cannot be immaterial in a way that would make it separable from the body.[19]

The soul of a human being is thus fragmented into two really distinct substantial forms, one of which has vegetative and sensitive powers. Ockham thinks that no further distinction is needed between those powers or souls, since the relation to bodily dispositions explains the variety of both the sensitive and the vegetative powers equally.[20] Whatever the vegetative powers and operations are exactly, and however they work in detail, this account makes it possible to treat those powers of human beings in the same way as those of other animals. Whether these vegetative powers are those of plants, animals, or humans, they are identical to a material soul, and no special explanation is needed for how they could be related to an immaterial soul. On this account, there are no vegetative powers which are immaterial in any sense. However, this comes at a cost: with its proliferation of substantial forms, Ockham's account faces the difficulty of how the unity of a human being can be preserved. After all, in his account, human beings have two different souls, and it is also debatable how he can explain the relationship of the immaterial rational soul to the body and its other functions.[21]

Let us turn to John Buridan, who thinks that two really distinct souls in a human being would be an absurd consequence. Like Ockham, Buridan distinguishes between powers of the soul as bodily dispositions and capacities of the soul itself which are identical to the soul; but like Aquinas, he maintains that there is only one substantial form in living beings, which in the case of human beings is an immaterial soul.[22]

---

[18] See Ockham 1980: *Quodl.* II q. 10 (ed. OTh 9, 156–161, esp. 159) and I q. 12 (68–71).

[19] Robert Kilwardby (ca. 1215–1279) expressed a similar argument: since powers are so different in nature, they cannot be rooted in one and the same form. See Silva 2012, 74.

[20] See Ockham 1981: *Quodl.* II q. 11 (ed. OTh 9, 164).

[21] On the debates on the unity vs. plurality of forms, see the detailed descriptions by Zavalloni 1951 and Schneider 1973. On the problem of the unity of substance in Ockham in particular, see Perler 2010.

[22] Buridan argues against a plurality of souls in forthcoming: *QDA(3)* II q. 4 (on the identity of the sensitive and vegetative soul) and III q. 17 (on the identity of the intellective and sensitive soul)

Buridan arrives at this conception by starting from a conceptual clarification of 'power', according to which every active or passive principle of some motion or operation just is an active or passive power. It is an active principle or power when it brings about a change in something else; it is passive when it is acted upon, and thus changed, by something else. Therefore, everything which acts is an active power, and everything which is acted upon is a passive power. Buridan concludes from this that the soul is identical to its powers because it is the principle of various operations, such as understanding and nutrition. Since, according to him, there is just one soul in every living being, the human soul is the principle of all human actions; thus, all of a human being's various intellective, sensitive, and vegetative powers are identical to the human soul. Seen from this perspective—the soul as the principle of all the operations of a living being—the soul is a single power able to operate in various ways. This also implies that powers are not distinct from each other. Only in a secondary sense can we say that the soul has distinct powers according to the different operations which it is able to perform. We may distinguish various powers from one another conceptually, but this does not imply a real distinction.[23]

However, there are also nutritive powers of the body which are really distinct from the sensitive powers and from the soul. For there are purely bodily activities to which the characterization of active and passive powers applies. They are nutritive powers because they are the active principles of nutritive processes. Since they are co-active with the soul's nutritive powers, Buridan is willing to call them powers of the soul, but only insofar as they serve as instruments for the soul to carry out its vital operations "as the blacksmith uses fire and a hammer". From the perspective of what it means for something to be a power, those powers are properly called powers in themselves, as Buridan seems to suggest; seen in relation to vital operations, they can be called powers of the soul, since they are dispositions for the acts of the soul.[24]

Buridan's account thus combines the *Unity Thesis* of Aquinas's account with the *Identity Thesis* suggested by Ockham. Recall that Aquinas had argued for a single immaterial soul in human beings (the *Unity Thesis*) which is distinct from its powers (the *Distinction Thesis*), while Ockham had claimed that powers are identical to the soul when not conceived of as dispositions for the soul to operate (the *Identity Thesis*), and that human souls consist of two really distinct substantial forms (the *Plurality Thesis*). Thus, Buridan's account can be summarized as the *Unity-Identity Thesis*:

(UIT) A living being is the essential unity of matter and only one substantial form, namely, its soul, which is identical with its powers.

(ed. Klima et al.).

[23] See Buridan forthcoming: *QDA(3)* II q. 5 (ed. Klima et al., nn. 18–22).

[24] See *QDA(3)* II q. 5 (ed. Klima et al., nn. 23–25).

Buridan sides with Ockham against Aquinas in distinguishing between powers as dispositions for the soul to act and powers as parts of the soul. Since those bodily dispositions serve to allow the soul to perform its operations, they are called *instrumental, dispositive,* or *remote powers*. They are distinct from each other and from the soul. Powers as parts of the soul, which Buridan calls *principal powers*, are distinct neither from the soul nor from each other; they are identical to the soul insofar as it performs an act. But against Ockham, Buridan defends Aquinas's claim that a human being has only one substantial form, namely, the soul, whose proper functioning can be explained in the same way for animals and for human beings. Together with the claim that the soul of human beings is immaterial, to which Buridan also subscribes, this implies that, just as Aquinas had claimed, all principal powers of the soul—vegetative, sensitive, and intellective—are immaterial and incorruptible. An immaterial soul of a human being does not lose those capacities which require a body to be functional, but simply does not exercise them for the reason that the bodily dispositions are not available.[25]

10.3 The Metaphysics of Generation, Nutrition, and Growth

It certainly sounds odd to conceive of vegetative powers as powers of an immaterial soul to which they belong, even without a body in which and through which they are exercised. Generally speaking, any metaphysics of powers should not only tell us where to place powers in our ontology but also give us an explanation of how they work. Regarding vegetative powers within the Aristotelian framework, this means asking how an organism manages to carry out processes of nutrition, growth, and generation *by its soul*. A pressing question for the vegetative powers in human beings, endowed with an immaterial soul, is how they can be postulated in accordance with the general principles of natural change. For this is what vegetative processes are: they are natural changes. The human body is nourished by assimilating food, which then serves for the growth and maintenance of the organism. Food is also transformed into sperm or menstrual blood, which serve for the propagation of new life by producing offspring through sexual reproduction. In this section I shall investigate how medieval thinkers accounted for the *metaphysics* behind the operations of the vegetative powers in human beings with immaterial souls, in particular when it comes to procreation and digestion.

[25] See Buridan 2010: *QGC* I q. 8 (ed. Streijger et al., 85) and forthcoming: *QDA(3)* II q. 5 (ed. Klima et al., n. 31) and III q. 17 (n. 14).

10.3.1 How to Produce a Human Being?

It seems reasonable to say that human beings would not be procreative if they did not produce new offspring. However, the human soul is immaterial, and this entails that it cannot be the product of natural generation; rather it must be created supernaturally. Medieval thinkers were therefore confronted with the following question: In what sense can human beings be called generative? After all, did not Aristotle hold that a human being begets a human being (*homo generat hominem*)?[26] Medieval thinkers agree that matter cannot be what is generated, since it is the substratum of substantial changes and becomes actualized by a new form; however, they disagree about whether it is this new form that comes to be in matter which is generated, or rather the substantial composite of matter and form on account of a new form inhering in matter.

But whether it is the composite of form and matter that is generated or just the form, sexual reproduction (whatever the biological details are) should somehow contribute to the coming to be of a new form, that is, the soul of the generated human being. How can this be reconciled with the commonly shared assumption that the human soul is not naturally generated but created and infused in the body by God? And what is it that human beings contribute to new offspring? I will compare two accounts: while Thomas Aquinas and John Buridan defend what can be called *dispositional procreation*, Henry of Ghent (c. 1240–1293) and Godfrey of Fontaines (c. 1250–c. 1305) argue for the proper *substantial generation of a body* as the product of human generation.

According to Aquinas, body and soul unite to form one thing, and are not separate things in the sense that the body has its own being and the soul is added to the body. It is the soul which brings the body into existence. Here, Aquinas faces the objection that generally things which are one in being do not result from different agents and actions; thus, if body and soul are one in being, they cannot be the results of different agents, and therefore they cannot be the product of both a natural agent (i.e., human semen) and a supernatural agent (i.e., God). If the human body arises ultimately from human semen, and if body and soul are one united thing, the soul too should be produced by the semen.[27]

Aquinas replies to this by scrutinizing the underlying premise that different agents bring about diverse actions resulting in different things. This is true only if the different agents are not properly ordered to one another; but if they are properly ordered to one another, they can bring about one thing and one united result. To defend this claim, Aquinas refers to the relation between the vegetative processes of the animated body. While the soul is the principle or agent of vegetative processes, and is thus properly said to be their cause, it cannot bring them about by itself, but needs to use the body as an instrument in order to bring about certain effects, such as digestion. The body, on the other hand, is a concomitant cause of this process.

---

[26] Cf. Aristotle 1984: *Phys.* II.2, 194b13.

[27] See Aquinas 1961: *SCG* II cap. 88 n. 1725 (ed. Marietti, 251a).

The soul is the principal cause and agent while the body, or the bodily dispositions, serve as an instrument. Aquinas then compares this relationship to the one between God and material dispositions: God, being the principal and primary cause of the human soul, uses those material dispositions in order to bring about the animated human body, that is, the union of body and soul.[28]

The question remains, however, exactly what human beings contribute to this joint production of a new human being. What is the point which Aquinas calls the *aliquid hominis*, or part of the new human being, at which the natural action of father and mother comes to a close? It cannot be the living body, since this would include the human soul as its substantial form. It seems that it cannot be the body without the soul either, since generally, according to Aquinas, the soul does not merely inhere in and give life to an already existing body, but brings a functional body into existence in the first place. To solve the problem, Aquinas resorts to matter—not prime matter, however, but what he calls "designated matter". The part which is produced by human beings within the process of generating new offspring is matter as temporally preceding a substantial form, that is, matter which is the potentiality to receive a form but does not yet possess a form. Matter in this sense can be understood as the "human body insofar as it is in potentiality to the soul", a body which, "as not yet having one, precedes the soul in time."[29]

This might explain how human beings are causally involved in sexual re-*production* and it might be enough to attribute reproductive powers to them. What this solution seems not to resolve, however, is the problem of how human beings can be described as producing other human beings in the technical or strict sense of *substantial generation*, that is, as bringing about a new substantial composite of matter and form. According to Aquinas, the "part of the human being" is not the definite human body, but is a material substance, namely, the embryo, which first has a vegetative soul, and then a sensitive soul which includes the vegetative powers. The procreative activities of human beings are not in vain, but cause a substantial change by generating a human body. However, this body has to develop and undergo several sequences of substantial generations and corruptions before it is sufficiently complex to be united with an immaterial soul which includes intellectual capacities. First, the embryo has a nutritive system, followed by sensory capacities. What happens at this stage is that the nutritive soul is corrupted and gets replaced by a sensitive soul, which then includes nutritive and sensitive powers, making for a more complex material substance. The principle of substantial change is also at work when God creates and infuses the rational soul after the embryo has been sufficiently developed to receive such a soul. The sensitive soul is corrupted, and is replaced by the rational soul, which includes the former vegetative and sensitive powers.[30]

[28] See Aquinas 1961: *SCG* II cap. 89 n. 1749 (ed. Marietti, 255b–256a).

[29] See Aquinas 1961: *SCG* II cap. 89 n. 1752 (ed. Marietti, 256a–b).

[30] See Aquinas 1961: *SCG* II cap. 89 nn. 1745 and 1757 (ed. Marietti, 255a and 256b–257a) and see Pasnau 2004, 100–104.

John Buridan draws a more radical conclusion regarding the question whether human beings can be properly said to bring about offspring. To the objection that if humans were composed of matter and immaterial soul only, they could not be considered the product of their parents' bringing about a substantial change, Buridan bluntly replies that neither males nor their semen are, strictly speaking, a proper agent in bringing about offspring; the male is hardly more than a sperm donor, as it were, since he could die immediately after ejaculation and hence would contribute nothing during the time of substantial generation. But the semen is also not the proper agent in substantial generation since it has a lower state of being which precludes it bringing about a higher state of being, that is, a substance. The male and his semen contribute to the production of a new offspring not substantially, but only instrumentally.[31] Buridan does not think that this is specific to humans with immaterial souls; rather, the soul, whether material or not, whether it the soul of a human or of a horse, produces the material dispositions necessary for an animated body in a remote way at most. Ultimately, any substantial form needs to be induced by what Buridan calls the principal generating agent, which he identifies with the Avicennian *dator formarum*.[32]

We can summarize Aquinas's and Buridan's account on human generation as *Dispositional Procreation*:

(DP) Human beings are involved in generative changes in bringing about the right material dispositions for the human soul, but properly speaking they produce neither the human body nor the human soul.

Both Aquinas's position and Buridan's have to face the following problem: If the created soul produces a different and new body (since the sensitive soul is corrupted when the created intellective soul is infused), how can we say that this generated being is partly produced by human beings? It seems rather that in the last stage of the production of a new human being everything that has been produced so far by human beings will be corrupted when the intellective soul is introduced. For this again is a substantial change: the former body, strictly speaking, will be destroyed, and a new body will come into existence on account of the intellective soul.[33] Thus, it seems ultimately to be God who produces the entire human being by infusing the soul that makes up not only the complete human being but also a completely new human body.

Henry of Ghent and Godfrey of Fontaines take up this worry by giving a different answer to the initial objection faced by Aquinas, namely, that diverse agents and

[31] Because both the male and the female produce the right material dispositions for the infusion of the soul, Buridan ascribes generative powers to both of them. On the biological details, see Beneduce 2020.

[32] See Buridan forthcoming: *QDA(3)* III q. 17 (ed. Klima et al., n. 23) and 2015: *QP(U)* II q. 5 (ed. Strijger and Bakker 2015, 274–6). Cf. Avicenna latinus 1980: *LPP* I tract. 9 cap. 5 (ed. van Riet, 488–494).

[33] In Aquinas, neither (designated) matter nor form nor "dimensive quantities" can guarantee the identity of the embryo from fertilization to birth; see Amerini 2013, ch. 5.

actions cannot result in a thing which is one in being. In order to reconcile the human and divine production of a human being, they conclude that the offspring must be the product of two essentially different actions each of which has its own specific product. Through substantial generation, that is, sexual reproduction, human beings bring about the product of the human body as a substantial composite of matter and form, but God introduces the human soul, thereby making this body an animated being.[34]

Thus, Henry and Godfrey argue for proper *Substantial Generation* of the human body:

(SG) In producing the substantial form of corporeity, the parents bring about the human body, which is then further informed by the immaterial soul as an additional substantial form of the human being.

The crucial difference between Henry and Godfrey on the one hand, and Aquinas and Buridan on the other, is that the former allow for the body to be a substantial composite on its own. Similar to Ockham's *Plurality Thesis*, and contrary to the *Unity Thesis* as defended by Aquinas and Buridan, their conception entails that a human being is the composite of two substantial forms: the form of corporeity which makes up the human body, and the intellect which makes this body an animated being. Human beings have proper vegetative powers of *generation*, since they bring about a new substance, namely, the body as a composite of matter and substantial form. But they do not properly produce a full human being: for the production of this, God is needed to animate the body by infusing the human soul.

10.3.2 How Is Food Digested?

As we have just seen, one problem for vegetative powers of human beings endowed with an immaterial soul is how they can produce a new substance when this composite is supposed to include an immaterial soul which is not naturally generated. A similar problem arises in nutrition and growth once this process is taken to be some kind of substantial change, namely, the corruption of food on the one hand, and the production of new parts of the living being (flesh and bones) on the other, which some authors call a partial generation or an "ad-generation" in the human being. Given that those new parts are also informed by the soul, medieval thinkers are again confronted with the problem of how something can be substantially generated in a human being when this substantial composite includes an immaterial form which is ingenerable. I shall present the accounts of Aquinas and Buridan, which do not involve a partial substantial generation on the part of the human being. The alternative account of Francis of Marchia (1285/90–c. 1344) argues that

[34] See Henry of Ghent 2011: *Quodl.* IV q. 13 (ed. Wilson/Etzkorn, 85–86.) and Godfrey of Fontaines 1904: *Quodl.* II q. 7 (ed. de Wulf/Pelzer, 96).

assimilation of food involves a partial generation of the body and makes an additional form of the body necessary.[35]

As Aquinas says in his commentary on Aristotle's *On Generation and Corruption*, the growth of every living body involves "some particular generation", because food is "converted" into that which is nourished, namely, the animated being which takes in food. Food is nourishing because it "produces increase" by becoming quantified flesh of the body.[36] Closely following Aristotle, Aquinas connects growth and generation only to a certain extent. Growth takes place as the result of the corruption of food, followed by the generation of flesh. Growth, however, differs from generation, for although food is converted into flesh, new portions of flesh are generated *into* already existing flesh, which thus grows. This process is an assimilation of something into something else whereby that which is assimilated to the other is entirely changed into what it is assimilated to. However, that which assimilates it is increased but not generated, Aquinas stresses, taking up Aristotle's analogy of water being poured into wine to illuminate this difference: if we add water to wine the water will be assimilated into the wine, which in turn will increase, but there is no new wine being generated, unlike when the juice of grapes is converted into wine.[37]

Thus, digestion and growth, as the absorption of food into the body, is a substantial change different from other substantial changes, since it involves the corruption of one thing (the food) without the generation of another (the body, or parts of the body, i.e., flesh). The crucial difference is that flesh is not generated separately from food (which would be generation properly speaking) but is assimilated into something that already exists. Nutrition is therefore a special kind of substantial change by means of which one substance is corrupted without a new substance being generated; nevertheless, if new flesh is not just generated but assimilated into already existing flesh, some new portion of the body is produced from something else, a new part of the body which will equally be informed by the soul. In fact, as Aquinas claims in the *Summa theologiae*, the food will be "converted into the substance of the one which is nourished"; that is, once the form of the food ceases to exist, the matter of food will "take on the form of the one which is nourished."[38]

Thus, Aquinas argues for what can be called *Transmateriation*[39]:

(TM) Nutrition involves no substantial generation on the part of the human being, even though the food is substantially corrupted and its matter is assimilated to the body and newly informed by the soul.

Aquinas's approach faces the problem that vegetative powers of the soul seem to bring about a change through nutrition, which falls outside the range of ordinary

---

[35] It is fair to say that Francis is probably a less well-known figure. On his life and work, see Friedman/Schabel 2006.

[36] See Aquinas 1886: *InDGC* prooem. n. 1 (ed. Leon. 3, 261b).

[37] See Aquinas 1886: *InDGC* lib. I cap. 5 lect. 16 n. 3 (ed. Leon. 3, 318a–b).

[38] See Aquinas 1906: *ST* lib. III q. 77 art. 6 (ed. Leon. 12, 202a–b).

[39] I owe this term to Brower 2014, cap. 11.2.

substantial changes since it involves only corruption but no generation.[40] At the same time, Aquinas subscribes to Aristotle's dictum in *De generatione et corruptione* that every generation or coming-to-be of one thing is the corruption or passing-away of another, and vice versa.[41] Yet in nutrition, only the substance of food is corrupted, without there being something generated in the living being. But this tension in Aquinas's account might turn out to be only apparent, given that from an Aristotelian perspective, generation, which is substantial change, and nutrition and growth, which is quantitative change, are two categorically different kinds of change which are interconnected in the corruption of food and simultaneous growth of parts of the body.

However, once one accepts that nutrition involves a substantial change on the side of what nourishes *and* on the side of what is nourished, the question arises of how to conceive of the latter. Applying Aristotle's principle that the generation of one thing implies the corruption of another and vice versa, authors after Aquinas spell out the substantial change in question in the following way: food being destroyed means that the substance of food is corrupted. This corruption of a substance involves the generation of another substance on the part of the living being by absorbing the food. Now, obviously this is not a case of an entirely new living being coming into existence (as in procreation); however, in the process of digestion the living being in question surely undergoes a certain change, which these authors call *partial* substantial generation on the part of the nourished animal. The fact that food will be changed into the substance of the nourished being means that the matter of the food takes on the substantial form of the nourished being, and thereby a new part of this already existing substantial form is generated. Putting the emphasis on the new form as the final result of the generative operation of nutrition—the endpoint (*terminus*), technically speaking—these authors conclude that what is generated must be a part of the soul of the nourished animal. To make this sound less odd, recall what Ockham thought of the vegetative-sensitive soul. According to Ockham, this soul has parts according to the different parts of the body in which it inheres. Thus, when such a part grows or is maintained by nutrition, a new part of the soul is generated. Again, the soul of a living being is not newly produced as a whole, but only a new part of it, by being "ad-generated" to the already existing soul.[42]

But if it is a part of the form which has to be generated, there again arises the difficulty of how to reconcile this with an ingenerable human soul. Buridan addresses this difficulty directly.[43] His solution is that in human beings, nothing new on the side of the soul comes about, and the immaterial human soul simply informs new portions of matter which had previously belonged to the food, but no new part of the soul is generated. Digestion turns out to be a special case for human beings, in

[40] See Brower 2014, 241–245. For more on Aquinas on nutrition, see Cadden 1971.

[41] See Aristotle 1984: *DGC* 318a23–25; 319a7–8, 21–22.

[42] On the question of the soul's extension and presence in the body see de Boer 2013, cap. 5.4.

[43] See Buridan forthcoming: *QDA(3)* III q. 17 (ed. Klima et al., nn. 3–4).

which only the material constitution of the body changes but not the soul. Yet Buridan is confident that this conception is still in line with biological explanations of nutrition and growth. He also thinks that between nutrition in humans and in all other living beings there is a crucial similarity which makes it possible to subsume both human and non-human nutrition under partial substantial change, for in both cases some portion of matter—in this case, the matter of food—is deprived of its form and takes on a new form, namely, the soul of the nourished being. But Buridan leaves unexplained how such a change comes about.[44]

Against this apparent sleight of hand on the part of Buridan, who seems simply to bypass the original problem of how human nutrition can be explained as a case of substantial *generation* and not just a sequence of forms in matter, other authors claim that in order to fully account for digestion in terms of a partial substantial generation of the nourished body, there must be a material form involved which will serve as the endpoint of this change. Since this form cannot be the human immaterial soul, we must posit another substantial form of the body as the point at which partial substantial generation arrives. As seen in the previous section, Henry of Ghent and Godfrey of Fontaines hold that a human being is composed not only of matter and soul, but also an additional form of corporeity. Against the background of generation as substantial change, they argue that in order to make it metaphysically explicable how humans procreate, we must assume that they produce a material form which makes up a new human body to be later informed by the soul.

A similar line of reasoning is employed by Francis of Marchia in explaining how the body of a human being is nourished and grows. Like Buridan, Francis holds that nutrition involves some kind of substantial change and that the natural generation of one thing entails the corruption of another thing, and vice versa. Since the form of food is corrupted, there must be a new form which is generated in the matter that was previously informed by the form of food. Again in line with Buridan, Francis rules out prime matter and the human soul being involved in this change; unlike Buridan, however, he holds that it must be some additional substantial form in the human composite that accounts for what is generated. Francis concludes that nutrition involves proper *Partial Generation* of the substantial form of corporeity:

(PG) For the assimilation of food into the human body to be considered a kind of natural generation, there must be an additional material form of the body which is partly generated from the matter of food.

Prime matter is the substratum of this change, and thus not what is generated; nor is the intellective soul a suitable option, since it is incorruptible and therefore does not need food to be maintained. The only necessary reason to take in food is that its recipient suffers from a lack which must be remedied in order for it not to perish. But if we take it for granted that the human soul is incorruptible it would be absurd to think that it needs this kind of refreshment: even if we starve to death, our soul

[44] See Buridan forthcoming: *QDA(3)* III q. 17 (ed. Klima et al., nn. 18–19). For more on Buridan on human nutrition, see Klein 2020.

will not be destroyed. Thus, to maintain that the soul is nourished by food contradicts the conception both of food as that which maintains what is nourished, and of the human soul as incorruptible.[45] Francis concludes that since neither prime matter nor the human soul can account for what is generated in nutrition, there must an additional form that does so.[46]

Francis pursues a reductive account. He seems to think that if we cannot identify one of the essential parts of the body as that in which nutrition is completed, it does not make much sense to take the body as a whole as the suitable object that undergoes the nutritive change; in Francis's technical language, we must be able to account for an essential part at which nutrition is terminated (*terminatur*). After all, we can metaphysically explain the corruption of food by pointing to its decomposition into its essential parts of prime matter and perishing form. We should equally be able to account for what is generated in a nourished being by focusing on the same metaphysical level of essential parts. Of course, Francis holds that what is in fact nourished is the body, as the composite of matter and form. But in order to be able to say that it is the human *body* which is nourished, Francis argues in favor of a metaphysical composition of prime matter together with not the immaterial soul but the additional substantial form of corporeity.

10.4 Conclusion

This paper has addressed two questions which arose in the late Middle Ages against the background of the immateriality of the human soul: Where should vegetative powers be placed in the metaphysical make-up of human beings? And how can the human vegetative powers account for generation, nutrition, and growth?

As to the first question, medieval thinkers debated whether the vegetative powers in human beings are powers of the body, of the soul, or of both. While all three authors presented—Aquinas, Ockham, and Buridan—agree that vegetative powers are *powers of the soul*, they operate with a certain ambiguity in the term "power of the soul". The ambiguity stems from the Aristotelian conception of living beings as soul-body composites, or animated bodies, in which the one component cannot be really understood without the other. On the one hand, there is a sense in which the soul is said to be active in vegetative operations by making use of the body; on the other hand, it is the composite of body and soul which is said to be that which is capable of performing vegetative operations. Aquinas, Ockham, and Buridan all make clear that the soul is not by itself capable of vegetative operations. This makes sense given that nutrition, growth, and procreation are processes that pertain to bodies. But these processes can happen in bodies only because bodies are animated. All three authors conceive of the soul as the actuator of vegetative operations carried out

[45] See Aristotle 1984: *DA* II.4, 416b17–20.

[46] Francis of Marchia 2012: *In II Sent. (Rep. IIA)* q. 38 nn. 22–24 (ed. Suarez-Nani et al., 121–122).

in the body and *by means* of the body. In Aquinas's account, this actualization is mediated in two senses: the soul acts (i) through its powers, which are distinct from the soul, and (ii) through the body, which the soul uses as an instrument. Ockham and Buridan deny (i) and reinterpret (ii) as involving an extended meaning of the term "power of the soul".

Ockham and Buridan deny that powers are additional intra-psychic entities which mediate between the soul and its activities. The soul, being identical to its powers, is the direct principle of a great variety of its acts. One could say that the soul is itself just a bundle of capabilities. Yet both authors retain a sense in which the soul is active by the mediation of bodily dispositions, which can be called powers of the soul in the extended sense; but since these dispositions are qualitative and quantitative parts of the body they are really distinct from the soul. They are powers of the soul only in relation to the soul's acting, as a means for the soul to act.[47] But are they dispositions simply of the body, or of the body *qua* ensouled? On this question there seems to be a difference between Ockham and Buridan.[48] Ockham seems to hold that there is a sense in which the vegetative powers of the soul *include* bodily dispositions, since the latter are powers of the soul insofar as they serve as a partial cause for the soul to act. Both senses of power of the soul entail the other, which is why Ockham also considers it impossible for an immaterial soul to have vegetative powers; in human beings, this makes an additional soul necessary. Buridan, on the other hand, holds that vegetative powers belong to the soul even separated from a body: they are immaterial and exist as the soul's capability even without the body. Buridan, like Aquinas, thus accepts a sense of vegetative power which does not include reference to the body. At the same time, he seems to accept a sense in which the vegetative powers are just the bodily dispositions themselves regardless of the soul, that is, not of soul and body together, but he hesitates to call them powers of the soul in the proper sense. They seem to be what Aquinas had called the actions of bodily instruments which the soul makes use of in performing vegetative operations as their principal agent.[49]

As to the second question about the vegetative processes performed by the vegetative powers in human beings, late medieval metaphysical investigations of those processes, whatever the biological details, concentrated on the precise nature of the changes involved and what has to be assumed in the metaphysical make-up of a human being in order to account for those processes. While all authors presented here took procreation to lead to a substantial change—namely, the generation of a new human being—they had obvious difficulties in attributing proper *generative* powers to its father and mother, given the theological proviso that it is God who creates the human soul. What they could agree on is that what parents do in sexual

---

[47] See de Boer 2013, 248 for this difference between instrumental powers and powers that emanate from the soul.

[48] Cf. the slight disagreement between Bakker 2019, 71 and de Boer 2013, 250.

[49] See Aquinas 1961: *SCG* II cap. 89 n. 1749 (ed. Marietti, 255b–256a). Thus, Buridan's and Aquinas's accounts of the soul using bodily dispositions as instruments ultimately do not seem to differ substantially, *pace* Wood 2017.

reproduction is at least to produce the material dispositions of the offspring that are suitable for the soul to be infused by God from the outside, as it were. They disagreed, however, on how to conceive of the ontological status of those material dispositions. Aquinas and Buridan hold that human beings are involved in generative changes in bringing about the right material dispositions for the human soul, but properly speaking they produce neither the human body nor the human soul. Against such an account, Henry of Ghent and Godfrey of Fontaines argue for the proper substantial generation of the body: in producing the substantial form of corporeity, the parents bring about the human body, which is then further informed by the immaterial soul as an additional substantial form of the human being.

A similar problem concerned nutrition. Medieval thinkers claim that food is properly corrupted in digestion and that its matter is incorporated to serve the growth and maintenance of the body. But how is the assimilation of food into the body to be explained? Aquinas and Buridan argue for transmateriation: nutrition involves no substantial generation on the part of the human being, although the food is substantially corrupted and its matter is assimilated to the body and newly informed by the soul. Francis of Marchia argues that mere transmateriation is not enough to explain the bodily change in nutrition. On the assumption that the generation of one thing implies the corruption of another, nutrition should involve a substantial generation of a new part of the animated body, that is, a composite of matter and form. Thus, according to this account, for the assimilation of food into the human body to be considered a kind of natural generation, there must be an additional material form of the body which is partly generated from the matter of food.

Ultimately, authors like Henry of Ghent and Francis of Marchia argue that the vegetative processes of generation and nutrition effect a substantial change, and thus bring about a material form, really distinct from the human soul, through a natural process. On such an account, in vegetative processes the soul acts on the body, which is a substantial composite in its own right. The soul seems to be acting merely *on*, but not *within* the body. One can take this also to be a radical consequence drawn from the view of vegetative powers as powers of the soul which make use of bodily instruments, the view maintained by Aquinas and, more radically, by Buridan. This is not yet a Cartesian mechanics of the body, though two theoretical steps towards it were taken in the late Middle Ages: the assumption of complex physical dispositions on which the soul merely acts, and the conception of those dispositions as a body in its own right, that is, a substantial composite of matter and a form which is not the soul. Descartes would later say: Let's study these bodily dispositions as a substance in its own right, irrespective of the soul.

Acknowledgments I am grateful to Paolo Rubini, Ian Drummond, and Can Laurens Löwe for their helpful comments on earlier versions of this chapter.

References

Amerini, Fabrizio. 2013. *Aquinas on the Beginning and End of Human Life*. Cambridge, MA: Harvard University Press.

Avicenna Latinus. 1980. Liber de philosophia prima sive scientia divina V–X, ed. Simone van Riet. Louvain: Peeters.

Aristotle. 1984. *The Complete Works of Aristotle: The Revised Oxford Translation*, ed. Jonathan Barnes. Princeton: Princeton University Press.

Bakker, Paul J.J.M. 2019. The Soul and Its Parts: Debates about the Powers of the Soul. In *Philosophy of Mind in the Late Middle Ages and Renaissance*, ed. Stephan Schmid, 63–82. London: Routledge.

Beneduce, Chiara. 2020. Filosofia naturale e medicina nella teoria buridaniana della generazione. *Rivista di Filosofia Neo-Scolastica* 112 (1): 165–186.

de Boer, Sander W. 2013. *The Science of the Soul. The Commentary Tradition on Aristotle's De anima, c. 1260–1360*. Leuven: Leuven University Press.

Brower, Jeffrey E. 2014. *Aquinas's Ontology of the Material World: Change, Hylomorphism, and Material Objects*. Oxford: Oxford University Press.

Buridan, John. 2010. *Quaestiones Super Libros De Generatione et Corruptione Aristotelis: A Critical Edition with an Introduction*, ed. Michiel Streijger, Paul J.J.M. Bakker, and Johannes M.M.H. Thijssen. Leiden: Brill.

———. forthcoming. *Questions on Aristotle's "On the Soul" by John Buridan: Latin Edition with an Annotated English Translation*, ed. Gyula Klima, John Peter Hartman, Peter G. Sobol, and Jack Zupko. Cham: Springer.

Cadden, Joan. 1971. *The Medieval Philosophy and Biology of Growth: Albertus Magnus, Thomas Aquinas, Albert of Saxony and Marsilius of Inghen on Book I, Chapter V of Aristotle's* De generatione et corruptione. Ph.D. thesis. Indiana University.

Dales, Richard C. 1995. *The Problem of the Rational Soul in the Thirteenth Century*. Leiden: Brill.

Francis of Marchia. 2012. *Quaestiones in secundum librum Sententiarum (Reportatio IIA): Quaestiones 28–49*, ed. Tiziana Suarez-Nani, William O. Duba, Delphine Carron, and Girard J. Etzkorn. Leuven: Leuven University Press.

Friedman, Russell L., and Christopher Schabel, eds. 2006. *Francis of Marchia, Theologian and Philosopher: A Franciscan at the University of Paris in the Early Fourteenth Century*. Leiden: Brill.

Godfrey of Fontaines. 1904. *Les quatre premiers quodlibets de Godefroid de Fontaines*, ed. Maurice de Wulf and Auguste Pelzer. Louvain: Institut supérieur de Philosophie de l'Université.

Henry of Ghent. 2011. *Quodlibet IV*, ed. Gordon Anthony Wilson and Girard J. Etzkorn. Leuven: Leuven University Press.

Klein, Martin. 2019. *Philosophie des Geistes im Spätmittelalter: Intellekt, Materie und Intentionalität bei Johannes Buridan*. Leiden: Brill.

———. 2020. Digestive Problems: John Buridan on Human Nutrition. In *Nutrition and Nutritive Soul in Aristotle and Aristotelianism*, ed. Giouli Korobili and Roberto Lo Presti, 259–284. Berlin: W. de Gruyter.

Pasnau, Robert. 2004. *Thomas Aquinas on Human Nature: A Philosophical Study of Summa Theologiae Ia 75–89*. Cambridge: Cambridge University Press.

Perler, Dominik. 2015. Faculties in Medieval Philosophy. In *The Faculties: A History*, ed. Dominik Perler, 97–139. Oxford: Oxford University Press.

Polansky, Ronald. 2007. *Aristotle's De anima*. Cambridge: Cambridge University Press.

Schneider, Theodor. 1973. Die Einheit des Menschen: Die anthropologische Formel ‚anima forma corporis' im sogenannten Korrektorienstreit und bei Petrus Johannis Olivi. Ein Beitrag zur Vorgeschichte des Konzils von Vienne. Münster: Aschendorff.

Silva, José Filipe. 2012. *Robert Kilwardby on the Human Soul: Plurality of Forms and Censorship in the Thirteenth Century*. Brill: Leiden.

Thomas Aquinas. 1886. *Commentaria in libros Aristotelis De caelo et mundo, De generatione et corruptione, et Meteorologicorum*, ed. Commissio Leonina. Opera omnia, vol. 3. Rome: Ex Typographia Polyglotta S. C. de Propaganda Fide.

———. 1889. *Pars prima Summae Theologiae a quaestione L ad quaestionem CXIX*, ed. Commissio Leonina. Opera omnia, vol. 5. Rome: Ex Typographia Polyglotta S. C. de Propaganda Fide.

———. 1906. *Tertia pars Summae Theologiae a quaestione LX ad quaestionem XC*, ed. Commissio Leonina. Opera omnia, vol. 12. Rome: Ex Typographia Polyglotta S. C. de Propaganda Fide.

———. 1961. *Liber de veritate catholicae fidei contra errores infidelium seu Summa contra Gentiles*, ed. Ceslai Pera. Turin: Marietti.

William of Ockham. 1980. *Quodlibeta septem*, ed. Joseph C. Wey. Opera theologica, vol. 9. St. Bonaventure: The Franciscan Institute.

———. 1981. *Quaestiones in librum secundum Sententiarum (Reportatio)*, ed. Gedeon Gál and Rega Wood. Opera theologica, vol. 5. St. Bonaventure: The Franciscan Institute.

———. 1982. *Quaestiones in librum tertium Sententiarum (Reportatio)*, ed. Francis E. Kelley and Girard I. Etzkorn. Opera theologica, vol. 6. St. Bonaventure: The Franciscan Institute.

Wood, Adam. 2011. The Faculties of the Soul and Some Medieval Mind-Body Problems. *The Thomist* 75: 285–636.

———. 2017. Aquinas vs. Buridan on the Substance and Powers of the Soul. In *Questions on the Soul by John Buridan and Others*, ed. Gyula Klima, 77–93. Cham: Springer.

Zavalloni, Roberto. 1951. Richard de Mediavilla et la controverse sur la pluralité des formes: Texts inédites et études critiques. Louvain: Éditions de l'Institut Supérieur de Philosophie.

Chapter 11
The Jesuit Cultivation of Vegetative Souls: Leonard Lessius (1554–1623) on a Sober Diet

Cristiano Casalini and Laura Madella

Abstract A useful prism through which one can examine the positive Jesuit attitude toward the human condition is that of the nature and operations of the vegetative soul. Rather than neglecting aspects that pertained presumably to the lowest functions of human life, Jesuit theologians dealt with issues of nutrition and generation in a variety of their works. Sometimes, Jesuit theologians even went beyond their own teaching interests to put the vegetative soul at the very center of their research. This contribution will inquire into a successful long seller during the seventeenth century as an example of such literature. In 1613, the Dutch theologian Leonard Lessius (1554–1623) published the *Hygiasticon seu vera ratio valetudinis bonae et vitae ad extremam senectute conservandam* (Antwerp), a peculiar treatise that was strictly focused on nutrition and the best ways to optimize it for the physical and spiritual health of human beings. Although such an outcome was the fruit of a moral habit rooted in human learning and observation, Lessius also connected it to traditional virtues such as "temperance" and "sobriety." By doing this, Lessius seemed to provide an implicit correction to the ranking of human functions–at least from the practical aspects of living a Christian life–, as nothing superior could be performed without rendering appropriate respect to the lower powers. As was true of any being in this world, humans were to be primarily gardeners of their souls.

C. Casalini (✉)
Institute for Advanced Jesuit Studies, Boston College, Chestnut Hill, MA, USA
e-mail: casalini@bc.edu

L. Madella
Department of Education, Parma University, Parma, Italy
e-mail: laura.madella@unipr.it

F. Baldassarri, A. Blank (eds.), *Vegetative Powers*, International Archives of the History of Ideas Archives internationales d'histoire des idées 234,
https://doi.org/10.1007/978-3-030-69709-9_11

11.1 Introduction

During the first century after the inception of the order (1540), the Jesuits had often to face theological accusations of holding overly positive views about post-lapsarian human nature—views that were often labelled by their enemies as "Pelagian," "semi-Pelagian," and the like. Among the most famous of the polemics branding the Jesuits with such a stigma were two parallel and roughly simultaneous controversies that occurred in Spain and Low Countries: in 1581, Dominican and Jesuit theologians in Spain began lashing out at each other regarding the soteriological doctrines of Luis de Molina, while at Leuven (1587), Jansen, Baianists, and the theological faculty accused Molina, Suárez, and Leonard Lessius, champions of Jesuit moral theology, of neo-Pelagianism.

Of course, a wide array of distinctions and details differentiate both these controversies and the doctrines of the individual Jesuit theologians involved. Yet some common traits can be identified and collected together under the more general rubric of a shared positive anthropology, one in which human beings are seen as not entirely corrupted in their post-lapsarian condition.

Compared to the Augustinian anthropologies that Protestants and, within Catholic culture, Jansenists and Baianists held, the balance between natural and supernatural that was characteristic of Jesuit theologians seemed to incline toward virtues connected to the creatural nature of human beings. That nature was one that, in the late sixteenth century, Jesuit theologians increasingly investigated the physiological aspects of, sometimes adopting patterns and methods that usually belonged to the domains of natural philosophy and medicine.

What were the implications of such speculations for the philosophy and the mentality of the Jesuits in general?

A useful prism through which one can examine the positive Jesuit attitude toward the human condition is that of the nature and operations of the vegetative soul. One might think that, as they constituted the humblest of the three levels of the soul's functioning, the vegetative powers of human beings would have been hardly considered by Jesuit theologians in matters that pertained to soteriology or even religious life. But on the contrary, issues dealing with nutrition and generation do appear in a variety of such works. Sometimes, Jesuit theologians even went beyond their own teaching interests to put the vegetative soul at the very center of their research. In 1613, for example, the Dutch theologian Leonard Lessius (1554–1623) published the *Hygiasticon seu vera ratio valetudinis bonae et vitae ad extremam senectute conservandam* (Antwerp: 1614),[1] a peculiar treatise that was strictly focused on nutrition and the best ways to optimize it for the physical and spiritual health of human beings. Although such an outcome was the fruit of a moral habit rooted in human learning and observation, Lessius also connected it to traditional virtues such as "temperance" and "sobriety."

---

[1] We quote from the first English translation (Lessius 1634).

In Lessius's account, the cultivation of a "sober life" was not a moral virtue because it mortified the flesh, as would have been implied by the traditional Pauline opposition between righteous spirit and sinful body. Rather, sobriety was a virtue because it was natural, in the sense that it obeyed the inner law of human nature.

In this contribution, we will inquire into Lessius's argument dignifying the supposedly lowest functions of the human soul and show how that ennoblement was enrooted in the Jesuit culture of the body, which had its origins in Ignatius of Loyola's own spirituality and involved a positive attitude toward the mainstream medical culture of humanism, particularly Galenic taxonomies of temperamental qualities.

11.2 Jesuits and Galenism

The influence of Galen and Galenism on early Jesuit culture cannot be overstated. Partly as an outcome of exposure to humanistic culture and partly as a useful tool for organization management, Galen's theory of humors, temperaments, and consequently psychology recur with distinctive frequency in Jesuit commentaries and treatises as well as in the bureaucratic documents of the Society, as is clearly exemplified by the content of the order's triennial catalogs (Casalini 2016; Ferreira Panazzolo and Massimi 2015).

Sometimes, Galen's influence surfaces in an indirect way, as in the case of Jesuit commentaries on the soul that (quite unusually) raise the question of whether or not the human soul coincides with temperament (Santiago de Carvalho 2016).[2] This had to be so, as any other solution could prove nothing but heterodox. Yet the very presence of such a question evidences the important role that temperament was seen to play in the functioning of human souls.

The core of Galen's humoral theory of medicine was that restoring the balance of humors in a body could remedy disease. An excess of one humor over the others resulted in—or at least accompanied—illness, with the kind of illness being determined by the humor that was in excess. This was especially true for those humors that were supposed to be produced as an evil waste of badly performed physiological processes. According to Galenists, the goal of the physician was to restore balance by offsetting or limiting the offending humor. There were three ways to achieve this, two of them based on the natural intercourse between bodies and their environment and one the product of "violent" human intervention. The natural ways were changing climate and diet, while the violent way was purging the body.

If the physician's goal was simply to restore balance, one could infer that everybody's perfect balance was exactly the same. But Galen's *Quod animi mores temperamenta sequantur* argued instead that each person was characterized by a

[2] One example of such is Manuel de Goís's commentary on *De Anima* (1598), which resolves the question by denying that the soul can be so reduced.

predominant disposition toward a particular humor. This predominance was reflected in a person's general physical complexion and—most importantly, from the psychological and moral points of view—in his or her peculiar character and behavior.

The determinism implied by such a theory was radicalized in the sixteenth century by the Spanish physician Juan Huarte de San Juan (1529–1588), whose *Examen de ingenios* (Baeza, 1575) connected Galen's theory of temperaments to the functioning of rational powers, thereby outlining a taxonomy of human talents based on the prevalence of particular humors in the body. As Casalini has argued elsewhere, Huarte's theory had a huge impact on Jesuit culture, and—despite the officially necessary rejection of it as potentially heterodox in its neglect of free will—many Jesuits adopted Huarte's taxonomic framework both in theory and in practice (Casalini 2016).[3]

This insertion of Galenism into Jesuit culture was probably made easier by the general attitude of the first Jesuits, and of Ignatius of Loyola in particular, toward the body and its role in human spirituality and religious life. After Ignatius's rejection of the fasts, vigils, and mortification of the flesh that he had initially practiced—a move John O'Malley rightly calls Ignatius's "reconciliation with the world" (O'Malley 2013, 112)—the care of the body came to permeate Ignatius's style of spiritual direction and, later on, his regulations for the daily routine of his fellow Jesuits. What had looked like a series of practices of self-abnegation by a humble creature before his God now seemed to Ignatius to be public and immodest exhibitions of sanctity that reeked of pride and bordered on scandal. Hence, his conversion to the rule of the "golden mean" (*aurea mediocritas*): for sake of human spirituality, any excess in behavior was to be avoided, including those practices that were ostensibly undertaken to honor God by mortifying the sinner's flesh. Rest, physical exercise, movement, and vacations counterpointed intellectual activities, spiritual retreats, and general pedagogy in Jesuit schools (Casalini 2014). And as Claudio Ferlan points out, appropriate nutrition became a tenet of Jesuit spiritual culture and the Jesuit mentality (Ferlan 2017).

Many times did Ignatius warn his fellows to limit their spiritual "intemperances," often recommending them to abstain from excesses in devotion that would weaken their bodies (Loyola 2006, 170). This was certainly intended to help the person so advised to become a perfect instrument of divinity—that is, someone called to cooperate with Providence by acting in this world for the greater glory of God (Van Ginhoven Rey 2013). Preserving the human temple, that is, the body as "spiritual experience's place" (Emonet 2006), keeping oneself healthy, and even exercising (a term that often recurs in relation to both spirit and body in Ignatius's writings), were elevated to a religious obligation.

[3] Huarte's theories reappeared in Antonio Possevino's *Coltura degli ingegni* (Vicenza: 1598)—drawn from the *Bibliotheca selecta*, in Antonio Zara's *Anatomia ingeniorum et scientiarum* (Venice: 1615)—and found application in the Jesuit triennial catalogs, where each member of the Society was briefly described according to his complexion and temperament.

In a letter he wrote to Francis Borgia on September 20, 1548, Ignatius tried to discourage Borgia's inclination toward extreme fasts and rigorous abstinence and clearly connected the care of the body to the service of God:

> I see that these fasts and abstinences keep the stomach from functioning naturally and from digesting ordinary meats or other foods which supply proper sustenance to the body. Instead, I would seek every possible means to strengthen the body, eating any permissible foods and [doing so] as frequently as you find them beneficial (barring scandal to the neighbor). For we ought to cherish and love the body insofar as it obeys and serves the soul, and insofar as with the body's help and obedience the soul becomes more fitted for the service and praise of our Creator (Loyola 2006, 255).

But, as even the Spiritual Exercises made clear, each person had a peculiar blend of physical structure, complexion, and character. Finding the right measure between preserving the body and challenging it for sake of devotional abstinence was a matter of attentive and individual discernment rather than a general rule that could be applied to every case. In other words, human beings were all different, and distinguishing between excesses and the right mean was a matter of individual judgment that echoed the etiological art of the physician.

11.3 The *Hygiasticon*

At the time when Lessius published the *Hygiasticon*, he had just left behind the most eminent chapter of his career as a professor of theology. After entering the Society of Jesus in 1572, he first taught philosophy at the English College in Douai. He subsequently left for Rome, where he attended the Roman College under the guidance of Robert Bellarmine and Francisco Suárez (Van Sull 1930, 328–341; Chamberlain 1939). Returning to Brabant in 1585, he took the chair of theology at Louvain, where he introduced pedagogical methods that would be emulated elsewhere in the Society, such as holding lectures on Aquinas instead of Peter Lombard (as tradition commanded) (De Smet 2001; Decock and van Houdt 2005; Decock 2009). Despite the unremitted waning of his frail constitution, during the first decade of seventeenth century Lessius wrote and published significant works on grace, free will, and ethics, which promptly reached a wide audience and anticipated his later productivity. His *De justitia et jure cæterisque virtutibus* (1605) is regarded as a milestone among the early European forerunners of modern economic thought (Gordon 1975; Decock 2016), while his *De gratia efficaci* (1610) engaged him in the notorious controversy on grace.

The *Hygiasticon* seems to abandon theological and speculative topics and mark a shift in Lessius's attention to the mechanics and habits of nutrition, a branch of knowledge that seemed mostly confined within worldly borders (examples of this literature include Calanius 1550; Boorde 1542; Rangoni [1550?]; Placotomus 1550;

Valverde de Amusco 1552; Bruyérin-Champier 1560; Pisanelli 1583).[4] This was made even more evident by the author's inclusion of an appendix entitled the *Treatise of a Sober Life* (1558) (Cornaro 1558), written by a Venetian layman and businessman named Luigi Cornaro (d.1566). Lessius claimed that Cornaro's ideas were coherent with the *Hygiasticon,* and they even possibly constituted a case study, as Cornaro scaffolded his doctrine with examples from his own experience.

Lessius's short presentation of Cornaro informs the reader that even though the Italian was a layman who lacked any doctoral certification in medicine, he was very well endowed with that practical wisdom that was required for the topic of Lessius's book. A wealthy landowner, Cornaro had thrived thanks to his skills in hydraulic engineering and his entrepreneurial spirit, while also cultivating a wide range of artistic and cultural interests and patronizing writers and musicians in his Paduan villa (Fiocco 1965; Gullino 1983; Menegazzo 1980; Puppi 1980; Bergdolt 1991; Milani 2014a, b). Over time, information about his age has been inferred from what he stated in his writings;[5] for a long time he was supposed to have been born in 1464 and dead at the venerable age of 102. Such extraordinary longevity served as evidence of his theories, lending the *Sober Life* great authority. Archival research recounts a different story, though, establishing 1484 as the most reliable year of Cornaro's birth.[6]

Due perhaps to Lessius's reputation, or perhaps to the power of its argument, or perhaps even to Cornaro's "human interest story," the *Hygiasticon* achieved immediate success in both Catholic and Protestant countries.[7]

Lessius laid out the *Hygiasticon* with three fundamental goals in mind: to provide an exact definition of a sober life, to inquire into the ways by which one could determine the just measure that should be observed in order to attain it, and to list the reasons for and benefits of doing so.

A concern to establish the epistemological basis of his work dominates in the background of Lessius's argument, as evidenced by his inclusion of declarations of approval from multiple physicians as forewords. The topic of how to live a long and healthy life must have seemed a very unusual one for an expert theologian—particularly a Jesuit, given the fact that the Society of Jesus neither ran schools of medicine nor even provided any teaching of medicine in its school network at the time.[8]

The origin of Lessius's felt need to justify his incursion into a typically medical field was twofold. On one side, he wished to show the general public that the content

[4] Authors of works on nutrition and health were mostly physicians. Tommaso Rangoni was a physician and philosopher who might have inspired Cornaro's *Sober Life* (Minuzzi 2013).

[5] In different writings, though, he declared different years of birth.

[6] Emilio Menegazzo stated 1484 to be the most likely because it appears on a paper signed by Cornaro before the Venetian Court (which made it an official document in which he was supposed to tell the truth).

[7] A 1614 print of *Hygiasticon* was even held in a Jesuit library of Bejing (Golvers 2011).

[8] For an overview of medicine in Jesuit education in sixteenth and seventeenth centuries, see: O'Malley 1993, 200–242; Grendler 2018, 23–37; Welie 2003; Sander 2014.

of his book was *scientifically* approved, and that he had the expertise needed to deal with his topic. On the other side, he sought to reassure potential theological readers who were acquainted with his soteriological works that the topic of the *Hygiasticon* was one that was of utmost importance from a moral and religious point of view (Lessius 1634, 10–11).[9]

It is in fact this latter point that is of greatest significance for this contribution. Dietetics as an art for living a long and healthy life was considered by Lessius to fall within the spectrum of moral theology, insofar as the observance of dietetic rules was fundamentally intertwined with the cultivation of the cardinal virtue of temperance and therefore most necessary for religious people seeking to live a life of contemplation and prayer. Moreover, both the structure and the dynamics of sobriety (i.e., proper and virtuous human nutrition) lent themselves to an institute of religious life in which self-restraint was intended to help the cleric (or monk) in the performance of his spiritual and intellectual duties by establishing a positive and healthy relation to his "fleshly" condition rather than by pushing him to mortify the flesh in order to root out the vice of gluttony. In other words, what Lessius sketches in his address to his religious readers is a perspective on religious practices that was quite consistent with the tradition of the Society of Jesus and with Ignatius of Loyola's own spirituality.

Lessius extends this perspective beyond the domain of religious orders, including clerics in his account for a sober life. In fact, the social status of intellectuals in northern Europe might have influenced Lessius here, as from the Middle Ages on intellectuals—who were mostly associated with universities—were usually clerics who professed simple vows. According to Lessius, a sober diet was quite necessary for those whose activities were prevalently spiritual and intellectual, by this meaning that in order to perform all the activities of the mind, attentive care in one's diet was required. The social distinction Lessius's argument evoked was one between intellectuals and "mechanical" people, defined by whether reason or movement was predominated in their life.

From a theological point of view, though, diet was strictly connected to soteriology, in the sense that—according to Lessius—it made a person apt to fulfill his or her mission as a perfect instrument of God. Rather than a mere act of moral self-restraint aimed at gaining merits for the afterlife, following a proper diet was an anthropological precondition for the accomplishment of God's unfathomable design. It is true that, as Julia Fleming argues, "never in the *Hygiasticon* does Lessius suggest that there is any moral obligation to pursue long life as a good in itself" (Fleming 2008, 110). Yet however "secondary" a goal living a long life might be, it was a predictable side effect of that which Lessius *does* enjoin as a necessary basis for something every Christian was obligated to provide—namely, a body suitable for cooperation with God's supernatural grace. Therefore, the *Hygiasticon*'s

[9] "I would not have any man to think strange of the matter, that I being a professed Divine should take upon me to write of this subject. For besides that I have long ago made some good progresse in the Theorie of Physick, this matter is on way discrepant to the profession of a Divine: in regard that it is the divine virtue of Temperance, which is chiefly in question."

public became theologically universal, and the cultivation of the supposedly lowest functions of the human soul got elevated in the rank of moral dignity to an indispensable precondition for any other virtue.

11.4 The Characteristics of Diet

Lessius believed that the key to living a long and healthy life was living soberly—that is, following a diet. The connection between longevity and health was neither absolutely necessary nor universal, as the theologian concedes that there were a few people who reached impressive ages despite being plagued by illness. But inasmuch as such a condition was clearly undesirable from a worldly point of view, even those people who happened to grow old in spite of what they ate and drank—whom Lessius declares to be exceptionally few—were models that good Christians should avoid, insofar as attaining a long life without sobriety was nothing but the endurance of affliction.

This moral rejection of distempered lives reveals a tenet of Lessius's doctrines regarding human psychology, as it posits an apparently circular relation between vegetative and rational powers (thus echoing the classic proverb *mens sana in corpore sano* that humanists also used to quote). From one side, a sober life was generally possible only when the superior functions of the soul led and constrained the lower ones. In this sense, Lessius follows the mainstream tradition of granting understanding and reason primacy over the lower faculties of the sensitive soul. From the other side, though, Lessius seems to state that without the proper functioning of the vegetative powers, and in particular without proper nutrition, neither memory, nor imagination, nor understanding could perform properly and attain their final goal (i.e., truth) (Lessius 1634, 154).[10] But a sober diet (despite Lessius's loose usage of the term 'abstinence' as a synonym) constituted a middle way between the overindulgence of gluttony and the excessive privation required by some medical regimens or spiritual devotions. Both modes of excess were not only irrational, as traditionally claimed, but also unnatural, as they ran contrary to the natural course of human nutrition and health. Therefore, the sovereignty that traditional spirituality and religious orders' ethics required to be exerted by the rational powers over (and yet somehow also against) the lower ones was, in Lessius's thought, not only limited but in fact reversed, becoming dependent to the natural functioning of "less dignified" powers.

We will return to these aspects of human psychology and theological anthropology (insofar as the question of what kinds of human powers are required for salvation is concerned) later on in this chapter.

The *Hygiasticon* reflects the general attitude of the Jesuits toward Galen's theory of humors and physiology, but—in the context of curing a distempered

[10] "The affections of the minde follow the apprehension of the Fancie: now the apprehension of the Fancie is conformable to the disposition of the Bodie, and to the Humours that are predominant therein."

balance—Lessius limits the factors that impacted the human body to diet and purges, thus excluding the influence of environment or climate that other Galenists and contemporary common sense held to be important as well. The reason Lessius excludes climate is a physiological bias, as he traces almost every sort of illness to digestive difficulties (i.e., difficulties of the stomach), perhaps because he identifies *pepsis* as the core issue (Galen 2003, 15).[11]

His definition of diet accordingly concerns only those substances that had physical contact with the stomach:

> We call that a Sober Life or diet, which sets stint not onely in drink, but also in meat: so that a man must neither eat not drink any more, then the constitution of his bodie allowes, with reference to the service of his minde (Lessius 1634, 14–15).

Lessius's definition makes it clear that the "right measure" that constituted a sober life ultimately depended on the condition of the stomach, with reference to each person's complexion, age, and activities (Lessius 1634, 16).[12] For different proportions (i.e., portions) were due to youths, adults, and the elderly, to the sick and the sound, to the phlegmatic and the choleric, and so forth. Such differences were necessary to keep in mind when looking for each person's right measure, as that measure was largely dependent upon the peculiar proportion of humors with which each person was naturally endowed.

Coherently with the idea that the natural balance of humors was unique to each person rather than universal, Lessius adopts the concept of *proportionality* in seeking the right ratio between humors (Lessius 1634, 99).[13] Although he repeatedly affirms the importance of considering the particular activities that each person usually performed, this came logically second in terms of finding the proper diet—for as we have seen in both Galen and Huarte, a person's activities and behavior usually followed their dominant temperament. Although Lessius does not explicitly mention it, this doctrine is clearly in the background when he makes statements such as "Phlegme is most of all other things contrary to the functions of the minde" (Lessius 1634, 150) and the like.

The particular balance of humors in each body determined not only its complexion but also the structure and functions of its internal organs. The physiology of the stomach especially was crucial, as that organ needed to be capable of perfectly concocting and digesting food "in the midst of any employment either of minde or

---

[11] "Pepsis was thought of in much the same light as cooking, and resulted from the action of the body's innate heat. For this reason it is translated here as concoction, rather than the digestion of some translators, as I discuss later. The stomach, Galen says, is the particular instrument (*organon*) of concoction."

[12] "This measure is not the same in respect of the quantitie in all sorts of people, but very different according to the diversitie of complexions in sundry persons, and of youth and strength in the self same bodie."

[13] "[A sober life] tempers the humours, and maintains them in an equall proportion, that they offend not any way either in quantitie or qualitie. Now where there is an agreeable proportionablenesse amongst the humours, there is no mater for sicknesse to work upon: inasmuch as the ground of health lies in this, That the humours be rightly and proportionably tempered in the bodie."

bodie" (Lessius 1634, 17). That function was vital for the sufficient nourishment of the body and had to occur without leaving any badly digested substances to spread into other organs and impede the smooth circulation of spirits (Lessius 1634, 29, 40).[14] Still, Lessius suggests that the work of the stomach differed between individuals, according to the fundamental temperament of each. A diet should therefore be attentive to the food each person was best suited for, as that criterion helped determine the proper measure for each particular stomach—that is, the amount it could process without harming the body and the superior powers of the soul.

Referring to the functions of the mind in order to determine the right measure of food and drink for the body implied a biological interaction between mind and body during the process of digestion. Lessius says that "the exercises and employments of the minde do very much hinder and disturb the concoction [digestion]" (Lessius 1634, 18). He goes on to identify two possible mechanisms for this: (1) the mind's abating and suspension of the inferior faculties when the whole force of the soul was called up, as when one prayed and didn't notice a clock ringing, and (2) the mind's withdrawal of not only the animal but also the vital and natural spirits from their proper service when it was intensely engaged (Lessius 1634, 19).

Any impediment or offense to the superior faculties arose from an overabundance of "vapors," which were sent up into the head chiefly out of the stomach; according to Lessius, who refers here to the common experience of his time, they would be but sparingly sent up if the right measure were not exceeded (Lessius 1634, 32). As Lessius said, the primary source of such vapors was meat, but they did not proceed merely from meat just taken (which immediately began to boil and concoct); they also proceeded from an abundance of blood and other humors in the liver, the spleen, and the veins, which together with the meat would send up a great quantity of fumes (Lessius 1634, 33–34).

A sober diet would gradually diminish this abundance of humors, abating their "ill moisture" and reducing them to their due proportions in both quantity and quality; as a result, noxious fumes would no longer arise upon eating. According to Lessius:

> When Nature doth perfectly govern all the humours of the bodie by the ministerie of the vegetative faculties, she doth so order and dispense all things, as neither any diseases arise in the bodie, nor any impediment follows to the superior offices and duties of the soul (Lessius 1634, 34–35).

Breaking with a long line of Renaissance physicians and natural philosophers, Lessius limits his concerns to the precise quantity of food that one should consume. Quality does not worry him, except for some particular meats that he deems blatantly harmful for the organism—specifically, those that he names "fat food." This latter concern arguably places his doctrine in the vicinity of those whose authors asserted

---

[14] "Now, that is alwayes excesse, which proves more in quantitie, then the stomack can perfectly digest, without leaving any crudities at all behind." Lessius, *Hygiasticon,* 29. And "ill humours do cloy up the muscles and the nerves, through which the spirits have their course and passage: whereby it comes to passe, that the animall spirits . . . cannot freely take their course, nor govern and order the bodies as they ought" (40).

the primacy of "subtle" food over "gross" meats. Huarte, Antonio Persio (1542–1612), and Girolamo Cardano (1501–1576) all devoted treatises to the concept of subtlety, a category and a concept that encompassed everything from fine abstract arguments (those that were both elegant and eloquent) down to the material aspects of human life, including the characteristics of the body and of food. Their emphasis on the quality of food was indeed consistent with Galen's theory of humors, insofar as they strictly connected that quality to the schemata of the four qualities (hot and cold, moist and dry) and the four elements (air, water, earth, and fire). Presumably, this should have had some relevance for dietetics. This is not the case in Lessius, though, as he took a person's familiarity with whatever kind of food they were currently eating to be a sufficient criterion for keeping that routine, allowing him to focus solely on the amount of food consumed each day (Lessius 1634, 57).[15]

One suspects, though, that Lessius might have waivered in his exclusive emphasis on quantity, for in his final definition a sober diet, he included "quality" among the criteria:

> [A good constitution of the body and health] consist in these two things, to wit, in the due proportion and symmetrie of the humours, both in respect of their quantitie and qualitie; and in a certain spongie kinde of disposition throughout the whole bodie, having no let nor impediment by obsructions, so that the spirits and bloud have their free passage and recourse through all parts (Lessius 1634, 111).

If what constituted a "proportioned disposition" was different for each temperament, one wonders how it could be possible to find the right balance for one's humors by varying merely the dosage of food. But even if quality of food remains secondary in Lessius's theory of dietetics, he does make some suggestions for anyone whose activities were mainly sedentary and intellectual. Meat, for example, was to be well-done and have no raw parts. Moreover, fat food was to be generally avoided (as noted above), but this was not a universally binding rule. Lessius does not in fact forbid one to indulge in "a little of Colewort, Onyons, Cheese, Beans, Pease, and the like; although they naturally breed melancholy, choler, slime, and windinesse" (Lessius 1634, 61). Lessius's diet was evidently rooted in the tradition of Continental cuisine, as meat and dairy are prevalent among his suggestions. Still, he admits that the diets of several nations, mostly exotic ones, were based on vegetables, and that this did not impede their people in growing old and staying healthy. Specifically mentioned are the Japanese, the Chinese, Africans of several regions, and the Turks, all of whom mostly fed themselves with rice and fruits. But the same could be said of "mechanical" people within Christendom, who ordinarily fed on "bread, butter, pottage, pulse, herbs, cheese, and the like, eating flesh very rarely" (Lessius 1634, 66).

It is curious to note that Lessius does not provide any further specification about the kind of meat one should prefer, as other dietologists of his time did, and that

[15] "Questionless a man may live long and healthfully on bread onely, with milk, butter, cheese, and beere; especially if he has from his childehood been used unto them."

even his own fellow Jesuits in charge of colleges were keen to do (Sumillera 2014, 295, 224, 305, 309).[16]

As an example of the latter, consider the case of Giuseppe Cortesono (c.1537–1571). In outlining the constitutions for the German college in Rome, he provides extensive recommendations about the kind of food that students should have in the refectory, offering along the way some insights into the underlying social divide in Jesuit boarding schools between students with different social and cultural heritages. Veal, for example, was to be cooked differently for noble and common students:

> To satisfy many nobles who gather there, there could be established a separate table where they would have the same size of serving but of some-what better quality; for example, the appetizers might be a bit elegant and they might get more tender veal roasted when the others have it boiled; when it comes roasted, in the later dish there should be the same consideration as with the appetizer, but additionally they would contribute more to the expenses than the others do (Cortesono 2016, 154).

Cortesono's document also makes mention of a dish that permeated the food culture of Jesuit schools at the time and greatly intrigues Lessius. That food was *panada*, or *panatella,* which was nothing but a traditional soup made by combining bread with water or broil and whose origins apparently traced back to the peasant cuisine of the central Italian countryside. This soup was generally meant to be curative but was in fact so appreciated by the students that—as Cortesono warns—they often indulged in fake illnesses so as to abuse it (Cortesono 2016, 149).[17] Lessius praises the excellent qualities of such a dish for weak and aged persons as well, pointing out that a diet based on panada, with the addition of one or two eggs, could help people live a long and healthy life.

The reasons Lessius looks so favorably on this sort of food are that it was light and easy to digest, most temperate in its qualities, and—as he remarks—"little subject to putrefaction and corruption" (Lessius 1634, 56). Moreover, it was supposed to produce an abundance of good blood. The fact that panada was a soup, and so a blend of dry and liquid food, challenged Lessius in his determination of the quantity of it that a daily sober diet should include. Indeed, he acknowledges that some might object to his prescription, since after deducting the water or broil, but a few ounces of food remained. But he replies that, just as when meat and drink are mingled, they have to be severely weighed and reduced to the just measures of the kinds to which they properly belong, so must the liquid portion of panada be treated as liquor, and the bread and the like as meat (Lessius 1634, 56).

---

[16] The first English translation of the Examen has recently been reprinted in Richard Carew. Huarte affirms that there are animals, like "hens, partridges, turtles, doves, thrushes, blackbirds and goats," whose meat, warm and dry, "impinges on the man's seed and make it apt to generate intelligent male offspring" (295); among them, "the stomach cannot suffer too much hens and partridges, because of their overly subtle substance" (224), while "cattles and swines are easy to digest but detrimental to the rational soul" (305, 309).

[17] "One should watch over those who might pretend to be ill in order to get out of classes, or to sleep a little more, healing them by means of *panatella* and *acquacotta*."

It is interesting to note that in the matter of the qualities of foods, Lessius suggests that he is simply relaying Luigi Cornaro's advice. This is true to a certain extent but needs some clarification. In the Latin translation of *Sober Life*, the first among Cornaro's treatises and the one Lessius appended to the *Hygiasticon*,[18] the Venetian does not mention panatella nor any other specific food but only the general bread, meat, eggs, and soup (Cornaro 1558, c. 10v). But in Cornaro's subsequent *Digest of the Sober Life* (1561), we find many more details regarding his diet and a recurring commendation of panatella, which he deems to be not only nutritious and easy to digest but also affordable for poor people (Cornaro 1561).[19] In addition, Lessius's translation of the original 1558 treatise incorporates some additional pages at the end, extracted from the so-called fourth *Discourse on Sober Life* (Cornaro 1591, c. 22v), in which we actually find panatella recommended for clergymen's diets.[20]

The example of panatella lets Lessius specify the golden rule of his diet. According to him, a sober diet should not exceed fourteen ounces of food per day, but more precise calibrations for each individual depended on a variety of aspects that were ultimately based on that person's temperament.

In fact, major variations in the course of a person's life significantly impacted his or her right measure of food and drink. Old age, for instance, was accompanied by a "consumption of the radical humor and native heat" (Lessius 1634, 151) and by a larger dispersion of phlegm throughout the body, which together reduced the ability of the brain to exercise its rational powers. Memory above all was hurt by phlegm, a cold humor possessing the brain, for it both stopped up the narrow passages of the spirits and "benumme[d] the spirits themselves, making them sluggish: whereby the apprehension of the minde bec[a]me slow, languid, and inconstant" (Lessius 1634, 164–165). Moreover, old age brought with it weaker physical powers, causing a decline in physical activity and movement, with the result that less food was required for the body's maintenance.

Another factor that would presumably have impacted the proper diet was the sex of the person following it, as all Galenists agreed that the qualities of females were generally cold and moist, while those of males were hot and dry. This distinction does not, however, appear at all in Lessius's account, as his addressees were supposedly all male.

Routine and costume were additional factors that Lessius believed to exert strong influence over the proper diet, as they apparently weakened the impact of nature on the whole person. He thus warmly recommends, for example, that the diet be

---

[18] After the positive response to his 1558 *Treatise*, Cornaro wrote at least three further texts on the same subject.

[19] "... & li miei cibi sono questi. Prima il pane, la panatella, o brodetto con ovo, o altre simili buone menestrine: di carne mangio, carne di Vittello, Capretto, e di castrato: mangio polli di ogni sorte, mangio Pernici, e ucelli, come è il Tordo: mangio ancora delli Pesci come e fra li salsi la Orata, simili: e fra li dolci il Lucio, e simili: questi sono cibi tutti apropriati al vecchio .... E quel vecchio, che per povertà non può avere di quelli, può conservarsi con il pane, panatella, & ovo."

[20] In the *Hygiasticon's* seventeenth- and eighteenth-century English versions, this passage was cut out.

inserted very carefully and gradually into a person's routine. For, he says, "that which is against Nature, so likewise that which is against long and inveterate Custome, is very grevious to be undergone, whilest the strength and power of Custome remains on foot" (Lessius 1634, 43–44).

The example of the holy fathers of the desert serves Lessius to praise the custom of having a sober diet, although he often refers back to them in order to stress to his contemporaries the importance of relaxing their own standards in pursuit of the same. For instance, while the ancients used to eat only once per day, Lessius recommends splitting one's food into two daily meals so as to ease the cycle of hunger (Lessius 1634, 33).[21] The significance of Lessius's reference to the ancient fathers is twofold. On one hand, it was a rhetorical device intended to inspire his religious public through the example of saints and extraordinary models. On the other, though, it was used to show that the behavior of such models was not so extraordinary that it could not be easily followed. The gift of a long and healthy life, after all, was not generally the fruit of supernatural intervention by God but rather a merit achievable through ordinary human powers (Lessius 1634, 127).[22]

Lessius's trust in natural power and merit for the cultivation of a sober life was so strong that it in fact moves him to provide evidence and examples from other religions:

> This Priviledge belongs not onely to Saints, but also to others: For the Brachmans among the Indians live exceedingly long by reason of their spare diet: and among the Turks, the Religious Professours of their Mahometicall superstition, who are very much given to abstinence and austeritie (Lessius 1634, 128).[23]

11.5 The Virtue of Nutrition as a Foundation for Rational and Spiritual Functions

Can we say, then, that according to Lessius, the sanctity of religious people was determined by the appropriate functioning of their stomachs? The answer seems to be yes. Lessius's arguments on the subject alternate between ethics and natural

---

[21] "Those holy Fathers of old, who eat onely once a day, did it so sparingly, as they were no whit at all thereby hindered in their performances of the functions belonging to the minde: How much more easily then may it be effected by them, who divide the quantitie, and twice a day use moderate refection?"

[22] "Nor can it be well sayd, that these whom we have recounted [a list of Saints and Blessed who lived long lives], lived to so great ages by the supernatural gift of God, and not by the power of Nature."

[23] Lessius does not forget to include the Jewish people in the list, by adding quotations from the Bible. This echoes an interesting chapter in Huarte's *Examen*, where the Spanish physician considers how the Jewish people's centuries-long custom of eating manna brought about a change in their fundamental temperament and talents, allowing them to produce many extraordinary physicians. Huarte used this example to emphasize the role of dietetic custom in the shaping of human talents (Sumillera 2014, 220–222).

philosophy. On the philosophical side, he says that people whose activities were mostly intellectual required less food, as their activities would otherwise be hindered by a slow and defective digestive process. From the moral point of view, he argues that following a diet that was even more restricted than one's order or society required (for which there was ample room, given that such rules merely established an upper limit) was more commendable than fasting and mortification of the flesh. This was for two reasons: first, because to keep one's diet restricted every day exhibited a more holy attitude than alternating between fasting and feasting, as demonstrated by the desert fathers; and second, because oscillating between radically different approaches toward food was difficult—not to mention dangerous—for both the stomach and the body in general. This latter point, though, was more of a medical argument, based on biology rather than moral theology.

It seemed to Lessius that any excess in diet would impede religious people in the performance of both their intellectual and their spiritual duties. Concoction stood in fundamental conflict with meditation, prayer, and study; a stomach that was too active would thus prevent the mind from fully performing its functions. The reverse, of course, was also true: an overly active mind would harm digestion by withdrawing power from the stomach. But since the mind was the superior faculty, the obvious solution was to restrict one's diet, rather than restricting one's intellectual and spiritual pursuits. To formulate this in a more radical way, it is possible to say that the prayer and study of people who did not follow a sober diet were defective.

In fact, we can say that a sober diet was the virtue of the vegetative functions of human beings. This virtue was of great importance to Lessius, as it allowed the superior functions of the soul (e.g., the sensorial and the rational) to perform their tasks. Contrarily, if one did not live a sober life, the vegetative part of the soul could cause "offence or hinderance to the operations of the superior faculties, to wit, of the Senses, the Imagination, the Understanding, or the Memory" (Lessius 1634, 31–32). Concerning movement and sensation, the core of the motive soul, Lessius concludes that if a man desired to be "always quick, apt, and ready to motion, and every other use of his senses" (Lessius 1634, 42) ill humors would have to be abated through a restricted diet. The same went for the governance of emotions and passions, which a sober life could facilitate, to the benefit of reason.

As for the rational powers, Lessius insists that failing to follow a sober diet damaged the proper functioning of memory, imagination, and intellect. The main mechanism by which memory was diminished was an excess of phlegm, specifically in the brain, which was provoked by bad digestion. The overly abundant phlegmatic humor would intercept the animal spirits upon which phantasy relied, rendering apprehension feeble and unreflective (if not causing it to cease altogether). According to Lessius, the apprehension of anything without reflection left an imprint insufficient for remembrance (Lessius 1634, 166).

The relation between imagination and a bad diet was bidirectional. From one side, a fervent imagination fixated upon fleshly appetites would, Lessius thought, move the soul toward gluttony, presenting the mind with delightful images in anticipation of the pleasure of eating. To cure this bad attitude, he suggests presenting the mind with awful and disgusting images when approaching meals, so as to abate the appetite (Lessius 1634, 70–71).[24] From the other side, a bad diet could derange the imagination by means of ill vapors entering the mind. When exacerbated by illness, a consistently immoderate diet could even produce a state of hallucination, driving people to psychosis and, in extreme cases, suicide.

Given the importance of a sober diet for memory and imagination, it seems as if it would have been easy for Lessius to conclude that the intellect, too, was dependent upon the cultivation of that fundamental virtue of the vegetative soul. Although he does not mention it, it is clear that Lessius agrees with Aristotle's gnoseology (which every Jesuit was bound to follow), according to which intellect worked on the phantasms that were presented by sensorial powers. Thus, any affliction or impediment arising from the body would diminish the capacity of the intellect to reach a true knowledge of things.

The upshot of all this is that Lessius's argument significantly inclines toward the foundationality of the vegetative powers for the natural functioning of the human soul:

> For both Nature and Reason exact, that the Vegetative part in a man (that is, that wherein the growth and conservation of the bodie consisteth) should be so ordered and cherished, as that there should arise no offence or damage thereby to the Animall and Reasonable parts of the Soul; in as much as the Vegetative part is ordained to the service of these other, and therefore ought to be of furtherance and help, and no ways of hinderance unto them in their severall functions and operations (Lessius 1634, 30–31).

Transitively, given the role that Lessius usually assigns to the intellect and human understanding in matters of faith, it would seem that a healthy stomach was foundational for religious life. Returning to the example of the ancient fathers, Lessius states that God granted them illuminations because they were always fresh in their minds, thanks to the fact they followed a sober diet (Lessius 1634, 172–173).[25] He adds, "No man without the assistance of Sobrietie can perform any such matter" (Lessius 1634, 173).

This conclusion might sound controversial, as sobriety seems to become the very basis for every aspect of human life, including the possibility of the cultivation of virtues—even those that were usually supposed to be beyond the scope of the natural powers of human beings. To avoid such an extreme conclusion, Lessius is forced to introduce a distinction that safeguards the primacy of faith over all other virtues. He says that faith is "the internall and primary foundation" of human virtues and functions, while sobriety "is an outward, secondary, and ministerial foundation,

---

[24] This symmetrical relation between imagination and appetite recurs when Lessius considers generation, another vegetative function of human soul, and the vice of lust that was connected to it: "The appetite doth not onely desire that which is necessarie to the conservation of the Bodie, but also that which may serve for the use of Procreation" (191). Also, "So, where there is no more sustenance taken in, then is sufficient for the nourishment of the bodie, there remains either nothing at all, or very little to be distributed for the increase of the seed" (193).

[25] "For having their minds alwayes lifted up and set on God, his Majestie vouchsafed to descend down to them, illuminating them wonderfully, (…) making them partakers of his secrets, and instrument of his miraculous works; that so the world know how acceptable their kinde of life was with God, and be provoked to the honour and imitation of them."

inasmuch as it removes those things which breed impediment to the exercises of Faith, and to the functions of the Intellectuall Facultie" (Lessius 1634, 175). He then calls sobriety the "the secondary Foundation of wisdom and spiritual progresse" (Lessius 1634, 178). However, the use of the term "secondary," if clearly understandable from an axiological and moral point of view, should not be taken to imply that sobriety was of mere accessorial value in the progress of the spirit toward the knowledge of God and, ultimately, its own perfection. For as Lessius says:

> All spirituall progresse doth depend upon the use of the Understanding, and of Faith which resides in the Understanding. For we cannot love any good thing, or profit in the love thereof, not hate any evil thing, or grow in the hatred thereof, except it be proposed by the Understanding, so as it may move the Affections: Whereupon he that is so disposed by heavenly Grace, as that heavenly matters are always in his mind ... will easily contemne all earthly things, and so by degrees, from a great measure of holinesse attained here below, mount up to the enjoyments of a glorious Crown of everlasting blisse in heaven... . From these grounds it is evident, that those things which hinder the functions of the mind ... are the things which in very truth debarre us from attaining to any great measure of perfection either in Learning, or in exercises of Religion, or in Sanctity of life (Lessius 1634, 176–177).

11.6 Luigi Cornaro's Idea of Diet and Temperance

Lessius's theological concerns and dietary conclusions can perhaps be better appreciated if they are compared with the doctrines of Luigi Cornaro, Lessius's own referenced author.

Unlike Lessius, Cornaro did not feel the need to substantiate his work from a scientific point of view nor situate his moral intent within a theological framework. His only religious concern seems to be that of professing himself a good Roman Catholic through a quick invective against the Lutherans.[26] English translators of the *Sober Life* suppressed this passage, as well as the last paragraphs mentioning panatella (this latter perhaps because they were expressly addressed to clergy). It's a matter of fact, however, that English translations and reprints of Cornaro's writings on temperance multiplied in the seventeenth and eighteenth centuries, both within the *Hygiasticon* and on their own (Milani 2014a, b; Cornaro 1743).[27]

In Cornaro's opinion, two natures acted in human beings: one that they were naturally endowed with and a second one provided by their habits, with the latter being so powerful that it could overwhelm the former. Intemperance was the most harmful habit for this second human nature because it affected the body directly and threatened to take people's lives before their times, possibly costing them their souls

[26] This was a quite common rhetorical pattern among Italian writers during the last phases of the Council of Trent. It should be noted that Cornaro's devotional assertions became more frequent in his later writings.

[27] Choosing the version closest to Lessius's original, we quote from Cornaro.

(because intemperance was a sensual sin) and certainly depriving friends, families and society of useful “supports” (Cornaro 1743, 6).[28]

The surest prophylactic against the diseases caused by intemperance was, according to Cornaro as well as Lessius, the observance of a sober diet:

> A Distemper, though fatal in itself, yet it may be easily prevented by a well regulated, or an orderly Diet (Cornaro 1743, 5).

“Distemper” refers here to an unbalance among the four humors, an invocation of the Galenic theory that was at the heart of medicine in Cornaro’s time. But Cornaro deals with the concept in a more commonsensical way, emphasizing its quantitative roots above all else. In fact, he is even stricter than Lessius in excluding quality from the categories to be considered:

> In short, he who is not offended at any thing, has the Quantity, and not the Quality for his Rule, than which nothing is more easy to be observed (Cornaro 1743, 24).

By eating only small quantities of food, a person wouldn’t feed the humors enough to develop the unbalances that trigger disease.

Cornaro believed that the stomach played the fundamental role in the mechanism of nutrition, and that it was up to each person to understand the quality and quantity of food that his or her own stomach could bear. Individuals had both the ability and the duty to tailor their own diets by gradually testing how much and what kind of food and beverage their stomachs could digest, as Cornaro himself had done (Cornaro 1743, 11). This takes us back to that circular relation between the vegetative and the rational soul, since human beings could afford a diet, as well as discern their own stomach’s needs, only through “solid and substantial reason” (Cornaro 1743, 6), while the facilitation of that reason was the most immediate result of a sober life. As Cornaro says of such a life:

> It makes blood sound and juyces sweet, it prevents the emissions of vapours to the head so to clean the brain and brighten reason, and give perfect use of all their faculties (Cornaro 1743, 25).

Although diet was the most essential instrument of temperance, other cautions also needed to be taken in daily life, such as avoiding whenever possible extreme situations (weather, climate) and efforts (movement, work, sexual intercourse). A serious concern for climate was logical for Cornaro, insofar as the Venetian lagoon had long since disclosed to him how noxious an environment it could be. Less obvious was the reason for his attention to housing—his insistence that functional spaces, bright rooms, and fair gardens eased his daily life by being regular and quiet, allowing him to make good use of both mind and body (Cornaro 1743,

[28] “How many Friends of a promising Genius, and lively Disposition, have I myself seen hurried off the Stage of Life in the very Sunshine of their Days, by this Plague of Intemperance? Who (if they had lived) might have been shining Ornaments to the World, and an inexpressible Comfort and Satisfaction to their Relatives and Acquaintance.”

29–30).[29] This observation was likely intended to celebrate the author's social achievements, as embodied in his estate properties, but it also served to recall (toward the end of the text) the important issue of harmony (Cornaro 1743, 17).[30]

Free from theological responsibilities, the *Benefits of a Sober Life* doesn't suggest any proper soteriology other than some common sense (the higher the virtue, the more likely God's mercy). Nonetheless, it provides a distinctive teleology, as the final purpose of diet and temperance went far beyond the conscious well-being of body and mind. Because of its very naturalness, a sober life granted men and women a peaceful death (Cornaro 1743, 36):[31]

> For I know (setting aside Casualties) I cannot die but by a pure Dissolution, my Regularity having left Death no other Way of destroying me. And that is an honourable and desireable Death, which comes upon us by no other Means, than by a total, or a natural Dissolution. Nature herself, who has linked the Bonds of our Life together, can easily loose them again, without the least Violence, and give Men a longer Respite than Disease usually do, which forcibly rend those Chains asunder (Cornaro 1743, 34–35).[32]

By attaching Cornaro's work as an appendix to his *Hygiasticon*, Leonard Lessius disseminated throughout Europe a remembrance of late Renaissance naturalism.

11.7 Conclusion

Most known for his involvement in the debate on Molinism, Leonard Lessius was consistent in his trust of human powers to help souls reach their perfection. In the preface of his *Treatise on Perfect Happiness*, he states that to obtain God's grace, prayer was the chief means, but that grace required our "unremitting cooperation," and that nothing was more important for that than the consideration of certain truths that could move us to practice Christian virtues (Lessius 1924). In light of the *Hygiasticon,* however, we must conclude that for Lessius, neither the understanding nor the practice of Christian virtues was possible without the foundation of a sober diet, which was a prerequisite of all moral and cognitive activities. Since nutrition was a virtue, a sober diet was meant not to mortify this natural function but to follow its natural course by rejecting excess in either direction (i.e., gluttony or starvation).

---

[29] Cornaro's age cultivated a strong interest in domestic and landscape architecture (Fischer 2014); the Venetians their own peculiar perspective thanks to the work of Andrea Palladio (1508–80).

[30] "If the Universe consist of Order, if our natural Life depend on the Harmony, or perfect Agreement of Humors and Elements, it is no Wonder, that Order should preserve, and Disorder destroy."

[31] Cornaro did, in fact, include women in his audience: "She [holy Sobriety] will be the faithful Guardian of every Man's Life, that will but embrace her; whether Rich or Poor, Young or Old, of what Sex soever."

[32] In a previous passage, there are echoes from the alchemical side of Renaissance natural philosophy: "[A sober life] softens the Pains and Agonies of Death, and when the radical Moisture is totally consumed, procures him a quiet Exit out of this World. Infine, it contains the imagined Virtues of Aurum Potabile, or the Philosopher's Stone, which Numbers have already sought after to no Purpose" (20).

At the same time, these two excesses were to be considered vices because they were unnaturally directed toward illness and the death of the body. Lessius thus seems to provide an implicit correction to the ranking of human functions, as nothing superior could be performed without rendering appropriate respect to the lower powers. As was true of any being in this world, humans were to be primarily gardeners of their souls.

References

Bergdolt, Klaus, ed. 1991. *Alvise Cornaro: Vom massvollen Leben*. Heidelberg: Manutius Verlag.

Boorde, Andrew. 1542. *Hereafter foloweth a Compendyous Regyment or a Dyetary of Helth*. London: Robert Wyer.

Bruyérin-Champier, Jean-Baptiste. 1560. *De re cibaria libri, vol. 12, Omnium ciborum genera, omium Gentium moribus et usu probata complectentes*. Lyon: Sébastien Honorat.

Calanius, Prosper. 1550. *Traicté excellent pour l'entretenement de santé*. Paris: Jean Bonfons.

Casalini, Cristiano. 2014. Active Leisure: The Body in Sixteenth Century Jesuit Culture. *Journal of Jesuit Studies* 3: 400–418.

———. 2016. Discerning Skills: Psychological Insight at the Core of Jesuit Identity. In *Exploring Jesuit Distinctiveness: Interdisciplinary Perspectives on Ways of Proceeding within the Society of Jesus*, ed. Robert A. Maryks, 189–211. Leiden: Brill.

Chamberlain, Cecil. 1939. Leonard Lessius. In *Jesuit Thinkers of the Renaissance*, ed. G. Smith, 133–155. Milwaukee: Marquette University Press.

Cornaro, Luigi. 1558. *Trattato de la vita sobria del Magnifico M. Luigi Cornaro Nobile Vinitiano*. Padova: Grazioso Percacino.

———. 1561. *Compendio breve della vita sobria del Mag. M. Alvise Cornaro: Con molte cose aggiunte a' vecchi sopra modo utili, e necessarie*. Padova: Percacino.

———. 1591. *Discorsi della vita sobria del Sig. Luigi Cornaro*. Padua: Miglietti. Originally published in *Amorevole Essortatione del Magnifico M. Alvise Cornaro, nella quale con vere ragioni persuade ogn'uno a seguir la vita ordinata & sobria …* (Padua: Percacino, 1565).

———. 1743. A Treatise of the Benefits of a Sober Life. In *A Treatise of Health and Long Life, with the Sure Means of Attaining It, in Two Books: The First by Leonard Lessius, the Second by Lewis Cornaro, a Noble Venetian*, trans. T. Smith. London.

Cortesono, Gioseffo. 2016. Constitutions for the German College (1570). In *Jesuit Pedagogy, 1540–1616: A Reader*, ed. Cristiano Casalini and C. Pavur, 107–156. Boston: Institute of Jesuit Sources.

De Smet, S. 2001. Lessius (Leys), Leonardus (Lenaert). In *Diccionario histórico de la Compañia de Jesús*, vol. 3, 2336–2337. Rome: Institutum Historicum Societatis Iesu and Madrid: Universidad Pontificia Comillas.

Decock, Wim. 2009. Lessius and the Breakdown of the Scholastic Paradigm. *Journal of the History of Economic Thought* 31: 57–78.

———. 2016. In Defense of Commercial Capitalism: Lessius, Partnerships and the Contractus Trinus. In *Companies and Company Law in Late Medieval and Early Modern Europe*, ed. B. van Hofstraeten and W. Decock, 55–90. Peeters: Louvain.

Decock, Wim, and Toon Van Houdt. 2005. *Leonardus Lessius: Traditie en vernieuwing*. Antwerp: Lessius Hogeschool.

Emonet, Pierre. 2006. Du bon usage du corps selon Ignace de Loyola. *Choisir* 47: 15.

Ferlan, Claudio. 2017. La pericolosità degli eccessi: Ignazio di Loyola e le regole della tavola. *Ignaziana: Rivista di ricerca teológica* 23: 3–16.

Ferreira Panazzolo, Lidiane, and Marina Massimi. 2015. Categorías antropológicas nos Catálogos Trienais da Companhia de Jesus. *JHS: Antiguos jesuitas en Iberoamérica* 3: 21–45.
Fiocco, Giuseppe. 1965. *Alvise Cornaro: Il suo tempo e le sue opere*. Padova: Neri Pozza.
Fischer, Sören. 2014. Die Fiktion von Aus- und Durchblicken auf eine gesteigerte Welt: Oktogon (Odeon Cornaro). In *Das Landschaftsbild als gerahmter Ausblick in den Venezianischen Villen des 16. Jahrhundert*. Petersberg: Imhof.
Fleming, Julia. 2008. When 'Meats Are like Medicines': Vitoria and Lessius on the Role of Food in the Duty to Preserve Life. *Theological Studies* 68: 99–115.
Galen. 2003. *On the Properties of Foodstuffs*. Trans. O. Powell. New York: Cambridge University Press.
Golvers, Nöel. 2011. The Jesuits in China and the Circulation of Western Books in the Sciences (17th–18th Centuries): The Medical and Pharmaceutical Sections in the SJ Libraries of Peking. *East Asian Science, Technology, and Medicine* 34: 47.
Gordon, Barry. 1975. *Economic Analysis before Adam Smith: Hesiod to Lessius*. London: Macmillan.
Grendler, Paul.F. 2018. *Jesuit Schools and Universities in Europe*, 1548–1773. Leiden: Brill.
Gullino, Giuseppe. 1983. Corner, Alvise. In *Dizionario biografico degli italiani* 29. http://www.treccani.it/enciclopedia/alvise-corner_%28Dizionario-Biografico%29/. (Consulted on 15 November 2019)
Lessius, Leonhard. 1613. *Hygiasticon, seu vera ratio valetudinis bonae et vitae una cum sensuum, iudicii, & memoriae integritate ad extremam senectutem conservandae: subjunctur tractatus Ludovici Cornari Veneti, eodem pertinens, ex Italico in Latinum sermonem ab ipso Lessio translatus*. Antwerp: Moretus.
———. 1634. *Hygiasticon; or, the Right Course of Preserving Life and Health unto Extream Old Age: Together with Soundnesse and Integritie of the Senses, Judgement, and Memorie*. Cambridge: Roger Daniel.
———. 1924. Preface. In *The Virtues Awakened: From the Treatise on Perfect Happiness*, trans. H. C. Semple. St. Louis: Herder Books.
Loyola, Ignatius. 2006. *Letters and Instructions*. Edited by J. W. Padberg and J. L. McCharty. Translated by M. E. Palmer. St. Louis: Institute of Jesuit Sources.
Menegazzo, Emilio. 1980. Alvise Corner: Un veneziano del Cinquecento nella terraferma Padovana. In *Storia della cultura veneta: Dal primo Quattrocento al Concilio di Trento, vol. 2*. Vicenza: Neri Pozza.
Milani, Marisa. 2014a. How to Attain Immortality Living One Hundred Years, or The Fortune of the Vita sobria in the Anglo-Saxon World. In *Writings on the Sober Life: The Art and Grace of Living Long*, ed. H. Fudemoto, 183–213. Toronto: University of Toronto Press.
———. 2014b. Introduction to Cornaro. In *Writings on the Sober Life: The Art and Grace of Living Long*, ed. H. Fudemoto, 3–71. Toronto: University of Toronto Press.
Minuzzi, Sabrina. 2013. Il medico Tommaso Giannotti Rangone (1493–1577) nell'economia della cura ovvero un trionfo di libri, segreti e regimen sanitatis. *Medicina e Storia* 13 (3): 61.
O'Malley, John W. 1993. *The First Jesuits*. Cambridge, MA: Harvard University Press.
———. 2013. *Saints or Devils Incarnate? Studies in Jesuit History*. Leiden: Brill.
Pisanelli, B. 1583. *Trattato de' cibi et del bere, con molte dotte et belle annotazioni*. Rome, Bartholomeo Bonfadino.
Placotomus, Jacob. 1550. *De tuenda bona valetudine*. Frankfurt: Christian Egenolff.
Puppi, Lionello, ed. 1980. *Alvise Cornaro e il suo tempo Catalogo della mostra*. Padova: Comune di Padova.
Sander, Christoph. 2014. Medical Topics in the De anima Commentary of Coimbra (1598) and the Jesuits' Attitude towards Medicine in Education and Natural Philosophy. *Early Science and Medicine* 19 (1): 76–101.

Santiago de Carvalho, Mario. 2016. Beyond Psychology – The Philosophical Horizon of the Coimbra Commentary on Aristotle's 'De anima' (1598). In *Cognitive Psychology in Early Jesuit Scholasticism*, ed. D. Heider, 67–95. Neunkirchen-Seelscheid: Editiones Scholasticae.

Sumillera, Rocio G., ed. 2014. *The Examination of Men's Wits*. London: Modern Humanities Research Association.

Valverde de Amusco, Juan. 1552. *De animi et corporis sanitate tuenda libellus*. Paris: Charles Estienne.

Van Ginhoven Rey, Christopher. 2013. *Instruments of the Divinity: Providence and Praxis in the Foundation of the Society of Jesus*. Leiden: Brill.

Van Sull, Charles. 1930. *Léonard Lessius de la Compagnie de Jésus (1554–1623)*. Paris: Éditions du Museum Lessianum.

Welie, Jos V.M. 2003. Ignatius of Loyola on Medical Education: Or, Should Today's Jesuits Continue to Run Health Sciences Schools? *Early Science and Medicine* 8 (1): 26–43.

Chapter 12
Nicolaus Taurellus on Vegetative Powers and the Question of Substance Monism

Andreas Blank

Abstract This article analyzes the treatment of vegetative powers in Nicolaus Taurellus's critical response to Andrea Cesalpino. Taurellus's interest in this topic derives from larger metaphysical and theological concerns. His concern is that Cesalpino's view that vegetative powers are due to a divine principle of activity inherent in natural particulars leads to a version of substance monism that is incompatible with the Christian doctrine of creation. Taurellus's critique can best be understood within the context of his defense of an immaterialist account of the basic constituents of reality, which is connected with an emergentist account of the origin of vegetative powers.

12.1 Introduction

Vegetative powers, especially the powers involved in the generation and nutrition of plants and animals, are prominent themes in Nicolaus Taurellus's *Alpes Caesae* (1597).[1] This work is a thousand-page book review of Andrea Cesalpino's *Quaestiones Peripateticae* (1571/1593). Cesalpino's work became influential at the University of Altdorf, where Taurellus held a professorship in medicine, and Taurellus's opposition to this influence cannot solely be explained by diverging views concerning the correct interpretation of Aristotle but rather are motivated by what Taurellus regarded as impious implications of Cesalpino's readings of Aristotle. As I will argue, Taurellus's detailed discussions of Cesalpino's account of the origin of vegetative powers can be understood as being motivated by Taurellus's opposition to Cesalpino's conception of God as a unique active intellect inherent in natural particulars. Taurellus used the analysis of vegetative powers to spell out a sense in which living beings can be regarded as substances, as opposed to the

[1] For an overview of Taurellus's life and works, see Blank (2016).

A. Blank (✉)
Institute of Philosophy, Alpen-Adria University Klagenfurt, Klagenfurt, Austria
e-mail: andreas.blank@aau.at

F. Baldassarri, A. Blank (eds.), *Vegetative Powers*, International Archives of the History of Ideas Archives internationales d'histoire des idées 234,
https://doi.org/10.1007/978-3-030-69709-9_12

substance monism that Taurellus saw as an implication of Cesalpino's natural philosophy.

Neither *Quaestiones Peripateticae* nor *Alpes Caesae* are easy readings. Taurellus does not get tired of pointing out how opaque many passages in Cesalpino's work are, and he is certainly right about that. But in spite of its enormous length, Taurellus's response to Cesalpino is itself often stenographic and replete with irony, puns and invectives (the title of the book—something like *The Alps Torn Down*—itself being a polemical pun on Cesalpino's name). Nevertheless, both in Cesalpino and Taurellus it is possible to discern coherent threads of thought concerning central themes in early modern natural philosophy, especially the theory of biological reproduction and its connection with issues such as the nature of elements, mixtures and the origin of vital powers. Taurellus's critique of Cesalpino is not an example of one philosopher talking past another. This is so because Cesalpino and Taurellus share the belief that vegetative powers cannot be reduced to the causal powers of elements. Their disagreements concern the question of how a reductionist position could be avoided. Cesalpino took up the medical tradition that regarded subtle matter (or "spirits") contained in plant and animal seeds as the carriers of vital heat that he regarded both as the origin of vegetative powers and as a divine principle inherent in nature. By contrast, Taurellus tried to eliminate such subtle matter from his ontology and, in contrast to Cesalpino and most of his contemporaries, even refused the reality of primary matter that is traditionally understood to be the substrate of substantial forms. As it turns out, the reasons for Taurellus's rejection of primary matter have to do with his worries concerning substance monism. And the implications of this rejection are far-reaching because it leads Taurellus to the view that the basic constituents of reality are immaterial entities that are bearers of active and passive forces that bring forth both corporeal beings and higher-order forms. Using such immaterial entities as the basis of the eduction relation is what makes Taurellus's account unique in the field of early modern theories of vegetative powers.

In what follows, I will support these claims by reading the relevant passages from *Alpes Caesae* against the background of three of Taurellus's other works. The first is his most comprehensive work in metaphysics, *The Triumph of Philosophy* (1573), which pursued the plan to provide a unified account of philosophical and theological truth and thereby to offer a philosophical defense of the doctrine of creation (Taurellus 2012, 220–222, 250–252; for overviews of Taurellus's metaphysics, see Petersen 1921, 219–258; Mayer 1959; Leinsle 1985, 147–165; Wollgast 1988, 148–153). Using a text such as *Philosophiae Triumphus* to elucidate a text that was written almost 25 years later requires some justification, but this justification is forthcoming easily since in *Alpes Caesae* Taurellus refers the reader back to his earlier treatment of the generation of humans (Taurellus 1597, 132; 134). Hence, Taurellus seems still to subscribe to the essentials in his earlier work (although, as we shall presently see, there is at least one significant modification to the account of elements as outlined in *Philosophiae Triumphus*). This is why it seems highly plausible to use *Philosophiae Triumphus* to elucidate some passages in *Alpes Caesae* that, due to their extremely sketchy nature, are difficult to interpret when taken in isolation. The second co-text considered here will be Taurellus's commentary on

Aristotle's *De vita et morte*. The chronological closeness of this text to *Alpes Caesae*, published a year later, as well as the thematic closeness of the issues discussed make this a natural point of reference.[2] The third co-text will be Taurellus's unpublished *De procreatione hominis theses medicae*, which is preserved in an undated manuscript at the Municipal Archive of Nuremberg.[3] The text was written for the doctoral disputation of one of Taurellus's students at the University of Altdorf, but, in line with academic practice in early modern Germany,[4] two references in *Alpes Caesae* identify Taurellus as the author of this work (Taurellus 1597, 819; 833).

12.2 Vegetative Powers and the Problem of Substance Monism

Cesalpino's vindication of a version of substance monism is what triggered some of the most interesting parts of Taurellus's polemical response. In his preface, Taurellus makes clear that one of the main targets of his criticism is Cesalpino's conception of God. As Taurellus points out there, in a crucial respect Cesalpino goes beyond Averroes:

> What [Averroes] said about the assisting intellect, Cesalpino extends to the souls of humans and of the other animals, and of the entire world, for Cesalpino asserts that a single soul exists by itself and is multiplied according to the bodies of living beings. And through participating in it, the bodies are animated and substances. (Taurellus 1597, 25).[5]

Taurellus is clear that such a conception implies that God is not separated from matter and also is not an efficient cause of things; rather, God is understood as a constituent cause (Taurellus 1597, 25–26).[6] What is more, Cesalpino's conception of God seems to imply substance monism not only with respect to all animate beings but also with respect to the world as a whole.

---

[2] I am aware of only a single, indirect reference in *Alpes Caesae* to this work: Taurellus says that he has discussed "elsewhere" the sense in which body and soul of humans could be understood to be immortal (Taurellus 1597, 294); see below, Sect. 12.4.

[3] Stadtarchiv Nürnberg D 16 Nr. 573/1, fol. 38r–46v. Passages from this manuscript are reproduced below in Sect. 12.5 with the kind permission from the Municipal Archive of Nuremberg.

[4] See Marti (1994); Freedman (2005); Füssel (2016).

[5] "Cesalpini de animabus humanis opinionem multo magis & absurdam, & impiam esse, quam fuerit Averrhois. Quod enim hic de intellectu assistente dixerat: hoc ad animas hominum, ceterorumque animalium, & ipsius adeo mundi totius transtulit Caesalpinus: unam asserens, per se esse animam pro corporum animatorum numero multiplicatam. Cuius scilicet participatione corpora sint animata, & substantiae."

[6] "Deus hic est Caesalpinianus: quem in tot discerpit particulas Caesalpinus: quot in hac rerum universitate putat esse substantias. Quo posito non tam Deus efficiens rerum caussa erit, quam constituens."

In fact, substance monism seems to be an implication of a combination of views held by Cesalpino: (1) the view that the active intellect is a substance that is capable of self-reflection; (2) the view that the active intellect "perfects everything"; and (3) the view that the active intellect "implants into things a striving for perfection insofar as it is intellection."[7] Thus, the substantiality of the active intellect is characterized as being due to its capability of self-reflection, and it is this specific activity that is understood as the origin of the activity of natural things, not only of minds. Due to the relation between the unique active intellect and a plurality of natural things, Cesalpino holds that there is more than one sense in which unity is the origin of multiplicity:

> Nature therefore is intellect, insofar as only one exists, and at the same time it belongs to many, while by itself, it belongs to a single disposition, as [color] belongs to whiteness and [at the same time] to blackness insofar as it is the removal of whiteness. ... Hence, from what has been said it is clear how a single intelligence contains all the acts of understanding things, for it is like the measure of all things. However, it is evident that things relate to each other as acts of understanding do; hence, it is not impossible that a multiplicity arises from the one. This also becomes clear from the reduction of entities to the one and from the way in which the kinds of substances are described according to addition and subtraction .... Hence, insofar as it is simply and is described with respect to the subtraction of all matter, there is a unique and simple substance. (Cesalpino 1593, fol. 35v–36r).[8]

Thus, one sense in which multiplicity arises from the one has to with the dependence of things on the divine understanding. A plurality of things comes about because there is a plurality of divine acts of understanding, to which the things correspond. Another sense in which multiplicity arises from the one has to do with the role of matter in bringing about plurality. This is the sense in which Cesalpino speaks of "addition" and "subtraction": "adding" matter to the single incorporeal substance leads to a plurality of natural beings, "subtracting" matter from the plurality of natural beings leads to the single incorporeal substance.

Concerning the question of whether the intellective, sensitive and vegetative aspects found in humans are qualities of one and the same immaterial substance, Cesalpino argues that there is an experiment with a non-human animal that can decide the question: If a worm is divided, all of the vegetative and sensitive powers of the soul can be observed that were observable in the whole, "as if they were inseparable from each other" (Cesalpino 1593, fol. 43r). By analogy, Cesalpino

[7]All three claims are packed into the following passage (Cesalpino 1593, fol. 35v): "[P]rimum omnium principium substantia est, & sui ipsius intelligentia. Quatenus enim perfectissimum ens est, caeteris esse distribuit & perficit omnia: quatenus autem intellectio est, perfectionis desiderium rebus indit."

[8]"Natura igitur intellectus est, ut unus existens, simul sit multorum, per se tamen unius id est habitus, ut albedinis, nigridinis autem quatenus albedinis remotio. ... Patet igitur ex dictis quo pacto unica intelligentia omnium rerum intelligentias contineat, est enim tanquam mensura omnium. Quemadmodum autem se habet intellectiones, sic etiam & res ipsas esse manifestum est: ab uno igitur multitudinem descendere non est impossibile. Patet etiam ex reductione entium ad unum, & ex modo quo substantiarum genera dicuntur secundum additionem & ablationem. ... Quatenus igitur simpliciter est, & per ablationem omnis materiae dicitur, unica & simplex est substantia."

suggests, in humans the vegetative and sensitive powers depend on the intellective powers; but if the intellective powers are understood as deriving from the participation of human beings in the unique active intellect, then humans are an example for how in organic beings, together with the material form, there is also an immaterial form—not in the sense of the presence of two distinct substances but rather in the sense that the unique active intellect communicates corporeal activities to the human body and intellective operations to the human soul. Both corporeal and intellective operations thus participate in the unique active intellect. As Cesalpino conjectures, this could be the sense of Thales' saying that everything is full of gods (see Aristotle, *De anima*, I.5, 411a8) (Cesalpino 1593, fol. 43r). And Cesalpino's argument for explaining vegetative powers through the influence of the unique active intellect is the consideration that, if vegetative powers have the function of securing the eternity of species, and if something eternal can only arise from something eternal, then the generation of living beings must be due to an active principle that itself is incorruptible (Cesalpino 1593, fol. 44r).[9] And, as Cesalpino argues, neither singular corruptible beings, nor all causes taken together, can be the cause of the eternity of species. This leads him to the conclusion that "[e]very soul, or part of a soul, of mortal beings therefore may seem to participate in the divine itself …" (Cesalpino 1593, fol. 44r).[10]

This account of the participation of natural particulars in the unique active intellect raises the question of the sense in which Cesalpino speaks of different "kinds of substances" that arise through the addition of matter. Earlier in the text, Cesalpino explains that he regards the concept of substance to be equivocal. Generally, he remarks that the different ways in which things are distinguished from each other depend on the different ways in which things "participate in being," that is, on the different ways in which things "descend from a unique substance" (Cesalpino 1593, fol. 9v).[11] Accordingly, he claims that substance is not a "univocal genus" but rather allows for equivocations. In fact, he argues that such equivocations occur in more than one sense. Cesalpino takes up the thought from Aristotle's *Metaphysics* VII.7 (1033a12–26) that not only form-matter composites, but also the matter and the forms that constitute beings such as animals and plants can be called substances (Cesalpino 1593, fol. 10r). As Cesalpino explains, the difference between these different senses of "substance" is that even if neither matter nor the composite are unities, they are directed towards something that is a unity, namely, form: matter is substance because it is form potentially, and the composite is substance because it possesses form (Cesalpino 1593, fol. 10r).[12] Thus, according to the first way of

---

[9] On the doctrine of the eternity of species, see Hull (1965). On Cesalpino's use of the doctrine, see Atran (1990), 138–142.

[10] "Omnis igitur anima seu animae pars mortalium divinum ipsum participare videbitur …"

[11] "alio modo quatenus ipsum esse diverso modo participant, ab uno quodam quae substantia est descendentia …."

[12] "[H]aec quamvis non secundum unum dicantur, ad unum tamen omnia sunt, scilicet ad eam substantiam, quae ut forma dicitur: ipius enim est ipsum quid est, & materia ideo substantia est, quia potentia est hoc, compositum autem quia actu id habet …."

disambiguating the notion of substance, there is a sense of substantiality that is derivative on the substantiality of forms—a sense of substantiality that denotes certain relations between other beings and forms. Moreover, disambiguating the notion of substance in this way allows for a plurality of substances, even if it should turn out that the number of substantial forms is exactly one.

This possibility is taken seriously in Cesalpino's explication of the second sense in which substances do not constitute a univocal genus. Here, Cesalpino takes up the thought from Aristotle's *Metaphysics* XII.5 (1070b36–1071a1) and XII.8 (1073a23–1073b1) that there are three kinds of substances: two natural, one immobile; two eternal, one capable of being generated and undergoing corruption. As Cesalpino understands it, the distinction is between the divine substance, mathematical substances, and physical substances (Cesalpino 1593, fol. 10r–v). Moreover, he explains the sense in which all these different kinds of beings can be understood as kinds of substances as follows:

> [T]hese three genera of substances are proportional to each other. For all of them underlie certain affections, and always the substance is prior and the cause of the others. For this reason, they come together in an analogous genus. (Cesalpino 1593, fol. 10v).[13]

This is the second sense in which, in Cesalpino's eyes, natural beings can count as substances, even if they owe their being to a single, divine substance: Once they have been individuated through the composition of a singular portion of matter and a part of the active, divine intellect, they become bearers of qualities and relations of their own. In these two ways, well documented through extended quotations in Taurellus's *Alpes Caesae*, Cesalpino uses insights from Aristotle's *Metaphysics* to disambiguate the notion of substance, which would allow reconciling the view that the active, divine intellect is a unique substance with the view that nevertheless there is a plurality of natural substances.

12.3 Elements and the Question of Substantiality

In response to Cesalpino, Taurellus defends a version of substance pluralism that does not derive from the view that the concept of substance is an equivocal concept. Rather, Taurellus defends the view that the concept of substance is a *univocal* concept—a concept allowing only a single meaning. His arguments are based on *ad-hominem* criticism of Cesalpino, but also on some deep layers of his own metaphysics. As to the *ad-hominem* criticism, Taurellus is quick to point out that the experiment that Cesalpino regarded to be decisive—cutting worms into parts—allows for alternative explanation. For instance, one could argue that the continuing life, sensation and motion of the worm parts are just caused by animal spirits, the

---

[13] "[T]ria ista genera substantiarum esse inter se proportionabilia, omnia enim affectionibus quibusdam subiciuntur, & ubique substantia prior est & caeterorum causa. Ideo conveniunt in genere quodam analogo."

continued presence of native heat, and favorable external influences (Taurellus 1597, 309). If such an explanation is successful, then the experiment does not show that animal souls are divisible, nor that vegetative and sensitive soul powers are present in all worm parts. On the contrary, Taurellus goes on to argue, the separation between vegetative, sensitive, motive and intellective powers can be observed in many body parts: some are nourished without being sensible; some are sensible without having motion; some have all three qualities without having intellection (Taurellus 1597, 309).

As to Cesalpino's central claim—the claim that in material forms there is at the same time something immaterial—it is evidently not enough to point to experiences of this kind since Cesalpino regards vegetative powers as something that participates in the divine substance even in the absence of sensation and intellection. On the level of diagnosis, Taurellus is clear that Cesalpino's conception amounts to the view that "the substance of God and of all other things are confused in such a way that each thing in some respect is said to be god himself" (Taurellus 1597, 309).[14] But what is wrong with regarding God as a constitutive cause of all things? Taurellus only remarks that he has written about these matters copiously in his metaphysics—that is, in *Philosophiae Triumphus* (Taurellus 1597, 309). And this is how some deep layers of Taurellus's metaphysics become relevant for his response to Cesalpino. In *Alpes Caesae*, he only alludes rather cryptically to these deep layers. One of his allusions concerns his view that, if matter and form are understood to be complementary entities that cannot exist without each other, then there is no reason to regard them as substances:

> If substance is said to be that which subsists by itself, then this genus will be univocal. But you may say that matter does not subsist by itself; nor does form; but matter and form nevertheless should be substances in such a way that they are parts of substance. Let me say what comes to my mind. I wonder whehter there is any substance or part of substance … that could not subsist by itself, separately, But about these matters elsewhere. (Taurellus 1597, 25).[15]

A second allusion concerns a consequence that, in his view, follows from this consideration, namely, that matter should not be understood as a constituent of natural things but should rather be thought of as the "nothing" out of which creation (in the sense of *creatio ex nihilo*) took place (Taurellus 1597, 26).[16] These are fascinating ideas that Taurellus developed in much detail in his earlier *Philosophiae Triumphus* (1573), and to which he refers frequently in *Alpes Caesae*.

[14] "Quid hoc quaeso est aliud: quam Dei: rerumque ceterarum omnium substantias ita confundere; ut unaquaeque Deus ipse dicitur esse aliqua ex parte?"

[15] "Si substantia dicatur id esse, quod per se subsistit: genus haec erit univocum. At inquies materiam non subsistere per se: neque formam: quas tamen ideo substantias esse oporteat: quod substantiae partes sint. Dicam quid mihi in mentem veniat. Miror an ulla sit vel substantia: vel substantiae pars … quae separata per se subsistere nequeat. Sed de hoc alias." For detailed discussion of these matters, see Taurellus 2012, 274–280.

[16] "Ita nobis occurrit: ut primam physicarum materiam non trina dimensione definiremus: quod ausus est Caesalpinus. Sed nihilum diceremus esse theologicum …."

As Taurellus argues in *Philosophiae Triumphus*, the relation between forms and primary matter raises a problem concerning the sense in which forms can be regarded as something substantial. Taurellus remarks that the Aristotelian principles "want matter which is nothing by itself to be something accidentally. For if this is true, it will be an accident, not a substance …" (Taurellus 2012, 240).[17] Something analogous holds for the conception of form in the Aristotelian tradition: "The natural philosophers have stated that natural forms cannot subsist without a subject; if this were admitted, they would not be substances but accidents, because substances are not in something or from something but subsist through their own force, while one says of accidents that they do not subsist but rather are in something" (Taurellus 2012, 318).[18] By contrast, a theory of immaterial forms is exactly what yields an analysis of the substantiality of elements. This is so because, once the assumption that form depends existentially on prime matter is given up, there is no need to regard form as an accidental being: "Tell me, does nature subsist by itself or does it subsist in something other? If you say that nothing underlies it, why shouldn't one be allowed to ascribe the same to form, such that it is nature and subsists without matter?" (Taurellus 2012, 376).[19]

For this reason, Taurellus eliminates prime matter from his ontology and accepts only a metaphorical usage of the notion of prime matter. As he puts it, with respect to God, prime matter is "the NOTHING" (in capital letters) out of which God creates the world (Taurellus 2012, 484). When he calls prime matter "nothing," what Taurellus has in mind is a strict identity statement. This is why it is a suitable metaphor for what is there before the divine act of creation, namely, no created being. Prime matter in this metaphorical sense is absolute non-being. What we are left with on the basic level of reality thus are immaterial, form-like entities. And since they are not accidents because there is nothing in which they could inhere, there is also no reason to assume that they could be accidents of a unique divine substance. This is crucial for Taurellus's characterization of the relation between God and the world: "Because the world is separated from God by its substance, it is not conjoined with him from eternity" (Taurellus 2012, 426).[20] The ensuing view of the relation between God and creatures accepts creation dependence of creatures on God but denies the dependence of creatures' continued subsistence on God.

If one compares the position taken in *Philosophiae Triumphus* with that taken in *Alpes Caesae*, one significant modification comes to the fore. In the later work, Taurellus comments upon Cesalpino's claim that passive qualities such as humidity and dryness constitute the forms of elements (Cesalpino 1593, fol. 50v–51r):

---

[17] "materiam quae per se nihil est, per accidens aliquid esse velint: Hoc enim si verum sit, accidens erit non substantia …"

[18] "Formas … naturales Physici sine subiecto subsistere non posse statuerunt, quod si concederetur non substantiae, sed accidentia forent, illae siquidem per se non in aliquo, vel ab aliquo, sed vi propria subsistant, haec vero non subsistere, sed inesse dicantur."

[19] "dic quaeso an per se natura vel in aliquo subsistat, quod si nihil ei subesse dixeris, cur non idem formae liceret adscribere, ut et natura sit ipsa, et per se sine materia subsistat."

[20] "Cum … mundus a Deo substantia separatus sit, ab aeterno ipsi nequaquam coniunctus fuit …"

> Why [do they] not [constitute] matter? It does not belong to form to be the subject but to matter. But heat and coldness are added to humidity and recede from it. Coldness informs humidity such that water arises; heat [informs humidity] such that air arises (Taurellus 1597, 405).[21]

But Taurellus is also not satisfied with a solution that would take passive qualities to be constitutive of matter and active qualities to be constitutive of forms: "If active qualities were forms of elements, they certainly would be accidental forms, not essential forms; because they can be separated [from passive qualities] without the corruption of their subject" (Taurellus 1597, 404).[22] From the consideration concerning the separability of active and passive qualities, Taurellus draws the conclusion that active and passive qualities cannot constitute a *single* form. For this reason, Taurellus gives up the view articulated in *Philosophiae Triumphus* according to which active and passive powers are inherent to the forms that function as elements (Taurellus 2012, 284; 292–294).[23]

But he also does not return to a traditional conception of prime matter. Rather, Taurellus uses the idea of a hierarchy of forms to explicate a sense in which one can speak about matter without being committed to matter as a constitutive principle of natural things. Already in *Philosophiae Triumphus*, his reinterpretation of secondary matter in terms of subordinate forms allows Taurellus to continue to talk about bodies (*corpora*) and corporeal beings (*entia corporalia*) (see Taurellus 2012, 160–162; 182; 352–354; 552–554). As he puts it, "if it is not understood as a less noble form, ... matter does absolutely not compose anything" (Taurellus 2012, 278).[24] In *Alpes Caesae*, Taurellus uses the subordination relation between forms to explicate the sense in which passive qualities can be said to be characteristic of what one could call the "matter" of elements. According to his revised view, forms, in so far as they function as forms, are always active. But in so far as forms can be subordinate to other forms, they can be said to be passive: "Form is passive not in so far as it is form but in so far as it is matter" (Taurellus 1597, 406).[25] Consequently, Taurellus still uses the notion of form to analyze the nature of elements, but he regards elements as composites of *two* forms one of which functions as the matter of the other (Taurellus 1597, 612). This is a significant modification in the ontology of the basic constituents of reality; but it is a modification that still operates within the framework of an immaterialist ontology. And if the point of immaterialism about the basic constituents of reality is to guarantee the substantiality of created beings, then the modified version of Taurellus's ontology still fulfils the same purpose.

---

[21] "Cur non potius materiam? Formarum non est, ut subiiciant, sed materiae. Caliditas & frigiditas ad humidum accedunt, & ab eodem recedunt. Humiditatem figiditas informat; ut fiat aqua: Caliditas, ut fiat aer."

[22] "Activae qualitates, si formae sunt elementorum: accidentales certe sunt: non essentiales; quia separantur citra subiecti corruptionem."

[23] For detailed analysis, see Blank (2014).

[24] "materia ... nisi pro forma ignobiliori sumatur ... nihil omnino componat."

[25] "Non enim forma ut forma, sed ut materia est, patitur."

12.4 Vegetative Powers and the Emergence of Substantial Forms

Taurellus's immaterialist ontology implies that composites can be composed only of immaterial beings. As Taurellus puts it: "Matter plainly does not exist, and nothing but forms can be and can enter composition" (Taurellus 2012, 280).[26] Evidently, this raises the question of where the vegetative powers characteristic of particularly complex composites come from. Here it is interesting to see that Taurellus integrates a version of emergentism into his immaterialist ontology. Emergentism—the view that, once material composites have reached some level of complexity, potencies arise that cannot be reduced to the potencies of the constituents (see Macdonald and Macdonald 2010)—was clearly articulated by some ancient thinkers, including Galen and the Aristotelian commentators Alexander of Aphrodisias and John Philoponus.[27] According to Alexander of Aphrodisias, the soul "is a power and form, which supervenes through such a mixture upon the temperament of bodies; and it is not a proportion or a composition of the temperament" (Alexander of Aphrodisias 2008, 104, 2012, 51). Jacob Schegk (1511–1587), Taurellus's teacher at the University of Tübingen, took up this idea. In his commentary on Alexander's *On Mixture*, Schegk describes the role of the tempering of elemental qualities in mixture as follows:

> Those entities that constitute a temperament are first divided and split up amongst each other into minute parts, then their activity is gradually diminished through the composition of minimal parts …, and third, as it were through some agreement, they jointly bring about a single form of the entire mixed body. (Schegk 1540, fol. 65r; see Todd 1976, 158 [*De mixtione* 233.2–5]).[28]

According to this account, a substantial form of the mixture arises through the tempering of elementary qualities. Similarly, Taurellus writes:

> [W]hen by mutual action and passion mixed things are changed in such a way that none of them remains entirely the same, but some new form arises out of them that relates to the forces of all of them, then without doubt there exist mixed forms that have the forces of many, and that bring about different effects, which is most evident in the changes of things and especially in the use of medicaments. (Taurellus 2012, 272).[29]

Taurellus maintains that what arises in genuine mixture is not only a composition of simple compounds but also a form. In his view, this form is simple "because it is not composed but rather generated" (Taurellus 2012, 42). Such a form differs from the mere composition of elementary forms: "Because through generation a really

[26] "materiam plane non existere, nilque praeter formas vel esse, vel componi posse …"

[27] On Alexander and Galen, see Caston (1997); on Philoponus see Ganeri (2010); Sorabji (2010).

[28] On Schegk's natural philosophy, see Hirai (2007); Blank (2018).

[29] "Cum … actione mutua, & passione res mixtae sic immutentur, ut earum nulla integra maneat, sed nova quaedam forma inde oriatur quae vires omnium referet, dubium non est quin compositae sint formae, quae multarum vires habent, differentesque proferunt effectus, quod in rerum mutationibus, tum maxime usu medicamentorum."

unique being arises, namely the substantial form, we hold that this is not simple with respect to conjunction or some other accident and also not composite, no matter whether it derives from a single being or many beings" (Taurellus 2012, 274).[30] Thus, there is a sense in which emergent forms are complex—they are bearers of a plurality of qualities without having parts from which they are composed, even if they arise from a composite that has such parts. Consequently, a form that emerges from the simple constituents of a composite cannot undergo a process of being split up, although it can perish when the basis from which it emerges is changed (Taurellus 2012, 274). This is why the simplicity and immateriality of emergent forms is compatible with their capability of being destroyed (Taurellus 2012, 44).

In Taurellus's commentary on Aristotle's *De vita et morte*, this line of thought is applied to the question of the origin of vegetative powers. There, vegetative powers in plants are treated differently from vegetative powers in animals, while the treatment of vegetative powers in brutes and humans is so strikingly similar that Taurellus faces considerable theological challenges, which he tries to solve within the framework of an emergentist ontology. As to vegetative powers in plants, Taurellus follows the Galenic tradition in holding that the nutrition and augmentation of plants has to be ascribed to natural principles other than souls (Taurellus 1586, sig. B3r; see Galen, *De naturalibus facultatibus* 1.1). He takes the postulation of vegetative souls in plant to be superfluous because nutrition is brought about by natural potencies (Taurellus 1586, sig. B3v). As he argues, this can be gathered from the fact that augmentation, diminution, attraction, alteration, and expulsion—the processes underlying vegetative powers—can be found also in inanimate beings (Taurellus 1586, sig. Cv).

However, Taurellus's rejection of plant souls does not imply that only the elements are operative: "We define nature not only as the temperament but also as the proper and essential form of each body; if you confuse this with the temperament, such that you reduce it to the elements, you go far astray" (Taurellus 1586, sig. G3r). In line with his theory of the emergence of substantial forms, he ascribes to each plant part a substantial form that accounts for some of the activities of the organic plant body (Taurellus 1586, sig. B3v)—activities that cannot be explained by the temperament of elementary qualities (Taurellus 1586, sig. G3r). As he summarizes: "We … maintain that in seeds there is a power by means of which they are converted into plants; and that the forms of plants arise from nothing else but out of seeds. But this power capable of generating plants is not a soul but a nature. For out of this own essential form plants produce seeds" (Taurellus 1586, sig. B5r).

As Taurellus argues, it is necessary to postulate the existence of such natural forms in plant parts because not all kinds of attraction relevant for operation of plants and their seeds can be explained by the agency of elementary heat (Taurellus 1586, sig. B5v). Thus, plant forms and seed forms possess new causal powers that go beyond the causal powers of the elements that function as the matter of the

[30] "quia generatione unum quid revera fit, forma scilicet substantialis, non coniunctione aut alicuius accidentis respectu, simplex id quoque non compositum seu ab uno, seu sit a multis esse iudicamus."

emerging form. These new causal powers explain the difference between fire, which continuously regenerates itself by using up fuel, and life: "Living substances do not live through subject matter, but by themselves, without any supply or change of substance, and are not generated by being alive" (Taurellus 1586, sig. F6v). As Taurellus puts it, life "is the power of a substance through which the other actions of a living being, as their foundation …, are triggered and animated" (Taurellus 1586, sig. F7r). This is why he takes life to relate both to primary actuality—in the sense that life depends on the presence of a substantial form—and to secondary actuality—in the sense that it manifests itself in characteristic activities found only in living beings (Taurellus 1586, sig. F6v).

As to vegetative powers of brutes and humans, Taurellus regards the substantial forms from which these powers are brought forth to be souls. As he argues, if the human soul is generated out of matter, then the seed is the most suitable portion of matter from which the generation of the soul takes its origin (Taurellus 2012, 12). This is so because not everything that proceeds from a corporeal seed is necessarily corporeal. For instance, the souls of brutes are incorporeal substances that proceed from their seeds (Taurellus 2012, 18). However, this does not imply that Taurellus would ascribe animal souls to animal seeds or human souls to human seeds. Rather, he defends the view that in the seed the features of an animal are not contained actually but potentially (Taurellus 1604, 20). Something analogous holds for the human seed: "Many things are in the human seed that is unknown to the nature and forms of each element. This is the essential form of the seed, due to which its corporeal bulk is easily transformed into the various parts of the human body" (Taurellus 1604, 19). Thus, as in the case of non-human animals, human seeds possess substantial forms of their own, which determine through formal downward causation the structure of bodily parts, from which subsequently the human soul emerges. From this ensues a circle of causation: the life of the soul is communicated to the body (Taurellus 1586, sig. Gr); but the life of the soul is also perfected by actions that it can only carry out by means of the body (Taurellus 1586, sig. Gv). Consequently, the soul perfects the body, the body perfects the soul (Taurellus 1586, sig. G2r).

Treating non-human and human souls in an analogous way obviously raises theological issues: Would not regarding non-human souls as immaterial beings render them as immortal as human souls were thought to be? Would not regarding human souls as emerging from organized matter render them as mortal as non-human souls were thought to be? In *Philosophiae Triumphus* Taurellus considers a creation theory of the origin of human souls (Taurellus 2012, 144), but he also voices doubt concerning such a theory. In particular, he cautions that an act of divine creation would render the imperfection of human souls inexplicable (Taurellus 2012, 166). Moreover, he argues that if the soul is infused from the outside, then humans would lack the capacity shared by plants and brutes to generate beings of the same kind. Otherwise, humans would give birth to human bodies, but God would generate the human soul. However, as he objects, producing a being of the same species is a most natural process (Taurellus 2012, 13). Also, in his view the imperfection of human souls speaks against a celestial or divine origin of human souls (Taurellus 2012, 166).

Accordingly, other passages in his commentary on *De vita et morte* point in a different direction. In these passages, he offers an integrated naturalistic account of the origin of the souls of non-human animals and humans. As he surmises, human souls have in common with the soul of brutes that "they necessarily have their essence and life in a body" (Taurellus 1586, sig. G4r; see Taurellus 1604, 11). Therefore, he claims that "the human soul by itself is capable of ceasing to be" (Taurellus 1586, sig. F4v). As in the case of all other natural forms, Taurellus holds that the human souls cannot undergo corruption in the sense of a dissolution into parts. However, since he holds that immaterial animal souls can cease to exist (Taurellus 1604, 21), he maintains that human souls by themselves are capable of ceasing to be (Taurellus 1586, sig. F4v). On first sight, this conclusion seems to be incompatible with the Christian doctrine of the immortality of the soul.

Taurellus is clear that, if body and soul perfect each other, there are two possibilities with respect to immortality: either the human soul is as mortal as the soul of brutes, or resurrection must include the body (Taurellus 1586, sig. G2r). However, he sees room to argue that in this respect there may be a dissimilarity between the souls of humans and non-human animals. In his view, this is so because, by Divine will, human beings are goals in themselves, while brutes are subservient to humans. Hence, brutes can fulfil their goal without immortality, but immortality is required for the fulfilment of the goal of humans (Taurellus 1586, sig. G4r–v). Thus, Taurellus uses theological considerations to show why the goal-directedness of the souls of brutes does not require assuming that they are immortal, while the goal-directedness of human souls requires assuming that human souls are immortal. Consequently, he does not challenge the theological idea of immortality. Rather, he argues that because humans are constituted not only by souls but also by bodies, felicity must be ascribed to soul and body together (Taurellus 2012, 562). This is why he believes that God does not want that the body perishes entirely (Taurellus 2012, 556). Taurellus concludes that if human beings are immortal, the relevant supernatural divine agency responsible for resurrection must relate to soul and body alike (Taurellus 1604, 26). Reinterpreting the idea of resurrection such as to include a human body from which the human soul emerges allows Taurellus to integrate an emergentist view of the human soul with the theological doctrine of immortality.

12.5 Vegetative Powers and Substance Pluralism

This, then, is the ontological framework in which Taurellus develops his critique of Cesalpino's account of the origin of vegetative powers. The univocal concept of substance that Taurellus defends is illuminating with respect to the question of substance monism since it allows for creation dependence—natural things would not exist if a divine act of creation did not take place—but also allows for activity independence—natural things are active without the influence of any continued divine agency—and for persistence independence—natural things do not need any divine agency in order to persist in being.[31] Taurellus's concept of substantiality contrasts sharply with Cesalpino's view that created beings possess substantiality only by participating in an inherent divine principle of activity. The reason for Taurellus's interest in Cesalpino's consideration of vegetative powers is easy to see. As it turns out, what Cesalpino says about the generation of living beings is closely connected with his view that the notion of substance is an equivocal notion.

Cesalpino assigns to living beings a status of being substances; however, because, in his view, substantiality depends on the presence of a divine principle of activity, he restricts the notion of substance to the notion of soul: As he puts it: "Separate elements … do not contain actually this divine being, but only potentially, but mixtures contain it actually; and for this reason, they are actually substances" (Cesalpino 1593, fol. 22r–v).[32] He derives support for this from a reading of Aristotle's *Meteorology* IV.3 (381a9–12), where putrefaction—a process that only bodies that contain moisture, such as nutriments, can undergo—is described as separation and aggregation of parts (Cesalpino 1593, fol. 23r).[33] If putrefaction is nothing but a separation and aggregation of parts, then the generation of inanimate composites, Cesalpino concludes, cannot be anything but the compositions of parts (Cesalpino 1593, fol. 23r). Consequently, metals and stones and all mixtures that lack animal power are not substances (Cesalpino 1593, fol. 23r–v). By contrast, Cesalpino holds that those composites that are substances owe their substantiality to the presence of vital heat, which he interprets as a divine principle. Cesalpino presents this view as a commentary on Aristotle's view that the heat of the fertile seed is not fiery but stands in an analogy to the heavenly bodies (*De gen. an.*, 736b29–727a7).[34] What is distinctive about Cesalpino's interpretation of this passage is his thesis that *all* animals that are generated sexually also could be generated spontaneously, without the intervention of seeds. As he argues, this is so because if celestial heat is what makes seeds fertile, then celestial heat could produce the same effect in other portions of matter as well (Cesalpino 1593, fol. 109r).

Taurellus raises a series of objections against Cesalpino's speculations concerning spontaneous generation. For instance, he asks why, if celestial heat is assumed to be everywhere, why does Cesalpino believe that putrefaction is necessary for spontaneous generation (Taurellus 1597, 102)? Also, he presses Cesalpino to explain why celestial heat should be capable of fulfilling the function of souls (Taurellus 1597, 102). As he notes, if celestial heat could fulfil the function of souls, then the world as a whole would be animate (Taurellus 1597, 102). And he objects that, if all mixtures were already animated, then there were no genuine generations of animals

[31] On the role of ontological dependence in the metaphysics of substance of Taurellus's *Philosophiae Triumphus*, see Blank (2014).

[32] "Elementa separata … actu non continent divinum hoc, sed potentia solum: in mixtione autem etiam actu: idcirco & actu sunt substantiae."

[33] On early modern alchemical interpretations of this passage, see Newman (2004), 200–206.

[34] On this passage, see Solmsen (1957). For critical discussion, see Preus (1970), 35–38; Freudenthal (1999), 19–29, 40–46, 114–129; Lennox, (2001), 229–249.

(Taurellus 1597, 102). Also, if spontaneous generation were due only to celestial heat, then it would become inexplicable why plants and animals are not generated out of whatever mixture (Taurellus 1597, 103). And finally, if everything were animate due to the presence of celestial heat, why should elements be thought to be inanimate? (Taurellus 1597, 103).

The internal difficulties of Cesalpino's account of celestial heat lend plausibility to Taurellus's alternative account: "The seed is one thing; the fertile seed is another thing. For the generation of an animal not only a seed but a fertile seed is required. Even if the sun can perhaps make a seed fertile—about which there is warranted doubt—the sun nevertheless cannot make a seed" (Taurellus 1597, 790).[35] As Taurellus argues in line with his emergentist views, heat cannot be enough to bring forth vegetative powers because a fertile seed requires an essential form as a primary agent, while heat can only be an assisting or instrumental cause (Taurellus 1597, 791). This can be inferred through the observation that sun heat is present in inanimate bodies: "The heat of the seed is not by itself fertile. For if it is in dirt, or some other matter, it has no fertile power" (Taurellus 1597, 791).[36] Consequently, Taurellus takes the sun to be a common cause, and if so, then something applies to the sun that applies to all common causes: they can have only common effects. For this reason, the sun heat is insufficient to explain the occurrence of a plant or an animal of a particular species, in contrast to the occurrence of a plant or an animal of a different species (Taurellus 1597, 791).

What is more, Taurellus takes seminal heat to be a quality that arises from the temperament of elements and, therefore, distinguishes it from celestial heat:

> [S]eminal heat and sun heat do not have the same power. For if the seed loses its native heat and receives as much heat as it had before only from the sun, it does not become fertile. Nor is native heat enough to make the seed fertile. For if the seed is hotter than would be just or more humid, or suffers from some other defect, it is infertile. Whatever we may say about solar heat, we cannot say the same about the essential form of the seed which only the testicles can bring about, since so many material conditions are required for fertility. (Taurellus 1597, 792).[37]

Thus, seeds that, due to a disbalance of their temperament, have become infertile cannot become fertile again only through the influence of sun heat. But also, the elementary heat inherent in seeds cannot be understood by itself as the origin of vegetative powers because any excess in elementary qualities leads to infertility.

---

[35] "Aliud est semen: aliud item est semen foecundum. Ad animalis generationem non modo semen: sed etiam foecundum semen requiritur. Licet ergo Sol forte semen foecundum facere possit: qua de re merito dubitatur: semen tamen facere non potest."

[36] "Seminis calor non est per se foecundus. Nam si sit in luto, vel alia quapiam materia: nullam habet foecunditatis vim."

[37] "Non eamdem esse vim, & naturam seminalis caloris, & solaris. Si namque semen nativum suum calorem amittat; & a Sole tantum accipiat caloris, quantum ante fuit, non fiet tamen ita semen foecundum. Neque caliditas nativa satis est, ut semen foecundum efficiat. Si namque iusto sit semen calidius: vel temperate quidem sit calidum: sed humidius tamen: aut aliud habeat vitii quidpiam: infoecundum est. Quid ergo de solari calore dicamus: cum tot materiae conditiones ad foecunditatem requirantur.

This is why Taurellus holds that vegetative powers must be the effect of a seed form that emerges from a particular structure of the mixture of the seed.

In Taurellus's view, these considerations have implications for the question of whether any living being that is generated sexually also could have been generated spontaneously. As Taurellus argues, in order for a human seed to be produced spontaneously, the same material conditions would be required outside the human body as they are found within the human body; and he takes it to be doubtful that the sun could bring these conditions about (Taurellus 1597, 793). Something analogous holds for the case of spontaneous generation where "some seminal power inheres in putrefying matter, of the kind that inheres in the perfect seed, but less perfect. Therefore, the reason of the generation of animals has to be attributed primary not to heat, but to this matter" (Taurellus 1597, 794).[38] Thus, Taurellus does not exclude the possibility of spontaneous generation but rather holds that in all cases of spontaneous generation the origin of vegetative powers must be explained by new causal powers that emerge from the specific structure of compounds.

Taurellus offers his emergentist views also as a remedy for the weaknesses of Cesalpino's account of sexual generation. Cesalpino still adheres to an Aristotelian one-seed theory of sexual generation according to which the female contributes only matter, while the male contributes the spirit that contains vital heat (Cesalpino 1593, fol. 112r–v). Taurellus offers the hypothesis of substantial forms emerging from the mixture of male and female seeds as an alternative to Cesalpino's hypothesis of celestial heat as the origin of vegetative powers. For instance, he holds after the mixture of seeds, seminal matter forms a unity of its own because it possesses an essential form from which the vegetative powers required in the generation of a living being arise (Taurellus 1597, 817). Already in *De procreatione hominis*, Taurellus denies that the temperament (and hence elementary heat) could be the bearer of generative power: "The seed certainly has its temperament; but this temperament does not have any generative power: Rightly so, for this power does not belong to the seed itself; therefore, the most potent forming power should not be ascribed to it; nor should it be ascribed to heat, be it inherent [in the seed] or coming from the outside" (§ 22).[39] And he offers the hypothesis of seed forms as an alternative to the hypothesis of heat as the origin of vital powers: "We therefore maintain that the form of the seed, in so far as it is a seed, is the primary efficient cause of our generation; very much as the temperament of the seed, and heat, and spirit (in case it exists), and very much as the uterus are not primary causes but assisting causes" (§ 23).[40]

[38] "seminalis quaedam vis in putrescente materia insit: qualis in semine est perfecto: sed imperfectior. Non ergo solari calori primo: sed huic materiae tribuenda est ratio generationis animalium."

[39] "Semini suum equidem est temperamentum: cuius nulla nequit esse vis ad generationem. Huic tamen: cum semini proprie non competat: potentissima vis efformatrix attribui non debet: uti neque calori seu sit insitus, seu sit influens."

[40] "Eam ergo seminis formam qua semen est, primariam censmus esse causam nostrae procreationis effectricem: ut temperamentum eius, et calor, et spiritus (si quis modo est), quemadmodum et uterus causae sint non primariae, sed adiutrices."

From this perspective, Taurellus rejects three possible explanations for the similarity between parents and offspring. As he argues, the only argument for explaining the similarity with respect to species membership through the role of blood is the observation that the offspring of parents belonging to two different species is more similar to the mother; however, Taurellus objects:

> But since this can happen due to some other cause, it is also evidently most absurd because species membership consists more in the spermatic body parts than in the sanguineous part: For the forming power … of blood does either not exist or is at least smaller than the one that resides in the seed. (§ 25).[41]

Taurellus also rejects the explanation of similarity with respect to sex through differences with respect to temperament between the testicles on the right side and those on the left side:

> Each genital vessel belongs to the same species and possesses the same power; after all, everyday experience shows that it is the same seed which has the same effect. Location does not allow us to make a distinction: For the mixed seed complements the total capacity of the uterus, such that it can be most conveniently fostered from each side. (§ 27).[42]

Finally, Taurellus rejects the explanation of similarity with respect to external traits through the workings of the imagination[43]:

> We do not want to neglect the conception of the mind (which is called phantasy): everyone knows very well how great its power is. But it is not yet sufficiently understood what the proximate efficient cause of this kind [of effects] is, if you say that it is the phantasy; for through the understanding of the intellect or the mind nothing other happens than that there occurs an understanding of itself. And certainly, phantasy does not possess forces as strong as it appears to have. (§ 35).[44]

In his critique of Cesalpino, Taurellus rejects the view that the male provides active force while the female provides nothing but passive matter (Taurellus 1597, 811); he uses an anatomic argument from the existence of female testicles (Taurellus 1597, 819) and argues that since both male and female seeds are infertile by themselves, both have to become more perfect through the interaction between their active powers (Taurellus 1597, 819). Already in *De procreatione hominis*, Taurellus defends a medical two-seed theory of animal generation: "But that females are less perfect

[41] "Sed cum hoc alia de causa fieri possit: etiam ob id absurdissimum esse liquet: cum in partibus spermaticis potius, quam sanguineis consistat species: Cumque facultas efformatrix, vel … sanguinis nulla sit: aut saltem eâ modô minor, quam semen obtinet."

[42] "Utriusque vasis genitalis eadem est species, eademque facultas: Quin est idem quidem semen, et eiusdem eius effectum esse quotidiana testatur experientia. Nec ullum nobis hic situs discrimen admittitur: Semen namque concretum omnem totius uteri capacitatem complet: ut omni ex parte foveri possit commodissime."

[43] On early modern imagination theories of trait acquisition, see Smith (2006); Blank (2009).

[44] "Nec animi conceptus (quem phantasiam vocant) negligi volumus: vis enim ei quanta sit, cuius cuivis est notissimum. Sed nondum tamen satis intelligitur, quanam sit huiusce generis effectivum causa proxima: si phantasiam esse dixeris: Cum nihil intellectu, vel animi compraehensione facit aliud: quam quod ipsius est intelligentiae. Nec certe convenit phantasiae tantae esse vires; quantas obtinere videtur."

than males—this comment can in no way be tolerated. In its kind, each is absolute. And if there is some difference in perfection, this does not depend on the first origin but rather on the various accidents of both sexes, such as an otiose life-style from which perhaps the female sex becomes weaker" (§ 30).[45] As he comments, "this weakness does not contribute anything to the generation of females" (§ 31).[46]

This is why in his response to Cesalpino, Taurellus ascribes both to male and female seeds active powers that offer mutual assistance; he takes only the composite into which the seeds are fused to belong to the species and essence of the animal to be generated (Taurellus 1597, 815). At the same time, Taurellus rejects the idea that the composite arising from the mixture of seed already contains preformed structures of body parts but rather holds that it contains structures that, through nutrition and other external influences, can be transmuted into body parts (Taurellus 1597, 815). A parallel passage from *De procreatione hominis* gives a clue to the reason why Taurellus rejects preformationism:

> Even if I see in the fetus most different parts: in the seed I do not see anything that has a relation to bones, body parts, and functional tissues. You may say that they all exist there although they are not perceived by our dull senses. This is the *diakrisis* of the atomists that Galen has often rejected. And certainly not without reason. For there would be no alteration and no generation, if everything remained always what it is. (§ 44).[47]

Thus, if the hypothesis that the seed actually contains the rudiments of all body parts, then living beings would come about through the mere composition (*synkrisis*) of parts and they would decay through the mere separation (*diakrisis*) of parts—exactly what the so-called "syndiacritic hypothesis" that was gaining prominence in early seventeenth-century natural philosophy claimed.[48] Preformationism is thus opposed to the claim that at a certain level of complexity composites could bring forth new causal powers. This is why Taurellus opts for the view that *none* of the structures of a living being are pre-formed in the seed: "We maintain that in the fetus there is nothing that was previously in the seed. For whatever is generated, did not exist before. And this splitting apart does not happen with the goal of generating bones. Because the seed consists of similar parts, everything can be generated out of the quality of a single part" (§ 45).[49]

---

[45] "Atqui vero faeminas esse maribus imperfectiores. Commentum hoc esse profitemur nullo pacto ferendum. In suo genere quidvis est absolutum. Et si quid sit in hac perfectione discriminis: id non a prima dependet origine: sed a variis utriusque sexus accidentibus: cuiusmodi est otiosum vita genus: quo ferè sexus muliebris evadit infirmior."

[46] "Non tamen haec infirmitas quicquam ad faemellarum generatione facit."

[47] "Equidem in faetu partes conspicor longe diversissimas: in semine nihil video, quod ossa: quod membras: quod parenchymata referat. Esse dicis haec omnia: licet sensibus obtusis non percipiantur. Haec Atomistarum est *diakrisis*: quam Galenus saepe reijcit: Et certè non immeritò. Nulla namque alteratio erit, nullaque generatio: si maneat idem perpetuo, quicquid est."

[48] For a detailed study of the development of this hypothesis, see Newman (2006).

[49] "Nos in faetu nihil esse dicimus, quod antea fuerit in semine. Quicquid enim generatur, id ante non fuit. Nec ad ossium generationem illa fit secretio. Cum enim semen sit homoeomeres: ex qualitate unius partis quidvis potest procreari."

Consequently, Taurellus takes the determination of sex to be a matter of the interaction between the active powers of the male seed and the active powers of the female seed. As he conjectures, the dominance of the male seed leads to male offspring, that the dominance of the female seed leads to female offspring, and that equality of powers between male and female seeds leads to hermaphrodites (Taurellus 1597, 815). He extends this pattern of explanation also to trait acquisition, without making the empirically implausible claim that the seed that is stronger in determining sex is also stronger in determining observable traits:

> But perhaps someone may hold that, if the seed of the father wins such that a male is generated, it also wins such that the male generated by the father is similar to the father. This is an admirable process. The tiny seed has almost infinite powers. For there have to be as many powers as there are effects. Through the conjoined powers of both seeds it can happen that the one wins with respect to this or that power, and the other is overcome. (§ 34).[50]

Taken together, these elements articulate a conception of animal generation according to which species membership, the determination of sex and trait acquisition are all explained through the differences between the structure of the constituent of mixtures and through the differences between the substantial forms emerging from different mixtures. These are exactly the differences for which Cesalpino's conjecture of a universal divine principle of activity as the origin of life does not offer any explanation. And this is why Taurellus holds that an emergentist explanation of the origin of vegetative powers provides a biological argument in favor of the substantiality of living beings and their constituents that does not involve a hypothesis of a divine cause inherent in living beings.

12.6 Conclusion

What motivates Taurellus's critique of Cesalpino's account of the generation of living beings is thus his opposition to the theological implications that Cesalpino's account brings with it. In particular, it is the connection between Cesalpino's analysis of vegetative powers and his version of a theory of a unique active intellect that makes Taurellus's critique an essential part of his own philosophical project. Contrary to Cesalpino's view of a unique active intellect that animates all living beings, Taurellus develops an emergentist view according to which the substantial forms of parts of living beings, as well as the substantial forms of living beings have causal powers that go beyond the causal powers of elements and their temperament but at the same time depend for their existence on mixtures of elements. Elements, in turn, are understood as form-like, immaterial principles that, because they do not depend on any other created beings, are substances, not modes of a divine being.

[50] "Sed existimaverit fortasse quispiam: si patris semen vincit: ut mas generatur: etiam vincere: ut mas hic a patre genitus patri sit similis. Mirandum hoc opus est. Semen exiguum vires habet propemodum infinitas: Tot enim oportet esse facultates: quot eius sunt effectus. His itaque seminis utriusque viribus coniunctis: fieri potest ut quod hac aut illa virtute vincet, alia superetur."

Substantial forms that emerge from the mixture of elements depend on composites but are not accidents because they are not composites themselves.

Emergentism, thus, is meant to defend the view that all natural beings act by means of their own natural powers. If so, then it is unnecessary to stipulate any divine principle of activity immanent in nature. Thereby, the degree of independence required by creation theory is upheld—natural beings depend for their existence on divine creation, but after the moment of creation, they act by means of their own powers alone. Taurellus's considerations concerning elementary substantial forms and emergent substantial forms thus provide philosophical support to the view that, after creation, God does not interfere with the course of nature on a regular basis. Taurellus's emergentist account of the origin of vegetative powers thus supports a version of substance pluralism that, in contrast to Cesalpino's theory of the substantiality of living beings, is incompatible with substance monism. And if neither elements nor living beings can be thought of as modes of divine substance, only the doctrine of creation offers an explanation for their coming into being.

Acknowledgments Work on this article has been supported by a research position (P 33429) at the Alpen-Adria-Universität Klagenfurt, funded by the Austrian Science Fund (FWF).

References

Alexander of Aphrodisias. 2008. *De l'âme*. Ed. and trans. M. Bergeron and R. Dufour. Paris: Vrin.

———. 2012. *On the Soul. Part I: Soul as the Form of Body, Parts of the* Soul, *Nourishment and Perception*. Translated by Victor Caston. London: Bloomsbury.

Atran, Scott. 1990. *Cognitive foundations of natural history. Towards an anthropology of science*. Cambridge: Cambridge University Press.

Blank, Andreas. 2009. Material souls and imagination in late Aristotelian embryology. *Annals of Science* 67: 187–204.

———. 2014. Nicolaus Taurellus on forms and elements. *Science in Context* 27: 659–682.

———. 2016. Nicolaus Taurellus. In *Stanford Encyclopedia of Philosophy*, ed. Edward N. Zalta, September 2016 ed. https://plato.stanford.edu/entries/taurellus. Accessed 10 July 2019.

———. 2018. Sixteenth-Century pharmacology and the controversy between reductionism and emergentism. *Perspectives on Science* 26: 157–184.

Caston, Victor. 1997. Epiphenomenalisms, ancient and modern. *Philosophical Review* 106: 309–363.

Cesalpino, Andrea. 1593. *Quaestionum Peripateticarum libri V*. Venice: Iuntas.

Freedman, Joseph S. 2005. Disputations in Europe in the early modern period. In *Hora est! On dissertations*, ed. Douwe D. Breimer, 30–50. Leiden: Universiteitsbibliotheek Leiden.

Freudenthal, Gad. 1999. *Aristotle's theory of material substance: Heat and pneuma, form and soul*. Oxford: Clarendon Press.

Füssel, Marian. 2016. Die Praxis der Disputation. Heuristische Zugänge und theoretische Deutungsangebote. In *Frühneuzeitliche Disputationen. Polyvalente Produktionsapparate gelehrten Wissens*, ed. Marion Gindhart, Hanspeter Marti, and Robert Seidel, 27–48. Köln, Weimar, Wien: Böhlau.

Ganeri, Jonardon. 2010. Emergentisms, ancient and modern. *Mind* 120: 671–703.

Hirai, Hiro. 2007. The invisible hand of God in seeds: Jacob Schegk's theory of plastic faculty. *Early Science and Medicine* 12: 377–404; reprinted as ch. 3 of Hirai, Hiro. 2011. *Medical humanism and natural philosophy: Renaissance debates on matter, life and the soul*. Leiden: Brill.

Hull, David L. 1965. The effect of essentialism on taxonomy—Two thousand years of stasis. *Part I. British Journal for the Philosophy of Science* 15: 314–326.

Leinsle, Ulrich Gottfried. 1985. *Das Ding und die Methode. Methodische Konstitution und Gegenstand der frühen protestantischen Metaphysik*. 2 vols. Augsburg: Maro Verlag.

Lennox, James G. 2001. *Aristotle's philosophy of biology*. Cambridge: Cambridge University Press.

Macdonald, Cynthia, and Graham Macdonald. 2010. Introduction. In *Emergence in mind*, ed. Cynthia Macdonald and Graham Macdonald, 1–21. Oxford: Oxford University Press.

Marti, Hanspeter. 1994. Disputation. In *Historisches Wörterbuch der Rhetorik*. Ed. Gert Ueding, 2: cols. 866–880. Tübingen: Niemeyer.

Mayer, H. C. 1959. Nikolaus Taurellus, der erste Philosoph im Luthertum. *Ein Beitrag zum Problem von Vernunft und Offenbarung*. Dissertation, University of Göttingen.

Newman, William Royal. 2004. *Promethean ambitions. Alchemy and the quest to perfect nature*. Chicago and London: University of Chicago Press.

———. 2006. *Atoms and alchemy. Chymistry and the experimental origins of the scientific revolution*. Chicago and London: University of Chicago Press.

Petersen, Peter. 1921. *Die Geschichte der aristotelischen Philosophie im protestantischen Deutschland*. Hamburg: Meiner.

Preus, Anthony. 1970. Science and philosophy in Aristotle's generation of animals. *Journal of the History of Biology* 3: 1–52.

Schegk, Jacob. 1540. *De causa continente. Eodem interprete Alexandri Aphrodisaei De mixtione libellus*. Tübingen: Ulrich Morhard.

Smith, Justin E.H. 2006. Imagination and the problem of heredity in mechanist embryology. In *The problem of animal generation in early modern philosophy*, ed. Justin E.H. Smith, 80–99. Cambridge: Cambridge University Press.

Solmsen, Friedrich. 1957. The vital heat, the inborn pneuma, and the aether. *Journal of Hellenic Studies* 77: 119–123.

Sorabji, Richard. 2010. Introduction. In *Philoponus and the rejection of Aristotelian science*, ed. Richard Sorabji, 1–40. London: Duckworth.

Taurellus, Nicolaus. 2012. *Philosophiae triumphus, hoc est, metaphysica philosophandi methodus*. Ed. and trans. Henrik Wels, Stuttgart-Bad Cannstatt: Frommann-Holzboog.

———. 1586. *De vita et morte libellus*. Nuremberg: Catharina Gerlach.

———. 1597. *Alpes caesae, Hoc est, Andreae Caesalpini Itali, monstrosa & superba dogmata, discussa & excussa*. Frankfurt: Palthenius.

———. 1604. *Theses de ortu rationalis animae*. Nuremberg: Paulus Kaufmann.

Taurellus, Nicolaus. Undated Ms. *De procreatione hominis theses medicae*. Stadtarchiv Nürnberg D 16 Nr. 573/1, fol. 38r–46v.

Todd, Robert B. 1976. *Alexander of Aphrodisias on Stoic physics. A study of the De mixtione with preliminary essays, text, translation, and commentary*. Leiden: Brill.

Wollgast, Siegfried. 1988. *Philosophie in Deutschland zwischen Reformation und Aufklärung*. Berlin: Akademie Verlag.

Chapter 13
Vegetal Analogy in Early Modern Medicine: Generation as Plant Cutting in Sennert's Early Treatises (1611–1619)

Elisabeth Moreau

Abstract This chapter examines the use of vegetal analogy in late Renaissance physiology through the case of the German physician Daniel Sennert (1572–1637). It is centered on Sennert's explanation of generation, in particular the transmission of life through the vegetative soul within the seed, as developed in his early works on medicine and alchemy, the *Institutionum medicinae libri V* (1611) and *De chymicorum...liber* (1619). This chapter first summarizes Sennert's account of generation and the seed's "formative force" according to Aristotle and Galen, as well as his appraisal of the medical debates on the origin of the seed's soul and form. Then, the next part explores Sennert's own interpretation of the origin of forms, for which plant physiology served as a common denominator of his medical, alchemical and theological inclinations. Finally, this chapter considers how Sennert attempted to harmonize his reasoning with the Paracelsian account of generation, seed and life.

13.1 Introduction

In Renaissance medicine, the vegetative soul was a central concept for the explanation of generation, growth and nutrition. In the case of generation, the faculties of the vegetative soul were considered as driving the development of the seed through its own "formative" force.[1] The origin of the soul within the seed, and more broadly, the transmission of life and physiological functions from parent to offspring, was one of the most difficult questions in the medical philosophy. Following Galen's

[1] "Generation" (*generatio*) here refers to reproduction as a physiological function common to all living beings. The same term could also designate the broader process of coming into being that applied to natural things, in reference to the Aristotelian physics.

E. Moreau (✉)
FNRS – Université libre de Bruxelles, Brussels, Belgium
e-mail: Elisabeth.Moreau@ulb.be

F. Baldassarri, A. Blank (eds.), *Vegetative Powers*, International Archives of the History of Ideas Archives internationales d'histoire des idées 234,
https://doi.org/10.1007/978-3-030-69709-9_13

The Construction of the Embryo, as well as Aristotle's *Generation of Animals*, many Renaissance physicians attempted to solve the origin of the vegetative soul by investigating the seed's "form" – in the Aristotelian sense of "essence" – as an active principle (Galen 1997; Hirai 2011a; Deer Richardson 2018; Roger 1997). Among them, the German physician Daniel Sennert (1572–1637) was constantly discussing this question throughout his medical and alchemical works.[2] A professor of medicine at the University of Wittenberg, Sennert was representative of the early modern German physicians learned in Aristotelian and Galenic philosophy, who strove to introduce the teaching of alchemy in the university program (Michael 1997, 2001; Stolberg 2003; Clericuzio 2000, 9–34). His works on natural philosophy, medicine and alchemy had a large audience in the seventeenth century as evidenced in the re–editions of his *Opera omnia* between 1641 and 1676.

Presented by historians as a precursor of van Helmont and a direct source for Boyle, Sennert has been the object of numerous studies on early modern science. Since the 1990s, his atomistic philosophy has garnered growing attention in the historiographical current on early theories of matter (Newman 2006, 85–156; Michael 1997, 2001; Lüthy 2005; Lüthy and Newman 2000; Clericuzio 2000, 9–34). It has been shown that Sennert merged the Aristotelian physics of elements and matter–form with atomistic concepts that he ascribed to Democritus. His theory of matter was grounded in late Renaissance natural philosophy as well as alchemical theory and practice, from late medieval Latin-Arabic authors to Paracelsus' followers and detractors.

Interestingly, Sennert's atomist philosophy had a medical side which has been explored regarding his theory of generation in the *Hypomnemata physica* (1636) (Stolberg 1993, 2003; Hirai 2011a, 151–172). In this treatise, Sennert considered seeds as living particles and atoms, which encapsulated the body's soul and superior form for their "multiplication" during generation. Following Galenic and Paracelsian accounts of generation, his interpretation was based on a theological conception of the transmission of the soul from parent to offspring, which stemmed from his Lutheran background.

While Sennert's theory of generation was imbued with medicine, natural philosophy, alchemy and theology, it also resorted to vegetal analogy to explain the phenomenon of reproduction. In this regard, Hiro Hirai has shown Sennert's analogy between spontaneous generation, the origin of forms and the formation of mushrooms in the *Hypomnemata* (Hirai 2011a, 151–172; Hirai 2015). However, prior to this treatise, Sennert already developed a vegetal explanatory model of generation in some of his medical and alchemical works. This vegetal representation was rooted in the longstanding medical concern with the vegetative soul. Among Sennert's sources on this theme, the Aristotelian and Galenic physiological texts abounded in the analogy between plants and the human being (Totelin 2018). As the most basic form of life, the vegetal realm was indeed the object of analogical reasoning in the medical tradition, as it provided visible and familiar evidence for otherwise obscure phenomena, especially in embryology (Holmes 2017).

---

[2] On this theme, see also Bigotti in this volume.

In addition to his vegetal model of generation, Sennert showed a broader interest in botany from his earliest works. For instance, his *Epitome naturalis scientiae* (1618) includes a book dedicated to the vegetative soul as well as the parts, differences and history of plants according to Aristotle and the Italian physician Julius Caesar Scaliger (1484–1558) (Sennert 1618, 409–494). Figuring prominently in Sennert's works, Scaliger applied an Aristotelian interpretation of matter–form to plant physiology, vegetal reproduction in particular, in his commentaries on ancient botanical treatises (Blank 2010, 53–72; Blank 2012, 503–523). These botanical works likely played an influential role in Sennert's medical philosophy along with Scaliger's *Exercitationes* (1557).

Moreover, from his alchemical "turn" around 1619, Sennert was familiar with the multiple vegetal metaphorical terms such as "seed", "root", "fruit" and "transplantation" in Paracelsian alchemy. One of his main sources, the Danish physician Petrus Severinus (c.1540–1602), claimed in his Paracelsian manifesto, the *Idea medicinae philosophicae* (1571), the importance of agriculture and the *res rustica* for the knowledge of nature (Severinus 1571, preface and 22; Shackelford 2004, 183 and 205). For this reason, Paracelsian alchemy also needs to be considered in the maturation of Sennert's view on generation and the vegetative soul.

In the context of his medical and alchemical interest in plants, Sennert began to develop his vegetal explanation of generation and seed propagation in two major treatises (Sennert 1611; Sennert 1619).[3] It was first presented in a physiological account of generation included in a systematic treatise on medicine, the *Institutionum medicinae libri V* (1611). This account was partially updated in a treatise on alchemy, *De chymicorum cum Aristotelicis et Galenicis consensu ac dissensu liber* (1619). As will be shown in this chapter, Sennert's interpretation of generation in these early treatises throws light on his medical theory of matter before his progressive "atomistic turn" around 1619.[4] Sennert indeed used the analogy with vegetal generation, an observable yet specific phenomenon, in order to explain the reproduction of all living beings at the level of their smallest components (Bailer–Jones 2002). Interestingly, his vegetal explanatory model of generation operated as a common denominator of the medical, alchemical and theological frameworks espoused in his works.

In this chapter, I explore the *Institutionum...libri* and *De chymicorum...liber* to elucidate Sennert's early theory of generation and his reception of the Paracelsian view on the seed and the transmission of life. This chapter first examines Sennert's medical account of generation and his appraisal of the main interpretations of the origin of forms. It then follows with his own explanation inspired from natural philosophy and horticulture. Finally, this chapter investigates how Sennert attempted to

---

[3] For the abridged version of these treatises, see Sennert, 1656, 1662.

[4] Sennert progressively adopted an atomistic view around 1619 as he considered elements as discrete fragments in the first edition of *De chymicorum...liber* (1619), while in the second edition of the same treatise (1629), elements were defined as discontinuous and intact units (Newman 2006, 85–156).

reconcile his account of generation and the origin of forms with the Paracelsian philosophy.

13.2 The Transmission of the Soul During Generation

Sennert provided his earliest medical account of generation in the *Institutiones medicinae* (1611). This systematic and didactic work consists of five books dedicated to physiology, pathology, "semiotics" (symptomatology), "hygiene" (dietetics) and therapeutics (Sennert 1611, 69a–86b). In structure and content, the *Institutiones* followed the example of late Renaissance Galenic *summae* along the lines of eponymous works by Leonhart Fuchs (1555) and Johan van Heurne (1592) (Fuchs 1555; Heurnius 1592). Sennert attempted to imitate both illustrious doctors from the reformed tradition by publishing his own *Institutiones* in Wittenberg, where he showed his mastery of Aristotelian physics and revealed his interest in alchemical pharmacology. In so doing, Sennert proposed his Aristotelian interpretation of debated notions, such as the union or "mixture" of elements and the "substantial form" of beings. His natural philosophy was indebted to that of Scaliger as an heir of the "Latin Pluralist" account of Aristotelian philosophy, which was promoted by the school of Padua in the Renaissance (Sakamoto 2016; Blank 2010, 27–52; Lüthy 2001). Following this current, Sennert posited that all beings were made of a hierarchy of substantial forms within discontinuous units of matter, also called "natural minima" (Michael 1997, 2001).

Sennert's treatise *De chymicorum...liber* (1619) was aimed to reconcile the Aristotelian–Galenic tradition with the "new" Paracelsian philosophy. In this work, Sennert maintained his stance on generation in the context of a chapter dedicated to the notions of substantial form and seed and to the Paracelsian account of "stars" (*astra*) (Sennert 1619, 189–230). Following the same interpretation as in the *Institutiones*, he added some brief remarks in reference to Scaliger and Paracelsian physicians, such as Petrus Severinus and Thomas Moffet (1553–1604). This led Sennert to posit a broader comparison between the Aristotelian and Paracelsian definitions of life. In the present and next sections of this chapter, I examine Sennert's discussion on the origin of forms in both the *Institutiones* and *De chymicorum...liber*, while the Paracelsian facet of his theory will be discussed in the last section.

In the first book of the *Institutiones*, Sennert explores the phenomenon of generation following Galen's *On Semen* (*De semine*) and Aristotle's *Generation of Animals* (*De generatione animalium*) (Galen 1992). He first states that generation aims at the conservation of living beings, including human, animal and plant species. Their seed contains a "generative" or "formative" force (*vis generatrix* or *formatrix*), through which they produce an animate being similar to themselves (Sennert 1611, 69b, 1619, 189). In the case of plants, this formative force is stimulated by solar heat and ambient air while, in the case of humans and "perfect" (achieved) animals, it comes from the seed emitted by the male in the female uterus. The seed is then mingled with menstrual blood, which serves as a material principle within the uterus. There, the seed is retained, warmed up and nurtured in order to be developed at the end of gestation or pregnancy.

Having established the general role of the seed in the generation of the fetus, Sennert further examines the seed's nature and composition following Aristotle's *Generation of Animals*. He begins by describing the seed as a subtle body, hot and moist in nature. It is produced from a "nobler" matter, which is abundant in the body parts during nutrition (Sennert 1611, 71b). During the last digestive phase or "third concoction", the body parts provide the necessary material for producing the seed, which the medical "spirits" send to the testes.[5] During the development of the fetus, it is the seed's formative force that prepares the seminal matter by assigning it a size, number, figure, order and position. The seed material is then subject to a "delineation" and a "signature", in other words, it receives the visible features of the fetus' body parts (Sennert 1611, 71b–72a; Kikuchihara and Hirai 2015).[6]

As Sennert explains, the key to the formation of animate beings during generation lies in the seed's formative force, which is a faculty of the vegetative soul. In reference to Galen's *The Construction of the Embryo* (*De foetuum formatione*), he recounts that the formative force has long remained an obscure and elusive concept (Galen 1997, 200–201; Hirai 2011a, 151–172). Galen deplored that philosophers had been discouraged to ever elucidate its nature as testified the various theories on the active nature of the seed, which was sometimes called "formative reason" or "plastic power" (Sennert 1611, 74b).[7] Nonetheless, Galen contended that the seminal cause of formation necessarily required the highest "skill" (τέχνῃ) and "intelligence" (σοφίᾳ) to produce an animal. In Sennert's view, such requirements for the formation of an animate being can only be operated by the soul itself through its faculties and its essence or "substantial form". Consequently, it is the origin of the seed's substantial form which needs to be investigated to understand the generation of living beings and the transmission of the soul from parent to offspring.

Before unfolding his position on the origin of the seed's form, Sennert appraises two main interpretations of this question within his medical sources. First, he examines the Platonic accounts of an "external", i.e., celestial, origin of forms. Then, he discusses the Aristotelian views on the "internal" origin of forms within the seed. The core of his argument is very similar to that of his later *Hypomnemata* (1636),

---

[5] Arist. *Gen. an.* 1.18, 726a16–726a28.

[6] Sennert 1611, 1.10, 71b–72a: '[...] sed [accipiendum est] quod illud, quod in partibus ad alendum superat et abundat, cum spiritibus ad testes mittatur, materiamque semini suppeditet. [...] Plerique enim seminis materiam a tertia coctione peti existimant; et quia a tertia coctione seminis materia decidatur et suppeditur, rudem quasi delineationem, signaturam et formam praecipuarum partium in se complecti statuunt. [...] quae tamen delineatio rectius soli animae et formatrici facultati tribuenda videtur.'

[7] Sennert 1611, 1.10, 74b: 'Alii enim in semine animam inesse, atque ab ea. omnes corporis partes delineari fabricarique statuunt: alii animam inesse negant in semine, et solum λόγον quendam πλαστικόν, seu δύναμιν πλαστικήν, et vires quasdam, potentiam et facultatem haec efficiendi in semine inesse dicunt. Verum alii aliter id explicant.'

which has been examined by Hiro Hirai (Hirai 2011a, 151–172). For this reason, I shall summarize the main steps of Sennert's appraisal before exploring his own explanation in the next section.

Sennert first considers the Platonic supporters of the celestial origin of forms in the medical tradition. One of his main sources, the French physician Jean Fernel (c.1497–1558), held this stance in his *Universa medicina* (1567), a systematic work on medicine, which was reedited many times in the early modern period (Sennert 1611, 74b–75a; Sennert 1619, 191–193). Fernel's medical philosophy was emblematic of the Renaissance Platonic response to the "materialistic" interpretation of Galen, which explained all physiological phenomena by the simple mixture of elements (Zanier 1987). In his treatise *On the Hidden Causes of Things* (*De abditis rerum causis*) (1548) included in *Universa medicina*, Fernel argued that all physiological functions, because they were related to the body's vital principle, had specific causes of celestial nature that were associated to their substantial form (Fernel 2005; Deer Richardson 1985). As Sennert points out, Fernel emphasized that the form of living beings had a celestial origin received by the seminal matter, which was well-disposed by the body's innate heat (Fernel 2003; Hirai 2011a, 46–79). According to this view, the heavens sent some "perfection" that stimulated life in the seminal species, whose matter was beforehand prepared. However, this reasoning leaves Sennert unconvinced.

In *De chymicorum…liber*, Sennert also alludes to the longstanding Avicennian account of celestial causation through a "giver of forms" (*dator formarum*). While Avicenna was an important authority in medical learning for his *Canon of Medicine*, he also provided an extensive work on Aristotelian natural philosophy. In his *Metaphysics*, he developed an account of celestial causation in the context of an emanationist cosmology. Accordingly, the "giver of forms" was a celestial emanation of the active intellect, which gave a form to the well-disposed matter of living beings. Sennert considers this notion as a subordinate deity, following Scaliger's criticism in *De plantis* (Sennert 1619, 190; Scaliger 1566, 29a).[8] In his view, forms are well and truly divine but cannot be transmitted by an external entity like the heavens because it would make the "equivocal" reproduction of "inferior" animals, that is spontaneous generation, impossible. This is the occasion for Sennert to recall the explanation of generation from *Genesis*. God created animals by endowing them with fertile forces within their seeds just as he did for plants, yet before creating the stars. With this reasoning, Sennert defends the internal origin of the seed's form, while maintaining its divine provenance following an interpretation compliant with the Scriptures.

Having expressed his doubts on the celestial "impression" of forms during generation, Sennert goes on to examine the Aristotelian view on the internal origin of

[8] Scaliger 1566, 1, 29a: 'Cum tamen ne inter primos quidem Philosophos satis constet: quis sit formarum dator, aut unde proficiscantur: ἐκτίνος ἐκμαγείου depromantur illae. Non enim facilis patere videtur ingeniis humanis aditus ad huiusce sacrarii penetralia. Adeo vero exagitati sunt sapientes, ut Deum quendam (loquar illorum more) sub alterum crearit Avicena tuus ille Scaliger: cuius Dei tum beneficio factae, tum officio datae formae reciperentur in materiam.'

the seed's form. In *De chymicorum...liber*, he begins by examining the theory of the "eduction" (*eductio*) of forms (Sennert 1619, 193–195). As he explains, this theory asserts the emergence of forms from the potency of matter, hence suggesting that the form of animate beings is subject to generation and corruption as it is coextensive of the body. Although Sennert correctly attributes this stance to the Aristotelian scholars of his time, it should be noted that scholastic philosophers rejected the "eduction" of forms applied to the rational (human) soul and affiliated it with the "materialistic" philosophy of Alexander of Aphrodisias (Pluta 2007). During the Renaissance, the philosophy of Alexander was, nonetheless, rediscovered and endorsed by some scholars from the University of Bologna and Padua, such as Pietro Pomponazzi (1462–1525), who claimed that Aristotle supported the mortality of the soul (Michael 2000). As for Sennert, he rejects the "eduction" of forms related to the vegetative soul because it assigns to matter the role of form as an efficient entity able to animate the seed. Since it presents matter as a "nobler" principle than the form, the theory of "eduction" subverts the Aristotelian physics of matter–form and needs to be dismissed.

Sennert then gives a closer look at the application of the "emergence" (*emersio*) of forms, following Scaliger's terms, to the physiological explanation of generation (Sennert 1619, 196; Scaliger 1557, 13v).[9] In his view, this theory implies that the form "draws" itself from the status of potentiality to that of actuality by preparing the seminal matter. At the same time, the progenitor's form moves from the status of "first actuality" to that of "second actuality" as a secondary instrumental cause able to vivify the seed (Sennert 1611, 75a–78a; Sennert 1619, 199–202). Sennert identifies this stance to that of the German physician Jacob Schegk (1511–1587), who was a professor at the medical faculty of Tübingen and a renown Aristotelian philosopher in the German intellectual world. In his *On the Plastic Faculty of the Seed* (*De plastica seminis facultate*) (1580), Schegk defined the seed's formative force or "plastic reason" (λόγος πλαστικός) as a "second actuality" related to the substantial form, hence potentially animate (Hirai 2011a, 80–103). Sennert deems this reasoning unsatisfying as it suggests that the seed's form is only an instrumental cause related to the spermatic moisture, which is still inanimate.

The main reason for Sennert's rejection of the Aristotelian theory of emergence is his understanding of the seed's formative force as the "formal agent" and "first actuality" pertaining to the soul. To support this view, he refers to Aristotle's *Generation of Animals* by stating that any form that accomplishes the operations of the soul is not only "noble" and superior but can only be the soul itself.[10] Such a form is an "efficient" entity that operates in actuality the vital functions within the seed in order to vivify and shape the fetus. Therefore, Sennert insists, the seed's

[9] Scaliger 1557, 6.5, 13v: 'Formam esse in semine canino: cuius in potestate dicitur esse, quia semen est potens dare formam, quam in se continet. Educitur autem de ea potentia remota, qui est actus primus, ad potentiam propinquam, qui est actus secundus: scilicet ut forma sit in eadem materia ad eum modum qui nullis egeat adminiculis: ut fine suo fruatur, ad quem comparatum est totum compositum. [...] Caeterum est emersio potius, quam eductio.'

[10] Arist. *Gen. an.* 2.3, 736a24–736b20.

form stems neither from the heavens as claimed by Platonic physicians, nor from a subordinate instrumental cause as established by the Aristotelian supporters of the emergence of forms.

In concluding his appraisal, Sennert finds it more relevant to consider the transmission of the progenitors' soul to their seed as a "multiplication" of forms. On this point, he relies on the Paduan philosopher Giacomo Zabarella (1533–1589) who asserted, following Albert the Great, that any form was "multiplicative" of itself to the extent that all animate beings generated their own kind by multiplying their form (Sennert 1619, 196–197; Zabarella 1590, 589; Spruit 2008). While Zabarella developed this argument in the context of an Aristotelian theory of perception, Sennert uses it in a physiological framework in order to overcome the aporias in his medical sources, namely Fernel, Avicenna and Schegk. This leads him to espouse his own interpretation of the transmission of life by animate beings, which is examined in the next section.

13.3 Multiplication of Forms and Horticulture

Although Sennert denies the celestial origin of the seed's form, he acknowledges that its initial provenance is somewhat divine. In his view, this "noble" origin, which has long remained unexplained, comes from the divine blessing of germinating plants and multiplying animals and humans as expounded in *Genesis* (Sennert 1611, 80a; Sennert 1619, 197).[11] God created the forms of living beings, which henceforth have been propagated by each progenitor's seeds, a phenomenon that Sennert calls "traduction" (*traductio*) of forms (Stolberg 2003; Vidal 2011, 21–57). Established in theological sources, the traducian theory stated the transmission of the soul by a portion of the parent's seed. Adopted by the Lutheran Church, traducianism was nonetheless in the minority and poorly used in medical treatises. Instead, the doctrine commonly adopted by the Catholic and Calvinist Churches was the "creationist" interpretation of the soul as infused by God during the development of the embryo.

According to Sennert, the "traduction" of forms implies that of the soul from a physiological point of view. The soul within the seed is "latent" to the extent that it is alive and able to germinate if stimulated by heat and moisture. As Sennert points out, this is manifest in the case of plants, whose seeds are preserved by the vegetative soul. Even at rest, vegetal seeds remain fertile for a certain duration, so that they are able to operate the actions of the soul if placed in a suitable material (Sennert 1611, 75b–76a, 1619, 197–198).[12] While the case of plant generation corroborates

[11] See *Genesis* 1:11, 22 and 28.

[12] Sennert 1611, 1.10, 75b–76a: 'Quodvis enim semen, ut in plantis manifestum est, vegetante anima conservatur et aliquandiu prolificum permanet: et quandiu integrum, et incorruptum est, in loco idoneo, et praesente alimento, ut vivens operatur et exercet suas actiones in eam, quae praesto est, materiam, non secus, ut ipsum vivens integrum omnibus partibus [...]. Nam eaedem

the presence of a "latent" soul in the seed, Sennert still has to show the relevance of vegetal generation for the rest of the living world, in particular, "perfect" animals, including human beings.

For this purpose, Sennert first considers plant reproduction by taking the example of the willow and rosebush. Both multiply by cutting, i.e., through the section of a stem planted in the earth in the same way as grafting and layering. According to Sennert, the botanical phenomenon of plant cutting, which the current horticulture calls vegetative "multiplication" or "propagation", is the explanatory model for the generation of all living beings – plants, animals and humans. Vegetative propagation is, indeed, the object of an analogical reasoning that Sennert extends to the reproduction of all living beings to build an interpretation of generation upon a familiar and observable phenomenon (Bailer–Jones 2002). From the ancient times, plant propagation had been known by naturalists and gardeners, while cutting was common horticultural practice (Ambrosoli 1997). If Wittenberg did not offer a botanical garden in the early seventeenth century, Sennert still had access to private gardens and longstanding botanical literature on vegetal reproduction, from Theophrastus and Pliny to Cesalpino (Bellorini 2016).[13]

In the late Renaissance, the analogy with vegetal reproduction went beyond the common biological framework as it stimulated original interpretations in medicine and natural philosophy. Plant grafting, in particular, raised the attention of scholars following the works of Giambattista Della Porta (1535–1615) on botany and natural magic. In this regard, Della Porta's theory and experiments concerning plant grafting inspired Gaspare Tagliacozzi on plastic surgery, William Gilbert on magnetic polarity and Francis Bacon on the prolongation of human life (Savoia 2017; Oppenheimer 1953; Rusu 2020). In the case of Sennert, it was the broader phenomenon of vegetative propagation that was at the center of his medical theory of generation in order to buttress his interpretation of matter–form and the vegetative soul.

Throughout Sennert's discussion on generation, the analogy with vegetative propagation allows to visualize and explain the essential mechanisms of embryological growth despite its hidden and complex character (Holmes 2017). What vegetal propagation reveals, according to Sennert, is the status of the plant's torn stem as a material which contains a "particle" of its soul and form. The latter, in turn, allows the cutting to grow (Sennert 1611, 78b; Sennert 1619, 202–203).[14] Although

operationes in semine, et in planta omnibus numeris integra conspiciuntur [...]. Eadem enim est omnino operatio, cum anima in semine latens ex attracta materia corpus plantae fabricat [...].'

[13] On ancient accounts of vegetative propagation, see for instance Theophr. *Hist. pl.* 2; Plin. *HN* 17.12. In the early modern period, Robert Sharrock, an English botanist and friend of Robert Boyle, dedicated a treatise on this topic in his 1660 *History of the Propagation and Improvement of Vegetables* (Webster 1966).

[14] Sennert 1611, 1.10, 78a: 'Atque animam in semine haec omnia efficere, neque formationem corporis animati alterius rei opus esse: satis quoque videtur probare plantarum nonnullarum generatio, quae ex parte a planta avulsa propagantur; cum scilicet particula animae cum parte materiae cohaerens avellitur, unde planta priori similis excrescit. [...] sicut in hoc propagationis modo ab animae parte cum parte corporis avulsa plantae formantur: ita etiam in semine vim formatricem partium non ulli alii attribuendam esse.'

Sennert does not state here that the form of an animate being is a divisible quantity, he believes that it is somewhat divided at the same time as matter during generation (Roger 1997 [1963], 84–86). The part of the soul that is detached with the seed's body makes an animate being which is similar to the parent. As Sennert acknowledges, reproduction by cutting does not occur in all plant species, nor does it exist in animals, either oviparous or viviparous. A similar phenomenon, however, is observed in the case of animals: their seeds or eggs act as cuttings to the extent that they are torn particles, which enclose a part of their soul. Sennert further supports the idea of a division of the soul during generation by quoting the Greek atomist philosopher Epicurus. As reported by Pseudo-Plutarch's *On the Opinions of the Philosophers* (*De placitis philosophorum*), Epicurus defined the seed as a "detached portion" or "particle" of body and soul (Sennert 1611, 70a, 1619, 204; Perseus Digital Library 2020).[15]

Sennert's accent on the form's ability to tear further reflects his theory of matter in the first edition of the *Institutiones* (1611) and *De chymicorum…liber* (1619). At that time, he was in the early stage of his Aristotelian matter theory following the "Averroist" account of elements and mixture (Newman 2006, 85–156; Michael 2001; Lüthy 2005). Before around 1629, Sennert adopted this stance along the lines of Zabarella and Scaliger as representatives of the "Latin Pluralism" promoted by Aristotelian philosophers of Padua in the Renaissance (Michael 1997, 2011). According to this view, elements were the smallest or "minimal" parts of bodies. When subject to a "mixture" for the constitution of a new being, they gathered as contiguous portions, which joined into a homogeneous compound. During mixture, their substantial forms tore and united in a plurality of subordinate forms, which constituted the new "median" form of the compound. Such a "tearing" (*refractio*) of forms in a range of diverse degrees reflected the hierarchy of beings as suggested by Aristotle's *Metaphysics*. Sennert fully supported this claim as he regularly insisted that the creation of natural beings obeyed a hierarchy of increasing degrees, which ended with the human being as the "noblest" living being.

Within a framework indebted to the "pluralist" Aristotelian physics of matter–form, Sennert's theory of matter was shaped by his progressive adherence to ancient atomism. Previous studies on Sennert's alchemical theories have shown that he later defined compounds as "atoms" endowed with a superior form (Newman 2006, 85–156). For this interpretation, he referred to Democritus, who was an important figure for the Renaissance "atomist revival" from the late sixteenth century (Lüthy 2000). Interestingly, in his early medical explanation of generation, Sennert rather alluded to Epicurus by way of *De placitis philosophorum*, hence providing an additional clue to his interest in atomistic explanations and authors. He openly shared Epicurus' approach to body and soul as discrete entities that were subject to division and tearing during the generation of living beings. This reasoning coincided with Sennert's "pluralist" stance that the matter of beings was made of contiguous parts, while their substantial form was subject to some "tearing". In this context, Sennert

[15] Sennert 1611, 1.10, 70a: 'Verum non incommode Epicurus, […] ψυχῆς καὶ σώματος ἀπόσπασμα esse dicebat. Viventia enim dum generant, aliquid de sua materia et sua forma largiuntur, semen exhibendo […].'

considered that Epicurus' account of the seed corroborated his own definition of seeds as torn particles, which contained a "latent" soul.

In claiming the presence of a latent soul within the seed, Sennert recalls Aristotle's description of the seed and its cause of formation in *Generation of Animals*. In this treatise, the seed was presented as a foamy body, which enclosed vital heat as a "spirit" (*pneuma*) whose nature was analogous to the element of the stars (Sennert 1611, 70a; Sennert 1619, 204).[16] In the late Renaissance, this statement was propagated by a major source of Sennert's medical philosophy: Jean Fernel. For his Platonic account of Galenic medicine, Fernel stressed the celestial and divine part of the living body as related to its soul and form (Hirai 2011a, 46–79; Walker 1958). At the physiological level, this celestial entity corresponded to "innate" heat, which served as an instrument of the soul to operate vital functions such as generation, growth and nutrition. For Sennert, the presence of innate heat within the seed makes the seminal matter suitable for the propagation of the soul and the generation of a similar animal. With the help of innate heat and the medical "spirits," the latent soul within the seed deploys its virtues and shapes the seed's material into a new animate being (Sennert 1611, 79a, 1619, 204–205).[17]

Because the soul makes for itself a suitable instrument to perform its duties, Sennert considers it as the "architect" of its own home (Sennert 1611, 89ab, 1619, 205). With this explanation, he seeks to comply with Aristotle's requirement of a principle of motion within the seed coming from the parent's form in actuality.[18] At the same time, Sennert's statement refers to Themistius, via Scaliger, who asserted that the seed's form had a most "noble" and "intelligent" virtue comparable to the architect of the Temple (Sennert 1619, 188; Scaliger 1557, 14r; Hirai 2011a, 151–172). In *De chymicorum...liber*, Sennert adds that the seed's formative force is an admirable aspect of divine providence (Sennert 1619, 205).[19] During the creation, God ordained the multiplication of such a "smallest" body, which was efficient enough for the conservation of species until the end of time. As Sennert explains, it was the ignorance of the Christian doctrine of creation that prevented Galen from being able to explain the origin of the formative force within the seed of animate beings.

[16] Sennert 1611, 1.10, 70a: '[...] estque semen corpus quoddam spirituosum, calidum et humidum, in testibus genitum, seu spiritu et θέρμῳ θειοτέρῳ τῶν καλούμενων στοιχείων, καὶ ἀναλόγῳ τῷ τῶν ἄστρων στοιχείῳ plenum, ad animae propagationem, et similis animalis generationem aptum.' See Arist. *Gen. an.* 2.3, 736b29–737a6.

[17] Sennert 1611, 1.10, 79a: '[...] ex semine ob hanc animam iam novum animal exoritur, dum anima in semine latens sese exserit et suas virtutes explicat, et operando sese manifestat, calidoque et spiritibus utens, omnia, quae ad animalis constitutionem necessaria sunt, fabricare exorditur, subiectamque materiam distinguit, disponit, ordinat, format, et effingit [...].'

[18] Arist. *Gen. an.* 2.1, 734b19–735a4.

[19] Sennert 1619, 9, 205: 'Neque quem exiguum illud seminis corpus offendat, sed potius Creatoris sapientiam, potentiam, bonitatem, qui cum exiguo corpore formas ad specierum conservationem ad finem usque mundi transferri et multiplicari in prima creatione et voluit et iussit, hic attentius admiremur: formarumque praeterea nobilitatem aestimemus, quae in minimo corpore aeque suam essentiam et potentias integras retinere possunt, ac in maximo.'

To support his interpretation of the "vegetal" multiplication of forms, Sennert had two possible and non–exclusive sources: Scaliger and Tertullian. As Sennert's priority source, Scaliger mentioned the case of the graft's transmission of its form but limited it to spontaneous generation, without adopting a traducian interpretation of the soul (Sakamoto 2016, 130–131). On the other hand, the Church father Tertullian (ca.160–225) used the same metaphor of vegetative multiplication to establish his view on traducianism – the Latin term *tradux* meaning "graft". This doctrine provided a corporeal interpretation of the soul by stating the transmission of the original sin by individuals of each generation as the "grafts", namely the offspring, of the previous generation.[20] In his treatise *On the Soul* (*De anima*), Tertullian explained that all shrub, stem and offspring contained the force of its soul and the things necessary for the generation of its own kind.[21] Setting aside the traducian interpretation of generation and ensoulment, the works of Tertullian raised discussion among early modern German reformed scholars, for instance Melanchthon in Wittenberg, in an essentially theological context (Fraenkel 1982).

In his turn, Sennert naturalizes Tertullian's metaphor to describe the propagation of the soul and form in a physiological framework. This allows him to clarify the mode of transmission of the seed's form by conforming to Aristotle's requirement of a form in actuality and to Galen's description of the formative power. Sennert merges this point with late Renaissance physiological theory of the body's vital heat as an instrument of the soul. He is careful, however, to clarify that his account is only centered on the vegetative soul present in the seeds of all living beings. Cautiously, he refrains from pronouncing on the origin of the rational and immortal soul that is specific to humans (Sennert 1611, 80b, 1619, 221).[22] However, this did not prevent Sennert from being the target of a controversy, in 1632–33, about his conception of the soul, form and innate heat as well as his atomistic matter theory (Clericuzio 2001, 30–32). Beyond his original interpretation of the Aristotelian and Galenic tradition, it was, overall, his status as a Paracelsian philosopher that was under attack. While the particulars of this controversy lie beyond the scope of this chapter, Sennert's appeal for Paracelsian alchemy in his early theory of generation is examined in more details in the following section.

---

[20] Traducianism was opposed to the doctrine of creationism as a divine infusion of the soul in each individual during generation, as was promoted by Lactantius (245–325) (Givens 2010, 99–128; Hirai 2011b).

[21] Tert. *De anim.* 19: 'Siquidem et illis necdum arbusculis, sed stipitibus adhuc, et surculis etiam nunc simul de scrobibus oriuntur, inest propria vis animae. [...] Aut unde mox illis et frutices inoculantur, et folia formantur, et germina inflantur, et flosculi inornantur, et succi condiuntur: si non in ipsis omnis paratura generis quiescit, et partibus promota grandescit?'

[22] Sennert 1611, 1.10, 80b: 'An vero haec, quae de anima in semine existente hactenus diximus, et quibus animae vegetantis, ut et sentientis, praesentiam in semine probavimus, de Rationali quoque anima intelligenda sint: hic non decidimus. Neque enim gravissimam illam quaestionem, de animae Rationalis, quae immortalis est, origine, hic discutiendam proposuimus, sed solum vim formatricem corporis animati inquirere voluimus.'

13.4 Forms, Seeds, and "Stars": A Paracelsian Reconnection

Sennert's works in natural philosophy and medicine took an alchemical "turn" in 1619 with the publication of *De chymicorum...liber*. The project of this treatise was to show the utility of alchemy for the preparation of medicinal drugs. Once a substance was subject to the extraction or "separation" of its alchemical principles, its powerful volatile part could be "fixed" into a moderate substance. For Sennert, this meant that poisonous minerals could be tamed into safe and efficacious remedies. Following this reasoning, he intended to reconcile Paracelsian alchemy with the medical tradition. In promoting such a "chemical compromise", Sennert nonetheless recommended to reappraise a series of debated ideas in the Paracelsian philosophy. In particular, he had in mind the Paracelsian penchant for neologisms, the excessive correspondence between macrocosm and microcosm and the reference to some religious vocation (Sennert 1619, 108–124).

Whereas Sennert did not deny the status of alchemy as an ancient source of knowledge, he rejected its sacred character promoted by Paracelsus and his followers, who at times considered themselves as "priests of nature". Moreover, the Paracelsian terminology reflected, in Sennert's eyes, a new "way" of acquiring knowledge, which diverged from the tradition based on reason and experience. As a typical Aristotelian philosopher, Sennert strongly believed that correct speech clearly expressed the meaning of thought, while erroneous discourse required a long process of deciphering and understanding. For this reason, he proposed to demystify the Paracelsian philosophy in light of Aristotle and Galen.

In *De chymicorum...liber*, Sennert recounts the concepts of "seed" and "star" at the center of the Paracelsian theory of generation (Sennert 1619, 178–182). His account mainly comes from Severinus' *Idea medicinae philosophicae* (1571), one of the earliest digests of Paracelsian medicine. Aimed at late Renaissance humanists, the *Idea* synthetized the works of Paracelsus with ancient philosophers, above all Hippocrates and Plato, hence making a major contribution to the diffusion of Paracelsianism in the early modern period (Shackelford 2002; Hirai 2005, 217–265; Bianchi 1982). In this treatise, Severinus considered "seeds" (*semina*) as the foundation of nature and knowledge but noted that the tradition gave them the restricted meaning of a fertile material involved in the reproduction of living beings. In contrast, Severinus more broadly defined seeds as the invisible and incorruptible principles of generation of all natural things, including mineral and celestial bodies.[23] He also called the seeds "stars" (*astra*) to the extent that they were at the origin of celestial cycles. Through their status of link (*vinculum*) between the higher and lower worlds, seeds as "stars" also influenced all beings of the terrestrial world

[23] The Paracelsian "generation" refers not only to reproduction but to the progressive coming into being and growth of natural things.

(Severinus 1571, 46–54). For this reason, they were considered as causing the regularity and durability of natural cycles.

As Severinus explained, seeds were incorporeal, dimensionless and only accessible by thought thanks to their link with the "elements". The latter were described as mere receptacles, envelopes, "abysses" and "matrices", which sheltered the seeds for generation (Severinus 1571, 54). For this reason, Severinus refused to base his philosophy on elements as the material components of natural change. He even disqualified the medical tradition by calling it "anatomy of death" for its approach to the living body in terms of perishable entities. For Severinus, the traditional elements established by the Aristotelian physics were "invalid" in the sense that they were devoid of any active properties.[24] In contrast, the Paracelsian "vital anatomy" that was based on the flow of the eternal seeds was believed to cause the fertile properties of beings at the origin of their active powers. Consequently, Severinus' account of generation diverged from the Galenic doctrine and proposed, instead, the idea of a "progression" of seeds following a Paracelsian interpretation.

During generation, the seeds incarnated in natural beings following a "progression" from their fundamental unity and perfection to the multiplicity of the world (Severinus 1571, 62; Hirai 2005, 249–261; Shackelford 2002, 180–185). They were first subject to incubation (*fomentatio*) in the elements as "receptacles" and "abysses," before progressing from a fundamental and obscure place called "darkness" and "Orcus" (Hades).[25] The latter was an underground reservoir, which constituted the starting and ending point of every being, where the cyclic flow of seeds took place. At the end of their progression from the elemental "abysses" to the "light" of the world stage, the seeds differentiated and separated to complete the multiplication of "fruits", i.e., the generation of natural beings. The seeds then assigned "signatures" to the new beings, namely their individual characteristics, such as size and figure (Kikuchihara and Hirai 2015; Bianchi 1987). In order to do so, the seeds deployed their own "knowledge" (*scientia*), that is a plan and internal know–how to develop bodies, which they received as a "gift" from divine providence.[26]

From this Paracelsian approach to seed and generation, Sennert seeks to establish a common lexicon with the Aristotelian and Galenic tradition. For this reason, he deems the Paracelsian notions of "seed", "star" and "root" as equivalent of the Aristotelian notions of form and soul (Sennert 1619, 181–182 and 222).[27] At first, Sennert is conciliatory toward this new terminology for a series of reasons. He agrees with the fact that "seeds" and "stars", as forms related to the soul, are dimensionless and incorporeal, and that they cause the life and powers of beings across

[24] See *De gradibus rerum naturalium et compositionibus remediorum* (Paracelsus 1589–1591, VII, 17–18).

[25] See *Philosophia de generationibus et fructibus quartet elementorum* and *De Meteoris* (Paracelsus 1589–1591, VIII, 54–159 and 206).

[26] See *Labyrinthus medicorum errantium* (Paracelsus 1589–1591, II, 215–220).

[27] Sennert 1619, 9, 222: 'Ut ergo ad institutum redeamus, appellant Chymici Recentiores, Astra Semina et Radices rerum, quae Philosophi et Medici hactenus appellarunt formas et animas.'

generations (Sennert 1619, 222–223). In addition, Sennert attempts to naturalize the Paracelsian notion of seed "progression" from the "abyss" towards the "world's scene" by analogy with vegetal physiology. As he explains, plants come up from seeds sown underground, which first grow in tiny leaflets. At this stage, they do not display a perfect structure as they need to draw from the earth an appropriate nutriment to thrive and acquire the "perfection" of their kind. This happens thanks to the plant's (vegetative) soul which shapes a suitable body for its nature. Outside of this reasoning based on the cyclic growth of plants, Sennert believes that the Paracelsian notions of "abyss" and "progression" are incomprehensible (Sennert 1619, 226).

Further to his project of conciliation, Sennert states that the Paracelsian concepts of "seed" and "star" correspond in many respects to the Galenic notion of formative force related to the seed's substantial form. As he explains in his *Institutiones* and *De chymicorum...liber*, the form attributes all the body's characteristics during generation. Through the functions of the vegetative soul, it enlivens and shapes the seed's body by assigning its size, figure, order, position and many other features. Sennert sees there a parallel with the distribution of "determined signatures" by the "internal star" that Severinus highlighted in his *Idea*. In the same way, he states that the formative force deploys something similar to the Paracelsian notion of "knowledge" (*scientia*) contained in the seeds for the development of beings. In this sense, the seminal knowledge is very close to the "art" (*ars*) and "wisdom" (*sapientia*) that Galen praised in his *Construction of the Embryo* (Sennert 1619, 185–187). Thanks to the divine providence, which introduced a formative force into them, the seeds are a divine instrument playing the role of God's "working hand". Sennert, thus, considers that the Galenic philosophy did emphasize that the soul and form contained the "knowledge" of making the body through the formative force. Nonetheless, the Galenic theory of generation had to be enlightened by Christian religion to show that this power was received during the divine creation.

Despite his compromising attitude, Sennert appears inflexible with some aspects of the Paracelsian theory of generation. In his view, Paracelsian philosophers are wrong in defining "seeds" and "stars" as celestial entities in the same way as Avicenna and Fernel were in positing the celestial origin of forms. Sennert more broadly considers that philosophers have tended to make the same conclusion because they have observed that living beings develop thanks to solar heat (Sennert 1619, 223). Moreover, because the movement of the heavens is regular and subject to cycles, philosophers have considered that the heavens are likely causing the "movement" of creatures living on earth. On this point, Sennert concedes that the Paracelsian philosophers are right to emphasize the "moments" and "terms" related to seeds as "stars" in the course of generations (Sennert 1619, 183). As he notes, it is well-known by botanists that plants produce flowers, fruits and seeds at regular and definite moments each year (Sennert 1619, 225).[28] In the same way, physicians are aware that physiological processes are subject to periodic times during

[28] Sennert 1619, 9, 225: '[...] et qui ignorat, plantas in producendis floribus, seminibus, fructibus; exarescendo quasi et rursum repullulascendo et revirescendo, stata tempora observare, rei botanicae plane ignarus est.'

digestion, generation, growth, pulse, menstruation and fevers. However, Sennert recalls, this does not imply the celestial nature of "seeds" and "stars". As the latter amount to the Aristotelian notion of form, they are actually transmitted through the seeds of living beings. Initially created by God, they contain a form and soul causing the order, regularity and periodicity of their vital functions (Sennert 1619, 225–226).

Sennert concludes his appraisal of the Paracelsian account of generation by comparing the Aristotelian and Paracelsian definitions of life. Aristotelian philosophers and Galenic physicians have held only the animate beings as living to the extent that life is a certain actuality and "vigor" of the soul. To be considered as living, one needs to display at least the operations of the vegetative soul, in particular nutrition (Sennert 1619, 226).[29] However, Sennert explains, Paracelsian philosophers have broadly extended the acceptation of life as they believe that celestial bodies, metals, minerals, gems and stones are alive too. Because the latter are endowed with a "seed" and "star", they allegedly possess an active power or a "spirit", which makes them alive (Sennert 1619, 227).[30] In this regard, Sennert refers to the English physician Thomas Moffet who, in his *Dialogus apologeticus* (1584) on the supremacy of Paracelsian drugs, defined life as the *energeia* – the Aristotelian notion of actuality – inserted in beings, in other words, their disposition to act (Moffet 1584, 29–30). With this reasoning in mind, Paracelsian philosophers have considered any active principle within beings as "vital" even if it is limited to sensory qualities or faculties. In the same way, they have considered as dead anything devoid of active and efficient powers.

For Sennert, the Paracelsian definition of life is not only excessive, it is based on a major confusion between living and acting. The latter, Sennert insists, is by far more general than living. For instance, physically dead bodies and substances subject to the alchemical process of "mortification" may still have sensory qualities and active powers. In Sennert's view, life is different and more specific than the *energeia* and disposition towards action. For this reason, he deems the Paracelsian definition of life as inappropriate and requiring reconsideration in light of the Aristotelian distinction between soul and nature (Sennert 1619, 229–230).[31] While bodies may

---

[29] Sennert 1619, 9, 226: 'Hactenus quidem Philosophi Aristotelici et Medici Galenici vitam tantum animatis tribuerunt, vitamque animae quendam actum et vigorem esse dixerunt: et nihil vivere concesserunt, nisi in quo aliqua animae, ad minimum vegetantis, operatio appareat; et quicquid vivit nutriri, et contra quicquid vere nutritur, vivere docuerunt.'

[30] Sennert 1619, 9, 227: 'Verum Chymici recentiores vitae nomen multo latius extendunt, et cum stellis vitam tribuunt, easque vitali seminum potestate perfundi, nec mortua esse corpora [...] docent; tum etiam metalla, mineralia, gemmas, lapides, vivere statuunt, et omnino quicquid semen vel astrum, quod appellant, in se continet et agendi vim habet, seu ut alii loquuntur, omne corpus, quod Spiritum habet, vivere dicunt.'

[31] Sennert 1619, 9, 229–230: 'Rectius vero Aristoteles et eius sectatores sentiunt, qui inter Naturam et animam distinguunt, et ab iis etiam in Natura rebus, quae animata non sunt, vi a Creatore indita, definitas et ordinatas actiones provenire statuunt; vitae vero principium tantum animam esse, quibus vitae principium est anima, et ubi anima non est, ibi vitam non esse docent [...].'

have particular virtues due to their nature, they can be considered as alive only to the extent that they are animate, in other words, that they hold a vegetative soul.

13.5 Conclusion

In his early physiological theory, Sennert proposed the model of vegetative propagation by cutting in order to explain generation and the transmission of life from parent to offspring. He tackled this question by discussing the transmission of the vegetative soul from the parent's seed. Following the Galenic and Aristotelian tradition, he stated that the seed was endowed with a formative force which operated the development of the embryo. To elucidate the origin of such a force, Sennert emphasized its relationship with the substantial form of the seed, which was "latent" though in actuality in order to perform physiological functions through the faculties of the soul. In support of this claim, he referred to *Genesis* by explaining that the seminal forms were initially created by God but remained immanent to the seeds in order to be subject to multiplication during generation.

With this interpretation, Sennert proposed an original account of the Aristotelian philosophy of matter–form by literally understanding the Epicurean notion of seed as a "detached portion" following the traducian view on the transmission of the soul. This led him to establish the propagation of forms as "detached particles" of the progenitor's soul in the same way as plant cuttings. During this process of propagation, the seminal form acted as an entity that was "torn" from the parents' substantial form within a material "minimum". In his *De chymicorum...liber*, Sennert applied his account of the substantial form within the seed to the Paracelsian notions of "seed" and "star" in order to enhance the form's active power, its relationship with the vegetative soul and its major role in physiological cycles.

In many regards, Sennert's early account of generation gives an insight into the maturation of his medical and alchemical ideas. First, it shows his longstanding view on body and soul, matter and form, as discrete entities which constitute the living body. Sennert's description of the seed's matter and form was nourished by the account of Epicurus, who thus needs to be included among his previously established sources: medieval alchemy, Democritean atomism and Renaissance "pluralist" Aristotelianism. In shaping his explanation of generation, Sennert merged a plurality of definitions of the seed and soul from diverse frameworks and contexts that all used the analogy with plant physiology. Most notably, he extended the vegetal analogy from the Aristotelian and Galenic views on the vegetative soul to the theological theory of traducianism and to the Paracelsian approach to generation. Such a "cross–pollination" of diverse epistemic frameworks allowed Sennert to provide a physiological explanation of generation that encompassed all living beings by taking into account their formal and material composition as well as their active alchemical properties.

In early modern medicine and natural philosophy, Sennert paved the way to a clear synthesis of Galenic and Paracelsian explanations of generation by harmonizing their most diverging views on the nature of life. In doing so, he provided a stimulating interpretation of matter–form through the notion of seed, which emphasized the celestial nature and powers of the substantial form while suggesting the atomic nature of matter. It is, therefore, unsurprising that Sennert's account of generation was widely read and discussed throughout the seventeenth century by an audience ranging from students of *Chymiatria* at the University of Marburg to experienced scholars such as Robert Boyle.

Acknowledgments This chapter was written at Princeton History Department with the generous support of a BAEF Hoover Foundation Postdoctoral Fellowship. I am grateful to William Theiss, Jennifer Rampling, Anthony Grafton, Keith Wailoo and all the participants in Princeton HOS Program Seminar for their helpful comments and suggestions.

References

Ambrosoli, Mauro. 1997. *The Wild and the Sown: Botany and Agriculture in Western Europe, 1350–1850*. Cambridge: Cambridge University Press.

Bailer-Jones, Daniela M. 2002. Models, Metaphors and Analogies. In *The Blackwell Guide to the Philosophy of Science*, ed. Peter Machamer and Michael Silberstein, 108–127. Malden/Oxford: Blackwell.

Bellorini, Cristina. 2016. *The World of Plants in Renaissance Tuscany: Medicine and Botany*. Routledge.

Bianchi, Massimo Luigi. 1982. Occulto e manifesto nella medicina del Rinascimento: Jean Fernel e Pietro Severino. *Atti e memorie dell'Accademia Toscana delle Scienze e Lettere La Colombaria* 47: 183–248.

———. 1987. *Signatura rerum: segni, magia e conoscenza da Paracelso a Leibniz*. Rome: Edizioni dell'Ateneo.

Blank, Andreas. 2010. *Biomedical Ontology and the Metaphysics of Composite Substances, 1540–1670*. Munich: Philosophia Verlag.

———. 2012. Julius Caesar Scaliger on Plants, Species, and the Ordained Power of God. *Science in Context* 25: 503–523.

Clericuzio, Antonio. 2000. *Elements, Principles and Corpuscles: A Study of Atomism and Chemistry in the Seventeenth Century*. Dordrecht: Kluwer Academic Publishers.

Deer Richardson, Linda. 1985. The Generation of Disease: Occult Causes and Diseases of the Total Substance. In *The Medical Renaissance of the Sixteenth Century*, ed. Andrew Wear, Roger K. French, and Ian M. Lonie, 175–194. Cambridge: Cambridge University Press.

———. 2018. *Academic Theories of Generation in the Renaissance: The Contemporaries and Successors of Jean Fernel (1497–1558)*. Cham: Springer.

Fernel, Jean. 2003. *The Physiologia of Jean Fernel (1567)*. Trans. John M. Forrester. Philadelphia: American Philosophical Society.

———. 2005. *Jean Fernel's on the Hidden Causes of Things: Forms, Souls, and Occult Diseases in Renaissance Medicine*. Ed. and Trans. John M. Forrester and John Henry. Leiden: Brill.

Fraenkel, Pierre. 1982. Mélanchthon, Beatus Rhenanus et Tertullien. *Bibliothèque d'humanisme et Renaissance* 44: 357–360.

Fuchs, Leonhart. 1555. *Institutiones medicinae...libri quinque*. Lyon: Thomas Guerin.

Galen. 1992. *On Semen*. Ed. and Trans. Philip De Lacy. Berlin: Akademie Verlag.

———. 1997. *The Construction of the Embryo*. In *Selected Works*. Trans. P.N. Singer. New York: Oxford University Press, 177–201.
Givens, Terryl. 2010. *When Souls Had Wings: Pre-Mortal Existence in Western Thought*. New York: Oxford University Press.
Heurnius, Johannes. 1592. *Institutiones medicinae*. Leiden: Officina Plantiniana/Frans van Ravelingen.
Hirai, Hiro. 2005. *Le concept de semence dans les théories de la matière à la Renaissance, de Marsile Ficin à Pierre Gassendi*. Turnhout: Brepols.
———. 2011a. *Medical Humanism and Natural Philosophy: Renaissance Debates on Matter, Life and the Soul*. Leiden: Brill.
———. 2011b. Justus Lipsius on the World-Soul between Roman Cosmic Theology and Renaissance *Prisca Theologia*. In *Hiro Hirai and Jan Papy*, ed. Justus Lipsius and Natural Philosophy, 251–273. Brussels: Royal Academy of Belgium.
———. 2015. Mysteries of Living Corpuscles: Atomism and the Origin of Life in Sennert, Gassendi and Kircher. In *Early Modern Medicine and Natural Philosophy*, ed. Benjamin Goldberg, Peter Distelzweig, and Evan R. Ragland, 255–270. Dordrecht: Springer.
Holmes, Brooke. 2017. Pure Life: The Limits of the Vegetal Analogy in the Hippocratics and Galen. In *The Comparable Body: Analogy and Metaphor in Ancient Mesopotamian, Egyptian, and Greco-Roman Medicine*, ed. John Z. Wee, 358–386. Leiden–Boston: Brill.
Kikuchihara, Yohei, and Hiro Hirai. 2015. Signatura Rerum Theory. In *Encyclopedia of Renaissance Philosophy*, ed. Marco Sgarbi. Cham: Springer. https://doi.org/10.1007/978-3-319-02848-4_405-1.
Lüthy, Christoph. 2000. The Fourfold Democritus on the Stage of Early Modern Science. *Isis* 91: 442–479.
———. 2001. An Aristotelian Watchdog as Avant-Garde Physicist: Julius Caesar Scaliger. *The Monist* 84: 542–561.
———. 2005. Daniel Sennert's Slow Conversion from Hylemorphism to Atomism. *Graduate Faculty Philosophy Journal* 26: 99–121.
Lüthy, Christoph, and William R. Newman. 2000. Daniel Sennert's Earliest Writings (1599–1600) and Their Debt to Giordano Bruno. *Bruniana & Campanelliana* 6: 261–279.
Michael, Emily. 1997. Daniel Sennert on Matter and Form: At the Juncture of the Old and the New. *Early Science and Medicine* 2: 272–299.
———. 2000. Renaissance Theories of Body, Soul, and Mind. In *Psyche and Soma: Physicians and Metaphysicians on the Mind-Body Problem from Antiquity to Enlightenment*, ed. John P. Wright and Paul Potter, 147–172. New York: Oxford University Press.
———. 2001. Sennert's Sea Change: Atoms and Causes. In *Late Medieval and Early Modern Corpuscular Matter Theories*, ed. Christoph Lüthy, John E. Murdoch, and William R. Newman, 331–362. Leiden: Brill.
Moffet, Thomas. 1584. *De jure et praestantia chymicorum medicamentorum dialogus apologeticus*. Frankfurt: Andreas Wechel.
Newman, William R. 2006. *Atoms and Alchemy: Chymistry and the Experimental Origins of the Scientific Revolution*. Chicago: University of Chicago Press.
Oppenheimer, Jane M. 1953. William Gilbert: Plant Grafting and the Grand Analogy. *Journal of the History of Medicine and Allied Sciences* 8 (1953): 165–176.
Paracelsus. 1589–1591. *Bücher und Schrifften*. Ed. Johannes Huser. Basel: Conrad Waldkirch, 10 vol.
Perseus Digital Library. 2020. Pseudo-Plutarch, Placita Philosophorum, Book 5, Chapter 3. http://data.perseus.org/citations/urn:cts:greekLit:tlg0094.tlg003.perseus-grc1:5.3. Accessed 1 September 2020.
Pluta, Olaf. 2007. How Matter Becomes Mind: Late-Medieval Theories of Emergence. In *Forming the Mind: Essays on the Internal Senses and the Mind/Body Problem from Avicenna to the Medical Enlightenment*, ed. Henrik Lagerlund, 149–167. Dordrecht: Springer.

Roger, Jacques. 1997 [1963]. The Life Sciences in Eighteenth-Century French Thought . Ed. Keith R. Benson. Trans. Robert Ellrich. Stanford: Stanford University Press.
Rusu, Doina-Cristina. 2020. Using Instruments in the Study of Animate Beings: Della Porta's and Bacon's Experiments with Plants. *Centaurus* 62: 393–405.
Sakamoto, Kuni. 2016. *Julius Caesar Scaliger, Renaissance Reformer of Aristotelianism: A Study of Exotericae Exercitationes*. Boston: Brill.
Savoia, Paolo. 2017. Nature or Artifice? Grafting in Early Modern Surgery and Agronomy. *Journal of the History of Medicine and Allied Sciences* 72: 67–86.
Scaliger, Julius Caesar. 1557. *Exotericarum exercitationum liber XV de Subtilitate*. Paris: Michael Vascosan.
———. 1566. *In libros de Plantis Aristoteli inscriptos commentarii*. Lyon: Jean Crispin.
Sennert, Daniel. 1611. *Institutionum medicinae libri V*. Wittenberg: Zacharias Schürer.
———. 1618. *Epitome naturalis scientiae*. Wittenberg: Caspar Heiden.
———. 1619. *De chymicorum cum Aristotelicis et Galenicis consensu ac dissensu liber*. Wittenberg: Zacharias Schürer Sr.
———. 1656. *The Institutions, or, Fundamentals of the Whole Art, both of Physick and Chirurgery, Divided into Five Books*. London: Lodowick Lloyd.
———. 1662. *Chymistry Made Easie and Useful. Or, the Agreement and Disagreement of the Chymists and Galenists*. London: Peter Cole.
Severinus, Petrus. 1571. *Idea medicinae philosophicae, fundamenta continens totius doctrinae Paracelsicae, Hippocraticae, et Galenicae*. Basel: Heinrich Petri.
Shackelford, Jole. 2004. *A Philosophical Path for Paracelsian Medicine: The Ideas, Intellectual Context, and Influence of Petrus Severinus (1540/2–1602)*. Copenhagen: Museum Tusculanum Press.
Spruit, Leen. 2008. Renaissance Views of Active Perception. In *Theories of Perception in Medieval and Early Modern Philosophy*, ed. Simo Knuuttila and Pekka Kärkkäinen, 203–224. Cham: Springer.
Stolberg, Michael. 1993. In Awe of Creation: *Tota Substantia, Calidum Innatum, Generatio Spontanea* and Atomic Form Teachings of Daniel Sennert. *Gesnerus* 50: 48.
———. 2003. Particles of the Soul: The Medical and Lutheran Context of Daniel Sennert's Atomism. *Medicina nei Secoli* 15: 177–203.
Totelin, Laurence. 2018. Animal and Plant Generation in Classical Antiquity. In *Reproduction: Antiquity to the Present Day*, ed. Rebecca Flemming, Nick Hopwood, and Lauren Kassell, 53–66. Cambridge: Cambridge University Press.
Vidal, Fernando. 2011. *The Sciences of the Soul: The Early Modern Origins of Psychology*, 21–57. London: University of Chicago Press.
Walker, D.P. 1958. The Astral Body in Renaissance Medicine. *Journal of the Warburg and Courtauld Institutes* 21: 119–133.
Webster, Charles. 1966. The Recognition of Plant Sensitivity by English Botanists in the Seventeenth Century. *Isis* 57: 5–23.
Zabarella, Giacomo. 1590. De sensu agente. In *De rebus naturalibus libri XXX*, 582–599. Venice: Paolo Meietti.
Zanier, Giancarlo. 1987. Platonic Trends in Renaissance Medicine. *Journal of the History of Ideas* 48: 509–519.

Chapter 14
Vegetative and Sensitive Functions of the Soul in Descartes's *Meditations*

Igor Agostini

Abstract According to Descartes, vegetative and sensitive activities, which in the Aristotelian tradition were considered dependent on the soul (the so-called lowest functions of the soul), are pure effects of corporeal dispositions. What is, however, the relationship between this doctrine and Descartes's metaphysical claim in the *Meditations*, namely, that the mind is the only cause of thought? In the *Meditations*, Descartes does not provide a doctrine of vegetative and sensitive functions, and he does not even establish the canonic Cartesian thesis that they are not dependent on the soul. In claiming that the lowest functions cannot be attributed to the soul, the Second Meditation is not making an ontological point. Nevertheless, in spite of this, the *Meditations* teach us something very important about the lowest functions: that both vegetative and sensitive functions depend on the soul is not, and never will be, a necessarily true proposition.

14.1 Introduction

In his metaphysics, René Descartes (1596–1650) notoriously challenged Aristotelian substantial forms and the appeal to souls, and claimed that both vegetative and sensitive activities are nothing more than effects of corporeal dispositions.[1]

Descartes had already formulated this doctrine in *L'Homme*:

> I desire, I say, that you should consider that these functions follow in this machine simply from the disposition of the organs as wholly naturally as the movements of a clock or other automaton follow from the disposition of its counterweights and wheels. To explain these functions, then, it is not necessary to conceive of any vegetative or sensitive soul, or any

[1] On the physiological study of the vegetative soul in Descartes and Cartesianism, see Des Chene 2001, Detlefsen 2016, Hutchins 2016, Baldassarri 2018, Baldassarri 2019 and Baldassarri, Chap. 15, in this volume.

I. Agostini (✉)
Dipartimento di Studi Umanistici, Università del Salento, Lecce, Italy
e-mail: igor.agostini@unisalento.it

F. Baldassarri, A. Blank (eds.), *Vegetative Powers*, International Archives of the History of Ideas Archives internationales d'histoire des idées 234,
https://doi.org/10.1007/978-3-030-69709-9_14

> other principle of movement or life, other than its blood and its spirits which are agitated by the heat of the fire that burns continuously in its heart, and which is of the same nature as those fires that occur in inanimate bodies. (*L'Homme*, AT XI 202; G 169)

This doctrine is later summarized in the Fifth Part of the *Discours de la méthode* (1637):

> So I contented myself with supposing that God formed the body of a man exactly like our own both in the outward shape of its limbs and in the internal arrangement of its organs, using for its composition nothing but the matter that I had described. I supposed, too, that in the beginning God did not place in this body any rational soul or any other thing to serve as vegetative or sensitive soul, but rather that he kindled in its heart one of those fires without light which I had alread explained. (*Discours*, V, AT VI 46; CSM I 134)

The same doctrine also appears in Descartes's correspondence. A famous letter to Regius of May 1641 establishes that there is only one soul in human beings, namely the rational soul, and that both alleged vegetative and sensitive souls (that is, vegetative and locomotive powers) should not be called 'souls':

> There is only one *soul* in human beings, the *rational soul;* for no actions can be reckoned human unless they depend on reason. The *vegetative power* and the *power of moving the body,* which are called the *vegetative* and *sensory souls* in plants and animals, exist also in human beings; but in the case of human beings they should not be called *souls,* because they are not the first principle of their actions, and they belong to a totally different genus from the *rational soul.* (Descartes to Regius, may 1641, AT III 371, CSMK III 182)

Descartes's polemical target is clear enough. His doctrine not only targeted Plato and his tripartite theory of the soul, which Descartes considers in contrast to the teaching of Christianity ("The first thing which I cannot approve is your saying that 'men have a threefold soul'. In my religion this is a heretical thing to say"),[2] but also Aristotle's conception of the soul. According to Aristotle, nutrition, growth, and reproduction, on the one hand, and sensorial perception, on the other hand, depend on the rational soul. And the rational soul is the principle both of these functions, the so-called 'lowest functions', and of intellectual activity, the higher function of the soul. For Descartes, on the contrary, the alleged lowest functions of the soul are nothing other than a consequence of corporeal dispositions and can be explained mechanistically, mainly according to the hydraulic model. In this way, Descartes eliminates the need for any soul other than the rational soul (AT XI 202, CSM I 108), that, being deprived of its lower functions, is reduced to pure thought.

For Descartes, thinking is the only function depending on that alleged immaterial life-giving principle traditionally labeled 'soul', which Descartes prefers to call, for this very reason, 'mind' (*mens*). Descartes's mind, unlike Aristotle's soul, is not the origin of the lowest functions, but is only the cause of thinking. In responding to Pierre Gassendi (1592–1655), Descartes develops this point at length:

> The next question you raise concerns the obscurity arising from the ambiguity in the word 'soul'. But I took such care to eliminate this ambiguity when it arose that it is tiresome to

---

[2] *Descartes to Regius*, May 1641, AT III 371, Cottingham III 181: "Primum itaque, quod ibi minus probo, est quod dicas *Animam homini esse triplicem*; hoc enim verbum, in mea religione, est haeresis". See, on this, Fowler 1999, 316.

> repeat myself here. I shall say only that it is generally the ignorant who have given things their names, and so the names do not always fit the things with sufficient accuracy. Our job, however, is not to change the names after they have been adopted into ordinary usage; we may merely emend their meanings when we notice that they are misunderstood by others. Thus, primitive man probably did not distinguish between, on the one hand, the principle by which we nourished and grow and accomplish without any thought all the other operations which we have in common with the brutes, and, on the other hand, the principle in virtue of which we think. He therefore used the single term 'soul' to apply to both; and when he subsequently noticed that thought was distinct from nutrition, he called the element which thinks 'mind', and believed it to be the principal part of the soul. I, by contrast, realizing that the principle by which we are nourished is wholly different – different in kind – from that in virtue of which we think, have said that the term 'soul', when it is used to refer to both these principles, is ambiguous. If we are to take 'soul' in its special sense, as meaning the 'first actuality' or 'principal form of man', then the term must be understood to apply only to the principle in virtue of which we think; and to avoid ambiguity I have as far as possible used the term 'mind' for this. For I consider the mind not as a part of the soul but as the thinking soul in its entirety. (*Fifth Replies*, AT VII 355–356, CSM II 246)

Now, since Descartes's thesis that the essence of the ego consists only in thought is formulated in the *Second Meditation*, interpreters sometimes claim that the Cartesian criticism of Aristotle's doctrine of the lowest functions is also contained in the *Second Meditation* as an essential component of his theory of mind.[3] And, indeed, Descartes alludes to – and criticizes – Aristotle's doctrine in the *Second Meditation* by discussing the question of the essence of self.[4]

However, things are much more complicated than they seem at first glance, and the sense of Descartes's criticism has to be correctly understood. In general, the *Second Meditation* has to be read in the context of the order of the *Meditations* which, as is known, follows the analytical *modus demonstrandi*. In the light of this order, in the *Second Meditation* there is only one proposition to which an ontological value can be attributed: I exist and – which is the same thing, according to Descartes – I am a thinking thing (*res cogitans*). No other statements attain this objective certainty. As we have seen, Descartes expresses this difference in the *Preface to the Reader* to the *Meditations* by distinguishing "the actual truth of the matter" (*in ordine ad veritatem rei*) from "the order corresponding to my own perception" (*in ordine ad meam perceptionem*) [AT VII 7, CSM II 8]; and, in the *Second*

---

[3] See the two classic studies on Descartes and the Aristotelian partition of the soul republished in *Descartes. Critical Assessments*: W. Soffer 1984, in *Descartes. Critical Assessments*, 1991, 193–204, at 200: "The lingering doubt concerning body provides the lever to move from the indubitable existence of the *cogito* to its manner of existence as non-bodily *res cogitans*. Here occurs the rejection of the Aristotelian soul as responsible for the life of the body"; S. J. Wagner 1984, in *Descartes. Critical Assessments* 1991, 205–222, at 218: "His conception of thought provides a roughly correct ground for rejecting Aristotle's partition of the soul. One of the most basic, best known elements in that conception of the unity of mental capacities: the claim that any thinking thing is capable of the full range of mental acts", AT VII, 28 (however, more correctly, Wagner also adds that "Descartes presumably knew that he lacked a real argument of the unity of mental capacities" [...], at 218)". Or, for instance, Rozemond 2014, 236: "a crucial aim of the Second Meditation is to supplant an Aristotelian concpetion of the soul with a Cartesian one".

[4] See, below, pp. 4–5.

Meditation, he claims that his aim is not to establish his nature considered "in reality" (*in rei veritate*) [AT VII 27, CSM II 18). This is all the more true as concerns the lowest functions. In the *Second Meditation*, Descartes is not making an ontological point about them: on the one hand, he maintains that vegetative and sensitive activities cannot be attributed to "thought"; on the other hand, he never affirms that they must be excluded from it.

In this chapter, I will first argue that the Cartesian assertion of the *Second Meditation*, claiming that "I am only a thinking thing", does not imply the exclusion of the lowest functions, which is neither stated in the *Second Meditation* nor elsewhere in the*Meditations*. Second, I will reconstruct both the aim and significance of the *Second Meditation*'s criticism of Aristotle's doctrine of the lowest functions. In truth, if this criticism is not meant to challenge Aristotle's doctrine, then what is?

14.2 The Pre-philosophical Notion of Man

Descartes's discussion of the lowest functions of the soul occurs in the context of the central question of the *Second Meditation*: Who am I?

Descartes's inquiry is conducted through the method of doubt, that now takes the form of a subtraction process removing all the doubtful properties included in the idea that the meditator had previuously formed of himself:

> I will therefore go back and meditate on what I originally believed myself to be, before I embarked on this present train of thought. I will then subtract (*subducam*) anything capable of being weakened, even minimally, by the arguments now introduced, so that what is left at the end may be exactly and only what is certain and unshakeable. (*Meditations*, II, AT VII 25, CSM II 17).

According to this former idea, Descartes observes, "I thought I was a man. However, it should be known, first and foremost, what a man is." Unlike in the *Recherche de la vérité*, in the *Meditations* Descartes does not discuss at length the traditional definition of man as a rational animal:

> Shall I say 'a reasonable animal'? No; for then I should have to inquire what an animal is, what rationality is, and in this way one question would lead me down the slope to other harder ones, and I do not now have the time to waste on subtleties of this kind. (*Meditations*, II, AT VII 25, CSM II 17)

In the *Meditations*, Descartes concentrates on what we might call *the pre-philosophical notion of man*, that is the thoughts which spontaneously spring to the meditator's mind, and which were not inspired by anything beyond his own nature alone, when he applied himself to the consideration of his own being (*quid sponte et natura duce cogitationi meae antehac occurrebat, quoties quid essem considerabam*).

The content of this pre-philosophical notion is described as follows by Descartes:

> The first thought to come to mind was that I had a face, hands, arms and the whole mechanical structure of limbs which can be seen in a corpse, and which I called the body. The next

> thought was that I was nourished, that I moved about, and that I engaged in sense-perception and thinking; and these actions I attributed to the soul. But as to the nature of this soul, either I did not think about this or else I imagined it to be something tenuous, like a wind or fire or ether, which permeated my more solid parts. As to the body, however, I had no doubts about it, but thought I knew its nature distinctly. If I had tried to describe the mental conception I had of it, I would have expressed it as follows: by a body I understand whatever has a determinable shape and a definable location and occupies a space in such a way as to exclude any other body; it can be perceived by touch, sight, sound, taste or smell, and can be moved in various ways, not by itself but whatever else comes into contact with it. For, according to my judgement, the power of self-movement, like the power of sensation or of thought, was quite foreign to the nature of a body; indeed, it was a source of wonder to me that certain bodies were found to contain faculties of this kind. (*Meditations*, II, AT VII 26, CSM II 17–18)

It is important to stop here for a moment in order to clarify the notion of the soul presented by Descartes in this passage. Of course, as he claims, it is a spontaneous and natural pre-philosophical notion. However, I must also be traced back to the Aristotelian tradition, according to which vegetal, sensorial and thinking functions are all dependent on the soul. The idea, here, is that the soul is the principle of both the vegetative – that is, self-motion and nutrition[5] – and the sensitive functions, as well as of "thought".[6] And seventeenth-century commentators on the *Meditations* did not hesitate to claim that in this passage Descartes refers to the old Aristotelian doctrine. For example, Johannes Schotanus, in his *Exegesis in Iam et IIam Meditationem*, writes:

> The whole thing is entirely due to the alleged universal knowledge of the essence of the soul, which many of the Peripatetic and Scholastic philosophers think they possess. (Schotanus 1687, 188)[7]

That the pre-philosophical notion of the soul overlaps with Aristotle's is justified by the fact that Descartes considers Aristotelian philosophy as a sort of conceptual elaboration on the commonsensical conception of the world: Aristotele's psychology, according to Descartes, actually originates from spontaneous and sensorial beliefs. In this sense, the pre-philosophical account of the soul described in the *Second Meditation* is, in fact, none other than a commonsensical version of the Aristotelian conception of the soul.

---

[5] See Des Chene 2000. Among the different vegetative functions traditionally attributed to the the soul, Descartes mentions two in the Second Meditation: nutrition and locomotion. With regard to Descartes's mechanical explanation of them, it is interesting to observe that *L'Homme* mostly deals with self-motion, and is concerned with nutrition only in the more general context of the circulation of the blood (Meschini 2015). However, in a note in the *Excerpta anatomica* entitled *De Accretione et nutritione* dated November 1637, Descartes offers an explanation in terms of internal alteration and arrangement of particles of accretion and nutrition. It is therefore certain that when Descartes composed the *Meditations*, he had already elaborated his own mechanical explanation of both self-motion and nutrition. See, on this aspect, Baldassarri in this volume.

[6] See Des Chene 2000.

[7] The original Latin is "Et profecto haec fere vel hujus farina est universa cognitio, quam plerique e Philosophis, tam antiquis quam recentioribus Peripatetics et Scholasticis, de animae essentia se habere profitentur." (My own translation).

Now, part of this pre-philosophical notion is the idea that the soul is the principle of both the vegetative and sensitive functions. Once again, seventeenth-century commentators on the *Meditations* will insist on this point:

> Indeed, they attribute to the soul, in addition to the faculty of thinking, the functions of nutrition and locomotion, as well as of sensation. They are persuaded that growing, locomotion and sensation are originated only from the soul, and not from the body. And, certainly, the Philosopher too attributed these properties to the rational soul. (Schotanus 1687, 188)[8]

Now, as we have seen in all his writings, including the *Replies* to the *Meditations*, Descartes strongly rejects this doctrine and the idea that the lowest functions depend on the soul. According to him, they are nothing other than effects produced by the corporeal disposition.

However, two points have to be made here:

1. That both vegetative and sensitive functions of the soul are nothing other than material dispositions is not a claim made explicitly by Descartes, neither in his *Second Meditation* nor elsewhere in the six meditations.
2. As concerns the *Second Meditation* in particular, there is no ontological claim on the nature of thought (and, therefore, on its relationship with its vegetative and sensitive functions).

Let us substantiate these points and concentrate on Descartes's treatment of vegetative and sensitive functions in the *Second Meditation*.

14.3 Abstraction and Functions of the Soul

In the passage of the *Second Meditation*, in which the notion of vegetative and sensitive functions is introduced, Descartes's goal is to show what remains of his pre-philosophical notion once he launches the subtraction process by applying the reasons for doubt of the *First Meditation* to the pre-philosophical notion of man:

> But what shall I now say that I am, when I am supposing that there is some supremely powerful and, if it is permissible to say, malicious deceiver, who is deliberately trying to trick me in every way he can? (*Meditations*, II, AT III 26, CSM II 18).

(1) To begin with, can I affirm that I possess the least of all those things which I have just said to pertain to the nature of the body? I pause to consider, I revolve all these things in my mind, and I find none of them pertaining to me. It would be tedious to stop and to enumerate them. It should be observed here that in eliminating the corporeity, Descartes eliminates both the ideas "I *am* a body" and "I *have* a body".

---

[8] The original Latin is: "Illi enim praeter cogitandi facultatem, nutriendi et loco movendi et sentiendi facultatem animae rationali ascribunt, putantes omnem augmentationem, motum ac sensum a sola anima, non autem a corpore fieri. Et has animae rationali proprietates ut actiones ascribebat quoque Philosophus". (My own translation).

(2) Let us pass to the attributes of the soul and see if there are any of them in me.

(A) The first attributes mentioned were the powers of nutrition and locomotion; however, if I have no body, I am capable neither of walking nor of being nourished:

> But what about the attributes I assigned to the soul? Nutrition or movement? Since now I do not have a body, these are mere fabrications. (*Meditations*, II, AT VII 26, CSM II 18)

The same conclusion is established on sensorial perception:

> Sense-perception? This surely does not occur without a body, too, and besides, when asleep I have appeared to perceive through the senses many things which I afterwards realized I did not perceive through the senses at all. (*Meditations*, II, AT VII 27, CSM II 18, modified)

Summing up, both vegetative and sensorial functions cannot be attributed to the self because, though dependent on mind, they "occur" in the body [*non fit sine corpore*], (AT VII 27, CSM II 18), which is still subject to the general doubt of the *First Meditation*, and as concerns sensorial functions in particular, to the dream argument.

(B) Things change if we consider the attribute of thinking. This is the only case in which I discover something which cannot be separated from myself:

> Thinking? At last I have discovered it – thought; this alone is inseparable from me. I am, I exist – that is certain. (*Meditations*, II, AT VII 27, CSM II 18)

Let us make a few comments on A and B.

(A) Descartes rejects the vegetative and sensitive functions insofar as they are subject to doubt. The idea is that, according to Aristotle, though these functions are dependent on the soul, they take place in the body. As a consequence, as long as doubt about the body subsists, they cannot be affirmed.
(B) With regard to B, by claiming that "thought is inseparable from me", Descartes makes the same identical statement as: *Ego sum, ego existo*. As a matter of fact, in asserting that even a supremely powerful deceiver "will never bring it about that I am nothing, so long as I think that I am something," Descartes implicitly argues that the deceiver will never be able to separate existence from thought. It is this inseparability which explains why both (and only) two statements "I am, I exist" (*Meditations*, II, AT VII 25, Cottingham II 17) and "I am a thing that thinks" (*Meditations*, II, AT VII 27, Cottingham II 18) are qualified not just as true, but as *necessarily* true.

I will return to this point below. Now let us go further into the text and consider Descartes's comments on the proposition "I am, I exist":

> I am, I exist – that is certain. But for how long? For as long as I am thinking. For it could be that were I totally to cease from thinking, I should totally cease to exist. At present I am not admitting anything except what is necessarily true. I am, then, in the strict sense only (*tantum*) a thing that thinks; that is, I am a mind, or intelligence, or intellect, or reason (*mens, sive animus, sive intellectus, sive ratio*) – words whose meaning I have been ignorant of until now (*prius*). (*Meditations*, II, AT VII 27, CSM II 18)

The last part of this passage seems clear enough. When Descartes affirms that the "*mens, sive animus, sive intellectus, sive ratio*" are words whose significance was unknown to me before (*prius*), he is referring to the pre-philosophical notion of the soul, according to which the names *mens, animus, intellectus,* and *ratio* designate the higher part of soul. According to him, this is wrong because these names express all that I am and not a mere part of me.

The *Replies* will make this point more expicit, as we have seen:

> I consider the mind not as a part of the soul but as the thinking soul in its entirety. (*Fifth Replies*, AT VII 355–356, CSM II 246)

Attention should be paid to the symmetry and complementarity of the *tota* of the *Fifth Replies* with the *tantum* of the *Second Meditation*. In fact, the *tantum* expresses not only the rejection of the body, but also the lowest functions, and so it permits the identification of the whole (*tota*) soul with thought.

What is the subtraction process of the *Second Meditation* for? The question here is much more complicated.

Indeed, some lines later, Descartes observes that he cannot exclude the possibility that the soul is corporeal:

> And yet may it not perhaps be the case that these very things which I am supposing to be nothing, because they are unknown to me, are in reality identical with the 'I' of which I am aware? I do not know, and for the moment I shall not argue on this point, since I can make judgments only about things which are known to me. (*Meditations*, II, AT VII 27, CSM II 18)

Descartes points out here that if, on the one hand, he cannot affirm that corporeity belongs to him, because body has been subtracted as doubtful from the notion that he has of himself, on the other hand, he cannot affirm that he is not corporeal: I know nothing about this (*nescio*), said Descartes. In other words, though claiming that the meditator cannot *know* that he himself is corporeal, Descartes prohibits him from claiming that he *is* not corporeal: the meditator is not authorized to make such a statement because "these very things" (that is, "all that I attribute to the body") are unknown (*ignota*) to me, and "I can make judgments only about things which are known to me". In spite of the subtraction of the body, I cannot exclude that I am a body.

There is a technical point at work here, which has to be traced back to the same Aristotelian framework that Descartes is challenging in these pages. The subtraction of the body is not an exclusion, but an abstraction: namely, what is called *praecisio* in the Scholastic tradition. The text is here explicit:

> I am, then, in the strict sense only (*praecise tantum*) a thing that exists. (*Meditations*, II, AT VII 27, CSM II 18)

However, if this is true, not only I am not in a condition to exclude the corporeity of the soul, but also to exclude the vegetative and sensitive functions. Indeed, the argument by which Descartes had subtracted the lowest functions from the pre-philosophical notion of the soul was based on the fact that they all need a corporal condition. If it is true – he had claimed – that I have no body, it is true likewise that I am capable neither of walking nor of being nourished: "Since now I do not have a body, these are mere fabrications" (*Meditations*, II, AT VII 26, Cottingham II 18). However, now I know that the subtraction of the body is nothing more than an abstraction. As a consequence, the subtraction of the lowest functions cannot be anything more than an abstraction as well.[9] My point is therefore that the *precise tantum* must be referred to anything which is rejected by the process of doubting, that is, not only to the body, but also to the vegetative and sensitive functions, insofar as, according to the very same Aristotelian doctrine, they take place in the body although being originated by the soul.

I think, consequently, that it is wrong to claim that in the *Second Meditation*, Descartes is stating that the lowest functions do not belong to the nature of the mind. Unfortunately, however, this is a very common claim among contemporary interpreters, as we have seen.[10] This claim has a long history too, which can be traced back to seventeenth-century commentators on the *Meditations*. In discussing the *Second Meditation*'s subtraction of the lowest functions of the soul on the basis of the argument that they "occur" in the body (AT VII 27, CSM II 18), Schotanus comments as follows:

> But he answers that these properties belong to the body, if a body exists, and not to the soul; moreover, he adds that as he has supposed, at this point, that he has no body, and that nourishing and walking are only fictions in him. (Schotanus 1687, 191)[11]

> Therefore, he does not think that walking and sensation belong to his nature because his nature is only a thinking thing and has no body. (Schotanus 1687, 191–192)[12]

All of this is false. In writing here that the lowest functions belonging to the body "occur" in the body (AT VII 27, Cottingham II 18), Descartes is not expressing his

---

[9] I agree with the remark by Gary Hatfield: "We must note exactly what Descartes does and does not claim for this finding about the nature of mind. He claims that the meditator's knowledge of herself is now restricted to thoughts and that these thoughts can be conceived without reference to body. As already mentioned (and further discussed below), he does not claim that she can know whether human thoughts or a human mind can exist apart from a body. Correspondingly, she presumably cannot know at this point whether the mind actually directs the processes of digestion or provides the active power that contracts the muscles. Such questions are beyond the *Second Meditation*" (Hatfield 2003, 119). With respect to Hatfield, I would be inclined to be much more categorical: the mind does not know *at all* whether at this point it actually directs the lowest functions or not.

[10] See, above, footnote n. 3.

[11] The Latin is: "Sed respondet has proprietates esse corporis, si quod datur; non autem animae, uti vulgus putat; imo addit, quoniam jam corpus suum nullum esse supposuit, nutriri et incedere, respectu sui non esse nisi figmenta".

[12] The Latin is: "Non igitur nutriri, incedere, sentire ad suam naturam, quia tantum res cogitans est, nec corpus habet, pertinent".

own position, according to which these functions *are* bodily dispositions. Descartes is just referring to the Aristotelian account of the soul, according to which these functions, though dependent on the soul, "occur" in the body. That is, they cannot *take place* without a body [*non fit sine corpore*] (AT VII 27, CSM II 18). In fact, Descartes is not at all concerned here with the exclusion of the lowest functions. He *could not* deny that the mind is body, just as we cannot deny the existence of the lowest functions.

In order to assess this point, one should consider that Descartes's perspective, at this stage of the *Meditations*, is not at all ontological. As he points out in the *Praefatio ad lectorem*, corporeity is excluded with respect not to "the actual truth of the matter" (*in ordine ad veritatem rei*), but only to "the order corresponding to my own perception" (*in ordine ad meam perceptionem*) [AT VII 7, CSM II 8]. This distinction is clearly evoked in the *Second Meditation*: it is possible that "these very things which I am supposing to be nothing, because they are unknown to me, are in reality [*in rei veritate*] identical with the 'I' of which I am aware," because "for the moment I shall not argue on this point, since I can make judgments only about things which are known [*nota*]" (AT VII 27, Cottingham II 18). The only conclusion that I can establish from this is that the knowledge [*notitia*] that I have of myself "does not depend on things of whose existence I am as yet unaware." (AT VII 27–28; Cottingham II 18–19).

However, this is not enough. It should be added that this epistemic limitation does not concern only the fact that in the *Second Meditation* I am not yet in the condition to guarantee the passage from the order of perception [*perceptio*] to the order of truth [*veritas rei*]: actually, it is in the very same order of perception (despite some retrospective statements made by Descartes himself)[13] that it is impossible to arrive at the exclusion of corporeity.[14] The point is that, at this stage of the *Meditations*, the knowledge [*notitia*] of myself as a thinking thing is not clear and distinct. The reason for this is that this knowledge, as noted previously, does not in fact *exclude* corporeity, but only *abstracts* from it.[15]

Descartes focuses on the difference between abstraction and exclusion in a letter to Mesland, where he claims that

> There is a great difference between *abstraction* and *exclusion*. If I said simply that the idea which I have of my soul does not represent it to me as being dependent on the body and identified with it, this would be merely an abstraction, from which I could form only a negative argument which would be unsound. But I say that this idea represents it to me as a

[13] See, in particular, *Fourth Replies*, AT VII 226.

[14] In spite of the old, and still dominant interpretation, claiming that in this passage Descartes is invoking his metaphysical doubt concerning the reliability of clear and distinct conception, Harry G. Frankfurt had correctly (in my opinion) argued that here Descartes is not at all concerned with the ontological problem of his essence. The only problem that Descartes is here dealing with is what characteristics may be properly ascribed to him (Harry G. Frankfurt 1970, 113–127). Unfortunately, Frankfurt is not clear in explaining for what reason the knowledge of myself as a thinking thing is not clear and distinct.

[15] See S. Landucci 2002, correcting this point in Frankfurt's interpretation.

> substance which can exist even though everything belonging to the body be excluded from it; from which I form a positive argument, and conclude that it can exist without the body. (*Descartes to Mesland*, 2 May 1644, AT IV 120; CSMK III 236. Italics in the text)

The second case is the one concerned in the *Second Meditation*: when the meditator affirms that he cannot attribute to himself corporeity and the lowest functions, he is not in a condition to represent himself "comme une substance qui peut exister, encore que tout ce qui appartient au corps en soit exclu", for the reason that he does not know whether corporeity and the lowest functions are actually excluded from the notion that he has of himself, or merely not included in it. To sum up, the *Second Meditation* cannot exclude corporeity and the lowest functions from the clear and distinct notion that the meditator has of himself, for they are subtracted not by exclusion, but by abstraction; and this is the sense of Descartes's claim that this notion is *praecise sumpta*, that is formed by abstraction.

14.4 What Is "Necessarily True"

The question that now arises is to know what is the subtraction process of the lowest functions for, given that not only it is limited to the *ordo* of perception,but it is a mere abstraction.

The sentence immediately preceding the *praecise tantum* is here telling:

> It could be that were I totally to cease from thinking, I should totally cease to exist. (*Meditations*, II, AT VII 27, CSM II 18)

By claiming that it is *possible* that I would cease to exist if I stopped thinking, Descartes is arguing that it is possible that anything not belonging (at this stage of the *Meditations*) to the *cogitatio* – viz., the body and the lowest functions of the soul – is not a part of myself. However, to say that *it is possible*, this also means that *it is not necessary*. If this is true, the assertion that it is possible "that these very things which I am supposing to be nothing, because they are unknown to me, are in reality identical with the 'I' of whom I am aware?" (*Meditations*, II, AT VII 27, Cottingham II 18) had actually been anticipated by the assertion that "it could be that were I totally to cease from thinking, I should totally cease to exist". (*Meditations*, II, AT VII 27, Cottingham II 18).

In my opinion, this has some decisive consequences for the interpretation of the proposition "*sum igitur praecise tantum res cogitans*", and in general, for the project of the *Second Meditation*. The difference between the Cartesian statement "I am only a *res cogitans*" and the two commonsensical statements dismissed in the *Second Meditation*, "I am a body" and "I am a soul as the principle of the vegetative and sensitive functions", cannot be merely expressed in terms of *truth*, as if the first statement was true and the second and the third were false. This is because, at this stage of the *Meditations*, it is still possible that the statements "I am a body" and "I am a soul as the principle of the vegetative and sensitive functions" are true; or, in other words, it is not possible to establish them as false. It will be only the *Sixth*

Meditation which will establish as true that "my essence consists solely in the fact that I am a thinking thing", and that "I am simply a thinking, non-extended thing" (AT VII 78, CSM II 54). And it will be only at this point that it is demonstratated as false that "I am a body" and, as a consequence of the real distinction, it is false that "I am a soul as the principle of the vegetative and sensitive functions".

What makes, in the *Second Meditation*, the proposition "*sum res cogitans*" a *unicum* is, first, that it is true *at this moment* (*nunc*). And there is much more than this: I am indeed certain that this proposition is not only true, but *necessarily* true. This is textually confirmed by Descartes's statement of himself as a *res cogitans*:

> At present I am not admitting anything except what is necessarily true (*necessario sit verum*). I am, then, in the strict sense only a thing that exists. (*Meditations*, II, AT VII 27, CSM II 18)

This necessity constitutes the distinctive mark of the proposition "*sum res cogitans*". With regard to the propositions"I am a body" and "I am a soul as the principle of the vegetative and sensitive functions", at this stage of the *Meditations* they are neither false nor true; and for this very reason, also in the (counterfactual) case in which they should be later demonstrated to be true, they will never be necessarily true.

However, "*sum res cogitans*" is not a necessary truth in a familiar logical sense: rather, it is a necessary truth in the sense that whenever the proposition "I am a *res cogitans*" is considered, it is true. The solution to the problem of the nature of the soul given in the *Second Meditation* (which is not yet a solution) is here perfectly symmetrical to the one of the *Ego sum, Ego existo*, which was established as necessarily true insofar as it is self-verifying,[16] or, in other words, it cannot be considered at all unless it is true.[17]

> This proposition, *I am, I exist* is necessarily true (*necessario esse verum*) whenever it is put forward by me or conceived in my mind. (*Meditations*, AT VII 25, CSM II 17).

To sum up, the proposition, *I am, I exist* is necessarily true because it cannot be formulated at all, mentally or verbally, unless it is true. This is because my existence resists all the reasons for doubt: I cannot formulate any reason for doubting unless confirming as true that I exist. The same for the determination of my nature as *res cogitans*: the proposition "I am a res cogitans" cannot be formulated at all unless it is true. The reason is, again, that it escapes all the reasons for doubt: no matter what reason for doubting I can excogitate, I will always be *cogitans*.

It is this kind of necessity, arising from the resistance to doubt, which explains the restriction *only* (*tantum*): I am only a *res cogitans* because the fact that I am *cogitans* is the only thing which is (now) and will be (always) necessarily true about

---

[16] Cf. respectively, Williams 1978, 59; Harry G. Frankfurt 1970, 134 and 137 f. This aspect of incorrigibility possessed by 'cogito' and 'sum' has been emphasized by many interpreters, and in order to indicate this aspect, Hintikka proposed his 'performatory' or 'performative' interpretation of the *cogito*, which, however, poses problems which have been widely discussed by scholars and will not be addressed here.

[17] See Harry G. Frankfurt 1970, 134 and 137 f.

me. This is a definitive result for the *Meditations*: all those truths which I will discover later about me will not be necessarily true: the proposition "I have a body", which will be demonstrated in the *Sixth Meditation* (AT VII 81, CSM II 81), will never become necessarily true. And this is also the case with the proposition "The bodies exist" (AT VI 78–80, CSM 74–75).[18]

The same conclusion has to be made on the vegetative and sensitive functions: what the *Second Meditation* teaches on this point is only that every proposition concerning the vegetative and sensitive functions is not necessarily true. Accordingly, it does not matter if I discover that they are dependent on the soul or that they are only dependent on the body: in both cases, indeed, this will not be a necessarily true proposition. In other words, what the *Second Meditation* does, concerning the vegetative and sensitive functions, is to exclude them from the domain of what is necessarily true. Whatever I will be able to affirm about them, when I am in the condition of making ontological statements, they will never be necessarily true statements.

14.5 Conclusion

In conclusion, not only do the *Meditations* fail to provide a doctrine concerning the vegetative and sensitive functions, but they do not establish the canonic Cartesian thesis that they are not dependent on the soul altogether. Nevertheless, in the *Meditations*, Descartes expresses something very important: the fact that the vegetative and sensitive functions depend on the soul is not necessarily a true proposition, and it will never be necessarily a true proposition.

References

Baldassarri, Fabrizio. 2018. Descartes' Bio-Medical Study of Plants: Vegetative Activities, Soul, and Power. *Early Science and Medicine* 23 (5–6): 509–529.

———. 2019. The Mechanical Life of Plants: Descartes on Botany. *British Journal for the History of Science* 52 (1): 41–63.

Des Chene, Dennis. 2000. *Life's Form: Late Aristotelian Conceptions of the Soul*. Ithaca: Cornell University Press.

———. 2001. *Spirits and Clocks: Machines and Organism in Descartes*. Ithaca: Cornell University Press.

[18] See this crucial passage of the *Synopsis*: "The great benefit of these arguments is not, in my view, that they prove what they establish – namely that there really is a world, and that human beings have bodies and so on – since no sane person has ever seriously doubted these things. The point is that in considering these arguments we come to realize that they are not as solid or as transparent as the arguments which lead us to the knowledge of our own minds and of God, so that the latter are the most certain and evident of all possibile objects of knowledge for the human intellect". (AT VII 15, CSM II 11).

Detlefsen, Karen. 2016. Descartes on the Theory of Life and Methodology in the Life Sciences. In *Early Modern Medicine and Natural Philosophy*, ed. Peter Distelzweig et al., 141–171. Dordrecht/Heidelberg/New York/London: Springer.

Descartes, René. 1964–1974. *Œuvres de Descartes*, ed. Charles Adam and Paul Tannery [Rev. ed. ed. Joseph Beaude et al.], 11 vols. Paris: Vrin. [AT in the text].

———. 1984–1985. *The Philosophical Writings of Descartes*, ed. John Cottingham, Robert Stoothoff, and Dugald Murdoch, 2 vols. Cambridge: Cambridge University Press. [CSM in the text].

———. 1991. *The Philosophical Writings of Descartes*, vol. 3: *The Correspondence*. Ed. and Trans. John Cottingham, Robert Stoothoff, Dugald Murdoch, and Anthony Kenny. Cambridge: Cambridge University Press. [CSMK in the text].

———. 2004. In *The World and Other Writings*, ed. Stephen Gaukroger. Cambridge: Cambridge University Press. [G in the text].

Fowler, Colin Francis. 1999. *Descartes on the Human Soul. Philosophy and the Demands of Christian Doctrine*. Dordrecht/Boston/London: Kluwer.

Frankfurt, Harry Gordon. 1970. *Demons, Dreamers and Madmen. The Defense of Reason in Descartes' Meditations*. Indianapolis/New York, The Bobbs-Merrill Company.

Hatfield, Gary. 2003. *The Routledge Guidebook to Descartes' Meditations*. London: Routledge.

Hutchins, Barnaby. 2016. Descartes and the Dissolution of Life. *The Southern Journal of Philosophy* 54 (2016): 155–173.

Landucci, Sandro. 2002. *La mente in Cartesio*. Milano: FrancoAngeli.

Meschini, Franco Aurelio. 2015. La dottrina della digestione secondo Descartes. Itinerari tra testi, contesti e intertesti. *Physis* 50: 113–164.

Rozemond, Marleen. 2014. The Faces of Simplicity in Descartes's Soul. In *Partitioning the Soul. Debates from Plato to Leibniz*, ed. K. Corcilius and D. Perler, 219–243. Berlin/Boston: De Gruyter.

Soffer, Walter. 1984. Descartes' Rejection of the Aristotelian Soul. *International Studies in Philosophy* 16: 57–69. [repr. In *Descartes. Critical Assessments*, ed. by G.D.D.Moyal, 3 vols. London and New York: Routledge, 193–204].

Schotanus. 1687. *Exegesis in Iam e IIam Meditationem ut et quaestiones metaphysicae in quibus Methodi Cartesii asseritur*. Franequerae: Ex Officina Johannis Gyselaar.

Wagner, Steven J. 1984. Descartes on the Parts of the Soul. *Philosophy and Phenomenological Research* 45: 57–68. [repr. In *Descartes. Critical Assessments*, ed. by G.D.D.Moyal, 4 vols. London and New York: Routledge, 3: 205–222].

Williams, Bernard. 1978. *Descartes: The Project of Pure Enquiry*. London/New York: Routledge.

Chapter 15
Failures of Mechanization: Vegetative Powers and the Early Cartesians, Regius, La Forge, and Schuyl

Fabrizio Baldassarri

Abstract René Descartes' mechanization of living activities lays bare a glaring lacuna that concerns vegetative functions, such as nutrition, generation, and growth: his cardiovascular framework affects any exhaustive explanation of these activities. When he mentions a mechanical vegetative power in his 1641 correspondence with Henricus Regius, this definition is unspecified, although it may be correlated to a few posthumous bio-medical notes. Descartes' mechanization of the vegetative soul remains puzzling. Early Cartesian scholars were thus obliged to fill this lacuna to produce a more exhaustive physiology. In this chapter, after reconstructing Descartes' own position in L'Homme and in his manuscripts, I explore the positions of Regius, Louis de La Forge, and Florent Schuyl, who dealt with Descartes' medical texts and physiological doctrine through various perspectives. Yet, while building a mechanization of vegetation upon Descartes' philosophy, their interpretations of a mechanical vegetative power uncover shortcomings, weaknesses, and failures.

In his attempt to explain all living functions, René Descartes (1596–1650) famously introduced the notion of the 'animal-machine', a methodological claim that brought about an ontological revolution in physiology. Descartes attributed all the functions of life and sentience to the body, mechanically reducing them to matter and motion, and thus dismissing any physiological role for souls, spirits, agencies, and substantial forms (Grene 1986; Hattab 2009). This simplification of living natures to extended matter compressed all faculties and forms to the same mechanical order. This entails a general question: What replaces these agencies in a mechanical interpretation of living bodies? And, would a mechanistic power merely substitute for the soul as a principle informing the living body, or would it be something different?

F. Baldassarri (✉)
Institute for Research in the Humanities, University of Bucharest, Bucharest, Romania
e-mail: fabrizio.baldassarri@gmail.com

F. Baldassarri, A. Blank (eds.), *Vegetative Powers*, International Archives of the History of Ideas Archives internationales d'histoire des idées 234,
https://doi.org/10.1007/978-3-030-69709-9_15

This question is especially striking in the case of vegetative functions. While Descartes' mechanistic reconstruction of animal motion, sensation, and other higher activities is well specified in *L'Homme* (1633, posthumously published in 1662 (in its Latin translation) and 1664 in French), *La Dioptrique* (1637), and the *Principia philosophiae* (1644), vegetative functions, such as nutrition, digestion, and generation, remain under-specified. In a few scattered passages of *L'Homme*, Descartes gives very slight attention to digestion and nutrition, and avoids speaking of generation altogether. Although scholars have considered this text a source of knowledge for these functions, Franco Aurelio Meschini's recent survey of Descartes' interpretation of digestion makes clear that this text has very little to do with vegetative activities (Meschini 2015).[1] The reason is simply stated: in *L'Homme*, Descartes mostly deals with self-motion (Part 2), sensation (Parts 3 & 4), and the functions of the brain (Part 5), and makes the centrality of blood the material agent of the machine. The few passages devoted to digestion and nutrition are indeed restricted within a cardiocentric perspective and lay bare an under-developed mechanization of these basic activities.

In this chapter, I first address the shortcomings of Descartes' discussion of vegetative activities in *L'Homme*, and then explore his later attempt to provide a more exhaustive explanation of them. This eventually results in a definition of a vegetative power that operates within the body. Yet, since this definition remains confined to an unpublished manuscript, a glaring lacuna affects Descartes' mechanical account of vegetative activities, and a contrast between his explanation of life and his understanding of living bodies surfaces.[2] Cartesian scholars were thereby compelled to fill this lacuna as they aimed to present a more exhaustive medical doctrine. In the second part of the chapter, I explore the case of three Cartesian scholars. The first is the renegade Henricus Regius (1598–1679), who personally discussed this matter with Descartes in 1640s, and then proposed a mechanization of the processes of vegetative functions in his natural philosophy. The second is Louis de La Forge (1632–1666), whose remarks on the French edition of *L'Homme* (1664) provide new support for Descartes' physiology. The third is Florent Schuyl (1619–1669), whose Introduction to the Latin translation of *L'Homme* (1662) contains a detailed mechanization of plants and vegetation. Nevertheless, all these attempts reveal shortcomings and flaws, unearthing the difficulties in defining a vegetative power for the body.

---

[1] Cf. Des Chene 2001, 18; Aucante 2006, 152–58. On Descartes see, Nadler et al. 2019

[2] I have recently discussed this issue in relation to Descartes' study of plants in Baldassarri 2018a. In the next two sections, I follow this article. For a reconstruction of the metaphysical investigation of the vegetative soul in Descartes, see Agostini, Chap. 14, in this volume.

15.1 Nutrition, Digestion and Growth in *L'Homme*

In *L'Homme*, Descartes explains all living functions, 'such as the digestion of food, the beating of the heart and arteries, the nourishment and growth of the limbs, respiration, waking and sleeping, the reception by the external sense organs of light, sounds, smells, tastes [...] the imprinting of ideas [...]' (Descartes 1964–1974, XI, 202; Descartes 1984, 108) and so on, pursuant to the analogy between living bodies and automata.[3]

Descartes describes digestion in article 3 of *L'Homme*, right after his claim that the activities of the body belong to the matter and arrangement (or disposition) of its organs. Digestion occurs by 'the force of certain fluids which, gliding between its parts, separate, shake, and heat them.' These fluids 'are brought from the heart through the arteries,' and their force consists of their rapidity and warmth (they 'must be very hot,' wrote Descartes.) In other words, these fluids are principally composed of blood and heated by the heart, the source of bodily heat and life, and also the source of digestion. At the same time, Descartes claims that digested food has 'such a nature that it can be broken down and heated up of itself' (Descartes 1964–1974, XI, 121; and Descartes 2004, 100). Descartes' interpretation thus unites (a) a mechanical corpuscular explanation of digestion, (b) iatrochemical claims, as he stresses a similarity between the operations of digesting juices and *aqua fortis* in breaking metals, and (c) a role for thermodynamic reactions, i.e., the fermenting process of self-heating particles. At this stage, Descartes does not institute any specific role for the stomach or for other organs of the abdomen. The first is just a container for food and a receptacle for fluids brought from the heart, and digestion occurs due to the heat brought by the blood. In other words, digestion is an effect of blood circulation and relies on the heat of the heart. This is not an original interpretation of digestion, as Thomas Hall has shown (Descartes 2003, 6–7, notes 12 and 13).

Descartes then continues, claiming that the agitation produced by the warmth of fluids, 'together with the agitation of the stomach [...] as well as the arrangement of the fibres from which the bowels are composed, cause these particles [...] to descend [...] and [...] encounter innumerable small holes through which they flow into the branches of a large vein that bears them toward the liver' (Descartes 1964–1974, XI, 121–122; and Descartes 2004, 100). In the liver, this whitish fluid 'is subtilized and elaborated, taking on the colour and form of blood' (Descartes 1964–1974, XI, 122; and Descartes 2004, 101).

[3] See also Descartes to Mersenne November or December 1632: '[...] j'entreprends d'expliquer toutes ses principales fonctions. J'ai déjà écrit celles qui appartiennent à la vie, comme la digestion des viandes, le battement du pouls, la distribution de l'aliment etc., et les cinq sens. J'anatomise maintenant les têtes de divers animaux, pour expliquer en quoi consistent l'imagination, la mémoire etc.' (Descartes 1964–1974, I, 263). See Antoine-Mahut and Gaukroger 2016. Cf Schouten 1974.

At this point in *L'Homme*, Descartes moves on to describe the heartbeat, respiration, the pulse, and arterial blood, 'that serves to nutrition' (Descartes 1964–1974, XI, 125), before speaking of nutrition itself. This latter consists of the attachment of particles conveyed by the arterial blood to 'the roots of certain little threads which [...] make up bones, flesh, membranes, nerves, the brain, and all the other solid parts depending on the different ways in which they are joined or interconnected' (Ibid., 126; and Descartes 2004, 103). These particles of digested food later refined into arterial blood enter the pores of the body and replace other particles, causing growth.

Descartes' cardiocentric physiology incorporates digestion into the movement of venous blood (which heats the stomach, and brings liquids to the stomach and the chyle to the liver,) and nutrition into the movement of arterial blood (Descartes 1964–1974, XI, 127; and Descartes 2004, 104). The stomach and other organs of the abdomen have little role, if anything, except for the haematopoietic function of the liver; and the main focus is given to the circulation of blood. Very little attention is provided to growth as well, which was one of the activities that differentiated between inert or dead and living bodies. Even the order of Descartes' explanation follows the movement of blood, as he separates the description of digestion and nutrition. He repeats a similar account in the *Discours de la Méthode* (1637), where he claims that the heat of blood first causes digestion, as some 'fluid parts of the blood [...] help to dissolve the food,' and then favours nutrition, as the blood rarefies and loses parts that occupy the pores according to their 'situation, shape, or minuteness' (Descartes 1964–1974, VI, 53; and Descartes 1984, 138).[4]

Despite being consistent with his mechanization of bodies, this account presents two main flaws. The first concerns the thread of causation. Blood circulation is not the principle of nutrition, digestion, and growth, but a mere circumstance or physiological condition concurring in these activities: blood only warms the stomach in the case of digestion and transports the particles of nutrients. The second is physiological. Despite his claim of focusing on the arrangement of the organs, the organic complexities and diversities of vegetative processes remain unexplored, and *L'Homme* mostly contains a description of the movement of fluids in the abdomen. As a result, his explanation is lacunose and under-specified; reducing the vegetative functions to mere effects of the heartbeat appears therefore unsatisfactory.[5]

---

[4] See Descartes 1964–1974, VI, 54; and Descartes 1984, 138.

[5] Stephen Gaukroger acknowledges the limitations of Descartes' explanation of vegetative activities, but also takes for granted this account: 'Of the functions traditionally ascribed to the vegetative soul, Descartes has little to say on digestion, respiration, and reproduction'; '[Descartes] has a deep interest in the movement of the blood (Gaukroger 2002, 182; 186; 192–196). See Des Chene 2000, 133–138 and Aucante 2006, 152–162. For a different interpretation, see Bertoloni Meli 2011, 132.

15.2 Descartes' Vegetative Power as *Immutatio*

In the *Discours*, Descartes acknowledges that fire without heat is an ineffective principle for vegetative activities such as generation, and affects his explanation of nutrition and digestion too.[6] By the end of 1637, however, he brought his medical education to completion by means of several observations on the organs of the abdomen, ultimately developing a physiological explanation of nutrition and growth. This work is partially collected in two biomedical texts. I examine in particular two notes dated 1637 in *Excerpta anatomica*, one of these texts, published posthumously in 1859–1860.[7]

The first note is entitled 'Compendium on the anatomical observation about the lowest parts of the abdomen' (*Observationum anatomicarum Compendium de partibus inferiori ventre contentis*). This contains observations on the formation and disposition of the organs of the abdomen, and their interdependency and connection with the circulatory system (Descartes 1964–1974, XI, 587–594).[8] Although this is not the first note in his manuscripts in which he discusses these organs, this Compendium has a more structured description and presents a more assertive tone. Here, Descartes describes the connection of these organs by means of the vessels that belong to the portal vein, and then describes these organs – the intestines, pancreas, spleen, gallbladder, liver and stomach – in detail.[9]

The explanation in this note presents echoes of, but differs from the explanation of *L'Homme*. The most evident difference is that there is no strict dependency between the heart and the organs of the abdomen. In contrast, an interdependency between the parts of the body shows the importance of blood circulation, but restricts its primacy. The *Compendium* thus reveals Descartes' attempt to conceive of the organs of the abdomen *per se* as working in a more organic way. Still, this note does not answer the questions about what causes vegetative functions.

In a second note entitled 'On Accretion and Nutrition' (*De Accretione et nutritione*) and dated November 1637, Descartes deals with these activities more directly. He distinguishes between accretion in inert bodies and nutrition in living bodies. The former is a mere aggregation of particles entailing a transformation through the substitution, expulsion or corruption of the parts, while the latter mostly consists in

---

[6] *Discours de la Méthode*, in Descartes 1964–1974, VI, 45: 'je n'en avois pas encore assez de connaissance, pour en parler du mesme style que du reste [...] en démontrant les effets par les causes, et faisant voir de quelles semences et en quelle façon la nature les doit produire.'

[7] This text has been known since Louis-Alexandre Foucher de Careil (1826–1891) published it from some notes discovered in the manuscripts of Gottfried Wilhelm Leibniz (1646–1716), who copied them in 1676 from the manuscripts Claude Clerselier (1614–1684) possessed and circulated among his contemporaries. This text is now edited in several editions of Descartes' and Leibniz's works. A critical study of this manuscript is still to be done. See Baldassarri 2018b, 63.

[8] Cf. Meschini 2015, 133–37. Baldassarri 2021a, 158–160.

[9] *Excerpta anatomica*, in Descartes 1964–1974, XI, 590: 'Vena portæ radices educit varias ex intestinis, ventriculo, mesenterio, omento, pancreate, liene et felle; itemque exiguam ex hepate; unam etiam, nempe vas breve, educit e ventriculo per lienem;' and in Descartes 1964–1974, XI, 592–94.

an arrangement of particles that relies on an internal change, or *partium immutatione*.[10] This internal change consists of a set of activities, namely a change, fragmentation, and mixing of particles that act upon each other, until they acquire a precise shape. Additionally, this activity regulates digestion, the change of food within the stomach, nutrition, the arrangement of particles within the body, and the self-regulation and growth of the body.

While in the Scholastic tradition these activities rely on the vegetative soul, Descartes emphatically characterizes them as a mechanical process of internal alteration and the arrangement of particles. The *immutatio* is the mechanical process that stands for the internal change particles undergo in transforming food into the body during digestion and nutrition, ultimately constituting the body.[11]

This definition is not inconsistent with what Descartes wrote to Regius in 1641.[12] In opposition to Regius' attribution of a soul to the vegetative functions, Descartes claims that these functions are achieved by a vegetative power. He defines the latter as 'nothing but a certain constitution of the parts of the body' (Descartes to Regius, May 1641, in Descartes 1964–1974, III, 370).[13] This corresponds to the *immutatio*, a mechanical process that regulates the conformation of the particles forming the body. Yet, since the *immutatio* specifies a set of operations that span from the conformation of digested particles to the composition of bodies, the identification of the vegetative power with the *immutatio* reduces the former to a precise mechanical system or a set of processes. As a combination of processes, the *immutatio* produces bodily effects without any need for a vegetative soul,[14] and consists of the conformation of the body. Although Descartes did not describe this operation in his further work, this is a crucial point: his dismissal of any principle animating living bodies corresponds to defining a vegetative power not as a mere principle of these activities, but as a set of mechanical functions coherently working within the body.

---

[10] *Ibid.*, 596: 'Accretio duplex est: alia mortuorum et quae non nutriuntur, fitque per simplicem partium appositionem, sine ulla earum immutatione […] Et fit etiam transmutatio ligni vel alterius corporis in lapidem per modum talis accretionis, dum partes lapidis poros ligni ingrediuntur, et praecendentes vel sibi assimilant, vel extrudunt, vel partim hoc partim illud. Alia accretio est viventium, sive eorum quae nutriuntur, et fit semper cum aliqua partium immutatione.'

[11] For a reconstruction of the scholastic use of *immutatio* in relationship to Descartes, see Baldassarri 2021b. Vanni Rovighi 2007, 84-ff. For a reconstruction of the scholastic use of *anima vegetativa* in relationship to Descartes, see Des Chene 2000, 133–38. Baldassarri 2021c.

[12] On this correspondence, see Bos 2002.

[13] See also Descartes to Regius, May 1641, Descartes to Regius, May 1641, in Descartes 1964–1974, XI, III, 371: 'Hæc autem *vis motrix* a *vi vegetativa* ne specie quidem differt; utraque autem toto genere a *mente* distat.' [Emphasis in the text.] Ibid., 372: 'Vis *autem* vegetativa … *nihil aliud est quam certa partium corporis constitutio*.' [Emphasis in the text.] Cf. Regius 1641a, 'XI. *Vis vegetativa* est certa corporis constitutio, qua substantiæ corporeæ calorifique perpetuam dissipationem […] conservamus, & ex semine nostri simile procreamus.'

[14] See Descartes to Plemp, 23 March 1638, in Descartes 1964–1974, II, 65.

15.3 Three Steps beyond Descartes: Regius' Vegetative Power

Despite these notes, a glaring lacuna affects Descartes' work that the scholars who knew, voiced, and spread his theories needed to bridge. Henricus Regius, a Dutch physician and professor of botany and medicine at Utrecht University, embodies the first attempt to fill that gap. On the one hand, Regius accommodated his medical knowledge within Descartes' mechanical framework, heavily borrowing or drawing conclusions from Descartes' premises.[15] On the other hand, he covered a broader territory than Descartes on medical and philosophical matters, sometimes substantially differing from him.[16] If anything, Regius was more than just an expositor of Descartes' theories and, despite the fact that he probably voiced a few of Descartes' unrecorded views, he could present a better explanation of some of them. In the case of vegetative activities in animals and plants, Regius' physiological explanation can be characterized as being three steps beyond Descartes' interpretation (Schmaltz 2019).

The first step is contained in Regius' 1641 medical disputation, *De illustribus aliquot Quaestionibus physiologicis*, the one Descartes edited through their 1641 correspondence. In this case, Regius followed the text of Descartes' letter, and wrote:

> The *vegetative power* is a certain constitution of the body, through which the perpetual dissipation of corporeal substance and heat is conserved by means of a fluid prepared in the heart and infused in the parts [of the body], and a similar body to ourselves is procreated by means of the seed (Regius 1641a).[17]

This text presents an original interpretation. While the first words repeat Descartes' comment, the following lines are not in Descartes' letter and, presumably, reflect Regius' own thoughts. He claims that the vegetative power is a self-preservation of the bodily constitution, achieved by means of the circulation of blood. In the next theses, Regius specifies that life consists in the distribution of the fluids adept at the conservation of the bodily heat from the heart, while the obstruction of this

---

[15] Descartes to Mersenne, 5 October 1646, in Descartes 1964–1974, IV, 510–511; Descartes to Mersenne, 23 November 1646, in Descartes 1964–1974, IV, 566–567. *Lettre-Préface*, in Descartes 1964–1974, IX-2, 19. See Schmaltz 2016, 73–4. On some possible collaborations between Descartes and Regius, see Baldassarri 2020.

[16] Descartes to Regius, November 1641, in Descartes 1964–1974, III, 443: 'Sed sane multa sunt in Thesibus tuis, quæ fateor me ignorare, ac multa etiam, de quibus si forte quid sciam, longe aliter explicarem quam ibi explicueris. Quod tamen non miror; longe enim difficilius est, de omnibus quæ ad rem medicam pertinent suam sententiam exponere, quod docentis officium est, quam cognitu faciliora seligere, ac de reliquis prorsus tacere, quod ego in omnibus scientiis facere consuevi.' See Bos 2017, 95–111.

[17] Regius 1641a, Theses 11: '*Vis vegetativa* est certa corporis constitutio, qua substantiae corporeae calorisque perpetuam dissipationem, per succum à Corde praeparatum, & in partes impulsum, conservamus, & ex semine nostri simile procreamus.' [Italics in the text.][Translation mine.] See Verbeek 1992, 15–16.

distribution is called death.[18] Two issues surface in this text: (a) Regius overlaps the vegetative power and the conservation of life, or life *tout court*. This power coincides with the circulation or distribution of fluid or blood prepared in the heart, the principle of life in Descartes' physiology. As a result, this power is just a condition of the body and not an activity properly speaking. (b) This fluid has little to do with digestion as such. Indeed, Regius speaks of coction and nutrition in the following theses, as separate issues, while in Descartes the vegetative power as *immutatio* is an activity regulating digestion and nutrition.

The second step was undertaken at the end of 1641, when Regius collected all his disputations in a volume entitled *Physiology or the Knowledge of Health* (*Physiologia sive Cognitio Sanitatis*), embedding his claims within a philosophical system. In this text, these passages are repeated literarily.[19] To this, Regius adds that the vegetative and sensitive powers '(which we could call *Nature* of the body) are nothing but the suitable temperament of the well conformed human body' and correspond to 'the multiple actions that are admired in clocks and other automata, accomplished by the sole conformation of the parts' (Regius 1641b, 15–6).[20] Regius reduces the substance composing the body to matter: truly a Cartesian claim.[21]

Still, he encompasses the explanation of living activities within an un-Cartesian framework. He presents the vegetative (and sensitive) power in a section entitled 'On Actions and the Human Soul,' where he conceives of it not only as a function of bodily processes, but also of human actions. Accordingly, these latter consist of both the operations performed by the body and the operations of the soul.[22] This is un-Cartesian insofar as Descartes himself firmly rejects the theory that the body is a source of human actions. In his 1641 letter to Regius, he writes: 'I do not agree with you when you define actions as *operations performed by man by the power of soul and body*. For I am one of those who deny that man understands by means of the

[18] Ibid., Theses 12: 'Succi illius apta ad corpus caloremque conservandum à Corde distributio est *Vita*; uti distributionis istius privatio *Mors* dicenda.' [Italics in the text.]

[19] Regius 1641b, 15: '16. *Vis* autem *vegetativa* in homine nihil aliud est, quam certa partium corporis constitutione, qua substantiae corporeae calorisque perpetuam dissipationem per succum à corde praeparatum, et in partes impulsum, conservamus, et ex semine nostri simile procreamus. Succi illius apta ad cordus caloremque conservandum à corde distributio, *vita* dicenda est: uti distributionis istius privatio, *Mors*.' (Italics in the text.)

[20] *Ibid*., 15–16: '18. *Hae duae* itaque (quae *Natura* corporis alleppari possunt) nihil aliud sunt, quam corporis humani apte conformati apta temperies: quandoquidem omnes illarum operationes ab hac ita fieri queunt, ut in horologio et aliis automatis plurimae actiones admirandae a sola partium conformatione peraguntur: ita ut non opus sit aliquam substantialem incognitamque formam hic vel alibi in similibus fingere, entiaque contra verissimum Philosophiae dictatum, multiplicare absque necessitate.' (Italics in the text.)

[21] It is however to note that Descartes advised Regius to avoid rejecting substantial forms openly. Cf. Descartes to Regius, January 1642, in Descartes 1964–1974, III, 491–92. *Discours de la Méthode*, in Descartes 1964–1974, VI, 42–3.

[22] Regius 1641b, 15: '*Actiones* sunt operationes ab homine vi animae humanae, vel corporis, vel utriusque factae. 14. *Anima humana* est actionum humanarum primum in homine principium …' (Italics in the text.)

body' (Descartes to Regius, May 1641, in Descartes 1964–1974, III, 374).[23] In contrast, following his interpretation of the mind-body distinction, Descartes stresses that human actions are only produced by the mind or soul (Descartes 1964–1974, III, 371).[24] Stating the contrary appears to Descartes as an attempt to introduce intelligence into bodily actions, which corresponds to the restoration of the souls. This was a prelude to what would go on to happen in the following years (Verbeek 1994).

The third step took place in 1646, when Regius published his main treatise on natural philosophy, the *Fundamenta physices*. This was considered a satisfying and exhaustive Cartesian natural philosophy, although Descartes had tried to prevent Regius publishing it.[25] The section 'On Living Bodies' (*De Corporibus vivi*) starts with a definition of 'living bodies [as those] whose parts are temperate and conformed in such a way that their bodily substance, which is perpetually dissipated, is conserved by means of the fluid prepared and infused from the inside, following the mixture of temperaments and the conformation of parts' (Regius 1646, 145).[26] Despite some slight changes, this text echoes Regius' words that define vegetative power in *Physiology*. Yet, he adds something more, as he writes that '[t]his arrangement of parts [*partium dispositio*] is the vegetative soul of these bodies: [the vegetative soul] is their principle, through which living bodies achieve their living functions' (Regius 1646, 145).[27]

Regius reintroduces the term soul to specify vegetation and, later, sensation in living bodies. Accordingly, plants [*Stirps*] live through the vegetative soul alone, and animals live through the vegetative and sensitive souls.[28] This is a problematic formula that Descartes had prohibited Regius using and should not be underestimated, as Marin Mersenne's (1588–1648) comment on the text reveals.[29] The question then is: What are the souls Regius speaks about?

---

[23] 'Non etiam tibi assentior, cum definis *actiones esse operationes ab homine vi animæ et corporis factas*; sum enim unus ex illis qui negant hominem corpore intelligere.' (Italics in the text.)

[24] '*Vis* autem *vegetandi*, et *corporis movendi*, quæ in plantis et brutis *anima vegetativa* et *sensitiva* appellantur, sunt quidem etiam in homine, sed non debent in eo *animæ* appellari, quia non sunt primum ejus actionum principium, et toto genere differunt ab *anima rationali*.' (Italics in the text.)

[25] Huygens to Mersenne, 21 August 1646, in Mersenne 1932–1988, vol. 14, 413: '*Fundamenta Physicae* [...] en fin sera un corps achevé ...'; Huygens to Mersenne, 12 September 1646, in Mersenne 1932–1988, 14, 450: 'Le livre de Regius vous contentera en sa methode.'

[26] 'Corpora viva sunt, quorum partes ita sunt temperatae & conformatae, ut corporea eorum substantia, quae perpetuo dissipatur, per succum praeparatum & in interiora impulsum, secundum temperiem & partium conformationem, conservetur.' [Translation is mine.]

[27] 'Haec partium dispositio est istorum corporum anima vegetativa: est enim principium, quo corpora viva actiones suas vitales perficiunt.' [Translation is mine.] The text has this writing in the margin: 'Anima eorum vegetativa.'

[28] *Ibid.*, 148: 'Stirps est corpus vivum sola anima vegetativa praeditum.' *Ibid.*, 154.

[29] See Mersenne to Huygens, end of September 1646, in Mersenne 1932–1988, 14, 496–497. Translation is from Fowler 1999, 150: Mersenne rhetorically asked Constantijn Huygens (1596–1688) whether 'he believes that Mr. Regius explains the movements of plants and animals without giving them souls, as the principles of Descartes seem to require?' While Fowler interprets this text as a confirmation of Regius' eradication of souls, I think this letter is much more problematic.

In *Fundamenta physices*, Regius explains plant and animal behaviour in exclusively mechanical terms and without any appeal to cognition.[30] Since these sections are nowhere to be found in Descartes' *Principia*, Regius' work is quite important. Accordingly, plants and animals achieve living functions by the sole arrangement of particles and bodily conformation; and vegetating corresponds to the movement of fluids. Thereby, attributing a soul to living bodies ought to adhere to this mechanical framework. Regius mechanizes the soul as well. This is well known for the case of the rational soul. In *Brevis explicatio mentis humanae* (1648), Regius conceives of the mind (or rational soul) either as a substance, or as a mode of a corporeal substance, or as an attribute co-existing with extension in bodies.[31] A similar materialization occurs for the vegetative soul, whose function consists in the motion of fluids. Regius considers it 'consisting in inner heat [*calore nativo*] [...] which helps in preparing the alimentary fluid...' (Regius 1646, 146) and so on.[32] According to Regius, the fluid 'is the instrument for the conservation [of the body] and is prepared by the vegetative soul' (Regius 1646, 146).[33] Notwithstanding the reduction of the vegetative soul to heat, this definition uncovers a new problem: in referring to inner heat, Regius uses a scholastic formula close to substantial forms (Regius 1641b, 12–13).[34]

Another question concerns the ways the vegetative soul prepares this fluid. In the next sections, Regius describes two actions of living bodies: nourishment and procreation, which supposedly produce the matter composing this fluid. Regius follows Descartes' mechanization in this description, both using the text Descartes suggested in their correspondence and the words and theory contained in Descartes' *Excerpta anatomica*. Regius dismisses the substantial transmutation of particles, because food remains of the same matter, and also a transmutation of particles, which according to Descartes corresponds to growth in inert bodies, and claims that food internally changes [*immutentur*] during digestion (Regius 1646, 146).[35] He endorses a mechanical account of nutrition. Nevertheless, the relationship between

---

[30] Regius 1646, 226; 241: 'Bestia est animal merum, seu tale, quod actiones suas sensitivas et motivas, per solam partium dispositionem, citra ullam cognitionem vel intellectum [...] perfecit.'

[31] Regius 1648, 7: 'mens possit esse vel substantia; vel quidam substantia corporea modus; vel, [...] Mens possit esse attributum quoddam eidem subjecto cum extensione in homine conveniens...'

[32] 'Eaque potissimum in calore nativo consistit; qui est ignis tantum calidus, in corporibus vivis a prima eorum productione genitus, cujus ope succus alimentarius, ad eorum conservationem, praecipue praeparatur, & per totum corpus distribuitur, eique agglutinatur.' (Translation mine.)

[33] 'Ipse autem ille praeparatus succus, quia est immediatum istius conservationis instrumentum, quodammodo etiam anima vegetativa dicit potest. Succi vero istius apta, ad corpus conservandum, distributio, vita est dicenda; & provatio istius dispositionis, mors.' (Translation mine.)

[34] Andrea Strazzoni has recently unearthed Regius' references to Jean Fernel and Vopiscus Plempius in his section on innate heat, but has failed to grasp its traditional reach; see Strazzoni 2018, 409.

[35] 'Itaque hic nulla intercedit substantialis transmutatio: cum substantia alimentorum, quae est ipsa materia, hic & ubique semper eadem maneat; sed accidentia essentialia, quibus, ex:gr: cibus à chylo, & chylus à sanguine differunt, in conctione tantum immutentur.'

the vegetative power conceived as a soul and these functions remains unspecified in Regius. While Descartes conceives of the vegetative power as a combination of bodily activities, i.e., as a power of the body itself, Regius attributes it a different role. The vegetative soul, material as it may be, is not a combination of mechanical operations, but a principle inhabiting and animating the body to achieve basic life functions.

Although Regius repeats Descartes' words and encompasses nutritive activities within a mechanical framework similar to that of Descartes, his definitions of vegetative power reveal a shift from the latter. Three steps characterize this change, which highlight three main problems. In his early disputations, Regius defines the vegetative power as a conformation of the body preserved by the circulation of blood. At this stage, the vegetative power is not a power at all, but a bodily condition maintained and prompted by cardiac heat. The second step consists in the definition of the vegetative power as a human action in the *Physiologia*, which allows for reintroducing the scholastic soul, something Regius did in the *Fundamenta physices*. The third step consists in invoking the inner heat, another scholastic formula close to substantial forms, and claiming the vegetative soul to be a principle to activities that animate living bodies. Indeed, Regius' vegetative power is not a set of mechanical functions, but a principle of organic activities. As long as he restores concepts from the scholastic tradition, the detachment from Descartes' mechanical interpretation of the vegetative power appears striking.

15.4 Two Problems Within Descartes: La Forge's *Remarks*

The second case is the French physician and philosopher Louis de La Forge, who collaborated with Claude Clerselier (1614–1684) in the posthumous edition of the *Traité de l'Homme* of 1664. La Forge especially provided lengthy footnotes to the text, published at the end of Clerselier's edition of *L'Homme* as *Remarks* (*Remarques*). In preparing this commentary, La Forge faithfully pursued Descartes' project, especially opposing Regius, and helped disseminate Descartes' physiology in France in favour of widespread support for Cartesian natural philosophy (Clarke 1989, 8).[36] Despite seeking to provide internal consistency to Descartes' physiology, two problems affect La Forge's interpretation of vegetative activities.

In the *Remarks*, La Forge aimed to fill the gaps, supplement, and correct Descartes' *L'Homme*, bringing his programme to completion. He thus read *L'Homme* in the light of recent discoveries and in the light of Descartes' complete work, considering Cartesian medicine as intrinsically valid, i.e., metaphysically

[36] Cf. Antoine-Mahut 2016, 17: La Forge 'presents himself in the role of disciple with a fourfold objective: *to supplement, correct, link, and apply*.' [Italics in the text.] On La Forge, see Drieux 2019. Cf. La Forge 1999.

grounded.[37] Finally, he implemented the study of the body with a study of the union of the mind and body in the *Treatise on the Human Mind, on its Faculties and Functions, and on its Union with the Body* (*Traité de l'esprit de l'homme, de ses facultez et fonctions, et de son union avec le corps. Suivant les principes de René Descartes*), written together with the *Remarks*, but published separately in November 1665, although the title page is dated 1666 (Manning 2012; Clarke 1989, 19). This text is original, since La Forge became aware of the lack of a Cartesian text specifically devoted to the mind, but still based as closely as possible on Descartes' work.

Let us start with the *Remarks*. When La Forge deals with digestion and nutrition in commenting on articles 3, 4, 8, 9, 10 and 11 of *L'Homme*, a first problem surfaces. In the first notes, La Forge actually speaks of a power regulating digestion, as he discusses the different rates of digestion, a subject under investigation at the time, and interprets it in the light of Cartesian mechanical physics, especially referring to Descartes' *Principia philosophiae* part Four, article 93.[38] Accordingly, in chemistry the particles dissolving bodies [sometimes] do not have the same shape [*figure*], or size [*grosseur*], or power [*force*] that are necessary to penetrate the pores of the other body and divide its parts (La Forge 1664, 178). Something similar occurs in digestion. As chemicals fail to dissolve certain bodies, some liquids need more time to dissolve some kinds of food. In other terms, digestion occurs by means of the particles of the liquids shaking, warming and dividing food. These actions rely on the shape, size and power of the particles of the liquids that transform the food. The result is a uniform substance called chyle (La Forge 1664, 179).

In commenting on articles 8, 9, 10 and 11, La Forge explains the formation and arrangement of the particles that constitute the body. Accordingly, 'the bodies of Plants and Animals are nothing but a group of small fibers, whose ways of maintaining and mixing serve to constitute the different parts of animals and plants' (La Forge 1664, 195). These ways of constituting the body reflect a mechanical system, established on the size, shape and movement of particles within the circulatory system. Two reasons explain why particles reach a precise spot in the body throughout nutrition: their movement and shape. Heat also plays an important role. The particles accumulated in the spleen are 'cooked [*se cuisent*], fermented [*s'y fermentent*] and digested [*s'y digerent*]' (La Forge 1664, 202) by the heat of the blood. The same activities occur in digestion. While La Forge attributes digestion in the stomach to the power of the acid liquid or stomach acid operating on food, he also claims that

[37] It is to be remembered that Descartes conceived *L'Homme* as a treatise in three main sections, the one on the human body, the second on the soul or mind, and the third on the union. Only the first is achieved, though partially. The *Meditations* could embody this second part. Reading *L'Homme* in 1664 cannot be isolated from the entire work. Cf. *L'Homme*, in Descartes 1964–1974, XI, 119–120; *Discours de la Méthode*, in Descartes 1964–1974, VI, 59.

[38] La Forge 1664, 178: 'Mais parce que nous voyons que nous ne digerons pas toutes sortes de choses avec une égale facilité, l'on peut demander pourquoy toutes sortes de menstruës (pour parler en Chymique) n'ont pas la Faculté de dissoudre toutes sortes de Corps.'

'this liquid comes from the arteries,' (La Forge 1664, 203) that is, from the heart.[39] In sum, La Forge embeds nutrition and digestion in the cardiac system. While he assimilates the power of the liquids transforming food into particles to a chemical power, he stresses that their force is related and originates in the heat of the heart.

In this sense, La Forge's *Remarks* are orthodoxly Cartesian, and his explanation of digestion and nutrition mostly follows Descartes' interpretation in *L'Homme*. For this reason, La Forge fails to fill the lacuna of Descartes' early physiology that restricted vegetative activities to the cardiac system. While following Descartes' physiology, the general question about a vegetative power regulating nutrition and digestion remains, and the problem of explaining this power endures.

La Forge opposes the scholastic uses of forces or occult powers, making clear that the cause of vegetative activities does not belong to the faculties; still, detecting their cause is complex (La Forge 1664, 217).[40] In the *Treatise of the Human Mind*, La Forge deals with causation more precisely, and an attention to vegetation surfaces. First, he claims that the life of the body 'consists in nutrition' (La Forge 1997, 9), and 'the principle of vegetation and nutrition [...] is distinct from the soul' (La Forge 1997, 15).[41] More precisely, he stresses that the life of animals and plants 'consists in the motion which the finer particles of their blood (in the case of animals) and of their alimentary juice (in the case of plants) gives to the less subtle parts of their bodies. They live as long as this blood or juice is able to move and maintain them in their natural condition' (La Forge 1997, 48). In other words, he conceives nutrition as the self-motion of particles, and claims this is the principle of life in animals and plants as long as it preserves their bodies.

However, while discussing his occasionalistic interpretation of causation and bodily motions (Clarke 1997, xix), a specific problem arises.[42] In Chap. 16 of this text, when speaking of a force moving living bodies, La Forge proves that 'the cause of motion of bodies is not therefore something which is as obvious as one might think, and [...] it is no more difficult to conceive how the mind moves the body than to know how one body moves another;' indeed, motion cannot 'be communicated only by impact or as if it were as easy to perceive how one body could move another as it is to see how it touches it' (La Forge 1997, 143). He acknowledges a difference between the motion and the force that makes something move, as this force is different from the body to which it applies and moves. Thus, he makes clear that

---

[39] 'Cette liqueur acide, à laquelle nous avons attribué la dissolution des viandes [...] nostre estomac dissout quelquefois le biscuit & le pain, sans que nous bevuions, ny le trempions, il faut qu'il y ait quelque liqueur dans son fond, qui s'insinuant dans les pores des viandes solides qua nous avons avalées, en separe les parties [...] Il y a grande apparence que leur entiere dissolution dépend d'une semblable liqueur [...] Et de plus les observations du docte Wallaeus [or de Waleus] font voir que cette liqueur vient des arteres.'

[40] '[e]st ce bein expliquer la cause d'une Diarrhée, par exemple, que de dire qu'elle vient, ou de ce que la faculté expultrice est irrité, ou de ce que la faculté retentrice des intestins est affoiblie; N'est ce pas en bon François dire je n'en say rien?'

[41] Cf. ibid. 19: 'the human mind is not the principle of vegetative and nutritive functions.'

[42] On occasionalism in La Forge, cf. Nadler 2011 and Carraud 2002.

'no body can have the power of self-movement in itself' (La Forge 1997, 145). Defining what produces self-motion such as digestion and nutrition appears to be difficult. Moreover, he stresses that 'not only can it not change its condition by its own power; [but also] there is no creature, spiritual or corporeal, which can cause change in it or in any of its parts [...] if the Creator does not do so himself' (La Forge 1997, 146–7). In claiming that this force 'belongs as much as thought does to a spiritual substance,' he also 'recognize[s] bodies and minds as particular causes of these same motions,' (La Forge 1997, 145, 148) therefore differentiating between a universal cause, that is God, and secondary causes. Yet, in the case of body-body causation, La Forge intends that 'bodies cannot bring about those effects on their own' (Sangiacomo 2014, 78). This interpretation is crucial insofar as it makes it more difficult to conceive of any autonomous vegetative power operating in digestion and nutrition as a cause of these functions, and any attempt to define a cause for them better fits a general system of motions, i.e., the cardiovascular framework, than a specific set of operations.

In the following chapter, La Forge tries to go a step further on the question of causation, as he insists that no functions depend on the soul or mind. In defining the essence of the mind, La Forge claims that 'the human mind is a substance which thinks and [...] it is [...] the human form or soul.' As such, the mind is the principle of what 'distinguish[es] human beings from what is not human' (La Forge 1997, 152). The human form is the principle of thought and there is no necessity for it to be the principle of all living functions. Against Aristotelian scholars, La Forge stresses that the Aristotelian definition of form, '*form is that which makes matter become a specific thing*', consists of something 'by which [...] animals differ from plants and plants differ from inanimate bodies' (La Forge 1997, 153, Italics in the text). Accordingly, form is what makes a specific thing to be, and it is the principle of all its properties. This entails that the mind, which is what makes humans to be a specific thing, cannot in principle be something that is common between bodies, such as vegetation. As a result, he is thus able to claim that 'even according to Aristotle's teaching [...] it is not necessary [for the human mind] to be the principle of vegetation, nutrition, and generation, nor of everything which pertains exclusively to the body in its animal functions' (La Forge 1997, 153).

Additionally, appropriating the Aristotelian concept of form (and conceding to Descartes himself on this point), La Forge individuates singular bodies, differentiating between inert and living bodies, but also between plants and animals. The consequence is striking. When he claims that 'the form of animals and plants should not be the principle of those functions which they have in common with each other, but only the source and origin of all the differences which are observable' (La Forge 1997, 153), it is easy to read this as saying that the form of plants cannot be what makes them vegetate, since all living bodies would share vegetative activities. This point is consistent with Descartes' philosophy,[43] insofar as La Forge characterizes

[43] Cf. Descartes to Regius, May 1641, in Descartes 1964–1974, III, 371.

vegetation as a set of functions within living bodies of the same genus as other functions. Yet, the result is that vegetation is part of a general system, while a specific principle is lacking.

La Forge does not appear to be interested in defining this principle. In stressing that the mind is not the principle of vegetation, he claims that 'none of the vegetative functions depends (except accidentally) on thought' (La Forge 1997, 153). However, this exception does not open up to any understanding of the mind as a principle of bodily operations, and has nothing to do with the study of the higher faculties with plants, nor with an exploration of the principle of vegetation. This latter remains undefined.

In sum, two similar problems overwhelm La Forge's texts. In the *Remarks*, he fails to provide a physiological explanation for an autonomous principle of vegetative activities such as digestion and nutrition, which remain restricted to the cardiocentric framework of Descartes' physiology. In the *Treatise on the Human mind*, La Forge's occasionalism prevents him from defining a vegetative power as the principle of some bodily activities. In both cases, while supporting and integrating Cartesian physiology and philosophy, La Forge's interpretation of a mechanical vegetative power remains unspecified.

15.5 Florent Schuyl: Plants and Vegetative Activities in *De Homine* (1662)

The third case is the Dutch physician Florent Schuyl, who in 1662 published a Latin translation of *L'Homme*.[44] This work, *De Homine*, presents a number of important issues, especially as the text differs from the French edition of 1664, and includes a large *Introduction*, 'Ad Lectorem,' in which Schuyl discusses the soulless condition of living bodies. In the introduction, he combines a new biological interest in the autonomy of living functions with a profound theological-philosophical interest in the exceptional status of the soul. Before moving to the text, it is important to note that Schuyl was a Cartesian *sui generis*, as the publication of *De Homine* marked the end of a process of intellectual transition from an Aristotelian to Cartesian position (Ruler 2008). In the *Introduction*, he discusses both the sensory and motor functions and the vegetative activities of plants, rejecting any vegetative soul and presenting a mechanization of vegetal bodies.

Schuyl advances two important claims: self-motion is a false proof of life, and, in contrast to the scholastic claim that a soul activates self-motion, the soul is not the principle of bodily motions, nor of life (Schuyl 1662, unpaged).[45] This lies at the origin of the scholastic misinterpretation of plants. Indeed, Aristotelians

[44] On Schuyl, see Lindeboom 1974.

[45] '[Q]uia veram illarum motus causam ignorans, [...] putat motum illum, fallax vitae indicium, ab anima pariter ac in hominibus determinari.'

[*Peripatismus*] provided plants with a soul, thus explaining their being alive, according to Schuyl. Yet, they consider this soul not merely as 'consisting of the arrangement of their parts, i.e., the cause of vegetation, but [...], ex arbitrio, as a mind [*mentem*] or a substance totally other than the matter or body of plants' (Schuyl 1662).[46]

Schuyl stresses that the soul is not the principle of life, motion, or vegetation. In contrast, he claims that the cause of vegetation is the arrangement of parts [*partium dispositione*], something that echoes Descartes' vegetative power, but fails to explain it in detail. However, Schuyl conceives the functions of plants as following the movement and arrangement of particles, which replace the vegetative soul (Schuyl 1662).[47]

He then describes nutrition and generation in plants. Grounded on observation through a magnifying glass [*quod vel vitrea arbuscula ad oculum demonstrari potest*], he claims that nutrition occurs in plants due to the movement of aliment or juices that enter the pores of roots, rise and move upwards within the trunk, then spread, ferment and turn around, and expand. The heating and rarefaction of particles produce these motions, as inspecting the blood of gourds [*sanguinis in cucurbitulas*] revealed (Schuyl 1662).[48] The similarity between juices and blood, a widespread belief at the time, is to be noted.

Plants grow from this movement and arrangement of particles caused by heat: knotted surfaces, leaves and branches develop, then seeds and fruit emerge. According to Schuyl, the diversity of particles composing the juices and the various pores that the particles fill produce the various parts of the plants (Schuyl 1662).[49] More generally, the variety of pores and the variety of juices that fill them determine the species of the plant, and belong to the plants or the stems that engender the seed (Schuyl 1662).[50] From the explanation of nutrition and growth, Schuyl moves to the

---

[46] 'Haud dispari imprudentia & prodigalitate Peripatismus praeter Creatoris intentionem, plantas anima vitaque donavit. Non ea. tantum, quae in partium dispiositione consistens, vegetationis causa est: sed & alia quam verae causae ignorantia, velut mentem, sive substantiam a plantarum materia sive corpore plane diversam, iss affinxit ex arbitrio.'

[47] 'Omnium enim Plantarum facultatum exercitia, nulla istiusmodi anima considerata, facile concipiuntur.'

[48] 'Etenim non est, quod quempiam lateat, qua ratione pro tempestatis, dierum noctiumque vicissitudine, succus sive alimentum in plantas ante calore rarefactas, subsequente frigore, non aliter quam sanguis in cucurbitulas, quod vel vitrea arbuscula ad oculum demonstrari potest, per poros radicum adigatur, sursum pellatur atque attollatur, quaqua versum distribuatur, fermentetur, & redeunte calore effervescat, & elaboretur.'

[49] 'Unde planta distendatur, & tandem oculi folia, rami, flores, semina, fructus ebulliant, condesentur, & velut crystallizentur. Diversa quidem pro diversitate succi, & pororum, per quos illi transitus est.'

[50] 'Quae quidem pororum diversitas, singulis plantarum speciebus propria, oritur ex diversitate pororum matris cujusque seminis, atque plantae: et succi materni illos perforantis, quemadmodum illos Naturae Fabricator primum condidit.'

explanation of generation, which he considers a kind of nutrition (Schuyl 1662).[51] This interpretation of the life of plants is Cartesian, as Schuyl emphasizes Descartes' principles of philosophy against the erroneous claims of a vegetative soul or, even, the soul-world principle.[52]

Schuyl's complete mechanization of plants operates in every aspect of vegetation, as the example of the tulip reveals. Thus, this plant, 'void of all cognition, automatically opens its flower to receive the morning sunlight, and closes it again in the evening to preserve its seed from the harmful cold of the night' (Schuyl 1662).[53] No soul animates plants, just the movement of the juices nourishing them. According to Schuyl, what the scholastic tradition called soul, and which in the case of animals is their blood, seed, flesh, and so on, in the case of plants is their juices, seeds, and matter.

If what replaces the soul of plants is their body, what replaces the vegetative soul as the source of these activities is more complex to articulate. Stressing that the movement and arrangement of particles make plants vegetate, as Schuyl does, is insufficient to explain what causes generation, nutrition, and growth. Yet, in Schuyl's text there surfaces a vegetative power consisting in a set of bodily activities, which regulate the constitution of the body. Indeed, Schuyl's detailed explanation of activities such as nutrition and generation in plants provides a clear analysis of their mechanical functioning, without the need attribute a mechanical cause or principle to these functions. Their cause lies in the plant, or in the animal, as it is possible to observe these phenomena.

This is something Schuyl confirms in a passage in *De veritate scientiarum et artium academicarum* (Leiden 1672), the text of an official speech he made in February 1667 at Leiden University's *dies natalis*. In rejecting a soul that endows plants with life and vegetation, Schuyl underlines that this soul fails to explain the ways in which nutrition, growth, and generation occur. In contrast, if one wants to achieve such knowledge, it is enough to observe the movement of juices, the arrangement of particles and the constitution of seeds, thus revealing vegetative functions (Schuyl 1672, 22).[54] What causes the vegetative activities is not a vegetative power endowing the body with life, but the bodily mechanistic system at work (Schmaltz 2016, 83). Mechanizing the vegetative soul thus consists in the mechanical analysis of nutrition, growth, and reproduction, which occur within living bodies in the ways Schuyl had observed.

---

[51] 'Nec est, quod quemquam forte remoretur abstrusior plantarum generatio, si nulla prorsus ratio ejusmodi animae habeatur. Etenim plantarum semn omnino nihil aliud esse, quam ramusculum, pulpa quadam, velut capsula, conclusum, autopsia constat. Unde sequitur eodem plane modo, quo rami, & semina produci, nutriri, atque crescere.'

[52] On Descartes' study of plants, see Baldassarri 2019.

[53] 'Hac Tulipa, licet omni propria cognitione destituta, folia sua matutino Soli explicat, quae, ne a nocturno frigore semini fiat ignuria, vesperi colligit, atque constringit.' [Translation is mine.]

[54] The text is reprinted with separate pagination in Lindeboom 1974, 125 ff: 'Creditum quoque fuit plantas quibusdam fuis animulis vivere & vegetari, jam vero, quicquid illis nutritionis, auctionis & generationis accidit, a sola fermentatione atque effervescentia oriri, non dicitud modo: sed etiam arte factis ex multiplici materia seminibus ad oculum demonstratur.'

15.6 Concluding Remarks

Descartes' mechanical definition of life conflicts with his under-specified account of the vegetative functions. In *L'Homme*, he restricts nutrition and digestion to his cardiovascular framework. This interpretation entails a few unanswered questions that concern (a) the cause of vegetative activities, (b) the organic functioning of the body, (c) the relation to blood circulation and the autonomy of these operations, and (d) the definition of a vegetative power endowing living bodies. Despite grappling with these questions in his biomedical notes, where he defines a set of operations regulated by the *immutatio*, later described as a vegetative power acting in the constitution of the body, this definition remains however absorbed in the primacy of the heartbeat, and receives little attention.

The early Cartesians who dealt with Descartes' physiology were obliged to deal with this issue, answering these questions and providing a more exhaustive definition of the vegetative power of the body. Yet, their results are not as satisfying as one may expect. In this chapter, I have explored the three main cases. The first is the renegade Henricus Regius, an early proponent of Cartesianism who personally discussed this issue with Descartes. In dealing with a vegetative power in his work, Regius travels an untrodden path, as he combines a mechanical interpretation of these functions with a restoration of scholastic formulas: he speaks of souls as principles of living functions, and claims the vegetative soul, though material, to be the inner heat, another scholastic formula. The second case is the orthodox Louis de La Forge, whose remarks on *L'Homme* significantly enhance the text and fills its gaps, but fails to deal with digestion, nutrition, and generation, as he affirms Descartes' cardiovascular framework. Additionally, in his explanation of causation, La Forge endorses an occasionalistic view which prevents him from defining a mechanical source for bodily activities, and to specify a vegetative power that regulates specific functions. The third is the newcomer Florent Schuyl, whose translation of *De Homine* carries through his transition from an Aristotelian to a Cartesian stance. His mechanical interpretation of vegetation relies on direct observation of plants. In rejecting the vegetative soul, Schuyl explains how nutrition, reproduction, and growth occur in plants (and animals). Still, he fails to define vegetative power, but claims that the body as a coherent system acts as source of power for the vegetative functions.

These positions unearth the challenges Cartesian natural philosophers faced while dealing with growth and vegetation in living bodies, as their interpretations either remain embedded within a traditional framework or fail to provide systematic knowledge of living bodies. In particular, Regius and La Forge restore the idea of a principle informing the matter of the living body, reverting to a different stage of the understanding of the body, while Schuyl does not elaborate on a vegetative power. Ultimately, the mechanization of vegetation uncovers the limitations, if not the failures, of Descartes' mechanistic programme to explain living bodies.

Acknowledgments Research for this chapter has been carried out with the support of the Land Niedersachsen Herzog August Bibliothek fellowship and by a grant of the Romanian National Authority for Scientific Research and Innovation (CNCS – UEFISCDI), project number PN-III-P1-1.1-PD-2016-1496, "The Overlooked History of Vegetal Life. From the Vegetative Soul to Metabolism in Early Modern Philosophy and Biomedicine." I would like to thank Igor Agostini, Vlad Alexandrescu, Delphine Antoine-Mahut, Gideon Manning, Steven Nadler, Michael Pickering, and Theo Verbeek for their helpful suggestions.

References

Antoine-Mahut, Delphine. 2016. The Story of *L'Homme*. In *Descartes'* Treatise on Man *and its Reception*, ed. Delphine Antoine-Mahut and Stephen Gaukroger, 1–29. Cham: Springer.

Antoine-Mahut, Delphine, and Stephen Gaukroger, eds. 2016. *Descartes'* Treatise on Man *and its Reception*. Cham: Springer.

Aucante, Vincent. 2006. *La philosophie médicale de Descartes*. Paris: PUF.

Baldassarri, Fabrizio. 2018a. Descartes' Bio-Medical Study of Plants: Vegetative Activities, Soul, and Power. *Early Science and Medicine* 23 (5–6): 509–529.

———. 2018b. I moti circolari nella meccanica della vita in Descartes: embriologia e nutrizione nella medicina e nella botanica. *Physis* 53: 57–74.

———. 2019. The Mechanical Life of Plants: Descartes on Botany. *British Journal for the History of Science* 52 (1): 41–63.

———. 2020. Descartes and the Dutch: Botanical Experimentation in the Early Modern Period. *Perspectives on Sciencie*, 28/6: 657–683.

Baldassarri, Fabrizio. 2021a. *Il metodo al tavolo anatomico: Descartes e la medicina*. Rome: Aracne.

———. 2021b. Immutatio. In *Nouvel Index Scolastico-Cartésien*, ed. Igor Agostini et al., forthcoming. Paris.

———. 2021c. Anima vegetativa. In *Nouvel Index Scolastico-Cartésien*, ed. Igor Agostini et al., forthcoming. Paris.

Bertoloni Meli, Domenico. 2011. *Mechanism, Experiment, Disease: Marcello Malpighi and Seventeenth-Century Anatomy*. Baltimore: Johns Hopkins University Press.

Bos, Erik-Jan. 2002. *The Correspondence between Descartes and Henricus Regius*. Utrecht: Zeno.

———. 2017. Descartes and Regius on the Pineal Gland and Animal Spirits, and a Letter of Regius on the True Seat of the Soul. In *Descartes and Cartesianism: Essays in Honour of Desmond Clarke*, ed. Stephen Gaukroger and Catherine Wilson, 95–111. Oxford: Oxford University Press.

Carraud, Vincent. 2002. *Causa sive ratio: la raison de la cause de Suarez à Leibniz*. Paris: PUF.

Clarke, Desmond M. 1989. *Occult Powers and Hypotheses: Cartesian Natural Philosophy under Louis XIV*. Oxford: Clarendon Press.

———. 1997. Introduction. In *Treatise on the Human Mind (1664)*. Trans. and Ed. Desmond M. Clarke, xiii–xxv. Dordrecht: Kluwer.

Des Chene, Dennis. 2000. *Life's Form. Late Aristotelian Conceptions of the Soul*. Ithaca: Cornell University Press.

———. 2001. *Spirits & Clocks: Machines and Automata in Descartes*. Ithaca: Cornell University Press.

Descartes, René. 1964–1974. In *Œuvres complètes*, ed. Charles Adam and Paul Tannery, vol. 11. Paris: Vrin-CNRS.

———. 1984. *The Philosophical Writings of Descartes*. Trans. and Ed. John Cottingham et al., 2 vols. Cambridge: Cambridge University Press.

———. 2003 [1972]. *Treatise of Man*. Trans. and Ed. Thomas S. Hall. New York: Prometheus Book.
———. 2004. *The Worlds and Other Writings*. Trans. and Ed. Stephen Gaukroger. Cambridge: Cambridge University Press.
Drieux, Philippe. 2019. Louis de La Forge on Mind, Causality, and Union. In *Oxford Handbook on Descartes and Cartesianism*, ed. S. Nadler et al., 319–331. Oxford: Oxford University Press.
Fowler, C.F. 1999. *Descartes on the Human Soul: Philosophy and the Demands of Christian Doctrine*. Dordrecht: Kluwer.
Gaukroger, Stephen. 2002. *Descartes System of Natural Philosophy*. Cambridge: Cambridge University Press.
Grene, Marjorie. 1986. Die Einheit des Menschen: Descartes unter den Scholastikern. *Dialectica* 40 (4): 309–322.
Hattab, Helen. 2009. *Descartes on Forms and Mechanism*. Cambridge: Cambridge University Press.
La Forge, Louis de. 1664. *L'Homme de René Descartes et un traité de la Formation du fœtus … Avec les Remarques de Louys de La Forge*. Paris.
———. 1997. *Treatise on the Human Mind (1664)*. Trans. and Ed. Desmond M. Clarke. Dordrecht: Kluwer.
———. 1999. In *L'Homme de René Descartes et un traité de la Formation du fœtus … Avec les Remarques de Louys de La Forge*, ed. Thierry Gontier. Paris: Fayard.
Lindeboom, Gerrit A. 1974. *Florentius Schuyl (1619–1669). En zijn betekenis voor het Cartesianisme in de geneeskunde*. Den Haag: Nijhoff.
Manning, Gideon. 2012. When the Mind Became Un-Natural: De La Forge and Psychology in the Cartesian Aftermath. In *Psychology and other Disciplines: A Case of Cross-Disciplinary Interaction (1250–1750)*, ed. Paul J.J.M. Bakker, Sander W. Boer, and Cees Leijenhorst, 131–153. Leiden: Brill.
Mersenne, Marin. 1932–1988. *Correspondance du P. Marin Mersenne*, ed. Corneliis de Waard et al., 16 vols. Paris: CNRS.
Meschini, Franco Aurelio. 2015. La dottrina della digestione secondo Descartes. Itinerari tra testi, contesti e intertesti. *Physis* 50: 113–164.
Nadler, Steven. 2011. *Occasionalism: Causation among the Cartesians*. Oxford: Oxford University Press.
Nadler, Steven, Tad M. Schmaltz, and Delphine Antoine-Mahut, eds. 2019. *The Oxford Handbook of Descartes and Cartesianism*. Oxford: Oxford University Press.
Regius, Henricus. 1641a. *Disputatio medica prima, De Illustribus aliquot Quæstionibus Physiologicis*. Utrecht.
———. 1641b. *Physiologia sive Cognitio Sanitatis*. Utrecht.
———. 1646. *Fundamenta physices*. Utrecht.
———. 1648. *Brevis explicatio mentis humanae sive Animae rationalis*. Utrecht.
Ruler, Han van. 2008. Substituting Aristotle: Platonic Themes in Dutch Cartesianism. In *Platonism at the Originis of Modernity: Studies on Platonism and Early Modern Philosophy*, ed. David Hedley and Sarah Hutton, 159–175. Dordrecht: Springer.
Sangiacomo, Andrea. 2014. Louis de La Forge and the 'Non-Transfer Argument' for Occasionalism. *British Journal for the History of Philosophy* 22 (1): 60–80.
Schmaltz, Tad M. 2016. The Early Dutch Reception of *L'Homme*. In *Descartes'* Treatise on Man *and its Reception*, ed. Delphine Antoine-Mahut and Stephen Gaukroger, 71–90. Cham: Springer.
———. 2019. The Curious Case of Henricus Regius. In *Oxford Handbook on Descartes and Cartesianism*, ed. S. Nadler et al., 434–449. Oxford: OUP.
Schouten, J. 1974. Johannes Walaeus (1604–1649) and his Experiments on the Circulation of the Blood. *Journal of the History of Medicine and Allied Sciences* 29: 259–279.
Schuyl, Florent. 1662. Ad Lectorem. In *De Homine*. Leiden, Unpaged.
———. 1672. *De veritate scientiarum et artium academicarum. Qua, demonstrata mentis & dei opt. Max existentia…* Leiden.

Strazzoni, Andrea. 2018. How Did Regius Become Regius? The Early Doctrinal Evolution of a Heterodox Cartesian. *Early Science and Medicine* 23: 362–412.
Vanni Rovighi, Sofia. 2007. *Filosofia della conoscenza*. Bologna: ESD.
Verbeek, Theo. 1992. *Descartes and the Dutch: Early Reactions to Cartesian Philosophy, 1637–1650*. Carbondale/Edwardsville: Southern Illinois University Press.
———, ed. 1994. *Descartes et Regius. Autour de l'*Explication de l'esprit humain. Dordrecht: Brill.

Chapter 16
Marin Cureau de la Chambre on the Vegetative Powers

Bálint Kékedi

Abstract Marin Cureau de La Chambre is generally considered to be a lesser figure on the intellectual landscape of seventeenth century France, best known for his theory of instinct, animal cognition, and the passions. In this paper, my principal aim is to examine Cureau's thoughts about the vegetative powers, based on his only work in which he touched upon this subject: *The System of the Soul* (1664). In that work, Cureau offers a general explanation of how cognition works in living creatures, and – rather unusually – he attributes cognition to every living thing, including plants. Even more unusually, perhaps, he offers an analogical explanation of how orderly and goal-directed inanimate processes might occur in nature, but he claims that here we are only dealing with "the shadow of cognition". In providing an account of Cureau's ideas about the vegetative powers, I shall outline his general theory of cognition, based on images; then I will contrast the workings of the vegetative soul with those of the higher order souls; and finally I shall explain the difference between the animate and the inanimate parts of nature, arguing for the conclusion that Cureau wanted to navigate between what was perceived as the Aristotelian tradition and the "moderns" of seventeenth century science, siding more with the latter in attributing a greater independence of the created universe from its Creator.

16.1 Introduction

Marin Cureau de La Chambre (1594–1669) is a rather neglected figure of the seventeenth century among historians of philosophy and science. Despite his preoccupation with classical philosophical issues – such as perception and cognition, for instance – historians of philosophy seem to agree that he did not exert a significant impact on the philosophical debates of his age. The domain where his ideas have prompted some scholarly interest at all is the history of psychology, for Cureau's

B. Kékedi (✉)
Independent Scholar, Budapest, Hungary

F. Baldassarri, A. Blank (eds.), *Vegetative Powers*, International Archives of the History of Ideas Archives internationales d'histoire des idées 234,
https://doi.org/10.1007/978-3-030-69709-9_16

theory of instinct was a highly original contribution to that topic.[1] In this paper, my aim is to present Cureau's conception of the natural cognition of the vegetative soul, and to contrast this type of cognition with the higher order faculties of living creatures, as well as with the seemingly "instinctive" behaviour of inanimate objects. The latter is a generally ignored question, but since my earlier research centred on a comparison between the operation of natural and artificial automata in Descartes' philosophy, I shall give a more detailed account here on the boundary between the living and the non-living to fill a gap in the existing Cureau scholarship. The argument will proceed along the following outline: first, I shall give a brief introduction to Cureau's conception of cognition in general; then I shall discuss the instinct of the vegetative soul, as opposed to that of the sensitive soul in humans and non-human animals; in the final part, I shall interpret the only place in Cureau's work where he puts forward a machine analogy to shed light on the working of natural cognitive processes in ensouled creatures, and I will examine the implications of that analogy from the perspective of our current inquiry.

16.2 On That Secret and Hidden Cognition: *Instincts*

Let us begin, then, by describing the most general characteristics of Cureau's theory of cognition. A key point has already been highlighted by Marcus Wild, namely that for Cureau cognition is first and above all an *activity* – he repeats this thesis several times throughout his work. Therefore, we should render the French terms "connaissance" and "connaître" with the English "cognition" and to "cognize" respectively, instead of "knowledge" and to "know", because "knowledge" has, or at least may admit more passive connotations than "cognition". But precisely what kind of activity is cognition? Here I shall quote Marcus Wild's summary of Cureau's position:

> It is the activity of forming representations or "images" as Cureau calls them. More precisely, the term "cognition" refers to the activity by which living beings acquire and process information from the environment or from within the living being. This information is stored in the form of images on which living beings can act. Cognition is the way in which living beings regulate both, the equilibrium inside their bodies and the navigation in their natural environments. This regulative activity is governed by acquired or innate images (Wild 2008, 445).

It is certainly not an accident that Markus Wild speaks about *information processing* in this passage, because if there is an apt synonym for Cureau's images, it is *information*. This can be seen from the fact that images need not imply any kind of pictorial resemblance or representation, and that Cureau makes an extremely extended use of images in his explanations of natural phenomena. For instance, in voluntary motion the brain communicates *images* with the muscles to order the sequence of their motion (Cureau 1664, 486), but images also serve to explain all of the sensory

[1] See Solomon Diamond's and Marcus Wild's works cited below.

modalities, the sense-independent vegetative functions, and even the orderly and sometimes goal-directed behaviour of inanimate bodies, such as the behaviour of magnetic compasses, or planetary motions, as we shall see.

Cureau maintains the Aristotelian tripartition of the soul in living beings, and believes that cognition in each part is always the activity of the relevant cognizing faculty. His treatise from 1664, called *Le système de l'âme* (*The System of the Soul* – henceforth: *The System*) contains six books, the first three of which are devoted to intellectual, sensitive, and natural cognition respectively. The temptation is strong to read those books as dealing with the rational, the sensitive, and the vegetative soul – as Markus Wild proposes –, but it does not seem to be the case that one could establish such strict correspondence between the types of soul and the types of cognition.

On the one hand, Wild is certainly right in emphasizing that the notion of *foreshadowing* – i.e. the idea that Nature makes rough sketches, so to speak, of higher cognitive faculties in lower level types of soul – is important for Cureau both as a principle relating the different types of cognition, and as a methodological starting point to get his analysis off the ground. Thus the estimative sense in animals foreshadows reason (it is a sort of practical reason according to Cureau), whereas certain actions of the vegetative soul foreshadows sensation – and sometimes even practical reason – in the sense that it makes its own kind of discriminations, which often surpass those of the senses. Cureau's example for the latter is our bodily reaction to the stinging of a bee, which the sense of touch cannot distinguish from the impact of a needle, but which provokes different reaction on the vegetative level from the simple sting of a needle. From a methodological viewpoint, foreshadowing is important, because it allows us to establish analogies between the workings of the different types of soul. Cureau holds that if we did not make this methodological presupposition, we could not possibly give an account of the workings of the vegetative faculty, because the latter is a completely impenetrable territory for the human mind.

It is equally important to note, however, that the higher level souls all have their own lower level layer(s) of cognition, and therefore when Cureau speaks about the operations of the intellect or of the sensitive soul in the book on natural cognition, he does not necessarily mean that it is the vegetative soul that interferes with the activity of our rational faculty or the senses. The way he introduces the notion of instinct in the third book is instructive:

> [...] one must observe that the actions of the vegetative soul come under the heading of instinct [...], and in fact the use of this term has been introduced to mark the cause of a motion which requires cognition, but one which is secret and hidden (Cureau 1664, 173 – if not stated otherwise, every quote from Cureau's works is my translation).

Even though the vegetative soul is eminently the domain of that secret and hidden cognition because all of its actions are performed solely through instinct, Cureau insists that this definition can be applied to actions performed by any part of the soul. Indeed, he will call instinctive not only the classic examples of animal behaviour (the purposes of which animals need not be aware of in the same sense as we are), but also the choices of the intellect which are not transparent to itself (e.g. its aesthetic preferences where one is not aware of the reasons of her attraction towards a certain person or work of art, etc.). Moreover, Cureau considers the orderly and goal-directed motions displayed by inanimate natural bodies to be the result of instinct, and he devotes a separate part of his treatise to the instinct of inanimate bodies. While all of these forms of behaviour are instinctive in Cureau's sense of the term, the different faculties have their own characteristic instinctive behaviour: the instinct of the intellect will show only an analogical resemblance to the instinct of the sensitive soul, just like the instincts of the animals will only remotely resemble the instincts of the vegetative soul, etc. The natural cognition of the vegetative soul is, therefore, only one subspecies of that "secret and hidden" kind of cognition, called instinct, which permeates all nature, including its non-living part.

16.3 The Nature of Vegetative Cognition

Of course, the instinct of the vegetative soul has its own characteristics. It is based primarily on innate natural images that are distributed all over the organic body; that is, they are not centralized in the way that, for example, the images of episodic memory are located in a given part of the brain. Because the innate and acquired images must be stored over time, Cureau maintains that the vegetative soul has its own kind of memory:

> [this lowest part of the soul must have memory] so that it could preserve the natural images as well as those that it produces while cognizing (Cureau 1664, 219).

The operations of the vegetative soul are slower than those of the higher faculties, on account of that lowest faculty's more material nature. Consequently, it is also less sophisticated than the sensitive or the intellectual soul: it can only detect the most general differences in its objects. The bee sting is, again, a good example: the vegetative soul detects something more than the sensitive faculty, but it only recognises the venom under the generic notion of poison, to which it reacts more or less the same way as it would to any other poisonous substance, trying to discharge it from the body. In that respect, Cureau, the otherwise very acute observer, compares the vegetative faculty to the sense of smell in a passage which makes one wonder whether he possessed that particular sense to full extent. He writes:

> [...] we could say that [the vegetative soul] does not know the differences of its objects better than our sense of smell knows the scents. The other senses discern every essential difference in their objects; our eyes can see the species of colours and our ears can detect those of the sounds. Smelling is so imperfect in humans that it can only distinguish scents as either being pleasant or unpleasant, good or bad. The vegetative faculty does not know more about its objects (Cureau 1664, 215).

We should note here that it is a characteristic tendency in Cureau's work that cognition of any kind is primarily modelled on perceptual cognition in that its main purpose is to enable organisms to distinguish what is beneficial and what is harmful to the organism (and this is even true for inanimate objects). He says, for example in the second book on the senses that "cognition [without any qualification] is designed to enjoy the goods and flee the harms."[2]

We may observe the same tendency in the phenomenon of bodily irritation, which Cureau portrays as the vegetative sensory modality *par excellence* in article 6 of Book III of his treatise. He writes:

> [...] we can see that bodily parts feel what is unpleasant to them, not only without the participation of the brain, which is the source and the main seat of the sensitive faculty, but even before the brain itself is formed, or the sensitive faculty has got any organs. For our latest observations teach us that the first lineaments with which Nature endows the animal body, at a time when there is nothing yet of the figure that the parts will have to assume, one cannot poke, no matter how gently, this unformed mass which takes the place of the future parts, and which is the very foundation of the entire piece of work, that it should not tremble and retract. Now, it is impossible that it should not feel what hurts it. And yet, there are no nerves, no brain, not even animal spirits that could cause this movement or the sensation that should precede it. It is necessary, therefore, that all this should originate from the natural faculty, which is the first of all the faculties in time, order, and function (Cureau 1664, 208–209).[3]

Cureau seems to be criticizing with his description of bodily irritation both the Aristotelian and the Cartesian position, and this attitude is generally characteristic of him.[4] Here, the implicit criticism consists in that the proponents of the mechanical hypothesis typically focus their attention on fully developed adult organisms, that is to say – in their conception – on fully operational machines whose parts are already in place, and this kind of analysis disregards the problem that nothing of the future main structures exists when the organism begins to show some primitive form of discriminative behaviour. Discrimination, again, implies the sorting of the objects into the categories of beneficial and harmful, and therefore must involve information [*image*] processing for Cureau.

One could object at this point, however, that only certain animal tissues can be irritated, plants cannot – with the exception, perhaps, of the mimosa, the sensitive plant which was widely discussed in the period, but which Cureau did not want to use as an example in his work, because it is too specific and ambiguous according

---

[2] "la Connoissance est destinée pour jouïr des biens & fuir les maux" (Cureau 1664, 69).

[3] I am grateful to Francois Duchesneau, Daniel Schmal and Charles T. Wolfe for calling my attention to the importance of not reading this passage in Cureau as an evidence of an early description of the notion of *irritability*. One must distinguish between *irritability*, which is a technical term that came into use only later in the works of Glisson and Haller (cf. e.g. Giglioni 2008), and *irritation* which had been a frequently used term in the medical tradition long before the importance of *irritability* was first emphasized. As a physician of the French court, who studied medicine in Montpellier, Cureau was obviously familiar with the medical uses of the term *irritation*.

[4] In general, he had more sympathy towards the *novatores*, and he even had a mutually cordial relationship with Descartes, despite his disapproval of the latter's extremely mechanical approach to the living (and despite Descartes' well-known impatience with respect to any criticism of the foundations of his philosophy – and the mechanical hypothesis was perhaps the most important foundation for him).

to him.[5] His reply to the objection is that there are, in fact, two types of vegetative soul: one in animals, and one in plants, and the former is more perfect than the latter, for it serves more functions. Besides nutrition and reproduction, the vegetative faculty of animals must also regulate the heartbeat, breathing, bodily temperature, the workings of the immune system, all of which are recurring examples in Cureau's treatise. They also contribute to the "economy" of the animal body, and the difference in perfection between the vegetative faculty of plants and animals may boil down to a difference in the complexity of the tasks related to the maintenance of the equilibrium of the organism.

Nevertheless, even the lower type of vegetative soul must actively process images in order to be able to perform its basic functions, which is why it is called a *faculty* in the first place (I shall return to this point later). The activity and the intelligence of this faculty, which is a constant source of wonder for human understanding, becomes most evident in a passage where Cureau draws a peculiar analogy between the vegetative faculty and the intellect. He says that the intellectual ideas that the human mind forms upon the reception of the phantoms[6] take on the properties of the original objects, and are of equal cognitive value than the original object would be, if it could unite itself with the intellect. This kind of "metamorphosis" of the intellect into its objects is no stranger, he says, than the inherent formative virtue of the semen that will develop the form of the bodily parts of living creatures. According to him, this is why we use the term *conception* both in the case of thought formation (i.e. cognition in the purest sense of the term) and the first formation of plants and animals. It is innate *images* (which are innate to the formative virtue) that will drive the process of generation towards a pre-determined goal, in other words, embryonic development is nothing but active image processing.

As far as the details of the cognition of the vegetative faculty are concerned, the *modus operandi* of this power is not very different from that of other cognizing faculties (such as the sensitive soul or the intellect), with the only exception perhaps that it is inaccessible to the senses and to a certain extent to reason too, although the latter may attempt to make sense of the operations of the vegetative soul by imagining how higher level functions might be foreshadowed in a lower level type of soul. Cureau condenses the reasons for thinking that we are dealing with a cognizing agent in the following passage:

> One must presuppose that there is a kind of cognition in humans, in which neither the senses, nor reason plays any part, and which can be observed primarily in the vegetative faculty. For it is impossible to observe so many different sorts of action, as well as the order and the measure that this faculty introduces in those actions without having to admit that there is a kind of cognition that regulates and conducts such a beautiful economy. Just considering the number, the figure and the situation of the parts of the human body in the process of its first formation – which are so right and so regular – should convince us that the

[5] It is ambiguous, because Cureau thinks his adversaries would say that it partakes in animal nature as well (cf. Cureau 1664, 208).

[6] In Cureau's vocabulary *phantoms* correspond roughly to what Descartes calls "corporeal ideas" in his early works, most notably in the *Regulae* and in *L'Homme*.

> cause directing the process is highly knowledgeable, and that it produces everything with more knowledge than reason itself could do, if it were to make such an endeavour on its own (Cureau 1664, 160).

Thus, Cureau justifies the inclusion of the vegetative faculty in the domain of cognition with the following characteristics of its actions: (1) the introduction of order and measure in the otherwise chaotic whirling of matter; (2) the creation and the maintenance of an equilibrium between the parts of the body (and between the organism and its environment, but Cureau's example here is confined to internal processes); and finally (3) the direction of the bodily processes towards a well-defined end. The first of these features is typically conceived as an authoritative act: a more noble entity (a God-created soul or faculty, or God itself) imposes order and regularity on dumb, inert, indifferent, or – even worse – reluctant matter. Matter is generally not supposed to be able to organise itself, so the presence of an apparent order in material things is a mark of an intelligent agent – one that has the power to impose its orders on its subjects. The second one, by contrast, supposes a bidirectional communication between the organizing entity and the organized parts: the maintaining of equilibrium requires feedback, which enables a continuous monitoring and control to balance out the inequalities when the state of the whole moves away from the state of equilibrium. The third feature need not imply feedback at all, and Cureau's description of the process indeed likens it to the mere execution of a program inscribed into the organism in the form of linked images. The second is a teleological process in a different sense than what the third point suggests, because the maintenance of an economy implies the keeping of the states of affairs within a narrow margin of error with respect to the equilibrium, where both the latter state and the deviation from it must somehow be represented – and therefore *known* – according to Cureau, if the vegetative faculty were to give meaningful responses. The third, by contrast, is only teleological insofar as we would recognise a deviation from an imagined "normal" state or outcome,[7] but it is a unidirectional procedure altogether, and is therefore simpler in terms of representational complexity than the creation and the maintenance of an equilibrium. It is also worth noting that when things go wrong or simply out of control from the point of view of the vegetative soul, it is always due to the failure of one of the three factors above, and therefore – as Cureau lengthily argues – it must never to be imputed to God. The vegetative faculty may lack the power to overcome the "disobedience of matter" (p. 169), the immune system might misfire or initiate false crises during serious illness, sometimes "monsters" are born, yet God is never the source of these "disorders".

[7] Cureau, as almost anyone in the period, discusses the cases of "monstrous children" who are defective because copies of the images issuing from the mother's imagination affect the formative virtue during the pregnancy (cf. Cureau 1664 107 – as is well known, this was a widely shared opinion, equally held by Descartes and his followers, most notably Malebranche).

16.4 The Shadow of Cognition in the Inanimate World

It is at this point that I would like to turn to what Cureau has to say about the inanimate world, for he introduces the *only* comparison in the *System* between living creatures and mechanical artefacts in order to justify God's exemption from the appearance of disorder in the world. He writes:

> I am not saying that things are not directed by Divine wisdom, but this direction is not in the actions; it is rather in the virtues with which God's wisdom endowed its creatures. [...] Or do you think that the direction given by a craftsman to his machines are in the movements of the machine? No, it is in the wheels and springs that produce the movement, for even if it turns out that these parts are never set in motion, they will still possess all the direction with which art can provide them. In the same vein, we may say that the faculties are like the springs that make things move [*agir*], and that strictly speaking they are the only one that are directed [by God]. This is why there is no error in God's direction of them, for He endows them with everything that is necessary to perform their actions, but these actions are not always conform to God's direction, because it does not extend to them (Cureau 1664, 170).

One should not look for intelligent direction *in the actions* performed by the arms of a mechanical clock, one should look for it in the machinery of the clock's wheels and springs (in other words: in their arrangement, or *disposition*, as Descartes would say). However, to someone who has not seen a mechanical clock before, it is the orderly motion of its arms that suggests that she should inquire into the causes directing the observed actions. Likewise, there is a range of natural phenomena in inanimate objects that made Cureau believe that they act out of instinct, that is, out of a "secret and hidden cognition". The most noticeable forms of behaviour in question are the orderly motions in nature (such as the circular motion of the celestial bodies) and the goal-directed behaviour of some objects, most notably of magnets, which strive to align themselves in the direction of the magnetic poles. These two types of behaviour – the simply orderly and the goal-directed – were not clearly distinguished in the early modern period, and I have argued elsewhere[8] that perhaps with the exception of Descartes and his followers, it was unanimously believed that any such behaviour requires intelligent direction either from the behaving entity or from its designer and creator.

Cureau is no exception to this rule. One of the main questions among those who held that intelligent direction is needed for purposeful action was where to draw the borderline between beings having a native intelligence complex enough to govern their own behaviour in an autonomous manner, and between those that are merely predetermined – in the strongest sense – to such and such purposeful action by a wholly external cause. Cureau will draw that demarcation between the animate and the inanimate parts of nature, and in so doing he perceives himself, interestingly, as

[8] In a yet unpublished paper, currently (4 March 2021) available at: https://www.academia.edu/41899593/SHEEP_AND_MAGNETS_TELEOLOGICAL_CONSIDERATIONS_IN_THE_EXPLANATION_OF_ANIMAL_BEHAVIOUR_IN_DESCARTES_NATURAL_PHILOSOPHY

being in opposition to the majority of the tradition, and as fighting the idea of occult virtues together with the moderns, like Descartes.

It is true that by allowing genuine cognition to occur even in the vegetative soul, Cureau extends the boundaries of the cognitive world beyond what would have been acceptable for the majority of the Scholastics. However, it is also true that a considerable group of the "moderns", namely the Cartesians, were moving in the exact opposite direction, which was to deny cognition to every non-human life form (at least here on Earth). When it came to explaining the behaviour of magnets and planets, both Cureau and the Cartesians employed a unified scientific methodology: the latter tried to explain these phenomena with the hypothesis of gigantic corpuscular vortices in the heavens or between the two magnetic poles of the Earth, whereas Cureau supposed that it was due to innate images, and not because of some occult sympathy and antipathy that the magnet turns toward the poles, or that it attracts iron, just as it is not because of some motive intelligence that the planets perform their circular motion. In his own words, it is "natural images imprinted in them by God, which act as the real intelligence that regulates the turn they have to perform, and that gives them the inclination and the willingness to execute it" (Cureau 1664, 227). In the most general manner, Cureau proposes the following principle:

> Effects are similar to their causes because they are the copies of the models that regulate and direct the active power of the causes. We may say, accordingly, that the arrow only goes towards its target because the string of the bow has imprinted on it the image of its movement: for this is how we might call that wonderful and inexplicable quality commonly referred to as *impetus*, which is the principle of every local motion (Cureau 1664, 205).

We can see that even inertial motion is the result of an image having been imprinted on the moving object from outside. The difference between the more or less complex behaviour of inanimate bodies is in the number and the complexity of the images imprinted in them.[9] By contrast, the difference between animate and

[9] Cureau never discusses the question of what exactly the clockmaker does when she fabricates and arranges the parts of a mechanical clock – he was probably not very interested in the question. From the example of the arrow, it seems clear that it would have to be the entire machinery – wheels, springs, etc. – that should imprint the *images* of their motion on the arms, but it is less clear what the clockmaker creates when she assembles the materials with which she works in a way that it should be able to bring about the desired effect. One would be tempted to say that it is the potential to produce the motions independently of their creator (after all, that is what defines an automaton), but that would bring that potential dangerously close to a faculty, and would open up the possibility for a reduction of the sensitive and vegetative faculties in living beings to mechanical causes in a more or less Cartesian fashion. Other options are equally problematic. What if the image of the motion imprinted on the arm of a clock is the result of the sum total of the motions of the parts? In that case, one would have to account for the fact that since cogged wheels and springs hardly ever occur naturally in the world, they have to be manufactured out of raw materials that do not have the same "innate images" to imprint the appropriate motion on another part of the device, and eventually on the arm, before they acquire their final shape, elasticity, and so on. But it would follow, then, that human craftsmen either create images (in Cureau's sense of the term) inside material objects – without being aware of it –, or the images of what a cogged wheel does in a clock are already contained in, say, copper, thanks to Divine providence. What the craftsman does in this case is like Michelangelo who famously claimed to have set free a pre-existing form from a piece

inanimate objects is that the former possess faculties that actively process images, so that the changes of behaviour should originate from within, so to speak.

Of course, the capacity to initiate change from within has been one of the principal marks of the presence of a soul since at least Plato,[10] unless we are willing to entertain the possibility of genuine – soulless – automata, like Descartes. The issue depends on the distinction between what one would call today active and passive behaviour. Active behaviour is such that the energy or the motive force invested in the output movement performed by the behaving object does not come directly from an input acting on it, but is largely supplied and directed by the object itself, whereas in the case of passive motion all the energy – and eventually the direction – derives from the input. The shooting of an arrow above is a good example of passive behaviour when we consider the motion of the arrow. Passive motion is also a constrained one, where the constraint is exercised by an external agent. In Cureau's model, that external agent is God in the case of inanimate natural agents. It is God who imprinted the motions of the planets and compasses on them from outside, like the archer imprinted its motion on the arrow.[11] The fact that Cureau did not discuss the case of clockwork and other automata of his age may indicate that he did not believe genuine automata could be fabricated (not even in theory), with the degree of spontaneity enabling them not having to be wound up by their maker; whereas Descartes' project was exactly to make that possibility conceivable.

If we put aside that fundamental difference between Cureau's and Descartes' thought, we shall recognise that both of them are arguing against the Scholastic position by allowing for a greater degree of autonomy in creatures than their Scholastic predecessors. Cureau makes this point explicit in a very interesting passage, in which he speaks about a limited notion of freedom that is applicable to the vegetative faculty in living beings:

> It is not that God could not direct the actions just as well as the faculties, and he frequently does so for particular reasons; but just as he gave freedom to Man's will, he also allocates a sort of liberty [*liberté*] to natural agents so that they could perform the functions to which they are determined. He permits that they act by themselves, without providing them with any other help than his general concurrence, which does not direct or determine them at all, contrary to what they say in the schools (Cureau 1664, 171).[12]

One could object that this assertion of a limited degree of liberty in the case of natural agents, like the vegetative faculty, contradicts Cureau's comparison of the faculties to the working of a mechanical device, where the effects follow necessarily

---

of marble to produce a sculpture. Perhaps it is needless to say that the mechanical hypothesis offers a more parsimonious and less problematic explanation of the working of artefacts, than any of the options above.

[10] See e.g. *Phaedrus* 245e.

[11] Descartes had a fully naturalized mechanistic account of both of these phenomena, see especially Part IV of his *Principles*.

[12] God's delegation of certain powers to its creatures comes up already on p. 166. Cureau introduces the topic with so many cautionary remarks that it makes even the current-day reader appreciate what was at stake in the early modern era in such a theologically delicate issue.

from the constitution of the machine. Moreover, Cureau makes it clear just a few pages earlier that the actions performed by the vegetative soul need not be directed by God, for "why should those actions be directed [from outside], if they can only happen one way?" (Cureau 1664, 168).[13]

The key to resolve the apparent contradiction in Cureau's text is to admit a different kind of spontaneity in the vegetative part of living creatures. That spontaneity would enable them to choose the right moment to initiate change without there being an obvious external trigger.[14] *Activity* understood that way, as opposed to *passivity*, constitutes the boundary between the living and the non-living. Cureau uses a special term in the *System* to try to shed light on this kind of activity, which is *alteration*. In the final part of this essay, we shall examine that notion.

Throughout his *System*, Cureau repeatedly explains cognition by saying that it is a kind of *alteration*, and he uses this term to explain the important distinction we have just discussed. Alteration occurs everywhere and for many reasons in nature. In inanimate objects, it is brought about by external factors, which may or may not be intelligent, depending on the case. However, alteration in the sense that it signifies cognition is like a self-induced qualitative change in organisms. Although the ontology of the images is not always clear in Cureau's work, he consistently claims that alteration begins with a quality and terminates in another quality – therefore it always refers to a qualitative change. Accordingly, he also says that images are nothing but qualities. He seems to allow for two kinds of images: spiritual and corporeal. Images of the first type are proper to the intellect; those of the second belong to the sensitive and the vegetative soul – the vegetative being even more corporeal in nature than the sensitive soul (cf. Cureau 1664, 104–108). Let us see what might be the best summary of Cureau's thoughts on alteration, with a particular focus on the vegetative soul and the boundary between the living and the non-living:

> No matter with which kind of vegetative soul we are dealing [plant or animal], it has the ability to cognize, because alteration, in which cognition consists, is never beyond its power: it is its proper motion, which distinguishes the vegetative soul from all other, purely natural forms. For inanimate things do not cognize properly speaking, all they have is the material and the shadow of cognition. Even though they do have the images that Nature has given them, they do not have any faculty to act on those images, or to paint their portraits, which constitutes the essence and the form of cognition, in a way that we may even say that *everything that lives cognizes and everything that cognizes is alive* (Cureau 1664, 216–217 – part in italics is Marcus Wild's translation).

---

[13] He goes on to compare the way that the vegetative faculty brings about its effects to the falling of a stone towards the centre of the Earth. In either case, events follow one another necessarily, and it could not be otherwise.

[14] There are cases, like the stinging of a bee, that requires immediate reaction, but if you consider sprouting, for instance, the temporal gap between the external trigger (humidity) and the beginning of the growth of a new plant is so large and differs from seed to seed that it suggests a degree of spontaneity to the observer. But even in the case of the bee sting, the point is rather that the triggering cause does not provide the organism with the energy to bring about the reaction, the latter is produced and directed solely by the internal resources of the living being.

Cureau is quite straightforward here: with the equation of even the most primitive forms of life and the capacity for cognition, his intention is to mark the boundary between the living and non-living, between beings having a certain degree of liberty because they have the faculty to determine to act on the stock of images with which they have been provided, and those whose seemingly orderly behaviour is utterly determined from outside – and therefore all they have is the 'shadow of cognition". Interestingly, he claims that *alteration* is the proper motion of the soul, which can only refer to the ability to initiate change from within, because images or chains of images are translated into motion by inanimate beings as well (only passively). The observation of self-motion thus defined is what the observer should recognize, according to Cureau, as the mark of cognition inside organisms.

But besides acting on images, Cureau also mentions the ability of any type of soul to "paint their portraits". This is a problematic point, as the supposed ability requires a representational faculty of some form, which produces a representational entity – "a portrait" – of natural images. That kind of activity is typically associated with the intellect or the sensitive soul, and indeed Cureau gives the most detailed discussion of this issue in the first book on the understanding, and then in the book on the sensitive soul (cf. Cureau 1664, 11–24 and 99–105 respectively). Just what the vegetative soul should represent to itself, and how exactly, is difficult to say even on Cureau's account, for he does not go into the details of vegetative representations. Considering his examples, there are only two possible candidates.

The first one emerges from Cureau's description of the formative virtue, which is common in plants and animals. According to a possible reading, because the generation and the first formation of living creatures is a teleological process, the formative virtue must possess a representation of the fully formed organism toward which the whole process should tend from the very beginning. We have seen that Cureau illustrates the knowledgeable aspect of the vegetative faculty by using the example of the marvellous working of the formative virtue, which requires more knowledge than what reason could contribute to the process. Nevertheless, no matter how tempting this reading might be, it goes against Cureau's description of the operations of any kind of instinctive behaviour. For he compares the execution of a program inscribed in the chain of linked images to the utterly blind following of orders by an officer who has no idea of the reasons behind the orders, nor of the goal that he should eventually realise by executing the commands he receives on paper (cf. Cureau 1664, 196).[15] We may conclude, then, that teleological processes need not be based on representations of the *telos*, they are rather blind with respect to the goal itself.

We can extract a second candidate from Cureau's discussion of certain vegetative reactions, most notably those of the immune system or of the digestive tract. The common feature of these bodily systems is that they can discriminate between what is beneficial and what is harmful to the body, and in so doing, their "knowledge"

[15] Cureau speaks there about the instinct of animals, namely the nesting of birds, but there is no reason to believe that the vegetative faculty should work differently in that respect.

seems to surpass by far that of the senses. As we have already seen in the case of the stinging of a bee, Cureau maintains that our bodily reaction requires cognition and the representation of the venom under the generic notion of "harmful substance" by the vegetative faculty. Cureau argues that it is not the bad quality of the venom that provokes the reaction by itself, for the reaction does not occur if the bee stings a lifeless part of our body (e.g. a gangrenous limb). The presence of the vegetative soul – that is: life – is therefore necessary, and seems as if that faculty would need to paint a generic portrait of the harm, in order to respond meaningfully. The problem with this form of behaviour, however, is that it is only available to humans and animals – with the exception, perhaps, of the mimosa, the sensitive plant, which Cureau refuses to discuss in his treatise. We should conclude, then, that Cureau did not have a clear and unified conception of a vegetative representational faculty – one that would be applicable across the domains of the living (plants, animals, humans) – or at least he could not articulate coherently such a theory. The most natural interpretation of the above-quoted passage is that the ability to act on images is a genuine and universal mark of cognition, and therefore of life – as opposed to inanimate objects –, whereas the ability to paint the portraits of images only belongs to higher order life forms.

16.5 Conclusion

By way of conclusion, we may say that in Cureau's conception the vegetative faculty possesses in itself the power of cognition, and is therefore capable of actively performing that secret and hidden kind of cognition which makes the behaviour it produces instinctive. Instinctive behaviour may occur everywhere in nature, even in its inanimate part, but there is an important distinction to make between living and lifeless creatures, namely that the cognition driving the observed intelligent or orderly behaviour is proper to living beings, but lies outside of inanimate things. The behaviour of the latter is completely passive, whereas even the lowest life forms have the ability to initiate change from within, thanks to a process which Cureau calls *alteration*. By making living creatures less dependent on God, than what was taught in the schools in Cureau's interpretation, and by fighting the idea of occult qualities, he saw himself as siding with the *novatores*, and thus as paving the way towards a science of a more modern flavour.

Acknowledgments The work on this paper was supported by the National Research, Development and Innovation Office, Hungary (grant number K 125012). I am grateful to Charles T. Wolfe for his comments on this paper, and for revising the final draft grammatically and stylistically.

References

Cureau de La Chambre, Marin. 1664. *Le Système de l'âme par le sieur (M. Cureau) de La Chambre*. Paris: Jacques d'Allin.

Diamond, Solomon. 1968. Marin Cureau de La Chambre (1594–1669). *Journal of the History of the Behavioral Sciences* 4 (1): 40–54.

———. 1974. Four Hundred Years of Instinct Controversy. *Behavior Genetics* 4 (3): 237–252.

Giglioni, Guido. 2008. What Ever Happened to Francis Glisson? Albrecht Haller and the Fate of Eighteenth-Century Irritability. *Science in Context* 21 (4): 465–493.

Wild, Marcus. 2008. Cognition of the Vegetative Soul: An Early Modern Theory of Instinct. *Vivarium* 46 (3): 443–461.

Chapter 17
Re-inventing the Vegetable Soul? More's Spirit of Nature and Cudworth's Plastic Nature Reconsidered

Sarah Hutton

For the Root and Soul of every Vegetable is the Spirit of Nature;
(More, *Immortality of the Soul*, 401)

The Plastick and Vegetative Life of Nature, [is] the Lowest of all Lives
(Cudworth, *True Intellectual System*, 163)

Abstract My paper explores the extent to which More's 'Spirit of Nature' and Cudworth's 'Plastic Nature' incorporated the functions of the Aristotelian vegetable soul, and how far, if at all, each was indebted to Aristotle. I argue that, although, on the matter of vegetable life there is some overlap between the functions of the Aristotelian vegetative soul and those ascribed by Cudworth to Plastic Nature and More to the Spirit of Nature, Cudworth and More were not simply reviving Aristotle in new dress. Both certainly drew on Aristotle when formulating their hypotheses, but these have a much broader range of functions than the Aristotelian vegetative soul. Even among those functions which concern the basics of organic life, both hypotheses have affinities with other the well-developed medico-physiological traditions of the late Renaissance. And, although this type of formative spirit is often compared to the Platonic World Soul, both hypotheses are indebted as much, if not more to Stoicism.

Henry More's Spirit of Nature (or Hylarchic Principle) and Ralph Cudworth's Plastic Nature are probably the most well-known of their philosophical views.[1]

[1] Hunter (1950), Henry (1990), Giglioni (1996), Lotti (2004), Reid (2012), Allen (2013). On Plastic Nature and consciousness, see Thiel (2011), Pecharman (2014), Lähteenmäki (2010). On Cudworth see also Andrault in this volume.

S. Hutton (✉)
Department of Philosophy, University of York, York, UK
e-mail: sarah.hutton@york.ac.uk

F. Baldassarri, A. Blank (eds.), *Vegetative Powers*, International Archives of the History of Ideas Archives internationales d'histoire des idées 234,
https://doi.org/10.1007/978-3-030-69709-9_17

Both are conceived as indirect causes, intermediaries between God and the natural world which account for the mundane operations of the physical universe, bringing together in a single agent efficient, final and formal causality. They were proposed not as certain truths but as plausible *hypotheses* about the functioning of the created world. Both More and Cudworth explicitly state that they formulated their respective hypotheses in the light of the poverty of scholastic Aristotelianism and the perceived inadequacies of contemporary natural philosophy, especially the mechanical philosophy of Hobbes and Descartes. Like many other thinkers of the period who rejected Aristotelianism, More and Cudworth held that mechanical models which sought to explain all things in terms of matter in motion could not satisfactorily account for all the phenomena of nature. *Inter alia*, the limitations of the new mechanical philosophies were thrown into sharp relief by their attempts to account for the complex combination of life functions formerly attributed to the soul by Aristotle. The Aristotelian vegetative soul was something for which the new 'machine model' of nature proposed by Descartes and Hobbes had no use (Baldassari 2019). The 'Spirit of Nature' or 'Hylarchic Principle', and 'Plastic Nature' can therefore be seen as filling a hiatus in the causal explanations of the natural world, consequent upon the demise of Aristotelian natural philosophy. Many other alternatives were proposed, some of which are being discussed at this volume: hylozoism, for example, which imputed life to matter, active principles, plastic powers and seminal forms. English natural philosophy is particularly rich in proposing possibilities. Robert Boyle, for instance, allowed the possibility of active matter, to explain phenomena not easily explicable in mechanical terms. Matthew Hale proposed a *vis activa* in matter; Francis Glisson developed a conception of natural perception, or *biusia*; Henry Power posited a material spirit or effluvium. Others looked to explain functions which Aristotle ascribed to the vegetable soul in terms of immaterial causes of one kind or another. Richard Burthogge posited a universal Plastic Spirit while the botanists John Ray and Nehemiah Grew each posited a 'vital principle'. The last three took their cue from Cudworth and More.[2]

In this chapter I discuss More's 'Spirit of Nature' and Cudworth's 'Plastic Nature' to assess how far they incorporated the functions of the Aristotelian vegetable soul, and how far, if at all, each was indebted to Aristotle. I first outline the philosophical context in which they framed their respective hypotheses. I then discuss the Spirit of Nature and Plastic Nature, highlighting their similarities and differences. After examining more closely their account of vegetative life, I argue that their sources indicate a strong debt to Stoic ideas, rather than Aristotle.

[2] Burthogge (1694), Burthogge (1699), Ray (1681), Grew (1671). See Henry (1986), Garrett (2003), Hirai (2011), Andrault (2014), Inglehart (2016), Hutton (2015), and chapter 5.

17.1 Context: The Challenge of Mechanism

It is frequently claimed that Cudworth and More's motivations for positing Plastic Nature and the Spirit of Nature were primarily theological. It is certainly the case that they held that philosophy should be compatible with the truths of religion, and they were concerned about what they saw as atheistical tendencies in Descartes' philosophy. However, theological concerns were the norm, not the exception for the period, as can be seen from Pierre Bayle's quarrel with Jean Le Clerc about Cudworth's Plastic Nature (Rosa 1994). Something too often over-looked in assessments of Cudworth and More is that theological concerns were not the only consideration in their evaluation of philosophical theories. They were equally interested in explanations which made sense in terms of the new natural philosophy. For this reason they repudiated scholastic Aristotelianism with its supporting cast of substantial forms and occult qualities, which More dismisses as 'that dotage of the confounded *Schools,* who have indued almost every different Object of our Senses with a distinct *Substantial form,* and then puzzle themselves with endless scrupulosities about the generation, corruption, and mixtion of them' (More 1659, 465–6). In Cudworth's view, 'He that asserts an *Occult Quality,* for the Cause of any *Phaenomenon*, does indeed assign no Cause at all of it' (Cudworth 1678, 154). Cudworth was in fact critical of Plato for repudiating atomism, which he did, according to Cudworth, because of his fears that it was tainted with atheism (Cudworth 1678, 8).

There were also philosophical motivations for their critique of Descartes' repudiation of final causes. This was a major bone of contention for many of Descartes' critics, Cudworth and More among them, and there is no question it was motivated by theological concerns, the main reason being that to repudiate final causes is to ignore the operation of divine providence. Cudworth, for instance, highlighted its damaging impact on the argument from design, and belief in divine providence. But teleology in the workings of nature is not just a matter of fulfilling God's providential design, it is also integral to the relationship of form to function in living things. The order of the universe which exhibits divine design is achieved by the structure of organisms and regularity in the workings of nature. Intentionality in the form or design of living beings serves the purposes of life, in plants and animals.[3] Cudworth denied that chance and laws of nature are sufficient to account for the structure of natural organisms or the regularity of natural processes: neither 'the Fabrick of the World,' nor 'the exact Organization, and successive Generation of Animals,' can be explained solely in terms 'the Turning round of a *Vortex* or Whirlpool of Matter ..., according to certain General Laws of Nature' (Cudworth 1678, 54).

Both Cudworth and More were initially drawn to Cartesianism because it offered an intelligible account of the phenomena of nature without recourse to scholastic obscurities, such as 'occult qualities' and substantial forms. Such a philosophy was

---

[3] Natural design was also held to serve human ends, for example to provide medicines or food for human beings (More 1671, chapter 24). On animals serving human ends see Muratori (2017).

plausible because, to use a seventeenth-century phrase, it 'saved the phenomena' of nature without the scholastic *paraphenalia* of occult qualities, intentional species and the like. With the new 'atomical philosophy' as Cudworth called it,

> there is no need of any thing else besides these simple Elements of Magnitude, Figure, Site and Motion (which are all clearly intelligible as different Modes of extended Substance) to salve the Corporeal *Phaenomena* by; and therefore, not of any Substantial Forms distinct from the Matter; nor of any other Qualities really existing in the Bodies without... nor of any Intentional *Species* or *Shews*, propagated from the Objects to our Senses. (Cudworth 1678, 7)

However, More and Cudworth increasingly persuaded that Descartes had not delivered on his own claims to have accounted for all the known phenomena of nature. They argued that even in its own terms, the mechanical philosophy could not satisfactorily account for all the phenomena of nature, or, for that matter, life itself. They were not persuaded that the structure, organisation and beauties of nature could result from the chance collision of particles of matter. That 'the whole System of this Corporeal Universe', could result from 'Fortuitous Mechanism, without the Direction of any Mind' (Cudworth 1678, 10). He charged further that the 'narrow principles' of the Cartesians 'who make the *Essence* of the *Soul* to consist in *Cogitation,* and again the *Essence* of *Cogitation* in Clear and Express *Consciousness*' cannot account for such phenomena as blinking, moving when asleep, respiration, heartbeat (Cudworth 1678, 161). More argued that, if the mechanical philosophy cannot account for such a simple phenomenon as the fall of heavy objects ('the falling of a stone to the Earth'), and still less 'the cause of sundry sorts of Plants and other things, that have farre more artifice and curiosity then the direct descent of a stone to the ground' (More 1659, 465–6). The function of Spirit of Nature therefore is for 'directing the parts of the Matter and their Motion, as cannot be resolved into meer Mechanical powers' (More 1659, 450).

Although both Cudworth and More were critical of Descartes, it makes a lot of sense to see Cudworth's Plastick Nature, and More's spirit of Nature as *modifications* of the mechanical philosophy: attempts to overcome their shortcomings by supplementing a basically mechanical view of the workings of nature with an incorporeal causal principle. More in fact discusses his Hylarchic Principle or Spirit of Nature as a corrective to the mechanical philosophy, 'at a certain distance the *Spirit of Nature* in some regards leaves the motion of Matter to the pure laws of Mechanicks, but within other bounds checks it,' (More 1659, 465). He agreed with Descartes, 'that nothing affects our Senses, but such variations of Matter as are made by difference of Motion, Figure, Situation of parts' but he disagreed that this could account for everything:

> I dissent from him in this, in that I hold it is not meer and pure mechanical motion that causes all these sensible Modifications in Matter, but that many times the immediate Director thereof is this Spirit of Nature ... one and the same every where, and acting alwaies alike upon like occasions. (More 1659, 465–6)

More nevertheless declared that he did not want 'to damp our endeavours in the search of the Mechanical causes of the Phaenomena of Nature but rather make us more circumspect to distinguish what is the result of the *mere Mechanical* posers of *Matter* and *Motion* and what of an Higher Principle' (More 1659, *Preface* 18–19).

On his account, the Spirit of Nature conserves an essentially Cartesian cosmos, in which it,

> keeps the Waters from swilling out of the Moon, curbs the matter of the Sun into roundness of figure, which would otherwise be oblong, restrains the crusty parts of a Star from flying appieces into the circumambulent Æther, carries along those large Regions of looser Particles of the third Element, together with the Comets, in their peregrinations from Vortex to Vortex, every where directs the magnetick Atoms in their right Rode. (More 1662, Preface General, xv)

In some ways the Spirit of Nature is just the laws of nature operating in a teleological manner ('*containing in it certain general Modes and Lawes of Nature for the good of the Universe*') (More 1662, Preface General xv).

17.2 Two Hypotheses

Cudworth's account of Plastic Nature is set out in his 'Digression concerning the Plastic Life of Nature' which is printed in the first part of *True Intellectual System* (Cudworth 1678, 146–72). The fullest account of the Spirit of Nature occurs in More's Preface to his *A Collection of Several Philosophical Writings* (1662) and his *Immortality of the Soul* (1659). In his *Enchiridion metaphysicum* (first published 1671), which collates arguments aired across More's earlier writings, the Spirit of Nature figures in the course of his most extensive and sustained arguments in favour of the existence of incorporeal substance in general culminating in a definition of spirit. Both Plastic Nature and the Spirit of Nature are conceived as immaterial substances. They are thus distinct from matter and not properties of matter. Their function is formative: a key idea in both is plasticity, to account for 'the Efformation and Organization of the Bodies of Animals, as well as the other Phenomena'. Their function also involves regulating and directing the operations of nature in order to sustain the mundane operations of the physical universe in an orderly fashion. Both work by the power of sympathy, and both relate in some way to mind and intellect, but both operate unconsciously. The Spirit of Nature is all encompassing: More conceives of it as a single universal spirit, 'an *Immaterial* Principle' which acts as the agent of God in the natural world, as, 'the Vicarious Power of God upon this great *Automaton*, the World' (More 1662, *Antidote*, 47). For Cudworth, too, there is only one Plastic Nature, but he allows the possibility of plural Plastic Natures: 'it is not impossible but that there may be other Plastick Natures also (as certain Lower Lives, or Vegetative Souls) in some Greater Parts of the Universe' (Cudworth 1678, 171).

In the Preface to his *A Collection of Several Philosophical Writings* (1662) More presents the Spirit of Nature in strongly Platonic terms, describing it in emanationist terms as '*the last* Ideal *or* Omniform Efflux *from God*' (More 1662, Preface General, xvi), and as a copy of the divine mind, 'a mute copy of the eternal Word (that is, of the Divine Wisedome)' which operates 'according to that Platform of which it is the Transcript, I mean according to the Comprehension and Purpose of those Idea's of things which are in the eternal Intellect of God'. In its turn, the Spirit of Nature is a repository of the ideas or forms of all things, it 'virtually contains all everywhere' (More 1662, Preface General, xv–xv). Cudworth too sees Plastic Nature in Platonist terms as a copy of divine wisdom.

> The difference betwixt *Nature* and Wisdom was before observed, that *Wisdom* is the *First* and *Highest* thing, but *Nature* the *Last* and *Lowest*; this latter being but an Umbratile Imitation of the former. (Cudworth 1678, 163)

Plastic Nature too is a repository of forms: it 'doth virtually contain within itself, the forms of all the several organical parts of animals, and displays them gradually and successively, framing an eye here and an ear there' (Cudworth 1678, 77).

An essential feature of both, is that they operate teleologically. Plastic Nature, 'acts *Regularly* and *Artificially*, … for *Ends,* or in Order to *Good* (Cudworth 1678, 172). And More notes that, as a copy of the eternal word of God, the Spirit of Nature executes its functions according to God's intentions ('Comprehension and Purpose') for the sake of 'the good of the Universe' (More 1662, Preface General, xvi). To convey its executive role in respect of God's designs, both More and Cudworth use the image of a servant to explain how each acts as an unthinking (unconscious), agent of the divine mind in nature. For Cudworth Plastic is God's executor ('Drudging Executioner'), the unthinking servant, or manual labourer ('manuary opificer') who carries out orders (Cudworth 1678, 172). The Spirit of Nature, too is a kind of servant of God. He calls it 'the great *Quartermaster-General* of divine Providence' (More 1659, 469). It too is unconscious (it '*need not be* perceptive *of itself*' (More 1662, Preface General, xv) as well as immaterial.

For both Cudworth and More, Plastic Nature and the Spirit of Nature are not merely 'add-ons' wedged into the causal gaps of mechanistic nature, but integral to a complex pneumatology of souls, spirits and incorporeal substances. For More, the Spirit of Nature is accompanied by a fourfold psychology consisting of four kinds of Spirits. In *Immortality of the Soul* (1659) he lists these as seminal forms, animal souls, human souls and angelic spirit:

> The *logoi spermatikoi* or Seminall Formes, the Soules of Brutes, the Humane Soule, and that Soule or Spirit which actuates or informes the vehicles of Angels. (More 1659, 49)

In this hierarchy of types of spirit, both animals and humans share basic 'vegetable' properties, as well as sensation and reason. Human beings, are further differentiated by having the additional property of reason. The human soul is thus a set of accretions to a basic vegetable spirit—hence 'More's poetic claim that 'The lower man is nought but a fair plant' (*Psychathanasia* 2.1.9 in More 1646, 108). In *A Brief Discourse of Enthusiasm* (1656) he conceives of the soul as having several faculties, the lowest of which are 'the *Vegetative* or *Plantall* faculties of the Soul' (More 1656, 2–3).

For Cudworth, Plastic Nature is one of many kinds of soul-substance in nature. For Cudworth the 'General Heads and Principles' of all being are matter and activity: '*Passive Matter,* and the other *Active Power, Vigour* and *Vertue*' (Cudworth 1678, 28) The active principle in all things is *energy*; a principle of 'self-activity' capable of initiating movement or thought. The latter constitutes life, and all life is *energy*. 'Plastic Nature' is just such a 'vital energy of nature' (Cudworth 1678, 163). Cudworth, like More, conceives of the human soul as a compound of life functions, more loosely organised as higher and lower. And he adopts the triad of rational, sensitive and vegetative life, among which he claims Plastic Nature is 'really the same thing with the *Vegetative*, which is Inferiour to the *Sensitive*' (Cudworth 1678, 162–3).

Although they have much in common Plastic Nature and the Spirit of Nature are not identical in all particulars. For example Cudworth countenances many plastic natures, but there seems to be only one Spirit of Nature. Also, in line with More's conception of spirit as *res extensa*, the Spirit of Nature is extended. By contrast, Plastic Nature is not extended: for Cudworth extension is a property of bodies ('resisting or antitypous extension') while immaterial spirit is un-extended ('internal energy and self-activity').[4] Cudworth draws attention to the life-function of Plastic Nature as a lower part or form of soul, which ensures that the continues to perform the functions which keep it alive, such as respiration and heartbeat. More discusses his Hylarchic principle largely in relation to the cosmos as a whole and the laws of physics. And each adduces a different set of arguments. For example, Cudworth indicates that Plastic Nature is an answer to occasionalism, since it explains God's immanence in the world, without requiring immediate divine intervention in the minutiae of day-to-day operations of nature. More emphasises that the Spirit of Nature can account for phenomena inexplicable by the mechanical philosophy (e.g. sympathetic vibration of strings and tidal movement). Cudworth is interested in the epistemological function of Plastic Nature for our knowledge of both our own bodies and the wider world. The operations of Plastic Nature makes the world intelligible, by imprinting disorganised matter with divine design. These differences notwithstanding, there is enough to show that the functions of both the Spirit of Nature and Plastic Nature exceed those of the Vegetable Soul.

17.3 The Plant Life of Nature

The scope of both the Spirit of Nature and Plastic Nature is evidently far wider than the scope of Aristotle's vegetative soul. But in so far as both govern the operations of created nature, they incorporate the operations which Aristotle ascribes to the vegetable soul: growth, reproduction and nutrition. In this respect the Spirit of Nature and Plastic Nature are very much to do with *life* at its most basic. This is referred to by both Cudworth and More as 'vegetative', especially in relation to plants. In *Immortality of the Soul* More gives more than a hint that the power of the

---

[4] Cudworth, *True Intellectual System*, p. 161.

Spirit of Nature is equivalent to the vegetable soul: as More puts it 'the *Root* and *Soul* of every Vegetable is the *Spirit of Nature*' (More 1659 402). Cudworth asserted that Plastic Nature is 'the same thing with the *Vegetative,* which is Inferiour to the *Sensitive*' (Cudworth 1678, 162–3). More sees the Spirit of Nature as responsible for 'shaping Vegetables into all that various beauty we finde in them' (More 1660, 458). As 'the Principle of all natural Generations, Growths and Perfections' (More 1660, 410), its functions are very much those of the vegetable soul.

Relative to other aspects of their accounts of Plastic Nature and the Spirit of Nature Cudworth nor More gives extensive attention to plant life. Cudworth's 'Digression' gives more attention to animal life than vegetable life, he does refer to plantal life, bracketing animal and plant life together within the scope of Plastic Nature. Arguments based on plants occur across More's writings, but proportionately less than other phenomena. For example, in his *Enchiridion metaphysicum* only one of twenty chapters devoted to demonstrating the existence of incorporeal substance deals with plants: Chapter 17 discusses the formation, propagation, system of nourishment, beauty and utility of plants figure as another example of phenomena, which cannot be accounted for by the chance motion of atoms (More 1995, 212–3; More 1671, 310–311). He attributes such phenomena as the growth of trees from tiny seeds to mighty trees, the rising of sap in trees, to laws of life in contained in the Hylarchic Principle, ('resolvitur in Leges Vitae Principio *Hylarchico* insitas').[5] In *Immortality of the Soul* he explicitly gives as a reason for positing the Spirit of Nature (construed as an '*Inferiour Soule of the World*') that it 'saves the phenomena' of plant life, because it accounts for a particular phenomenon of plant life: the capacity to be propagated by means of grafting and cuttings

> 'tis exceeding rational that all Plants and Flowers of all sorts (in which we have no argument to prove there is any particular Souls) should be the effects of this Universal Soule of the World. Which Hypothesis ... is also very serviceable for the preventing many hard Problems about the Divisibility of the Soules of Plants, their Transmutations into other Species, the growing of Slips, and the like. Which is the common and onely *logos spermatikos* of all Plantal appearances, or of whatever other Phaenomena there be, greater or smaller, that exceed the pure Mechanical powers of Matter. (More 1659, 467–8)[6]

There is some ambiguity here, since elsewhere in the same book, More also ascribes vegetative functions to the afore-mentioned seminal forms, which he makes responsible for giving life to matter for forming plants ('organizing duely prepared matter into life and vegetation proper to this or the other kind of plant') (ibid. 50). By 1662, he had concluded that there need be only one seminal form, i.e. the Spirit of Nature–

> *That there need be no other logoi spermatikoi, or* Seminal forms, *then this one, which virtually contains all everywhere, and is therefore rightly styled* The Universal Spirit of Nature. (More 1662, xv–xvi)

[5] English translation by Alexander Jacob adjusted. This translation should always be checked against the Latin. An idiosyncracy of his edition is his re-arrangement of More's chapters.

[6] John Ray concurs, rejecting the idea of an individual vegetative soul in plants because to account for the capacity of plants to propagate from cuttings, that soul would have to be divisible.

He even speculates as to whether plants 'have any distinct [souls] from the Universall Spirit of Nature' (More 1659, 306–7).

For Cudworth, as for More, a major difference between the plants and animals is that animal have individual souls, but plants do not. It is, he writes, 'not reasonable to think, that every Plant, Herb and Pile of Grass, hath a Particular Plastick Life, or Vegetative Soul of its own, distinct from the Mechanism of the Body' (Cudworth 1678, 171). But he allows that there may be other kinds of Plastic Nature for special purposes: 'it is not impossible but that there may be other Plastick Natures also (as certain Lower Lives, or Vegetative Souls) in some Greater Parts of the Universe'. And he goes on to conjecture that there could be a single Plastic Nature governing plants:

> there may possibly be, for ought we know, one Plastick Nature or Life, belonging to the whole Terrestrial (or Terraqueous) Globe, by which all Plants and Vegetables, continuous with it, may be differently formed, according to their different Seeds. (Cudworth 1678, 172)

This suggestion comes pretty close to positing a vegetative soul for all plants. As a single, all-encompassing spirit or energy, which constitutes the lowest common-denominator of life, both More's hylarchic spirit and Cudworth's Plastic Nature could be said to be the vegetative equivalent of Averroes' *intellectus agens*—as it were a '*natura agens*' that operates at some level in all living things.

17.4 Sources

In view of the evident overlap between the functions of the Spirit of Nature and Plastic Nature and those of Aristotle's vegetable soul, the question naturally arises how far, if at all, More and Cudworth drew on Aristotle, and the Aristotelian notion of the vegetable soul when framing their respective hypotheses. And if so, whether, their attempt to avoid errors of mechanism, they are after all open to Bayle's charge against Cudworth that he had re-invented substantial forms. So I turn now to the question of sources.

17.4.1 Sources: More

Henry More's conception of the Spirit of Nature developed over time (Reid 2012). A continuous thread of association between the Spirit of Nature and vegetable or 'plantall' life can be traced from his early to his mature works (2014, chapter 9). Initially, in his early poems he posited a multitude of seminal principles which he called 'seminal forms' or 'plantall souls', the function of which was limited for the basics of vegetable or plantal life. But, as we have seen, seminal forms figure in relation to his conception of the soul in later writings, until, in 1662, when he rolls them into a single universal spirit. Thus the Spirit of Nature, as a concept develops from earlier accounts of plantal and vegetative life. Along the way, More is not in the habit of giving his sources. But there are, as we have seen, echoes of the Aristotelian vegetable soul in his conception of it. However there is no clear evidence that More specifically links the Spirit of Nature to Aristotle. On the contrary his use of the term, 'seminal form' in its evolution alerts us to non-Aristotelian sources, while his explanation of the workings of the Spirit of Nature by the power of 'sympathy' is decidedly non-Aristotelian. In fact his earliest musings on the vegetative life strongly suggest non-Aristotelain sources at the origins of his conception of the Spirit of Nature. In his *Philosophical poems* vegetable life is treated in terms of seminal forms or 'logoi spermatikoi', which are endowed with formative or plastic powers. He glosses 'sperm' or seed, as 'the *logos spermatikos*, the *ratio seminalis,* or the invisible plasticall form that shapes every visible creature'. 'Spermaticall' is glossed as having even wider scope, anticipating those of the Spirit of Nature, since it has plantal, plastic and magnetic powers. It 'belongs properly to Plants, but is tranferred also to the Plasticall power in Animalls, I enlarge it to all magnetick power whatsoever that doth immediately rule and actuate any body' (More 1647, 434). An obvious source for *logoi spermatikoi* as plastic powers governing natural life is, as Cudworth reminds us, Plotinus (Cudworth 1678, 163). And Plotinus may indeed have been a source for More, who had a syncretic tendency to treat seminal forms as common currency across philosophies In *Divine Dialogues*, for example, he invokes the notion of seminal form as a universal principle of life among Platonists and Pythagoreans) (More 1668, 1, 36–38). However, More's definition of 'Spermaticall' in the poems additionally glosses it as *physis* or nature construed as vegetable: 'this world is but one Plant, one λογοσ σπερματικοσ giving it shape and corporeall life'. This strong pointer to Stoic rather than Aristotelian sources, is borne out in Cudworth's discussion of analogues for Plastic Nature.

17.4.2 Sources: Cudworth

Cudworth's 'Digression concerning the Plastic Life of Nature' is the only source for his conception of Plastic Nature, so it is not possible to track how his thinking about it developed over time. However, Cudworth reveals the sources of this thinking in a wide range of ancient philosophy in the extensive taxonomy of philosophy which takes up most of his *True Intellectual System of the Universe*, sorting different philosophies into cognate groups of atheistic or theistic philosophies. In his 'Digression' Cudworth cites a number of philosophers, notably Plotinus. But Aristotle figures prominently as well. In fact Cudworth tries to recruit Aristotle as a proponent of Plastic Nature—an ancestor, so to speak, of his own hypothesis. In so doing, he makes no mention of the Aristotelian vegetative soul *per se*. Although he claims Plastic Nature is the equivalent of 'the *Vegetative Soul*', what he highlights are not its nutritive and generative functions (as in Aristotle), but its *formative* function. Instead of *De anima* (which he doesn't cite) his Aristotle sources are *De partibus animalium* and *Physics*, an indication that his interest is not in Aristotle's psychological taxonomy of souls, in but in the Aristotelian conception of nature as art—hence Cudworth's stress on Plastic Nature as 'artificial' nature (e.g. Cudworth 1678, 154). Above all, in 'A Digression', Cudworth focuses on Aristotle's teleological conception of nature, as a match for his own conception of nature as purposive, designed, contrived (that is nature as guided by Plastic Nature). He presents Aristotle's conception of nature in terms that echo his account of Plastic Nature:

> *Aristotle*'s Nature is no Fortuitous Principle, but such as doth Nothing in *Vain*, but all for *Ends*, and in every thing pursues *the Best*; and therefore can be no other than a Subordinate Instrument of the Divine Wisdom, and the *Manuary Opificer* or *Executioner* of it. (Cudworth 1678, 54)

Since Aristotle, by contrast with Descartes, retains final causality, so making nature an instrument of divine wisdom, '*Aristotle*'s System of Philos ophy seems to be more consistent with Piety,' than the Cartesian *Hypothesis*'.

> We add further, that *Aristotle*'s System of Philosophy seems to be more consistent with Piety, than the Cartesian *Hypothesis* it self, which yet plainly supposeth Incorporeal Substance. (Cudworth 1678, 54)

Aristotle is, of course not the only ancient philosopher recruited by Cudworth in support of this conception of Plastic Nature. As already noted, Plotinus is an important source. But Cudworth's syncretic view of philosophy allows him to blend together conceptions of plastic nature with the vegetative soul. Although he does in a general way identify Plastic Nature with the Aristotelian vegetable soul, he treats it as a common currency among 'Latter *Platonists* and *Peripateticks* [who] have unanimously followed their Masters herein, whose *Vegetative Soul* also is no other than a *Plastick Nature*' (Cudworth 1678, 153).

However, for the genesis of Cudworth's conception of Plastic Nature as a *vegetable* principle, we should look not to Aristotle, but to the Stoics. On this point there is an important passage in his discussion of the classifications of types of atheism shortly before his 'Digression'. This occurs in Cudworth's discussion of the variety of atheism which he calls *Cosmo-plastick Atheism*. According to Cudworth, atheists of the Cosmo-plastic variety posit a view of nature which is cognate with Plastic nature. A Cosmo-plastick view of nature, he writes, 'supposes one kind of *Plastick* and *Spermatick, Methodical* and *Artificial* Nature... to preside over the whole World, and dispose and conserve all things, in that Regular Frame in which they are' (Cudworth 1678, 131). On this view

> ... the whole World is no *Animal*, but as it were, one *Huge Plant* or *Vegetable*, a Body endued with one *Plastick* or *Spermatick Nature*, branching out the whole, Orderly and Methodically, but without any Understanding or Sense. (Cudworth 1678, 132)

A key feature of this 'cosmo-plastick' conception of nature is that it does not result from chance, but exhibits regularity and orderliness: 'it does not suppose Nature to be a mere *Fortuitous* ... thing'. It

> quite excludes *Fortune* or *Chance*, subjecting all things to the Regular and Orderly Fate, of one Plastick or Plantal Nature, ruling over the Whole. (Cudworth 1678, 131–2)

For this reason Cudworth differentiates this view of nature from the 'hylozoism', which attributes life to matter. Nevertheless he holds that the 'cosmoplastick' view of nature is atheistic, because it lacks a guiding mind, or 'any clearly Intellectual Principle or Conscious Nature' (Cudworth 1678, 132). Leaving the issue of atheism aside, the source which Cudworth cites is Seneca. He arrives at this view of the *Cosmo-plastic* conception of the entire world as a giant vegetable endowed with one plastic pature ('a *Great Plant* or *Vegetable*, having one *Spermatick Form*, or *Plastick Nature*') via a reading of a passage in Seneca's *Naturales quaestiones*, 3.29:

> Sive Animal est Mundus, sive Corpus Naturâ Gubernante, ut Arbores, ut Sata;

Cudworth translates this as follows

> Whether the whole World be an Animal (i.e. endued with one Sentient and Rational Life) or whether it be only a Body Governed, by (a certain Plastick and Methodical, but Sensless) Nature, as Trees, and other Plants or Vegetables. (Cudworth 1678, 131)

The conception of nature which Cudworth gets from this is a conception of nature which is very close to his own conception of Plastic Nature. On Cudworth's interpretation, the passage in question describes, 'one kind of *Plastick* and *Spermatick, Methodical* and *Artificial* Nature', where

> the whole World is no *Animal*, but as it were, one *Huge Plant* or *Vegetable*, a Body endued with one *Plastick* or *Spermatick Nature*, branching out the whole, Orderly and Methodically, but without any Understanding or Sense. (Cudworth 1678, 132)

Cudworth's reading is not unproblematic: aside from his textual interventions (e.g. translating 'anima' as 'animal'), he acknowledges that this Cosmo-plastic view of nature is not explicit in Seneca, even that it is 'doubtful', and that his interpretation is only 'hinted to us'.[7] Nevertheless, this does not deter Cudworth from comparing the Cosmo-Plastic Nature with Plastic Nature. One important difference between this and Cudworth's version of Plastic nature is that Stoic nature is material (Cosmoplastic nature 'supposes the Whole World … [has]nothing but Body in it)'. And, crucially for Cudworth, this Stoic account of 'Spermatick' or 'Vegetable' Nature is not subject to an over-arching intellectual principle of any kind, 'it does not derive the Original of things in the Universe, from any clearly Intellectual Principle or Conscious Nature'. Nevertheless, this vegetable world of the Stoics is closer to Plastic Nature than Aristotle's vegetable soul (see also Sellars 2011).

17.5 Conclusion

Both the Spirit of Nature and Plastic Nature are conceived as basic forms of life *life* and are treated by both Cudworth and More as a 'vegetative' function of their respective theories. Both we contain echoes of ancient philosophy, Platonism,

[7] For a fuller discussion of Cudworth and Stoicism, see Sellars (2011). See also Hirai (2007) for Justus Lipsius use of the same passage in *Physiologia stoicorum*.

Aristotelianism and Stoicism, Plato's *anima mundi*, Stoic *pneuma*, as well as the Aristotelian vegetable soul. The danger of highlighting the classical sources is that they make them appear to be mere throw-backs to ancient philosophy. However, they were not *revivals*, but *reformulations* designed to meet contemporary explanatory needs, among them the inadequacies of the new mechanical philosophies of nature for explaining life. Although these life functions requiring explanation correspond to those covered by Aristotle's vegetable soul, Cudworth and More looked to account for them differently from Aristotle. The roots of their thinking lie not with the vegetable soul of Aristotle, but in Stoic vegetable nature. Here they were in the good company of seventeenth-century humanist medicine (Distelweig et al. 2016, Hirai 2011 and 2016). Seen in this light, the Spirit of Nature and Plastic Nature appear less as airy metaphysical or theosophical notions as has sometimes been suggested (Duchesneau 1982), but rooted in more earthly natural philosophy. This also helps to explain their appeal to natural philosophers like Grew and Ray, not merely for their compatibility with religious doctrine, but as theories which accounted for vegetative life in terms which met the demands of natural philosophy.

References

Allen, Keith. 2013. Cudworth on Mind, Body, and Plastic Nature. *Philosophy Compass* 8 (4): 337–347.

Andrault, Raphaële. 2014. What is Life? A comparative study of Ralph Cudworth and Nehemiah Grew. In *The Life Sciences in Early Modern Philosophy*, ed. Ohad Nachtomy and Justin E.H. Smith, 29–46. Oxford: Oxford University Press.

———. 2017. *Life as Vegetation. Limit cases and theological problems for seventeenth-century thinkers'*. Unpublished paper given at BSHP-2017 conference.

Baldassarri, Fabrizio. 2019. The mechanical life of Plants: Descartes on botany. *British Journal for the History of Science* 52 (1): 1–23.

Burthogge, Richard. 1694. *Essay upon Reason and the Nature of Spirits*. London.

———. 1699. *Of the Soul of the World, and of Particular Souls*. London.

Cudworth, Ralph. 1678. *The True Intellectual System of the Universe*. London.

Distelzweig, Peter, Benjamin Goldberg, and Evan Ragland, eds. 2016. *Early Modern Medicine and Natural Philosophy*. Dordrecht: Springer.

Duchesneau, François. 1982. *La Physiologie des lumières: empirisme, modèles, et théories*. The Hague: Martinus Nijhoff.

Garrett, Brian. 2003. Vitalism and Teleology in the Natural Philosophy of Nehemiah Grew (1641–1712). *The British Journal for the History of Science* 36 (1): 63–81.

Giglioni, Guido. 1996. Anatomist Atheist? The "Hylozoistic" Foundations of Francis Glisson's Anatomical Research. In *Religio Medici. Medicine and Religion in Seventeenth-Century England*, ed. Andrew Cunningham and O.P. Grell, 115–135. London: Scolar Press.

Grew, Nehemiah. 1671. *Cosmologia Sacra. or a Discourse of the Universe. As it is the Creature and Kingdom of God*. London: W. Rogers, S. Smith, and B. Walford.

Henry, John. 1986. Occult Qualities and the Experimental Philosophy: Active Principles in Pre-Newtonian Matter Theory. *History of Science* 24 (4): 335–381.

———. 1990. The Spirit of Nature and the Nature of Providence. In *Henry More, Tercentenary Studies*, ed. Sarah Hutton, 55–76. Dordrecht: Kluwer.

Hirai, Hiro. 2007. Justus Lipsius on the World-Soul. Between Roman Cosmic Theology and Renaissance '*Prisca theologia*'. In *Justus Lipsius and Natural Philosophy*, ed. Hiro Hirai and Jan Papy, 63–79. Brussels: Koninklijke Vlaamse Academie van Belgie voor Wetenschappen en Kunsten.
———. 2011. *Medical Humanism and Natural Philosophy: Renaissance Debates on Matter, Life and the Soul*. Leiden: Brill.
———. 2016. Mysteries of Living Corpuscles: Atomism and the Origin of Life in Sennert, Gassendi and Kircher. In Distelweig et al. *History, Philosophy and Theory of the Life Sciences* 14 (2016): 255–269.
Hunter, William B. 1950. The Seventeenth Century Doctrine of Plastic Nature. *Harvard Theological Review* 43: 197–213.
Hutton, Sarah. 2015. *British Philosophy in the Seventeenth Century*. Oxford: Oxford University Press.
Inglehart, Ashley I. 2016. Boyle, Malpighi and the Problem of Plastic Powers. In Distelweig et al. *History, Philosophy and Theory of the Life Sciences* 14 (2016): 295–321.
Lähteenmäki, Vili. 2010. Cudworth on Types of Consciousness. *British Journal for the History of Philosophy* 18 (1): 9–34.
Lotti, Brunello. 2004. *Ralph Cudworth e l'idea di Natura Plastica*. Udine: Campanotto Editore.
More, Henry. 1647. *Philosophicall Poems*. Cambridge.
———. 1653. *Antidote against Atheism*. London.
———. 1656. *A Brief Discourse of Enthusiasm*. London.
———. 1659. *Of the Immortality of the Soul*. London.
———. 1660. *An Explanation of the Grand Mystery of Godliness*. London.
———. 1662. *A Collection of Several Philosophical Writings*. London.
———. 1668. *Divine Dialogues*. London.
———. 1671. *Enchiridion Metaphysicum*. London.
———. 1995. *Henry More's Manual of Metaphysics: A translation of the Enchiridium metaphysicum (1679). Introduction and Notes [by] Alexander Jacob*. Hildesheim: Olms.
Muratori, Cecilia. 2017. In Human Shape to Become the Very Beast: Henry more on Animals. *British Journal of the History of Philosophy* 25 (5): 897–915.
Pécharman, Martine. 2014. Cudworth on Self-consciousness and the 'I Myself'. *Vivarium* 52 (3–4): 287–314.
Ray, John. 1681. *The Wisdom of God in the Works of Creation*. London: Samuel Smith.
Reid, Jasper. 2012. *The Metaphysics of Henry More*. Dordrecht: Springer.
Rosa, Susan. 1994. Ralph Cudworth in the République des Lettres: The Controversy about Plastick Nature and the Reputation of Pierre Bayle. *Studies in Eighteenth-Century Culture* 23: 157–160.
Sellars, John. 2011. Is God a Mindless Vegetable? Cudworth on Stoic Theology. *Intellectual History Review* 21 (2): 121–133.
Thiel, Udo. 2011. *The Early Modern Subject: Self-Consciousness and Personal Identity from Descartes to Hume*. Oxford: Oxford University Press.

Chapter 18
Margaret Cavendish and Vegetable Life

Justin Begley

Abstract This paper traces how the seventeenth-century poet, playwright, and natural philosopher, Margaret Cavendish, developed her ideas on plant life in three major publications: her 1655 *Philosophical and Physical Opinions*, 1664 *Philosophical Letters*, and 1666 *Observations Upon Experimental Philosophy*. As I seek to demonstrate, Cavendish's mostly overlooked ruminations on vegetality were indispensable to her formation of a substance theory according to which all matter is endowed with life, vitality, knowledge, and, in her mature works, perceptivity. Over the course of this paper, I also suggest that her visions of matter and plant life were largely grounded in Galenic and Aristotelian insights that she extracted and adapted from a range of sources, not least from William Harvey's field-changing writings on the circulation of blood and the generation of animals. In building upon this material, Cavendish was driven to uphold the innate livelihood of plants against mechanical accounts and to champion a "natural" mode of classification that focussed on the diversity of plant attributes rather than foregrounding specific features that humans deemed valuable or significant.

18.1 Introduction

Recent years have witnessed a flurry of interest in the seventeenth-century poet, playwright, and natural philosopher, Margaret Cavendish. Among scholars who have examined her natural philosophical output, some have now considered the bearing that her so-called "material vitalism" had on her understanding of animals.[1] Yet, while Cavendish was at least as preoccupied with investigating the nature of vegetable as animal life, her ideas on this front have been given surprisingly short shrift. My paper seeks to fill in this lacuna by surveying her views on vegetality as they were expounded in three major publications: her 1655 *Philosophical and*

[1] See, for example, Boyle 2018, 189–214 and Broad 2002, especially 54–5.

J. Begley (✉)
Herzog August Bibliothek, Wolfenbüttel, Germany
e-mail: justin_s_begley@live.com

F. Baldassarri, A. Blank (eds.), *Vegetative Powers*, International Archives of the History of Ideas Archives internationales d'histoire des idées 234,
https://doi.org/10.1007/978-3-030-69709-9_18

Physical Opinions, 1664 *Philosophical Letters*, and 1666 *Observations Upon Experimental Philosophy*. I suggest that this significant facet of Cavendish's thought largely developed out of her engagement with Galenic and Aristotelian insights, which she extracted from a range of sources, not least William Harvey's field-changing works on generation and the circulation of blood. By analysing and contextualising Cavendish's understanding of the vegetal, this paper aims to illustrate the crucial bearing that ruminations on plant life had on her formulation of a substance theory according to which matter is endowed with life, vitality, knowledge, and, in her mature writings, perceptivity.

This chapter begins with a brief outline of Cavendish's evolving substance theory, which was inseparable from her wider conception of matter, motion, and knowledge. It then delves into her case for vegetable sensitivity and rationality, and sketches out its implications for seventeenth-century debates about how to differentiate plants from animals and minerals. As we will see, Cavendish drew attention to an assortment of marvellous plant traits to counteract what she regarded as nebulous and imprecise claims about their inferiority to animals. In closing, I turn to how her outlook on matter and vegetality fed into her hypothesis about natural perceptivity, which served to complicate dichotomous notions of humans as subjects and plants as objects. Throughout we will see that Cavendish upheld the innate livelihood of plants against mechanical accounts and advocated a "natural" mode of classification that was not predominantly constructed around aptitudes that humans identified as valuable or significant.

18.2 Galenic Matter and Knowledge

While Cavendish's substance theory morphed incrementally, all of her natural philosophical treatises—from her 1653 *Philosophicall Fancies* to her 1668 *Ground of Natural Philosophy*—defend the existence of three qualitatively discrete substances: rational, sensitive, and dull matter. In her pre-1664 volumes, rational and sensitive matter are styled as "spirits" that possess innate motion and incessantly move dull matter. In framing her discourse thus, Cavendish made it plain that her chosen terminology of "rational and sensitive spirits, [is that] which others call the animal, and vital" (Cavendish 1655a, 138). The latter distinction was rooted in the Galenic physiology that continued to be taught in all European universities and practised by most educated physicians throughout the seventeenth century.[2] Within this tradition, *pneuma zotikon* (known as *spiritus vitalis* or vital spirit) was said to shuttle heat and vitality through the body, and *pneuma psychikon* (known as *spiritus animalis* or animal spirit) transferred this vitality into psychic activity and transmitted the higher functions of the soul through the nerves.[3] But, in a bid to account for the dynamic

[2] See Frank 1997 and Webster 2002, especially 15–22 and 308–14.

[3] See Galen 1916, II.98 and xxxiv–v. On whether Galen also posited natural spirits (*pneuma physikon*), see Temkin 1977, 154–161.

symbiosis between living beings and the external world, Cavendish expanded the role of these spirits from overseers of internal functions to the basic constituents of nature. One of the inspirations for this manoeuvre was almost certainly Francis Bacon who at times also depicted a natural world teeming with material spirits, and whom Cavendish lauded as "Learned, Eloquent, Witty, and Wise" having pored over his mythological and philosophical output some years earlier (Cavendish 1664b, 146).[4] Outstripping even Bacon, however, she modified the ordinary lexicon of "vital" and "animal" to reflect her conviction that all natural spirits are vital and by no means restricted to animal bodies.

As she fleshed out the nature and capacities of her spirits, Cavendish devised a matter-motion-knowledge triad. From the premise that matter can move, she reasoned that it must know how to move, concluding that all matter with its "*innate motion*, is *knowing*: but according to the several motions, are several knowledges made; for *knowledge* lives in *motion*, as motion lives in *matter*" (Cavendish 1655a, 52). She held, in short, that knowledge does not require a knowing subject. With her stress on material rationality or knowledge, Cavendish again strayed from Bacon who had ascribed appetites or passions to matter, but had stopped short of endowing it with *knowledge*.[5] Yet Bacon was never her prime target. Instead, at various junctures in her career, Cavendish explicitly pitted her knowing matter against, *inter alia*, Henry More's immaterial force that pervades nature and René Descartes' association of knowledge not only with God but also the immaterial human soul.[6] While subsequent sections touch on Cavendish's critique of Descartes, it is noteworthy here that where he coalesced the rational and immaterial souls (and reserved the latter for humankind alone), Cavendish retained an immaterial soul that lacked a physiological purpose and substituted the conventional rational soul for diffuse matter. On her account, the knowledge proper to each natural being—whether plant, animal, or human—is a function of the configuration of this pervasive yet nonconscious material rationality.

Cavendish made it clear that vegetal considerations were indispensable to her matter theory when she proposed, in the opening pages of *Philosophical and Physical Opinions*, that "*Vegetables* and *Minerals* may have some of those *rational spirits*, which is a *minde* or *soul* in them, as well as *man*" (Cavendish 1655a, 21). It is important to clarify at this stage that sense and reason were, for Cavendish, by no means equivalent to traditional conceptions of animal perception or human rationality, and were rather natural capabilities of matter with observable expressions that could differ conspicuously across species (as we will see). Even with this proviso, her expansion of rationality not only posed a challenge to Descartes, but also to Aristotle's famous division of life forms in *De anima*, according to which the vegetative soul facilitates plant nutrition and reproduction, the sensitive soul empowers

[4] On Bacon's notion of spirits, the classic studies are Walker 1972 and Rees 1984, and, for some recent updates, see Anderson 2016 and Rusu 2018.

[5] See Giglioni 2010 and Andersson 2016.

[6] See Hutton 2003 and Semler 2012.

animals to perceive and respond to their environments, and the rational soul gives rise to human intellection.[7] But Cavendish's contention that human intelligence is an extraordinary feature that emerges from the non-conscious knowledge of matter did have a venerable Aristotelian lineage that primarily branched out of Alexander of Aphrodisias's commentaries.[8] Alexander was able to generate his reading of Aristotle because—in stark contrast to the Platonic standpoint that something intellectual must be derived from a higher-order being—Aristotle's *Historia animalium* presented a bottom-up picture of the *scala naturae* according to which "Nature proceeds little by little from things lifeless to animal life in such a way that it is impossible to determine the exact line of demarcation" (Aristotle 1910, 588b3–6).[9] With her notion that only degrees of intricacy distinguish matter itself from the features of minerals, plants, and animals—which she labelled "nature's three kingdoms"—Cavendish took the impossibility of differentiating "things lifeless" from "animal life" more seriously than even Aristotle, while simultaneously foregrounding the individual characteristics of plants and animals rather than livelihood as such (Cavendish 2001, 211).

One of Cavendish's techniques for complicating customary boundaries between life forms was to establish similarities between the internal functions of plants and animals. To this end, she again turned to Galen but now she drew on the four auxiliary capacities of the *facultas naturalis*—which he had posited as common to all living beings—to argue that the rudimentary internal motions that spirits know how to enact are attractive, retentive, digestive, and expulsive.[10] Mapping these onto observable vegetal functions, Cavendish went on to contend that "*Attractive* is that which we call *Growth*, or *youth*. *Retentive*, is that we call *strength*. *Digestive* is that we call *Health*, that is an equal distribution of parts to parts, and agreeing of those spirits. *Expulsive* is that which we call *Death*, or *decay*" (Cavendish 1655a, 7). As should be clear, she accepted, in broad strokes, both Galen's matter theory, according to which motion is inherent to spirits, and the corresponding capacities that he ascribed to plants and animals.

Following her speculative ruminations on matter and her adaptation of Galen's four auxiliary capacities, Cavendish turned to more anatomical and physiological matters, pointing out that "*rational spirits* seem most to delight in *spungie soft* and *liquid matter;* as in the *Blood, Brain, Nerves,* and in *Vegetables*" (Cavendish 1655a, 19). Her list of blood, brain, and nerves as the *loci* where spirits are concentrated resonates with Galen's notion that *pneuma psychikon* principally flows from the brain (along with the exhalations from useful blood) and *pneuma zotikon* from the

[7] On the reception of the Aristotelian notion of the soul during the early modern period, see Serjeantson 2011; Park 1988; and McCracken 2000.

[8] For this longer history, see Pluta 2007.

[9] The classic source on the *scala naturae* is Lovejoy 1936. Also see Wolff 1947; Kuntz and Kuntz 1987; and Lewis 2012.

[10] For a discussion of these motions, see, for example, Galen 1916, III.4–8. More generally, see Debru 2008, especially 273–4.

heart.[11] While it might appear peculiar, at first glance, that she added vegetables to this list, since *pneuma zotikon* was customarily deemed to function in plants while *pneuma psychikon* was confined to animals, Galen had stated that the vegetables which humans consume are of an essentially liquid constitution (similar to blood, the brain, and nerves), and he even surmised that they generated blood in the process of assimilation.[12] Once Cavendish granted rational spirits dominion outside of animal bodies, she quite naturally supposed that they would suffuse vegetables, given that vegetables share matter and internal motions with animals. All told, while much has been made of the idiosyncrasies of Cavendish's substance theory, a closer analysis reveals that her views on matter and the basic capacities of living beings stemmed in large part from her engagement with the Galenic tradition.[13]

18.3 William Harvey and Vegetable Rationality

The Aristotelian-Galenic ideas that captivated Cavendish also served as a launchpad for much of Harvey's groundbreaking research in physiology and comparative anatomy, and there are solid reasons to believe that Harvey's historic tomes strongly informed her interpretation and appropriation of ideas from these ancients.[14] As early as 1655, Cavendish eulogised the "most Learned Physician, Doctor *Harvey*" for having "found out the Circulation of the Blood, by his industrious study" (Cavendish 1655b, 180).[15] It is notable in relation to this discovery that she not only vaguely imbued vegetables with rational spirits, but specified that "the *Sap* in *Vegetables* may be of the same *substance,* and *degree* of the *Brain*", which she had previously coupled with the blood and nerves (Cavendish 1655a, 23). Harvey is best remembered for discovering that the heart's active motion was the forceful systole instead of the diastole. But, when it comes to Cavendish's association of sap with the animal organs and substances where active spirits are concentrated, it is pertinent that Harvey had proposed (already in his 1616 *Prelectiones anatomiae universalis*) that the blood, as a vital force, is the material and final cause of circulation. Not only did Harvey conceive of blood as self-moving, but he also intermittently described this power as a vegetative force, such as when he explained that blood circulates so that the "first innate warm moisture might be retain'd in life, and in their own vital and vegetative essence, and perform all their functions, whilst (as the Naturallists say) they are sustain'd and actuated by natural heat, and vital spirits" (Harvey 1653b, 17). With her vital and indeed "rational" sap, Cavendish built upon

[11] See Temkin 1977, 154–161.

[12] See, for instance, Galen 2003, II.626.

[13] For a fuller discussion, see Begley 2016, 90–119 and Boyle 2018, 215–37.

[14] There is now a large body of research on Harvey's Aristotelianism, but some of the best studies are Wear 1983; Schmitt 1984; and Lennox 2002.

[15] For a broader discussion of Cavendish and Harvey, see Begley 2016, 117–44 and Goldberg 2017.

the alignment of the circulation of the blood in animals with the movement of sap through plant "veins" that gained traction in England after Harvey's *De motu cordis* appeared in 1628 and only picked up speed following Marcello Malpighi's detection of the capillaries in the 1660s.[16] It was even reckoned in some quarters that plant sap must be more puissant than blood because it could flow from deep within a tree's roots to the tips of its branches (and back again) without a centralised mechanism such as a heart.[17]

In *Philosophical Letters* (1664), Cavendish went on to grapple at some length with Harvey's other hugely significant publication: the 1653 translation of his *Exercitationes de generatione animalium* (initially published in 1651).[18] In this work, Harvey complicated the standard four-element theory of the Aristotelian tradition and imputed an even greater set of powers to seeds and blood than he had formerly, arguing that they contain a "more divine substance then fire is" and that their spirits are "answerable in proportion to the *element* or *substance* of the *Stars*" (Harvey 1653a, 452). Harvey considered blood to pre-exist the heart, but, significantly, he also rendered it the vehicle of the soul that directed the embryo as it developed (in the case of humans) from a vegetative to a sensitive and finally a rational being.[19] One of his chief objectives in this work was thus to penetrate into "the abstruse nature of the *Vegetative Soul*" that the blood (or seed in the case of plants) originally spawned, and which he lavishly praised as the most deific force in nature, not least because it exhibited the power of godly rather than mere human reason (Harvey 1653a, 2). By this, Harvey meant that God had sagaciously crafted matter with the power to form the vegetative soul—as an internal, teleological principle—that in turn carried out His plans for the continuation of life.[20]

While Harvey (*à la* Aristotle) saw the soul as the principle that guided the activity of organised matter, we have seen that Cavendish went further with her contention that matter itself performs the traditional functions of not only the vegetative and sensitive souls but also the rational one. At the same time, ever wary of encroaching on the territory of divines, she underscored the natural character of matter rather than following Harvey in casting it in a quasi-spiritual light. Notwithstanding these divergences, Cavendish accepted the crux of Harvey's argument that a primordial rationality—which she framed as "the *spirits* that are in the *Seed*" in her early publications—permeated all manner of life (Cavendish 1655a, 9). She came to this conclusion because (like Harvey) she passed off the supposition that a non-living and unknowing principle could engender a living and knowledgeable one as an insuperable paradox. Her slant was that anything with the capacity to beget life must, at some level, be knowing, and anything with knowledge is necessarily alive.

[16] See Roos 2011, 151–66.

[17] See Hartley 2010, 239–44 and Begley 2021.

[18] For her explicit citation of *Exercitationes generatione de animalium*, especially see Cavendish 1664a, 419–432.

[19] See White 1986 and French 1994, 302–5.

[20] "Exercitation 50" is most important here, especially Harvey 1653a, 265–268. For a fuller discussion of this passage, see Goldberg 2013.

Within this framework, the parent of an animal or the seed of a plant are only secondary or instrumental causes, subordinate to the innate potency of matter. To convey the sentiment that living creatures are ultimately products of pre-existing life, Cavendish consistently deployed the antimetaboles "the *spirits* of *life*, or *lively spirits*" and "the spirits of knowledge, or the *knowledge* of *spirits*" (Cavendish 1655a, 8 and 17). Whereas Harvey (as a physician) principally dwelt on the nature of matter as a means to comprehend animal generation and physiology, Cavendish (as a philosopher) tended to gear such insights towards speculative ends, and, in the process, set much store by plants as uniquely capable of pointing towards the inscrutable livelihood of matter itself.

18.4 Redefining Plant Life

As she tried to make sense of why most philosophers, physicians, and theologians who tackled questions about the nature of substance, mind, and life dismissed out of hand the proposition that plants (let alone matter) possess an essential rationality, Cavendish pinpointed the elevation of language as a major roadblock.[21] She probably had at the forefront of her mind Descartes' well-known proclamation that animals "could never use words, or put together other signs, as we do in order to declare our thoughts to others", which he isolated as an attribute that proved the rationality of humans and raised them above other species (Descartes 1985, 140). Ventriloquising this line of reasoning, Cavendish remarked that since vegetables and other animals "do not prate and talk as Man, Man believes they have not so much wit as he hath" (Cavendish 1664a, 43).[22] Just because plants do not vocalise their knowledge, however, she denied that they were therefore utterly bereft of rationality. Indeed, Cavendish mocked philosophers who yoked reason to language at the expense of other abilities as blind to the foundational truth that it is not "speech that makes sense and reason, but sense and reason makes them" (Cavendish 1664a, 43). In formulating this critique, she was being somewhat facetious, given that most philosophers of her day agreed that reason preceded and enabled human speech.[23] Yet, in extending the remit of rationality, her broader point was that an intelligent speech act (let alone mindless gossip) was only one rational activity among many and it could plausibly be said to pale in comparison to the astonishing work of matter that daily brought forth and sustained all manner of natural life.

All of this indicates that an *aporia* lurked behind Cavendish's treatment of plants: she argued that vegetables enjoy rationality and knowledge while maintaining that their manifestations of these attributes depart drastically from human ones and are

---

[21] For the supposition that speech was perhaps the fundamental marker that distinguished humans from the rest of nature, see Thomas 1983, 32.

[22] Also see Cavendish 1655a, b, 21–22.

[23] See Fudge 2006, 41–8.

virtually inaccessible to us. Aware that her explanations for why and how vegetable life is rational and knowing could come across as particularly speculative, Cavendish pointed to drugs and foodstuff in a bid to furnish her readers with concrete instances of vegetable and mineral knowledge. She pronounced at one moment, for example, that although vegetables cannot "slice, pound or infuse, as man can, yet they can work upon man more subtilly, wisely, and as sensibly either by purging, vomiting, spitting, or any other way" (Cavendish 1664a, 43). Based on her affiliation of knowledge with motion—and against the backdrop of the popular Hippocratic adage that nature is a wise healer—she accordingly saw no reason why humans should be said to *know* how to prepare vegetables if it is denied that plants *know* how to purge humans.[24] As she put it, "Vegetables will as wisely nourish Men, as Men can nourish Vegetables" (Cavendish 1664a, 43). A substantial difference between vegetables and humans is of course that plants have a natural and inborn knowledge of how to nourish and ameliorate human health, whereas humans tend to take recourse in oral and book learning to understand how to achieve the same ends. When Cavendish discussed the knowledge, rationality, or wisdom of vegetables, she therefore did not mean that they had abstract thoughts about the processes and consequences of nourishing, which she quite conventionally saw as peculiar to humans or at least higher animals, but she was nonetheless adamant that simply because physicians happen to enjoy this distinctly human instantiation of rationality, they should not be credited more than the plant itself when a prescribed herb successfully treats a disease.

Cavendish's extension of a primordial rationality to all natural bodies was partially a result of her decision to embrace the language of "figure" in place of the more abstract and potentially rigid Aristotelian-Galenic concept of "soul".[25] Many of her atomistic and mechanically-inclined contemporaries preferred to discuss "figures" because this term was shorn of the implications of teleology and internal cohesion that weighed down the Aristotelian vocabulary of souls or forms.[26] Owing to her extension of rationality and sensitivity to matter and her rejection of substantial forms, Cavendish's understanding of plant and animal life as distinguished only by degrees of complexity could seem to render her view the equal and opposite of the Cartesians, who regarded sufficiently complex man-made automata and living beings as virtually indistinguishable. But Cavendish was careful to clarify that machines operate "as a *hand* that is guided by another, and not of its own strength: that is the reason, *Arts* have not so much perfection as *nature*" (Cavendish 1655a, 8). While "figure" sometimes narrowly signified the physical shape of a natural body, Cavendish also clarified that there remained "particular natures in every creature, which are the innate, proper and inherent interior and substantial forms and figures" by which one being is "distinguished from the other" (Cavendish 2001,

[24] On the more commonplace notion, see Newton 2015.

[25] A useful source on this transition is Joy 2006.

[26] On questions of the relationship between "form" and "figure", see Pasnau 1982 and Lüthy and Newman 1997.

197). Here she made it palpable that each being has a distinctive nature due as much to internal features as external markers, but, more importantly, she retained the age-old Aristotelian notion that life forms differ from man-made automata inasmuch as they are invested with the inborn capacity for continuous generation (along with growth and nutrition) as well as an internal principle of change and rest.[27] By focussing on these traditional vegetative functions, she circumvented the conclusion that all bodies, whether natural or artificial, are equally alive because matter is, in a sense, living.

While Cavendish appreciated that all vegetables involuntarily work to attain their proper ends, she was obviously cognisant of the fact that most vegetables (save "sensitive" and "carnivorous" plants) imperceptibly grow and spread their roots over protracted stretches of time, and even their nutrition is a mostly internal process that consists of sucking up nutrients from the earth. One of the main reasons why Cavendish viewed machines such as those that inspired Descartes as unconstructive models for organic life was that they spotlighted immediately visible locomotion in a manner that was designed to invoke wonder but threatened to draw philosophers and physicians away from diligently studying the complex, slow, and imperceptible workings of other bodies, particularly vegetables.[28] Despite being sessile, she regarded plants as active agents, entirely capable of responding to environmental cues. At the same time, Cavendish deemed mechanical models of motion to overstate the speed at which natural actions occur, noting that some vegetables "are but in their prime after a hundred years", which would no doubt necessitate a radically different relationship to time (Cavendish 1655a, 39). She accordingly took the sessility of plants, their virtually invisible functions, and their frequently glacial motions as prerequisites for deeply enviable capacities such as longevity. In this sense, the fact that the standard vegetative functions of growth, nutrition, and reproduction were difficult, if not impossible, to replicate with automata only spoke in favour of plant subtlety and complexity for Cavendish.

Perhaps owing to an appreciation of the sophistication of vegetables, many mechanists—with Robert Hooke foremost among them—opted to elucidate the nature of plants and seeds not so much by calling on mechanical analogies (like Descartes) but rather by using mechanical devices. In *Observations*, Cavendish responded directly to Hooke's somewhat poetic declaration in his 1665 *Micrographia*, which accompanied his hyper-realistic copper-plate engravings, that seeds are "the Cabinet of Nature, wherein are laid up its Jewels" with the insight that "if seeds of vegetables did lie never so long in a storehouse, or any other place, they would never produce, until they were put into some proper and convenient ground" (Hooke 1665, 152; Cavendish 2001, 68–9). While some scholars have aligned Cavendish and Hooke by arguing for their shared amazement at the intricacy of miniature bodies as viewed through microscopes, it should be emphasised that their approaches

---

[27] But see the various kinds of automata surveyed in Giglioni 1995.

[28] On automata and wonder, see Marr 2006.

diverged in two striking ways.[29] First, an Aristotelian preference for plain-sight observation prompted Cavendish to deny that Hooke's sketches elucidated the "true" nature of seeds, and she pointed out that merely scrutinising their texture from an alternative frequency told us nothing about their underlying matter or fundamental natures.[30] Second, she held that because microscopes detached seeds—and other minuscule bodies—from their natural environments they tended to foreground the complexity of particulars at the expense of a broader and more accurate picture of natural relations and causes.[31]

Although Hooke drove Cavendish to admit that seeds relied on soil for their continued existence, she defied the presumption that this dependence uniquely hampered plants, since "animals can live no more without air, then fishes can live without water, or Vegetables without ground" (Cavendish 1666, 56). Indeed, she ventured that humans are perhaps more overtly enmeshed in webs of reciprocal subsistence than plants insofar as they rely not only on the classical four elements but also plants for health and continued existence. As these deliberations indicate, Cavendish considered the human reliance on vegetables to run far deeper than the vegetable dependence on humans: humankind would soon perish if all plants were wiped out, whereas plants could easily flourish in a world devoid of humans. With all of this in mind, she quipped that as "Man makes use of other Creatures, so other Creatures make use of Man, as far as he is good for any thing [...] being not so profitable for use, as apt to make spoil" (Cavendish 1664a, 147). On her account, humans are uniquely inclined to destroy and consume more than they conserve and produce.

18.5 Generation and Growth

One of the principal reasons that Cavendish imputed rationality and sensitivity to matter was because she judged the natural world to be a closed system. Whereas she surmised in her early works that spirits are scattered in a vacuum, she later postulated a material continuum on which rational, sensitive, and dull matter are thoroughly intermixed. Grounded in the notion of continually shifting and recycled matter, this later theory had direct implications for her vision of generation. According to Cavendish, when a man, for example, "is dissolved, his parts dispersed, and joined with others, we may say his former form or figure of being such a particular man is buried in its dissolution, and yet liveth in the composition of other parts" (Cavendish 2001, 69). In the extant historiography, this conceit has been confused with the theory of palingenesis, which was a naturalised version of Pythagorean metempsychosis that was derived from Pseudo-Paracelsus's *De natura*

[29] For a partial alignment of Cavendish and Hooke due to their emphasis on "wonder" and "fantasy", see the relevant chapter in Campbell 1999, 181–220 and Linden 2001.

[30] On Cavendish's critique of Hooke's microscopy, see Lawson 2015; Wallwork 2001; and Aït-Touati 2011, 174–190; and Begley 2016, 214–41.

[31] For her criticism of Pliny for overemphasising the minute particular, see Cavendish 2001, 200.

rerum, whereby a deceased plant could be reconstituted from its ashes.[32] But, highly critical of such chymical positivism, Cavendish's contention was the more pedestrian one that the decomposing matter of, say, a human cadaver re-joins the continuum and inexorably serves, over time, as material for the generation of inorganic bodies or other living beings. It is significant that these newly created beings need not possess self-conscious rationality, and certainly lack access to the memories of dead organisms.

Having established that organic life springs from the unconscious rationality of matter, Cavendish went on to assert that the generation of "maggots, and several sorts of worms and flies, and the like, which have no generator of their own kind, but are bred out of cheese, earth and dung" is natural and perfectly explicable. Rather than seeing those substances out of which organisms emerge as passive incubators, she stressed that cross-species generation is an active process whereby "vegetables can and do breed animals, and animals, minerals and vegetables" (Cavendish 2001, 66–67). That one species could spawn another was in fact a seventeenth-century commonplace thanks to the all-pervasive ideas of Aristotle and Pliny.[33] The Aberdeen-born physician who emigrated to Paris, William Davisson, for instance, wrote to William Cavendish in a letter conserved in a medical manuscript co-produced by William and Margaret that "often wormes generate in young Peoples Bodys".[34] In an article on early modern spontaneous generation, Justin Smith situated Cavendish within a Lucretian genealogy when he argued that her intimation that "a creature could be generated *de novo* from a new and fortuitous arrangement of existing matter" in *Observations* "is tantamount to acknowledging the reality of species transmutation".[35] Yet her stance was actually far more orthodox, for she clarified that there was not "want or decay of general kinds of creatures, but only a change of their particulars" (Cavendish 2001, 132). While she in no way denied the existence of one-off mixed species such as the mule—and indeed reflected with delight on her experience of having eaten an apple that tasted like a cabbage—she was sure to note that such productions were infertile, and that the apple-cabbage in particular was a preternatural or "artificial" product of selective cross-pollination (Cavendish 1664a, 273). Cavendish, then, held fast to two broadly Aristotelian positions: the essences of species are fixed, but, in suitable environments, some beings and substances can bring forth a particular specimen of an eternally existing kind.[36] To her mind, cases of seedless generation were possible only because life pre-existed even the seed.

[32] The discussion here opposes the loose usage of "palingenesis" in Thell 2016. On palingenesis more generally, see Secret 1979; Roling 2014; and Marx 1971.

[33] On how Aristotle developed his views, see Lennox 1981 and Zwier 2018. For early modern debates, see Farley 1977; Parke 2014; and Muratori 2013.

[34] Nottingham University Library, PwV 90, ff. 22v. Benjamin Goldberg and I will publish a scholarly edition of this medical manuscript with Palgrave Macmillan later in 2021. For a fuller discussion of worms and generation during this period, see MacInnes 2012.

[35] Smith 2009, 95–6.

[36] See Aristotle 2014a, 415a27-b9 and Aristotle 1910, 539a23–5.

While cross-species generation fascinated Cavendish, she simultaneously identified certain generative capacities as *sui generis* such as when a plant clones itself from a severed part or cutting. As she recorded, if the "boughs, arms or branches" of some trees "be lopt or cut from a lively stock, those boughs or branches will many times remain lively, according to the nature of the figure whose parts they were, for a good while; nay, if they be set or planted, they will grow into the same figure as the stock was" (Cavendish 2001, 224). Cavendish summoned this observation to shore up her argument that an essential rationality animates plants in their entirety, which, once again, was a partial outgrowth of the post-Harveian conception of sap. As a result of this quality being distributed throughout its parts, she also realised that a plant could heal and continue to flourish even after much of it was destroyed. By contrast, a human is of course incapable of regenerating after most of its body has been severed, or, for that matter, of cloning itself from a dismembered arm or leg (Cavendish 2001, 224). But it would never be supposed that humans are, for this reason, inferior to vegetables. Despite aligning plants with animals (and against minerals) on the grounds that they are organic bodies with manifold parts, Cavendish realised that certain plants can only enjoy these generative capacities inasmuch as they consist of relativity few organs when compared to higher animals (Cavendish 1655a, 45). This simplicity was sometimes regarded as making them inferior, but she appreciated that it also undergirded one of the most spectacular vegetal traits: the capacity to regenerate a whole from a part.

18.6 Perception and Patterning

It was in *Observations* that Cavendish finally afforded an explanation for the nature of sense perception that attempted to account for how and why humans struggled to grasp the knowledge and functions of plants. While we have seen that her early works allied rational and sensitive matter, she explicitly refers to (and closely relates) "sensitive perception" and "rational perception" in *Philosophical Letters* and *Observations*, rendering perception the major category that wedded these endogenous traits.[37] Cavendish tellingly began to discuss these two kinds of perception at the very moment that she relinquished individualised spirits in favour of a material continuum. Whereas she had treated material sensitivity and rationality as effectively selfish forces—fuelled by self-motion and self-knowledge—she deemed perceptivity to allow for the communication and attendant organisation of matter along the continuum. As Cavendish put it, perception "proceeds from an interior principle of self-knowledge, yet it is nothing else but an observation of exterior parts or actions" (Cavendish 2001, 144). It was accordingly the mechanism that facilitated continuity between the external world and internal bodily economies. At least in the case of most animals, she argued that sensitive perception patterns "out the

[37] The fullest discussion of Cavendish's ideas on perception in *Philosophical Letters* is Clucas 2014.

figure or action of an outward object in the sensitive organ", whereas internal rational perceptions make "a figure of the same object in their own substance" (Cavendish 2001, 150). But because Cavendish saw natural bodies as composed of perceptive matter—with feeling and sensation emerging thanks to specific material configurations—she contended that the diversity of bodily compositions necessarily engendered differences in perception.

While Cavendish insisted that special bonds connect certain species or even genera (such as humans and dogs or bees and flowers), she considered the diverse configurations of perceptive matter to be one of the reasons why understanding and communication were typically most effective between members of the same species or those of a similar nature and complexity. In dwelling on differences in perception, she contended that a "vegetable, or mineral" could not "have an animal knowledge or perception, no more than the eye patterning out a tree or stone, can be said to have a vegetable or mineral perception" (Cavendish 2001, 141). Observing a plant, in other words, only supplies information about its external contours, but tells us nothing about its internal nature, never mind what it is like to be a plant.[38] More fundamentally, Cavendish's ascription of perception to plants and even minerals posed a challenge to Aristotle's notion—most comprehensively outlined in *De anima*—that animals actively obtain perceptual forms whereas lower beings (such as plants and minerals) are passive objects in this process.[39] To establish the rival position that various life forms could engage in perceptive acts without necessarily having the capacity to reflect on the contents of those perceptions (or even their ability to perceive), Cavendish made the point that when "one man, perceives another, he doth not perceive his knowledge; for it is one thing to perceive the exterior figure of a Creature, and another thing to perceive its interior, proper, and innate actions" (Cavendish 1666, 161–2). By extension, when a human feels the rough bark of a tree, the tree may concurrently pattern out the shape of the human hand, yet nothing about human perception illuminates the experience of the tree, just as the tree is presumably ignorant of what bark feels like to a human. In this sense, Cavendish's distinction between external figures and internal motions allowed her to extricate the perceptions of an observer (which tend to consist of figures and locomotion) from the immanent motions or "perceptions" of particular bodies.

Cavendish's theory of natural perceptivity brought her towards the fringe of mid-seventeenth-century attitudes towards substance and vegetality. But it is here that her ideas came to resonate most closely with those of Francis Glisson, the Cambridge Regius Professor of Physic from 1636–1677, three-time president of the Royal College of Physicians, and founding member of the Royal Society. His views converged with Cavendish's on numerous fronts: he not only endowed matter with life and self-motion, but also an unconscious perceptivity that coordinated the natural world. These ideas were most thoroughly adumbrated in his monumental *De natura*

[38] On the limits of inspecting exteriors, see Richter 2016.

[39] The *locus classicus* for this discussion is Aristotle 2014a, 416b33–34. On Aristotle and active perception, see Spruit 2008.

substantiae energetica of 1672.[40] In spite of their significant overlaps, however, Cavendish did veer away from Glisson insofar as she layered perception and sensation, rather than categorically distinguishing unconscious perception from conscious sensation. That is, instead of exclusively tying perception *qua* sensation to species with specific organs, she thought that this capacity was essentially explicable in terms of the nature of matter. While her parallel stress on the perceptive capacities of both matter *and* higher animals was another step toward problematising the *scala naturae*, she nonetheless retained a hierarchy according to which vegetables sit somewhere between the wholly unconscious perception of matter and the fully conscious perception of humans.

As a consequence of imputing perception to matter, Cavendish also came to complicate the agent/patient relationship. To her mind, when a human observer obtains an image of a tree through an intentional perceptive act, the matter composing that tree has a role to play in actively conveying its likeness to the human. In the process of evaluating ancient philosophies in *Observations*, she pointed out that her view, in this regard, could appear to differ markedly from the standard Scholastic account according to which forms are divorced from matter during a perceptive act and create a pictorial representation in the human soul (Cavendish 2001, 271–272). Yet she reminded her readers that, even for Aristotle, "matter and form are but one thing", which should render such a separation impossible (Cavendish 2001, 270). Further, on Cavendish's account, this divorce procedure would require the form of an object as a totality (rather than the relevant parts) to enter the sense organ, which she thought was as impossible as the atomist's stance that particles glide off the surface of a body yet retain the same position and order as they proceed to bombard the eye (Cavendish 2001, 265–266). While she defended the trustworthiness of human sense perception against sceptics of all stripes, Cavendish realised that perception always occurs from a particular vantage point, and that each observation therefore only captures a particular angle on an object as opposed to the true nature of the thing itself. In striving to square these ostensibly competing impulses, she concluded that the sensitive, perceptive matter of the object relays, through the material continuum, images to the equivalent matter in the observer, and it was this immediacy that served as a natural safeguard against deluding observations. Even so, experience led her to believe that a perceiver could never fully or directly access the internal workings of the rational, perceptive matter within another being. Along with her arguments for the natural livelihood and knowledge of matter, this conception of perceptivity transformed any confident assertion that vegetables lacked sensitivity or rationality into an epistemologically unjustified leap.

[40] On Glisson's substance theory, see Giglioni 1996 and 2002. For more on the relationship between the ideas of Cavendish and Glisson, see Begley 2016, 138–44.

18.7 Conclusion

It is notable in concluding that the eminent physician and FRS, Nehemiah Grew, read and took extensive notes on Cavendish's ideas about matter and life as spelt out in *Philosophical and Physical Opinions*.[41] Suggesting Grew's interest in Cavendish's emphasis on the shared features of plants, animals, and minerals, he specifically alighted on her point, which he slotted under the heading "Vegetables", that there is "A question whether they [vegetables] have not sense since they may possibly have with Animals the same Matter & Motion".[42] Grew's sustained engagement with Cavendish's substance theory and reflections on plant sensitivity testifies to their lasting relevance and appeal. Yet it is notable that even this so-called "Father of Plant Anatomy", who was renowned for complicating distinctions between life forms, felt compelled to re-organise the contents of her volume so that it better reflected the Aristotelian *scala naturae*.[43] As the tension between the structure of his compendium and the thrust of his extracts indicate, establishing a new, fine-grained, and less value-laden classification of natural bodies (particularly plants) was a gradual and even circuitous process.

While Cavendish regularly castigated natural historians who failed to ascend from minute particulars to general truths, she was philosophically committed to studying plants in their own right instead of making wholesale judgements of superiority or inferiority rooted in soul-based divisions. On her appraisal, the human capacity for conscious self-reflection and deliberation almost guarantees that "man thinks himself to have the Supreme knowledge" yet this self-aggrandisement was only justified insofar as human knowledge was set *a priori* as the benchmark against which to assess the capacities of all other life forms (Cavendish 1655a, 40). If one were to read *De anima* alone, it would be easy to pin the blame for such anthropocentric taxonomical tendencies on Aristotle, but, in *De partibus animalium*, Aristotle had in fact refused to define genera and species in terms of any one essential feature.[44] It was these competing Aristotelian inclinations that gave rise to a heated late seventeenth-century conflict over whether taxonomists should pursue a "natural" method, duly considering the nature and features of each plant species, or an "artificial" one, which differentially weighed specific attributes or functions, traditionally with an emphasis on plant fructification.[45] In the years leading up to these debates, Cavendish not only handled results procured using mechanical apparatuses with a hearty dose of scepticism and resisted the application of mechanistic

[41] For a fuller discussion of this manuscript, see Begley 2017.

[42] Grew Sloane MS 1950.

[43] On this as a common way to organise early modern textbooks, see Blair, *The Theater of Nature*, 30–39 and 65–81.

[44] Aristotle 2014b, 643b15. For a fuller discussion, see Lloyd 1961 and Lennox 1980.

[45] The classic essay on this topic is Sloan 1972, but, for some helpful correctives, see Müller-Wille 2013.

metaphors to natural bodies, but she also championed a natural approach to the categorisation of living beings.

Cavendish gestured at her "natural" approach in *Philosophical and Physical Opinions* when she asserted that "an oak is not like a tulip, or roses; for trees are not like flowers, nor flowers like roots, nor roots like fruit, nor all flowers alike, nor all roots alike, nor all fruits alike, nor all trees" (Cavendish 1655a, 39). As this implies, she deemed there to be almost as much variety within natural kingdoms as across them, and, as any reader of Cavendish's books knows well, the endeavour to capture this natural diversity often left her no choice but to list and catalogue exhaustively the traits and motions of living beings before delving into the precise causes of natural functions. Just because they could be labelled "vegetative life", she saw as little reason to lump together a blade of grass and a majestic oak as a worm and an ape. To Cavendish's mind, it would be absurd to suppose that the smallest worm is, in some sense, of a higher order than the tallest oak simply because it is an animal rather than a plant. One consequence was that she revelled in the differences, as much as similarities, between natural bodies, even underlining that plants enjoyed many capacities—such as self-cloning, unaided healing, regeneration, and, in some cases, remarkable longevity—that humans could only dream of. This is not to mention the nutritional and curative properties of plants that humans had always prized. Cavendish's hope was that, by drawing attention to these exceptional vegetable attributes, she might alert fellow natural philosophers to the self-love that was apt to taint their outlooks on the natural world at large.

Acknowledgements Many thanks for the edifying questions and discussions at the 2018 "Vegetative Powers" conference in Padua where an earlier version of this paper was presented, and to Benjamin Goldberg for his extremely helpful comments on a draft of this chapter.

References

Aït-Touati, Frédérique. 2011. *Fictions of the Cosmos: Science and Literature in the Seventeenth Century*. Chicago: University of Chicago Press.

Anderson, Miranda. 2016. Francis Bacon's Flux of the Spirits and Renaissance Paradigms of Hybridity and Adaptation. In *Francis Bacon on Motion and Power*, ed. Guido Giglioni, James Lancaster, Sorana Corneanu, and Dana Jalobeanu, 113–152. Dordrecht: Springer.

Andersson, Daniel. 2016. Appetites, Matter and Metaphors: Aristotle, *Physics* I, 9 (192a22–23), and Its Renaissance Commentators. In *Francis Bacon on Motion and Power*, ed. Guido Giglioni, James Lancaster, Sorana Corneanu, and Dana Jalobeanu, 41–59. Dordrecht: Springer.

Aristotle. 1910. *Historia Animalium*, ed. and trans. D'Arcy W. Thompson. Oxford: Clarendon Press.

———. 2014a. *On the Soul; Parva Naturalia; On Breath*, ed. and trans. Walter S. Hett. Cambridge, MA: Harvard University Press.

———. 2014b. *Parts of Animals; Movement of Animals; Progression of Animals*, ed. and trans. Edward S. Forster and Arthur L. Peck. Cambridge, MA: Harvard University Press.

Begley, Justin. 2016. *Margaret Cavendish, the Last Natural Philosopher*. Unpublished DPhil thesis, University of Oxford.

———. 2017. 'The Minde Is Matter Moved': Nehemiah Grew on Margaret Cavendish. *Intellectual History Review* 27: 493–514.

———. 2021 (forthcoming). Kenelm Digby, Nathaniel Highmore, and Plant Life. In *Interpretations of Life in Heaven and Earth*, ed. Hiro Hirai. Leiden: Brill.

Blair, Ann. 1997. *The Theater of Nature: Jean Bodin and Renaissance Science*. Princeton: Princeton University Press.

Boyle, Deborah. 2018. *The Well-Ordered Universe: The Philosophy of Margaret Cavendish*. Oxford: Oxford University Press.

Broad, Jacqueline. 2002. *Women Philosophers of the Seventeenth Century*. Cambridge: Cambridge University Press.

Campbell, Mary Baine. 1999. *Wonder & Science: Imagining Worlds in Early Modern Europe*. Ithaca: Cornell University Press.

Cavendish, Margaret. 1655a. *The Philosophical and Physical Opinions*. London.

———. 1655b. *The World's Olio*. London.

———. 1664a. *Philosophical Letters, or, Most Reflections upon some Opinions in Natural*. London: Philosophy.

———. 1664b. *CCXI Sociable Letters*. London.

———. 1666. *Observations upon Experimental Philosophy*. London.

———. 2001. *Observations Upon Experimental Philosophy*, ed. Eileen O'Neill. New York: Cambridge University Press.

Cavendish, William and Margaret Cavendish. Nottingham University Library. Pw V 90.

Clucas, Stephen. 2014. 'A Double Perception in all Creatures': Margaret Cavendish's *Philosophical Letters* and Seventeenth-Century Natural Philosophy. In *God and Nature in the Thought of Margaret Cavendish*, ed. Brandie Siegfried and Lisa Sarasohn, 121–139. Farnham: Ashgate.

Debru, Armelle. 2008. Physiology. In *The Cambridge Companion to Galen*, ed. R.J. Hankinson, 263–282. Cambridge: Cambridge University Press.

Descartes, René. 1985. *The Philosophical Writings of Descartes*, Vol. 1–3, ed. and trans. John Cottingham, Robert Stoothoff, and Dugald Murdoch. Cambridge: Cambridge University Press.

Farley, John. 1977. *The Spontaneous Generation Controversy from Descartes to Oparin*. Baltimore: Johns Hopkins University Press.

Frank, Roger G. 1997. Medicine. In *The History of the University of Oxford: Volume IV, Seventeenth Century Oxford*, ed. Nicholas Tyacke, 505–558. Oxford: Oxford University Press.

French, Roger. 1994. *William Harvey's Natural Philosophy*. Cambridge: Cambridge University Press.

Fudge, Erica. 2006. *Brutal Reasoning: Animals, Rationality, and Humanity in Early Modern England*. Cornell: Cornell University Press.

Galen. 1916. *On the Natural Faculties* (*De naturalibus facultatibus*), ed. and trans. Arthur John Brock. London: W. Heinemann.

———. 2003. *On the Properties of Foodstuff* (*De alimentorum facultatibus*), ed. and trans. Owen Powell. Cambridge: Cambridge University Press.

Giglioni, Guido. 1995. Automata Compared: Boyle, Leibniz and the Debate on the Notion of Life and Mind. *British Journal of the History of Philosophy* 3: 249–278.

———. 1996. Anatomist Atheist? The 'Hylozoic' Foundations of Francis Glisson's Anatomical Research. In *Religio Medici: Medicine and Religion in Seventeenth-Century England*, ed. Ole Peter Grell and Andrew Cunningham, 115–135. Aldershot: Scolar Press.

———. 2002. Francis Glisson's Notion of Confœderatio Naturæ in the Context of Hylozoistic Corpuscularianism. *Revue d'histoire des sciences* 55: 239–262.

———. 2010. Mastering the Appetites of Matter: Francis Bacon's *Sylva Sylvarum*. In *The Body as Object and Instrument of Knowledge: Embodied Empiricism in Early Modern Science*, ed. Charles Wolfe and Ofel Gal, 149–168. Dordrecht: Springer.

Goldberg, Benjamin. 2013. A Dark Business, Full of Shadows: Analogy and Theology in William Harvey. *Studies in History and Philosophy of Biological and Biomedical Sciences* 44: 419–432.

———. 2017. Epigenesis and the Rationality of Nature in William Harvey and Margaret Cavendish. *History and Philosophy of the Life Sciences* 39: 1–23.

Grew, Nehemiah. 1950. British Library. Sloane MS.

Hartley, Beryl. 2010. Exploring and Communicating Knowledge of Trees in the Early Royal Society. *Notes & Records of the Royal Society* 64: 229–250.

Harvey, William. 1653a. *Anatomical Exercitations, Concerning the Generation of Living Creatures*. London.

———. 1653b. *Two Anatomical Exercitations, Concerning the Circulation of the Blood*. London.

Hooke, Robert. 1665. *Micrographia, or, some Physiological Descriptions of Minute Bodies Made by Magnifying Glasses*. London.

Hutton, Sarah. 2003. Margaret Cavendish and Henry More. In *A Princely Brave Woman: Essays on Margaret Cavendish, Duchess of Newcastle*, ed. Stephen Clucas, 185–199. Aldershot: Ashgate.

Joy, Lynn. 2006. Explanations from Formal Causes to Laws of Nature. In *The Cambridge History of Science, Vol. 3: Early Modern Science*, ed. Katherine Park and Lorraine Daston, 70–105. Cambridge: Cambridge University Press.

Kuntz, Marion Leathers, and Paul Grimley Kuntz. 1987. *Jacob's Ladder and the Tree of Life: Concepts of Hierarchy and the Great Chain of Being*. New York: Peter Lange.

Lawson, Ian. 2015. Bears in Eden, or, This Is Not the Garden You're Looking For: Margaret Cavendish, Robert Hooke and the Limits of Natural Philosophy. *British Journal for the History of Science* 48: 583–605.

Lennox, James. 1980. Aristotle on Genera, Species, and "The More and the Less". *Journal of the History of Biology* 13: 321–346.

———. 1981. Teleology, Chance, and Aristotle's Theory of Spontaneous Generation. *Journal of the History of Philosophy* 19: 219–238.

———. 2002. The Comparative Study of Animal Development: William Harvey's Aristotelianism. In *The Problem of Animal Generation in Early Modern Philosophy*, ed. Justin Smith and Michael Ruse, 21–46. Cambridge: Cambridge University Press.

Lewis, Rhodri. 2012. *William Petty on the Order of Nature: An Unpublished Manuscript Treatise*. Tempe: Arizona Center for Medieval and Renaissance Studies.

Linden, Stanton J. 2001. Margaret Cavendish and Robert Hooke: Optics and Scientific Fantasy in *The Blazing World*. In *Esotérisme, Gnoses & Imaginaire Symbolique: Mélanges Offerts à Antoine Faivre*, ed. Richard Caron, Joscelyn Godwin, Wouter J. Hanegraaff, and Jean-Louis Vieillard-Baron, 611–624. Louvain: Peeters.

Lloyd, G.E.R. 1961. The Development of Aristotle's Theory of the Classification of Animals. *Phronesis* 6: 59–81.

Lovejoy, Arthur. 1936. *The Great Chain of Being: A Study of the History of an Idea*. Cambridge, MA: Harvard University Press.

Lüthy, Christoph, and William Newman. 1997. 'Matter' and 'Form': By Way of a Preface. *Early Science and Medicine* 2: 215–226.

MacInnes, Ian. 2012. The Political Worm: Invertebrate Life in the Early Modern English Body. In *The Indistinct Human in Renaissance Literature*, ed. Jean E. Feerick and Vin Nardizzi, 253–273. Basingstoke: Palgrave Macmillan.

Marr, Alexander. 2006. 'Gentille curiosité': Wonder-Working and the Culture of Automata in the Late Renaissance. In *Curiosity and Wonder from the Renaissance to the Enlightenment*, ed. R.J.W. Evans and Alexander Marr, 149–170. Aldershot: Ashgate.

Marx, Jacques. 1971. Alchimie et Palingenesie. *Isis* 62: 275–289.

McCracken, Charles. 2000. Knowledge of the Soul. In *The Cambridge History of Seventeenth-Century Philosophy*, ed. Daniel Garber and Michael Ayers, 792–832. Cambridge: Cambridge University Press.

Müller-Wille, Staffan. 2013. System and How Linnaeus Looked at Them in Retrospect. *Annals of Science* 70: 305–317.

Muratori, Cecilia. 2013. The Earth's Perilous Fertility: Telesio on Spontaneous Generation and the Continuity of Living Beings. In *The Animal Soul and the Human Mind: Renaissance Debates*, ed. Cecilia Muratori, 131–151. Pisa-Rome: Fabrizio Serra.

Newton, Hannah. 2015. 'Nature Concocts & Expels': The Agents and Processes of Recovery from Diseases in Early Modern England. *Social History of Medicine* 28: 465–486.

Park, Katherine. 1988. The Organic Soul. In *The Cambridge History of Renaissance Philosophy*, ed. Charles Schmitt, Quentin Skinner, Eckhard Kessler, and Jill Kraye, 464–484. Cambridge: Cambridge University Press.

Parke, Emily C. 2014. Flies from Meat and Wasps from Trees: Reevaluating Francesco Redi's Spontaneous Generation Experiments. *Studies in the History and Philosophy of Biological and Biomedical Sciences* 45: 34–42.

Pasnau, Robert. 1982. Form and Matter. In *The Cambridge History of Later Medieval Philosophy: From the Rediscovery of Aristotle to the Disintegration of Scholasticism, 1100–1600*, ed. Norman Kretzmann, Anthony Kenny, Jan Pinborg, and Eleonore Stump, 635–646. Cambridge: Cambridge University Press.

Pluta, Olaf. 2007. How Matter Becomes Mind: Late-Medieval Theories of Emergence. In *Forming the Mind: Essays on the Internal Senses and the Mind/Body Problem from Avicenna to the Medical Enlightenment*, ed. Henrik Lagerlund, 149–167. Dordrecht: Springer.

Rees, Graham. 1984. Francis Bacon and *Spiritus Vitalis*. In *Spiritus: IV Colloquio Internazionale del Lessico Intellettuale Europeo*, ed. Marta Fattori and Massimo Bianchi, 265–282. Rome: Edizioni dell' Ateneo.

Richter, Virginia. 2016. Exterior Inspection and Regular Reason: Robert Hooke's and Margaret Cavendish's Epistemologies of the Senses. In *The Five Senses in Medieval and Early Modern England*, ed. Annette Kern-Stähler, Beatrix Busse, and Wietse de Boer, 96–116. Leiden: Brill.

Roling, Bernd. 2014. Exkurs ins Pflanzenreich: Die Rose des Paracelsus. Die Idee der Palingenesie und die Debatte um die natürliche Auferstehung zwischen Mittelalter und Neuzeit. In *Zoology in Early Modern Culture*, ed. Karl A.E. Enenkel and Paul J. Smith, 263–300. Leiden: Brill.

Roos, Anna Marie. 2011. *Web of Nature: Martin Lister (1639–1712), the First Arachnologist*. Leiden: Brill.

Rusu, Doina-Cristina. 2018. Same Spirit, Different Structure: Francis Bacon on Inanimate and Animate Matter. *Early Science and Medicine* 23: 444–458.

Schmitt, Charles. 1984. William Harvey and Renaissance Aristotelianism: A Consideration of the Praefatio to 'De Generatione Animalium' (1651). In *Humanismus und Medizin*, ed. Rudolf Schmitz and Gundolf Keil, 117–138. Weinheim: Acta Humaniora.

Secret, François. 1979. Palingenesis, Alchemy and Metempsychosis in Renaissance Medicine. *Ambix* 26: 81–92.

Semler, Liam. 2012. Margaret Cavendish's Early Engagement with Descartes and Hobbes: Philosophical Revisitation and Poetic Selection. *Intellectual History Review* 22: 336–342.

Serjeantson, Richard. 2011. The Soul. In *The Oxford Handbook of Philosophy in Early Modern Europe*, ed. Desmond M. Clarke and Catherine Wilson, 119–137. Oxford: Oxford University Press.

Sloan, Philip R. 1972. John Locke, John Ray, and the Problem of the Natural System. *Journal of the History of Biology* 5: 1–53.

Smith, Justin. 2009. 'The Unity of Generative Power': Modern Taxonomy and the Problem of Animal Generation. *Perspectives on Science* 17: 78–104.

Spruit, Leen. 2008. Renaissance Views of Active Perception. In *Theories of Perception in Medieval and Early Modern Philosophy*, ed. Simo Knuuttila and Pekka Kärkkäinen, 203–224. Dordrecht: Springer.

Temkin, Owsei. 1977. *Double Face of Janus and Other Essays in the History of Medicine*. Baltimore: Johns Hopkins University Press.

Thell, Anne M. 2016. "Lady Phoenix": Margaret Cavendish and the Poetics of Palingenesis. *Early Modern Women* 11: 128–136.

Thomas, Keith. 1983. *Man and the Natural World: Changing Attitudes in England, 1500–1800*. London: Allen Lane.

Walker, D.P. 1972. Francis Bacon and *Spiritus*. In *Science, Medicine, and Society in the Renaissance: Essays to Honor Walter Pagel*, ed. Allen Debus, vol. 2, 121–130. New York: Science History Publications.

Wallwork, Jo. 2001. Margaret Cavendish's Response to Robert Hooke's *Micrographia*. In *Women Writing, 1550–1750*, ed. Jo Wallwork and Paul Salzman, 191–200. Bundoora: La Trobe University.
Wear, Andrew. 1983. William Harvey and the 'Way of the Anatomists'. *History of Science* 21: 223–249.
Webster, Charles. 2002. *The Great Instauration: Science, Medicine and Reform, 1626–1660*. Bern: Peter Lang.
White, John. 1986. Harvey and the Primacy of the Blood. *Annals of Science* 43: 239–255.
Wolff, Emil. 1947. *Die goldene Kette: Die Aurea Catena Homeri in der englischen Literatur von Chaucer bis Wordsworth*. Hamburg: Hansischer Gildenverlag.
Zwier, Karen. 2018. Methodology in Aristotle's Theory of Spontaneous Generation. *Journal of the History of Biology* 51: 355–386.

Chapter 19
Plantanimal Imagination: Life and Perception in Early Modern Discussions of Vegetative Power

Guido Giglioni

Abstract Relying on works by Plotinus, Galen, Ficino, Cesalpino, Kepler and Harvey, this chapter introduces the notion of 'plantanimal' imagination to explore the ways in which early modern philosophers and physicians conceptualized the elusive notion of vegetative perception. According to Plato, this perception was characteristic of plants (*Timaeus*, 77B). By concentrating on a series of interrelated notions that helped shape the category of vegetative perception (such as Ficino's inborn *intuitus*, Cesalpino's natural soul, Harvey's unconscious tactility and Kepler's telluric imagination), I will show how early modern thinkers manifested the need to expand the otherwise too narrow concept of animal and nervous reactivity. The process of nutrition will be at the centre of my argument. I will argue that the authors examined in this chapter laid the groundwork for a new idea of digestive and irritable self, through which living beings, including plants, were supposed to be endowed with a sense of natural discernment, capable of recognizing and altering their surrounding reality without losing any of their own identity.

Vegeto, ergo sum, I am alive, therefore I am. We all know that history – at least that sort of crystalline province of history which we like to call the history of ideas – went in another direction, for now we prefer to think that we exist insofar as we think, that is, insofar as we are aware of ourselves and present to ourselves. And yet we really cannot embrace this Cartesian fantasy entirely because we also are – exist – in a larger sense, that is, we are insofar as we live, and live as part of a greater living organization.[1] Let's call this organization the world of vegetative life. This kind of life, though not a cogitative affair, is knowledge all the same. I argue that life

[1] This point is now recognized on many levels. See Descola 2013 [2005]; Kohn 2013.

G. Giglioni (✉)
Università di Macerata, Macerata, Italy
e-mail: guidomaria.giglioni@unimc.it

F. Baldassarri, A. Blank (eds.), *Vegetative Powers*, International Archives of the History of Ideas Archives internationales d'histoire des idées 234,
https://doi.org/10.1007/978-3-030-69709-9_19

is knowledge because its basic operations – nutrition, growth, self-repair and reproduction – presuppose primordial acts of recognition stemming from a self-focused sense of identity and otherness. But, if life is knowledge, what kind of knowledge would it be? Can we describe the immediate act of metabolic recognition as a form of awareness? Philosophers like the Stoics in antiquity, and Bernardino Telesio (1509–1588) and Tommaso Campanella (1568–1639) during the Renaissance, would certainly answer in the affirmative. Other thinkers – philosophers, but especially physicians – would combine elements of Aristotelian and Galenic natural philosophy to provide a more nuanced answer: the sense of vegetative identity, while unaware of its operations, has an immediate grasp of what is necessary to do to stay alive. What is lost in terms of self-knowledge is gained in vital immediacy and urgency. The grasp of the vegetative power is indeed so immediate that in fact it is already action, prior to any division into subject and object of knowledge. It's a way of expressing things while doing them: Galen called this seamless act of knowledge and modelling a plastic power (διαπλαστική δύναμις).[2]

Aristotle consistently ruled out the possibility that the vegetative power required knowledge, let alone awareness, in order to be effective. At most, he envisaged the existence of ontological hybrids having both plant-like and animal attributes, such as sea anemones, corals, sponges and sea nettles (*Parts of Animals*, IV, 5, 681ab; *History of Animals*, 588b). These hybrids were called zoophytes, which in Latin were rendered as *plantanimalia*.[3] In this chapter, I am going to explore the characteristics that define this ability to sense metabolic identity and otherness. I propose to call this non sentient faculty of representing reality 'plantanimal imagination'. To put it brutally – and to challenge Aristotelian science – do plants, sponges, livers and embryos imagine, that is, have representations, *species*, of what they do while they do it? Or is the life of plants, sponges, livers and embryos so fully immersed in matter that their sensing things is already an act of transformation? The question is whether the vegetative power is capable of representing the object of its action, and more deeply, whether the transformative work that the vegetative faculty exercises on an object while making and fashioning it requires the ability to represent that object.

Since antiquity, to investigate the nature of the vegetative soul meant to ask questions about the material bases of life. The specific philosophical and scientific question I would like to address in this chapter lies at the intersection of anatomy and psychology, taking here psychology in the old philosophical sense of a doctrine concerning the soul and its faculties. The term *psychologia* was coined and used during the early modern period by such different authors as Rudolph Goclenius the Elder (1547–1628) and Christian Wolff (1679–1754), but in fact the discipline had long been created by Aristotle to collect together a set of interdependent inquiries

[2] Galen 1916, 18, 158. On the notion of plastic power in the early modern period see Hirai 2005; Hirai 2007.

[3] On Aristotle's zoophytes, see Granger 1985; Lloyd 1996, 67–82. On plants and animals in Aristotle, Sprague 1991; Falcon 2015.

dealing with biology, metaphysics and natural philosophy.[4] The tensions between anatomy and psychology do of course mirror the complex relationship between body and soul, but the specific issues related to a possible *anatomy of the soul* are even more complex, as is witnessed by the history of both medicine and philosophy. When the question of the soul is tackled from the point of view of anatomy, it doesn't matter too much whether we consider the soul and its various levels through the lens of Platonic, Aristotelian or Galenic models – that is, whether we divide the soul into reason, emotion and appetite (Plato), or into mind, the senses and the vegetative functions of life (Aristotle), or into reason, spirit and nature (Galen). The anatomical side of the story brings to the fore specific and notorious difficulties of its own, such as whether the mind is a section of the brain, whether the senses and the motions related to the senses are reducible to the cardiac or nervous system, or, finally, whether the liver and the digestive and venous apparatuses are the source of the natural – that is, involuntary – operations of an animal.

And yet, if we approach this matter from the point of view of the vegetative life, the difficulties are even greater. During the Renaissance there were anatomical research programmes of a Galenic or an Aristotelian kind, or eclectically Aristotelian-Galenic in which the functions of nutrition, growth and reproduction were examined in the hope of shedding new light on the mystery of the formative faculty – indeed the mystery of any Galenic faculty, as satirized by Molière in *Le malade imaginaire* (1673). As we will see towards the end of this chapter (Sect. 19.3), it was William Harvey (1578–1657) who, moving from strongly Aristotelian premises, laid the foundations for an original anatomical project centred on the notion of vegetative power. Francis Glisson (1599?–1677), relying on both Francisco Suarez and Francis Bacon (indeed, quite an unusual pairing), would then develop the anatomy of the vegetative power into a system of biological philosophy premised on a comprehensive theory of irritability.[5] To emphasize the significance of Harvey's anatomical project, in the first two sections of this chapter I will focus on Marsilio Ficino (1433–1499) and Andrea Cesalpino (1519–1603), who during the fifteenth and sixteenth centuries redefined a few key pieces of the puzzle through their original reading of, respectively, Plotinus and Galen. Section 19.4 of the chapter, finally, will briefly concentrate on Johannes Kepler and his attempt to link the activity of the vegetative soul, understood as an instinctual response, to a broader power in the cosmos, i.e., 'sublunary nature'.

[4] See Goclenius 1590; Wolff 1732; Wolff 1734. On the *Parva naturalia* as the part of the Aristotelian encyclopedia in which issues of plantanimal life were especially investigated during the Renaissance, see Bydén 2018. On the early modern evolution of the discipline dealing with the 'soul', see Vidal 2016 [2006].

[5] On Glisson's contribution to the early modern notion of vegetative power, see Temkin 1964; Henry 1987; Hartbecke 2006; Giglioni 2008, and Daniel Schmal in this volume.

19.1 A Plant's-Eye View of the Universe: Plotinus and Ficino

One of the most perceptive and sophisticated analyses of what being vegetative means can be found in Plotinus's *Enneads*, more specifically in Ennead 3, Treatise 8.[6] Here Plotinus, though well acquainted with the Aristotelian metaphysics of the intellect, follows Plato's puzzling position in the *Timaeus*: plants are animals, and like animals, they perceive (77B).[7] While developing the Platonic thesis, Plotinus maintains that the capacity to contemplate – θεωρία – is diffused at every level in nature. The difference between animals and plants consists in different degrees of intuitive clarity. Plants, Plotinus seems to say, while incapable of knowing things, are able to attain a glimpse – an 'image' (μίμησις καὶ εἰκών) – of θεωρία. The vegetative operations of nature indicate therefore that a different kind of perception and understanding is at work. Plotinus calls this perception 'σύνεσις τις ἢ αἴσθησις', some sort of sagacity and sensation (Plotinus 1966–1988, III, 370).[8]

The fifteenth-century Italian philosopher Marsilio Ficino, who first translated Plotinus's *Enneads* into Latin between 1484 and 1486, addressed the question of plant sagacity in his Commentary on the work published in 1492. He described nature, understood as a 'universal and primary' power, as a 'substantial and vital form' which transforms matter by applying to it 'a certain natural art', that is, a template of specific patterns of rhythm and proportions which regulates the development of motions and qualities (*moderamen* and *ratio seminalis*). According to Ficino, the works of vegetative productivity occur without any need for nature to rely on conscious decisions (*electio*).[9] This activity of transformation is a 'living and natural art', which acts internally in the substratum to be shaped while being ontologically independent of matter. Life shapes bodies because nature conveys and imprints intelligible patterns in them without requiring the intervention of deliberate and conscious thought. By virtue of its own power, life presides over bodies: it is 'kindred' (*cognata*) to matter – in this respect, matter is *sua*, not *aliena* with respect to life – but the origin remains transcendent, for life draws on a repository of intelligible archetypes.[10]

---

[6] On Plotinus on nature and the vegetative soul, see Wildberg 2009.

[7] On Plato's position on plants in the *Timaeus*, see Skemp 1947; Carpenter 2010.

[8] Plotinus 1966–1988, III, 360–362. Plotinus's enquiry concerns the level of θεωρία which is accessible to plants, trees and even the earth: 'νῦν δὲ λέγωμεν περί τε γῆς αὐτῆς καὶ δένδρων καὶ ὅλως φυτῶν τίς αὐτῶν ἡ θεωρία'.

[9] Ficino 1576, II, 1723. On Ficino and Plotinus see Garin 1974; Allen 1992; Saffrey 1996; Celenza 2002; Gersh 2017.

[10] Ficino 1576, II: 1723: 'Ars haec vivens atque naturalis omnes in se secundum actum naturales continet formas, non aliter quam ars humana possideat artificia. Materia enim quae in potentia formabili semper est ad omnes, non aliter ad actum perducitur omnium quam per vitam materiae quidem praesidentem, sed cognatam, in se omnes actu comprehendentem. Natura quidem haec artifex omnium, neque materia quadam tali vel tali, neque ipsa simpliciter materia ad suam indiget existentiam'.

Natural forms, which Ficino simply characterizes as *rationes* (i.e., specific self-active patterns of regulated causality), are therefore paradigms of action and transformation. They are vital *intelligentiae* and *conceptus*, i.e., natural minds devoid of self-awareness, and yet liable and ready to operate when material circumstances allow them to do so. The works of vegetative production are a manifestation of 'natural intelligence', which for both Plotinus and Ficino is just another way to describe the unconscious art of nature.[11] In this complex discussion, the world of human technology and culture offers a possible model to understand the type of production that is enacted by the vegetative power. According to Ficino, human art in all its manifestations provides examples of artificial productivity in which ideas find their material equivalents increasingly attuned to the intelligible order of things: paintings, sculptures, songs and poems, but also, more simply, images, sounds and words.[12] If these are all fitting embodiments of the expressive faculties of the human mind, the results of nature's craftsmanship are even more outstanding. Nature acts as a vegetative artist every time it enacts vital processes that correspond to tacit and unconscious instructions laid out by a superiour intelligence (*rationes*).[13] The vital *rationes* underlying vegetative activity, however, are very different from all forms of human thinking. Plants do not adjust to their world in the same way in which we conceptualize reality through sensible impressions, mental images and abstract notions. They shape themselves and their immediate surrounding through a full immersion into the countless folds of matter. The more we know with our minds, the more we distance ourselves from nature; plants, by contrast, and all the functions we share with them, remain closer to their matter. We objectify matter; matter, on the other hand, is never an object for plants, but a living component of their own very being.

In this sense, for both Plotinus and Ficino, the difference with human art is that the vegetative power of nature does not postulate a separation between the productive cause and the planned result, between intention and fulfilment, between subject and object. That element of sentience which marks a duality between idea and image in human and non-human animals is absent in the operation of the vegetative faculties, where the immediacy of action takes precedence over the detached waverings of reasoning. Natural forms and functions are self-active images of original ideas that do not require an extra level of cognitive expertise to be implemented. In this sense, these images are unconscious, and are unconscious because they are

[11] Ficino 1576, II, 1724: 'Formae igitur naturales in primis id habent, ut et appellentur et sint intelligentiae quaedam conceptusque expressi ipsius intellectualis naturae sive naturalis intelligentiae. Natura enim ipsa rerum artifex est potestas quaedam viva intellectualis animae, figurans materiam, non alienam, ut ars, sed suam: neque sicut ars extrinsecus, sed intrinsecus: neque electione tali, sed essentia tali ac sine instrumento atque semper'.

[12] Ficino 1576, II, 1724: 'Artificia sane ad visum spectantia dici solent intelligentiae quaedam ingeniaque artificis; multo magis soni musici; magis insuper cantus, quoniam cogitationem expressiorem affectumque vivum secum ferunt; maxime verba sive prolata sive scripta'.

[13] Ficino 1576, II, 1724: 'in quolibet animato non qualitas elementalis, sed vitalis quaedam ratio principium est eorum quae in corpore fiunt'.

active states of being: 'any *ratio* contained in nature is not a mental concept (*notio*), which at some time or other one conceives out of an argument or investigation, but it is an essential attribute of being and the very activity of being, which does not differs from being in such and such a way'.[14] Ficino calls this knowledge immersed in matter and yet capable of expressing things *intuitus*, a 'viewing'. *Intuitus* is also the way in which Ficino translates Plotinus's technical term θεωρία. It's a 'view' that brings with it a certain level of perceptual acquaintance with reality (*quidam sensus*). And like Plotinus before him, Ficino acknowledges that plants may shed light on this kind of vegetative *intuitus*:

> In the seminal nature there is therefore a certain *ratio* of that which originates from it. And this *ratio*, although it is an essential act of nature and its very being, nevertheless, through some kind of translation and transfer is called a view (*intuitus*), which is not acquired at a certain point in time, but is naturally affixed to it. This act of viewing does not look for something, but owns that something from the beginning; it does perceive something in itself, but it is almost senseless, as is the case with people who are in a state of stupefaction, or perhaps with that type of perception (*sensus*) which some people would attribute to plants.[15]

Plotinus had suggested that plants were able to share in the universal θεωρία of nature by elaborating their own 'images' (μίμησις καὶ εἰκών) of this contemplative activity.

Ficino further pursues this line of investigation by scrutinizing the characteristics of vegetative 'contemplation':

> And indeed this nature is the vast plant of the universe which enjoys its own life and conceives natural beings through a peculiar perception that is silent and substantial. And in his hymn to nature, Orpheus assigned to nature this perception without noise (*sine strepitu sensus*), that is, a perception in which the self is not separated from what is different from the self (*distractio sui ad aliud*). No doubt, to view (*intueri*) is the same in nature as being such and such a thing and making such and such a thing. And since nature does this by means of a certain viewing (*intuitus*) and sight (*spectamen*) as it were, that which is generated in matter can be called a spectacle (*spectaculum*), and indeed it can be called a spectacle of nature (*spectamen naturae*), as we are accustomed to say in a child: this mark that is born with him is a kind of desire and imagination by the mother.[16]

---

[14] Ficino 1576, II, 1724: 'Sed ut ad naturam redeamus, id est, vitam ab anima dependentem, in hac sunt rationes omnes formarum naturalium productrices, ipsaque in ea ratio qualibet non notio quaedam est, ex argumento vel indagine quandoque concepta, sed essentialis proprietas atque actus essentiae non differens ab esse tali quodam atque tali'.

[15] Ficino 1576, II, 1724: 'Est igitur in natura seminalis quaedam ratio eius quod inde gignitur, quae quamvis sit actus essentialis naturae ipsius ipsumque suum esse tale, tamen translatione quadam nominatur intuitus non acquisitus aliquando, sed naturaliter illic affixus, non quaerens aliquid, sed ab initio possidens, nec animadvertens in se aliquid, sed quasi stupidus, qualis in attonitis esse solet, ac forte talis quidam sensus, qualem nonnulli plantis attribuebant'.

[16] Ficino 1576, II, 1724: 'Nam et haec est communis universi planta per suum quendam quietum substantialemque sensum sua vita fruens naturaliaque concipiens; atque hunc sine strepitu sensum, id est, sine distractione sui ad aliud Orpheus naturae tribuit in hymno naturae. In natura quidem intueri nihil aliud est quam esse tale et tale quiddam facere. Et quoniam quasi quodam intuitu et spectamine facit, ideo quod gignitur in materia spectaculum dici potest, atque ita insuper

When nature contemplates, the difference between the act of knowing and the known thing disappears. Seminal reason is the illustrious name given to this vegetative sense of identity and otherness, an 'essential and effective viewing (*intuitus*)' that alters things while looking at them, just as the gaze of the pregnant mother may affect the development of the embryo.[17]

19.2 The Inner Touch of the Vegetative Soul: Galen and Cesalpino

At the beginning of my chapter, I suggested that we may call plantanimal imagination the primal coalescence of viewing (θεωρία, *intuitus*) and fashioning that is at work in all the natural phenomena of nutrition, growth and reproduction, evident in plants but also in certain areas of animal physiology. The plantanimal entity – half plant, half animal – helps us conceptualize this type of knowledge that is the knowledge of life. As we have seen, both Plotinus and Ficino say that this knowledge is a kind of perception – σύνεσις τις ἢ αἴσθησις, i.e., *sensus quidam*. One of the keys to interpret this complex issue lies in the sense of touch. Of all the senses, touch is the one which is the most immersed in matter and the most immersive in the way in which it probes the suitability and unsuitability of a potential object, starting with food as the primal object. As such, touch is the threshold between *vegetatio* and *sensatio*. According to the late Aristotelian commentator John Philoponus (c.490–c.570), to be provided with the sense of touch does not make a living thing an animal. Zoophytes who have the sense of touch are more plants than animals, and yet they have the sense of touch.[18] They need this sense to deal with the way in which they adjust themselves to their environment, but also to process their food. In this respect, the act of eating is an act of original touch.

In antiquity, Galen was the author who came the closest to provide a possible definition of this hybrid perception, a plantanimal and tactile acquaintance with reality. His treatise on the multiple operations of the vegetative power – *De naturalibus facultatibus* – is a comprehensive account of nutrition and digestion, with a large number of observations concerning the natural and involuntary operations of life. Galen's most interesting remarks on the possibility of a vegetative perception, however, are to be found in three texts he composed at the end of his life: the final two chapters of the autobiographical treatise *De propriis placitis* ('On My Own

spectamen appellari naturae, ut solemus in puero dicere signum hoc huic ingenitum est quaedam concupiscentia et imaginatio matris'.

[17] Ficino 1576, II, 1724: 'Intellectus autem et natura, quae a substantia intellectuali dependet, naturaliter possident, imo sunt illa ipsa quae intuentur et faciunt intuendo, quem quidem essentialem efficacemque intuitum appellat seminariam rationem'. On seminal reasons in the Renaissance, see Hirai 2005.

[18] Philoponus 2005, 14: 'Indeed, even the very sharing in sense does not thereupon make an animal. Things, at least, that share only in the sense of touch are not animals, for zoophytes share in touch'.

Opinions'), which became known during the Middle Ages and the Renaissance with the Latin title of *De substantia naturalium facultatum* ('The Substance of the Natural Faculties'); a short introduction (a 'synopsis') to Plato's *Timaeus*; and a commentary dealing with the medical sections of the same Platonic dialogue, a work known in antiquity as *On the Medical Statements in Plato's Timaeus*. From these works, it is apparent that in the last part of his intellectual career Galen was becoming increasingly concerned with the problem of how to explain the elusive nature of this hybrid *sensus*.[19]

While Galen's synopsis of the *Timaeus* survived in a medieval Arabic translation, we have only fragments in Greek and Arabic of the Timaeic commentary. Working on a Greek manuscript which is now lost, the Venetian humanist Agostino Gadaldini (1515–1575) translated the extant fragments of the commentary into Latin, which he published in 1550. The excerpts from Galen's *De propriis placitis* had instead already been translated at the beginning of the fourteenth century by Niccolò da Reggio. Then, during the sixteenth century, new translations appeared by Johann Winter von Andernach (1505–1574) and Vittore Trincavelli (1496–1568).[20] Of this material, two Galenic loci are particularly relevant for the topic of this chapter. One is from the fragment on the *Timaeus*, in Gadaldini's translation. Here Galen agrees with Plato that plants have the ability to 'recognize' (*agnoscunt*) what is known (*familiare*) and what is foreign (*alienum*), and this ability, says Galen, can be called 'a perception of a specific and peculiar kind' (*genere proprio ac peculiari sensus*).[21] The other relevant excerpt is in *De substantia facultatum naturalium*. In the passage, Galen distinguishes between two types of perception: one is a kind of knowledge that specifically belongs to the senses (*idipsum cognoscere ad sensitricem facultatem pertineat*); the other is a *vis*, a power, that discriminates suitable from unsuitable substances in the process of nutrition (*vis quaedam qua congrua ab alienis discernantur*). In his translation from the original Greek, Trincavelli significantly uses the phrase *naturalis anima*, 'natural soul', to denote the type of knowledge that, in Galen's opinion, is different from the conscious power of the five external senses, and identifies and attracts that which is suitable to life while transforming it into a part of the body.[22]

---

[19] On Galen's *De substantia naturalium facultatum*, see Nutton 1987; Nutton 2003; Fortuna 2014. On his synopsis and his commentary on the *Timaeus*, see Das 2013.

[20] Galen 1528; Galen 1550a; Galen 1934; Galen 1951. On Agostino Gadaldini, see Garofalo 2004; Petit 2007; Fortuna 2012, 399–401.

[21] Galen 1550b, 111: 'Ostensum enim a nobis est in commentariis de facultatibus naturalibus stirpes facultatem eam possidere qua tum familiares substantias a quibus nutriuntur, tum alienas a quibus laeduntur agnoscunt, ideoque familiares quidem attrahere, alienas vero avertere atque propulsare. Ob id ergo et ipse Plato stirpes et genere proprio ac peculiari sensus participes esse dixit. Nam et familiare et alienum agnoscunt'.

[22] Galen 1550a, 177–178: 'Nam naturalis anima nequit cuiuspiam alterius cognitionis, quae ad sensus spectet, esse participes, utpote quae visiles qualitates, vel auditiles, vel gustatiles, vel olfactiles, vel tactiles nequaquam possit apprehendere, sed illas tantum quae his alere possunt, vel secus, insunt. Id siquidem ea attrahit quod possit illi esse alimento, id continet, id concoquit, id demum in illud transmutat quod affine sit et familiare rei quae alitur: ea vero quae alere nequeunt

'There is a different kind of perception in plants' ("Ετερόν ἐστι γένος αἰσθήσεως ἐν τοῖς φυτοῖς)', thus Galen summed up his position in *De propriis placitis*.[23] The thesis became a most important harbinger of new ideas when Renaissance authors began to interpret this part of the Galenic canon. An Aristotelian who in the late Renaissance recognized the importance of mediating Plato (the 'Timaeic' Plato) with Galen (Galen as a commentator of Plato) was Andrea Cesalpino. His main contributions belong to the fields of medicine and botany. In his *Quaestionum medicarum libri duo* ('Two Books of Medical Questions'), published in 1593, Cesalpino explicitly referred to Galen's *De substantia naturalium facultatum* to emphasize the link between touch and nutrition.[24] He argued that there cannot be life – and above all, that life would not be able to continue – without assuming that a sense of touch was always at work in all types of digestive and vegetative operations.[25] Contravening one of the most stringent Aristotelian principles, Cesalpino declared that the capacity to perceive was necessary to account for the works of nature (*sensus igitur ad opera naturalia necessarius est*). What was decisive in Cesalpino's argument, however, was that this *sensus* had to be *naturalis*, i.e., unaccompanied by reflective cognition and conscious will, and was a type of perception capable of sensing food (*sensus alimenti*) and all the operations required to process it.[26]

---

repellit. Et propterea recte profecto mihi Plato dixisse videtur stirpes esse sensu praeditas, eo nempe qui congruorum incongruorumve sit, ut hac deinde ratione non absurde animalia nuncupari possint, cum non omnino eo motu careant qui ex ipsis est'.

[23] Galen 1821–1833, IV, 764.

[24] Cesalpino 1593, 226v: 'Tetigit hanc rationem Galenus in libro de Substantia facultatum naturalium in fine, ubi testatur Platonem ob id etiam plantis sensum quendam tribuisse diversum a caeteris, quo familiaria et aliena dignoscant'.

[25] Cesalpino 1593, 226rv: 'Cum nutritio quatuor facultatibus naturalibus perficiatur, scilicet attractione, retentione, concoctione alimenti atque expulsione excrementorum, attractio autem et expulsio sine sensu tactus non fiant, perempto omnino hoc necesse est et nutritionem perire. Quod autem attractio alimenti et expulsio rei inutilis sensu tactus egeat, patet. Non enim attrahit natura alimentum, nisi id familiare esse sentiat, neque expellit aliena, nisi ab iis irritetur. Haec autem sine sensu tactus non contingunt, est enim tactus sensus alimenti 2. De anima, tex. 28. Ob id conclusit Aristoteles 3. De anima, tex. 67. Animalia solo hoc sensu privata mori necessarium esse'.

[26] Cesalpino 1593, 226v: 'Si enim membra nutritionis sensu non dignoscerent quae molestiam exhibent, brevissimo tempore animalia labefactarentur: nam ab excrementis, quae in ipsa quotidie acervantur, ulcerarentur ac putrescerent… Nutritio igitur plantarum absque ulla dignotione alimenti peragitur. Animalia autem nutriri non posse absque sensu alimenti, id est calidi frigidi humidi et sicci, quibus aluntur omnia viventia, et quorum est sensus tactus, quamvis superius ostensum est, iterum ex iis quae in medicina apparent, exactius ostendamus'.

19.3 Perception as the Pulse of Undifferentiated Matter: Harvey

The complex tangle of Platonic, Aristotelian and Galenic motifs underpinning early modern discussions about the cognitive abilities of the vegetative power is at the core of Harvey's discussion of animal reproduction in his *Exercitationes de generatione animalium* ('Exercises on the Generation of Animals'), published in 1651. In this work, he explored the initial stages in the development of life, what he called the *primordium vegetale*, i.e., the original vital principle (Harvey 1766 [1651], 476–477, 479, 482–483). The unique characteristic of this power manifested itself above all in the process of reproduction, which, in line with Aristotelian and Galenic tenets, was for him a seamless combination of nutrition, growth and self-differentiation.

As mentioned earlier in this chapter, Aristotle, Plotinus, Galen and all their numerous followers in late antiquity, the Middle Ages and the Renaissance had recognized the 'artistic' nature of this formative force; Harvey's original contribution lies in his argument that life's ability to replicate itself by growing and shaping itself betrays a surplus of knowledge rather than being the mere actualization of a teleologically pre-determined potentiality. These reactions, Harvey insists, belong to the category of motions and actions that physicians had long called 'natural'; they cannot be controlled by the will or the mind, and yet, Harvey points out, they do not take place without involving a certain degree of sentient engagement, for they are stimulated, irritated and altered by the sense. The consequence is clear in Harvey's mind: 'Everything that through its various motions reacts against any source of irritation and distress, must be endowed with sense', and 'everything that is completely devoid of sense does not seem to be in any way capable of being irritated or stimulated to produce motions and actions'.[27] Such elusive phenomena as inflammation, infection, nausea and vomit are for him incontrovertible evidence that there is in the intestine a kind of cognitive capacity (the *sensus aliquis* of Plotinus, Galen, Ficino and Cesalpino) that does not depend on the brain; in the case of vomit, for instance, a form of obscure perception (*sensus quidam*) located in the stomach is clearly able to discriminate between harmful and healthy substances. Harvey calls this ability 'a certain kind of sensation' that occurred without the exercise of the senses (*sensatio quaedam... sine sensu*):

> Just as there are certain actions and motions whose control and direction do not depend on the brain, and therefore are called 'natural', so we should also say that there is a certain sense of touch (*dari sensum quendam tactus*), which does not refer to the common sense,

[27] Harvey 1766 [1651], 450–451: 'Quicquid enim contra irritamenta et molestias motibus sui diversis utitur, id sensu praeditum sit necesse est... Quicquid enim sensus plane expers est, non videtur ullo modo irritari, aut ad motum actionesque aliquas edendas excitari posse'. I have discussed the nature of Harvey's *sensus* in Giglioni 2016. On the arrival of the 'sensitive plant' on the scene of early modern experimental science, see Giglioni 2018b.

> nor does it communicate in any way with the brain. In a sense of this kind we therefore don't perceive that we feel.[28]

The complementary notions of feeling unaware of its sentient faculty (*sensatio sine sensu*) and primal tactile responsiveness (*sensus tactus*) characterize Harvey's account of vegetative power, which he defines as a nexus of nutrition, growth and reproduction. This activity of plastic construction is not exposed to the decisions of sentient perception because it comes before the very formation of the nervous system. This is what Cesalpino, while commenting upon Galen, had called *sensus alimenti*:

> in any process of nutrition and growth, the juxtaposition of the parts and the digestion and distribution of nourishment to be applied to the parts are equally necessary, and we should not regard digestion and distribution as less of a true nutrition than juxtaposition, for both processes occur through the reception, apposition, gluing and transformation of fresh nourishment.[29]

This incessant shaping of matter presupposes a type of perception that is prior to any organized state of the body. Matter is therefore alive and sentient (not consciously sentient) before it is shaped into organs. In arguing along these lines, Harvey goes against one of the hallmarks of the Aristotelian theory of life, i.e., the necessity for matter to be structured ('organic') in order to be a proper substratum of vital function. Speaking of the vital outline that is first to appear in the developing embryo, i.e., the *colliquamentum*, he credits this particular section of matter with feeling and irritability. Then on the fourth day in the development of the chick, a major transformation (*metamorphosis*) takes place, when inside the egg a shift occurs from plant to animal life (*de vita plantae ad animalis vitam sit transitus*), a shift that is marked by the emergence of a 'leaping' dot of blood (*punctum sanguineum saliens*): 'So small is the beginning of animal life!', Harvey exclaims, and this beginning is activated by 'the plastic force of nature' through principles that are barely visible.[30] The interplay of *colliquamentum* and *punctum saliens* in the formation of the embryo is the anatomical counterpart of the relationship between vegetative power and sentient faculty in the succession of the vital functions. The most extraordinary fact for Harvey is that 'the blood seems to exist before pulsation' (Harvey 1766 [1651], 233). In other words, experimental observations seem to confirm the hypothesis that the plastic and constructive abilities of the vegetative power precede the existence of any specific vital organ, including sense organs. Life is therefore radically fluid; homogeneous, and yet capable of organizing itself.

---

[28] Harvey 1766 [1651], 451: 'Quemadmodum tamen actiones et motus quidam sunt quorum regimen sive moderamen a cerebro non pendet atque ideo naturales appellantur, ita quoque statuendum est dari sensum quendam tactus qui non referatur ad sensum communem, nec cerebro ullo modo communicetur, ac propterea in eiusmodi sensu non percipimus nos sentire'.

[29] Harvey 1766 [1651], 216: 'in omni nutritione et accretione, aequaliter necessaria est partium iuxtapositio et applicati alimenti concoctio ac distributio, neque haec minus quam illa vera nutritio censenda est, siquidem utraque novi alimenti accessu, appositione, agglutinatione atque transmutatione contingit'.

[30] Harvey 1766 [1651], 249: 'Tantillum nempe est vitae animalis exordium quod tam inconspicuis initiis molitur plastica vis naturae!'

It is no accident that in Harvey's descriptions of embryonic development the language is pervasively botanical. The fetus is slowly produced out of fluid matter through the intervention of a plastic power which shapes all parts in the same way in which plants grow out of their seeds. The argument, though, is not simply linguistic and analogical. Harvey maintains that, in order to understand the primordia of life, we need to start from plants and devise a kind of plantanimal anatomy. What he therefore asks the reader to do in the middle of his analysis is only seemingly rhetorical: 'Would you like to have a specimen of these processes? Imagine a little plant with your mind'.[31] More than proposing an exercise in comparative anatomy, Harvey is suggesting that the beginning of life, the *primordium vegetale*, emerges out of an original condition of undifferentiated matter and it is at its most vegetative precisely in that stage of development. If one wants to shed light on the anatomy of the vegetative soul, the path to follow is therefore plant anatomy, for plants hold the secret of animal life.[32] The gradual spreading of ramifications and filaments in the *ovum* is what marks the beginning of the animal, for the future animal unfurls from a plant-like tangle of roots and branches developing inside the *ovum*. In Harvey's model of plantanimal anatomy, the physiological processes that take place in animal bodies follow the patterns of vegetative expansion and vital continuity characteristic of plant anatomy: no specific motor organs that expel, transmit and draw matter can be detected in the way in which the *ovum* naturally drops from the racemus, settles in the womb, grows as a vitellus and covers itself with albumen.[33]

Besides revealing that animals grow like plants, anatomical and embryological observations confirm Harvey's thesis that the vegetative power is not a neatly individuated faculty, but a source of diffuse energy. In this sense, beyond fecund animals and growing plants, it is the earth as a whole that in the first place appears to be endowed with the ability to grow and to shape homogeneous matter. Harvey acknowledges that it is necessary to establish a link between the autonomous power

[31] Harvey 1766 [1651], 191: 'Cupis harum rerum [i.e., *ovarium*, *infundibulum*, *racemus*, *vitella*] specimen aliquod? Finge animo pusillam plantam'.

[32] Harvey 1766 [1651], 251–252: 'Hoc certo constat futuri foetus nihil omnino hoc die apparere praeter sanguineas lineas et punctum saliens venasque illas quae omnes ab uno trunco (quemadmodum iste a punto saliente) propagantur et per totum colliquamentum plurimis fibrarum ramificationibus sparguntur; quae postmodum vasa umbilicalia constituunt, quibus longe lateque disseminatis, foetus demum, prout augetur, ex albumine et vitello sibi alimentum haurit. Harum venarum earumque propaginum vivum exemplar videas in arborum foliis, quorum fibrae omnes a pedunculo oriuntur et ab uno trunco per totum folium diffunduntur'.

[33] Harvey 1766 [1651], 296: 'Multo vero magis constabit animam ei [i.e., the *ovum*] inesse consideranti quo pacto quove motore vitellus rotundus et amplus, a vitellarii racemo abruptus, per infundibulum (exiguum nempe tubulum tenuissima membrana contextum nullisque fibris motoriis instructum) descendat viamque sibi aperiendo, uterum per tantas angustias adeat, ibidemque sese nutriat, augeat, albumine cingat. Cum interea nullum organum motorium in vitellario reperiatur, quod expellat, aut in infundibulo quod transmittat, aut in utero, quod attrahat: neque ovum utero per venas iungatur, ut in ovario; nec per umbilicum ab eo dependeat; ut Fabricius vere asseruit, et clare ad oculum liquet. Quid superest igitur tam ingentia opera cernentibus, nisi ut cum Poeta dicamus: Spiritus intus alit, totamque infusa per artus/Mens agitat molem'. See Fabrizi d'Acquapendente 1621, 9–10.

of vegetative transformation and a wider background of vital energy.[34] More importantly, Harvey's account of the developing embryo presupposes a type of anatomy in which processes prevail over structures, where the *contenta* (the fluids 'contained' inside the organs) are more important than the *continentia* (the 'containing' organs). Indeed, it is a paradoxical anatomy in which the *contenta* are already in place – the blood, for instance – when the *continentia* – the heart and the vascular system – are not formed yet. In Harvey's anatomical model the inner force shapes the outer form from within.

From this point of view, Harvey's embryological model presents the vegetative soul as a power that, by processing food and shaping it into a corporeal substratum, builds its own body without being aware of the plan it is executing (the plan is 'epigenetically' built step by step). Does this soul have some knowledge of the task to perform during the activity? To answer this most puzzling question, Harvey puts forward a daring hypothesis in a supplementary *Exercitatio* devoted to the topic of reproductive fertilization: let's try to imagine that the womb works like the brain, that roots are similar to nerves and that the functions of generation have something of a perception, of a 'conceiving' act: 'that which in the brain is a representation or appetite, this very thing or at least its analogue is stimulated in the womb by coitus, and the result is the generation of the *ovum*'.[35] In other terms, the womb activates the equivalent of the appetitive *phantasma* – some sort of perceptual and discerning activity – and turns that into *ova*, into principles of generation, in the same way in which the brain activates *phantasmata* into *cogitationes* and *intellectiones* (Giglioni 1993).

19.4 The Vegetative Imagination of the Earth: Kepler

There is an evident paradox in Harvey's treatment of the vegetative power: epistemologically speaking, there cannot be instinct, for no knowledge is inborn (*nulla cognitio nobis innata est*: 1766 [1651], 173), and yet the vegetative soul is *indocta sapientia*, i.e., effective knowledge that requires no training. As we have just seen, in the first actions of nature *sensus* is the most necessary ability to discern and separate life from death; awareness, however, can only grow out of memory, and only a certain number of animals can rely on this faculty. What in nature is ontologically

[34] Harvey 1766 [1651], 304: 'Quinetiam foemina potiore iure efficiens videatur... Idque etiam exemplum a terra deductum magis confirmat: peregrina enim semina ad postremum pro terrae natura redduntur. Unde videretur probabile, foeminam plus in generatione efficere quam marem: nam in universo quoque, naturam terrae, quasi foeminam matremque statuunt; coelum autem et solem, et reliqua generis eiusdem, nomine genitoris patrisque appellant. Atque etiam terra sua sponte plurima generat sine semine.

[35] Harvey 1766 [1651], 594: 'quod phantasma sive appetitus est in cerebro, isthuc idem vel saltem eius analogum a coitu in utero excitari, unde generatio, sive procreatio ovi contingat'. On Harvey's epigenesis, see Pagel 1967, 233–248; Pagel 1976, 21–23; Roe 1981, 3–4.

undivided is split into perception and action in the human mind. The great majority of living beings rest on the *indocta ars* and *indocta sapientia* of nature, for nature knows better what to do than any human faculty; above all, nature already knows what it needs to know. This is the paradox that the English anatomist Francis Glisson addressed in *De natura substantiae energetica* ('On the Energetic Nature of Substance', 1672) by introducing his idea of a natural perception. For other physicians and philosophers at the time, instinct did not represent a problem. Indeed, the notion was part of a glorious tradition whose antecedents can be traced to Stoic philosophy. One of the authors who made ample use of the concept of instinct to explain the scope and limits of vegetative power in nature is certainly Johannes Kepler. In his opinion, the moment of vital self-recollection in the construction of harmonious patterns in nature – what in the *Harmonice mundi* ('The Harmony of the World', 1619) he characterized as 'recognition' (*harmoniarum agnitio*) – was performed by a representative faculty that was not entirely aware of its activity. Kepler called this faculty 'sublunary nature'.[36]

In Kepler's discussion of the vegetative power, the scene is broadened beyond the limits of the natural bodies to include both the Earth and the surrounding cosmos. In *De stella nova in pede Serpentarii* ('On the New Star in Ophiuchus' Foot), the very framework of celestial influence (*influentia coelestis forma*) is sufficient evidence for Kepler that sublunary beings are endowed with a receptive disposition. This *passio* betrays the presence of sense perception or a faculty that is very similar to it (*est sensitiva vel sensitivae simillima*). The interplay of action and reaction demonstrates that 'there is a kinship (*cognatio*) between the souls and the faculties of sublunary things, on the one hand, and the nature of the sky, on the other'. Kepler describes this network of influences in terms of 'affinity' (συμπάθεια); it extends from one end of the cosmos to the other, from celestial motions to voluntary faculties, manifesting itself at three levels: heat, motion and reason. The faculty which on the Earth corresponds to the light of the heavens is the natural heat within living things (*vegetabiles res*), able to preserve their life for a sustained period of time. With respect to motion, both celestial bodies and sublunary living things are endowed with moving faculties, including some hint of voluntary (*animalis*) causation.[37] This begenning of 'animal' purposiveness is the instinctual response to celestial influence that was originally embedded in the work of divine creation:

---

[36] Kepler 1937–, VI, 277: 'Nam ea quae videntur oculis, serviunt discursui: harmoniarum agnitio sine discursu est. Sic innatam esse pullo gallinaceo ideam milvi, sciscit I. C. Scaliger; non eam quidem simplicem, sed cum nota fugiendae perniciei.' On Julius Caesar Scaliger and his contribution to early modern discussion on vegetative life, see Giglioni 1999.

[37] Kepler, 1937–, I, 314–315: 'Statuta est enim omnis influentiae coelestis forma, non in actione coeli [...] sed in receptione, et sic in passione naturae sublunaris, quae vel est sensitiva, vel sensitivae simillima. Animis quippe et facultatibus rerum sublunarium cognatio intercedit cum natura caeli; eaque triplex, caloris, motus, et rationis. Primum enim lux caelorum est caloris effectrix, et animae facultatesque rerum vegetabilium suum quaeque calorem in corpore quod informant excitant eumque pro viribus tuentur. Sic stellis additae sunt facultates quae illas movent, ut plurimum quidem materiales seu cum corpore dividuae, tam insitae quam adventitiae (qualis quidem hic

> just as the motive faculties of the stars somehow participate in the mind, for they almost understand, imagine and aim at their path, certainly not by reflecting on it, as we human beings do, but following an inborn power (*ingenita vis*), which was instilled (*instincta*) in them at the very beginning of the creation, so the voluntary (*animales*) faculties of natural things get some understanding of their purpose (by all means, without reflecting on it), and they direct all their actions towards this purpose.[38]

Moving from different disciplinary perspectives, both Harvey and Kepler suggest that we think of vegetative powers in terms of an unconscious formative imagination. However, while Harvey's anatomy of the vegetative functions relies on embryology, Kepler's anatomy of the vegetative soul is grounded in geology and meteorology. In that long section in the *Harmonice mundi* which is a digression on the essence and properties of the vegetative power ('On sublunary nature'), the plastic faculty of the Earth is described in terms of imagination (*animae Telluris imaginatio*).[39] In a similar way, in *De stella nova* Kepler had already compared the body of the Earth reacting to astral influences to the body of a human being whose vegetative functions are momentarily altered by the images of a dream:

> What can be more similar to this phenomenon than the fact that there is a faculty in the earth, capable of perceiving the celestial aspects (*aspectuum coelestium perceptrix*), and that this faculty, after it has been stimulated by some aspect, exudes rainy vapours? In both cases, an immaterial species, which comes across the imagination (*obiecta phantasiae*), sets genital matter in motion.[40]

Kepler elaborates a complex theory of representation in which the time-honoured *species* of human knowledge are the crucial intermediary agents between the intelligible realm of the immutable archetypes (παραδείγματα) and the changeable

---

penes nos est magnetica) sed tamen etiam animales: et hic quoque in terris similiter animae facultatesque rerum vegetabilium corporibus suis insunt vitae motusque causa'.

[38] Kepler, 1937–, I, 315: 'Denique ut facultates illae stellarum motrices sunt mentis quodammodo participes, ut suum iter quasi intelligant, imaginentur, affectent; non ratiocinando quidem, ut nos homines, sed ingenita vi et quae in prima creatione ipsis est instincta: sic facultates animales rerum naturalium obtinent quendam intellectum finis sui (sine quidem ratiocinatione) in quem omnes suas actiones dirigunt. Tanta cum sint cognatio, nihil amplius est cur mireris eam συμπάθειαν, inter caelorum motus et facultates animales, quam supra manifestis experimentis confirmavi cap. VIII. et X.'.

[39] Kepler, 1937–, VI, 269.

[40] Kepler, 1937–, I, 317: 'In Telluris globo gemina est facultas (quod meteora attinet), altera attrahendi aquas marinas in occultas concoctionis sedes, altera expellendi vapores concoctos per quendam quasi sudorem; aut si mavis, compara hanc expulsionem excretioni seminali animantium. Harum facultatum alteram expultricem, stimulari ab aspectibus planetarum harmonicis certissimum est: qua stimulatione efficitur, ut continua quidem (in fluminum generatione) at non perpetuo uniformis sit expulsio, sed multis intervallis interrupta eius vehementia, prout aspectus ordine astronomico inciderint. Et me hercule non absurde quis huic excretioni etiam voluptatem suam adiunxerit: ita multa terrae cum animantibus conveniunt. Eatenim docent medici, si quando humore genitali tument venae, facile vel per somnum, imagine dulci obiecta, nullo etiam contactu accedente, fieri excretionem. Quid vero huius rei similius quam quod constat inesse in terra facultatem, aspectuum coelestium perceptricem, quae stimulata aliquo aspectu, exsudet vapores pluvios? Utrinque species immateriata, obiecta phantasiae, ciet materiam genitabilem: quid impediet igitur utrinque ex perceptione speciei et expulsione materiae existere voluptatem?'.

domain of sublunary likenesses (εἰκόνες).[41] What is significant here is that for Kepler, even at the level of vegetative power, the mediation of *species*, the representations, is necessary to convey and transmit information from the intellect (including astral intelligences) to the natural functions of the soul (including the soul of the world), from disembodied ideas to the embodied processes of life on earth.

With Kepler's cosmic anatomy of the vegetative soul, it becomes evident that, while the principal referents are still plants, Kepler's plants, even more than Harvey's, need a soil, that is, the earth on which plants feed and from which they grow. Kepler's contribution to the history of early modern vegetative power demonstrates how this function is firmly grounded in varieties of botanical and telluric hylozoism (vital materialism). Together with William Gilbert (1544–1603), Kepler is the other great explorer of the soul of the earth during the early modern period. Indeed, I would say that a true hylozoistic – consistently hylozoistic – account of the vegetative soul needs a cosmological backup. The explanatory framework cannot be limited to animals, plants or plantanimals. It has to be a function of astro-biological life.[42]

19.5 Conclusion

Starting with Galen in late antiquity, a number of philosophers and physicians began to investigate the operations of the vegetative soul by commenting upon the section in the *Timaeus* where Plato had provided plants, too, with a perceptual ability (77B). In this chapter, I have taken advantage of this commonplace in philosophical botany to embark on an investigation into the nature of plantanimal imagination. By this I mean a way of sensing and representing reality that animals are deemed to share with plants when both fulfil their basic tasks of nutrition, growth and reproduction. I have shown how interest in the unconscious operations of life in plants and animals intensified during the Renaissance and culminated in Harvey's theory of reproduction in *De generatione animalium*, a true watershed in the history of the notion of vegetative power.

Going back to the question which opened this study (if life is knowledge, what kind of knowledge is it?), we may now qualify life's knowledge in a better way by referring to the phenomenon of nutrition. Life presupposes nutrition, nutrition presupposes selection, selection presupposes recognition. Recognition is what allows living organisms to incorporate or reject the selected material. Nutrition also presupposes a sense of touching, since the kind of 'contemplation' (θεωρία) lurking behind recognition and selection is engulfed in matter. As a result, this knowledge is a view of things that is necessarily drowsy and blurry. Plantanimal *sensus*, the

[41] Kepler, 1937–, VI, 270.

[42] On issues concerning life and astro-biology in Kepler, see Boner 2008; Boner 2013; Regier 2014; Christie 2019.

natural *sensus*, is primarily and primordially a *sensus alimenti*, understanding by *alimentum* the original matter that is being metabolized and shaped by a living principle. Likewise, the silent and inarticulate life of plants – blind, deaf and invisibly reactive – is paradoxically closer to the inner senses of the imagination than the external senses.

As pointed out by both Plotinus and Ficino, the actions of the vegetative power are still occurrences of θεωρία and *intuitus*, with the difference, of course, that the imaging of plants is plastic and confined in matter, more similar to immediate reactions of pain and pleasure than conscious sensations of sight and hearing. When plants 'imagine', they grow by assimilating nutriment from the soil and the air; when animals – including the human animal – imagine, they release themselves from the pressure of the soil and the air. These are two very different ways of imagining: one is a path to abstraction and freedom from matter, the other is a path to immersion and concentration into matter. Plato claimed that plants were animals. Harvey, by relying on a new and original way of combining Aristotelian science with Galenic medicine, intimated that plants were superior animals, for they were pure, unmediated enactments of vegetative life. As such they were showing the way towards a possible overcoming of the dichotomies of the intellect. Harvey extolled 'the higher and divine nature of the vegetative soul' and praised its extraordinary works, which surpassed the power of our intellect as much as the gods surpassed human beings.[43] Gods were immortal, and like the gods, plants had the power to regenerate themselves indefinitely. According to Harvey, though, this was not the main title of honour for the vegetative soul. The immersive knowledge of plants and the perfection of all vegetative operations involved an ever greater connection between nature and the mind, free from the uncertainties and qualms of animal perception. Here it may be worth reminding ourselves that the adjectives 'vegetal' and 'vegetative' do not indicate the same thing. Cabbages and cats share the property of being alive – *vegere*, *vigere* and *vegetare* in Latin, i.e., to arouse, invigorate, enliven, quicken and animate (cf. ὑγιής, i.e., 'sound', in Greek) – but vegetables are more vegetative, i.e., more *alive* (*vegeti*) than animals. Compared to animals, they seem to have a greater ability to regenerate. The Italian idiomatic phrase 'vivo e vegeto' means that one is more than simply alive: one is full of energy.

This chapter has shown us that one of difficulties in dealing with the nature of the vegetative power is that to delve into the meanders of plantanimal imagination means to part with the familiar world of consciousness. Plotinus, Galen, Ficino, Cesalpino, Harvey and Kepler all demonstrate in different ways that any discussion concerning the possibility of a plantanimal imagination tests the limits of perception in general. It is as if, besides the traditional division into external and internal senses, along with their well-known series of functions (sight, hearing, smell, taste and

[43] Harvey 1766 [1651], 297: 'Verum hac de re postea uberius dicetur, cum de anima embryonis generatim disputabimus et de vegetativae animae praestantia et numine ex admirandis ipsius (non sine providentia, arte et intellectu divino) operibus agemus, quae profecto intellectum nostrum non minus quam dii homines superant adeoque omnium consensu admirabiles sunt, ut obscura mentis nostrae acies lumen tam impervestigabile nullo modo penetrare queat'.

touch, on the one hand, and common sense, imagination, memory, estimative and cogitative powers, on the other), we discover the existence of a parallel series. Rather than emerging from matter and evolve towards the light of the intellect, this serie goes deeper into matter and reaches the point where the difference between animate and inanimate natures establishes itself. These are the inner senses of nutrition, which in fact can only be labelled 'senses' in an equivocal way because they go underneath the very threshold of sense perception. As already noted, all the authors who tried to enquire into this type of representative activity inevitably resorted to the customary landscape of conscious perception to describe this submerged life of the senses: θεωρία, *intuitus*, *tactus*, *conceptus*, *sensus quidam*, *sensatio sine sensu*. Even when dealing with the unconscious power of instinct, Kepler mentioned the *species* of the imagination. The fact is that the inner senses of nutrition are inarticulate; they betray obscure feelings of pleasure and pain, nausea, stupefaction and drowsiness.[44] Kepler is the author who went so far as to ascribe a sort of telluric imagination even to the Earth in order for the planet to regulate by itself its complex physiology of meteorological transformations. Interestingly, Kepler's radical stance resulted from embracing the gist of Plotinian arguments concerning life and the intellect. The plant's-view of the universe (θεωρία, *intuitus*) allowed him to advance the possibility that the *species*, the representations of things, had a much wider role than the animal *species* characteristic of animal perception. In this respect, Kepler's *species* are the same as Ficino's *rationes*. Moreover, the link with the soil is reinforced and becomes crucial: it's not just the plant that imagines; perhaps even the Earth as a whole complex organism has the power to form representations of what is necessary for its own survival. And like Harvey at the end of his career, Kepler suggested that life is grounded in primal forms of conception, suitable for both the mind and the organs of generation.

References

Allen, Michael J. B. 1992. Summoning Plotinus: Ficino, smoke and the strangled chickens. In *Reconsidering the Renaissance*, ed. Mario A. Di Cesare, 63–88. Binghamton: Center for Medieval and Early Renaissance Studies. (repr. in M. J. B. Allen, *Plato's third eye: Studies in Marsilio Ficino's metaphysics and its sources*. Aldershot: Variorum, 1995).

Boner, Patrick J. 2008. Life in the liquid fields: Kepler, Tycho and Gilbert on the nature of the heavens and earth. *History of Science* 46: 275–297.

———. 2013. *Kepler's cosmological synthesis: Astrology, mechanism, and the soul*. Leiden: Brill.

Bydén, Börje. 2018. Introduction: The study and reception of Aristotle's *Parva naturalia*. In *The Parva naturalia in Greek, Arabic and Latin Aristotelianism: Supplementing the science of the soul*, ed. Börje Bydén and Filip Radovic, 1–50. Cham: Springer.

[44] I have discussed the relationship between pain and unconscious imagination ('raw imagination') in Elijah Montalto's *Archipathologia* (1614) in Giglioni 2018c, the view of pain as a vital response in Cardano in Giglioni 2018a, and the connection between nausea, natural perception and imagination in Glisson's theory of irritability in Giglioni 2008.

Carpenter, Amber D. 2010. Embodied intelligent (?) souls: Plants in Plato's *Timaeus*. *Phronesis* 55: 281–303.

Celenza, Christopher S. 2002. Late antiquity and Florentine Platonism: The 'Post-Plotinian' Ficino. In *Marsilio Ficino: His theology, his philosophy, his legacy*, ed. Michael J.B. Allen, Valery Rees, and Martin Davies, 71–97. Leiden: Brill.

Cesalpino, Andrea. 1593. *Quaestionum medicarum libri duo, in Quaestionum peripateticarum libri quinque. Daemonum investigatio peripatetica. Quaestionum medicarum libri duo. De medicamentorum facultatibus libri duo*. Venice: Giunta.

Christie, James. 2019. From influence to inhabitation: The transformation of astrobiology in the early modern period. Cham: Springer, 2019.

Das, Aileen R. 2013. *Galen and the Arabic traditions of Plato's* Timaeus. PhD thesis. Warwick: University of Warwick.

Descola, Philippe. 2013 [2005]. *Beyond nature and culture*. Trans. Janet Lloyd. Chicago: The University of Chicago Press.

Fabrizi d'Acquapendente, Girolamo. 1621. *De formatione ovi et pulli*. Padua: Luigi Benci.

Falcon, Andrea. 2015. Aristotle and the study of animals and plants. In *The frontiers of ancient science: Essays in honor of Heinrich von Staden*, ed. Brooke Holmes and Klaus-Dietrich Fischer, 75–91. Berlin: De Gruyter.

Ficino, Marsilio. 1576. *Opera, et quae hactenus extitere, et quae in lucem nunc primum prodiere omnia, omnium artium et scientiarum maiorumque facultatum multifaria cognitione refertissima*, ed. Adam Henricpetri, 2 vols. Basel: Heinrich Petri.

Fortuna, Stefania. 2012. The Latin editions of Galen's *Opera omnia* (1490–1625) and their prefaces. *Early Science and Medicine* 17: 391–412.

———. 2014. Le traduzioni di Galeno di Niccolò da Reggio: Nuove attribuzioni e datazioni. *Galenos. Rivista di filologia dei testi medici antichi* 8: 79–104.

Galen. 1528. *Libri tres. Primus, de facultatum naturalium substantia. Secundus, quod animi mores, corporis temperaturam sequuntur. Tertius, de propriorum animi cuiusque affectuum agnitione et remedio*. Trans. Johann Winter von Andernach. Paris: Simon de Colines.

———. 1550a. *De naturalium facultatum substantia*, Trans. Vittore Trincavelli. In *Galeni Pergameni de naturalibus facultatibus libri tres*, ed. Thomas Linacre and Jacques Dubois, 170–178. Lyon: Guillaume Rouillé, Philibert Rollet and Barthélemy Frein.

———. 1550b. *Fragmentum ex quatuor commentariis quos ipse inscripsit de iis quae medice dicta sunt in Platonis Timaeo*. Trans. Agostino Gadaldini. In *Claudii Galeni aliquot opuscula nunc primum Venetorum opera inventa et excusa*, 107–134. Lyon: Guillaume Rouillé and Philibert Rollet.

———. 1821–1833. *Opera omnia*, ed. Karl Gottlob Kühn, 20 vols. Leipzig: Carl Cnobloch.

———. 1916. *On the natural faculties*. Trans. Arthur John Brock. London/Cambridge, MA: Heinemann and Harvard University Press.

———. 1934. In *Platonis Timaeum commentarii fragmenta*, ed. Heinrich Schröder. Leipzig: Teubner.

———. 1951. *Compendium Timaei Platonis, aliorumque dialogorum synopsis quae extant fragmenta*, ed. Paul Kraus and Richard Walzer. London: The Warburg Institute.

Garin, Eugenio. 1974. Plotino nel Rinascimento. In *Plotino e il Neoplatonismo in Oriente e in Occidente*, 537–553. Rome: Accademia Nazionale dei Lincei. (repr. in E. Garin, *Rinascite e rivoluzioni: Movimenti culturali dal XIV al XVIII secolo*, Bari: Laterza, 1975).

Garofalo, Ivan. 2004. Agostino Gadaldini (1515–1575) et le Galien latin. In *Lire les medecins grecs à la Renaissance: Aux origines de l'edition medicale*, ed. Veronique Boudon-Millot and Guy Cobolet, 284–321. Paris: De Boccard.

Gersh, Stephen. 2017. Analytical study of the Commentary on *Ennead* III. In *Commentary on Plotinus*, ed. Marsilio Ficino, vol. 4, xi–ccxxxi. Cambridge, MA: Harvard University Press.

Giglioni, Guido. 1993. *Conceptus uteri/conceptus cerebri*: Note sull'analogia del concepimento nella teoria della generazione di William Harvey. *Rivista di Storia della Filosofia* 48: 7–22.

———. 1999. Girolamo Cardano e Giulio Cesare Scaligero: Il dibattito sul ruolo dell'anima vegetativa. In *Girolamo Cardano: Le opere, le fonti, la vita*, ed. Marialuisa Baldi and Guido Canziani, 313–339. Milan: Angeli.
———. 2008. What ever happened to Francis Glisson? Albrecht Haller and the fate of eighteenth-century irritability. *Science in Context* 21: 465–493.
———. 2016. Sentient nature and the great paradox of early modern philosophy: How William Harvey and Francis Glisson reinterpreted Aristotelian φύσις. In *Natureza, causalidade e formas de corporeidade*, ed. Adelino Cardoso, Marta Mendonça, and Manuel Silvério Marques, 9–28. Ribeirão (Vila Nova de Famalicão): Edições Húmus.
———. 2018a. If you don't feel pain, you must have lost your mind': The early modern fortunes of a Hippocratic aphorism. In *Et amicorum: Essays on Renaissance humanism and philosophy in honour of Jill Kraye*, ed. Anthony Ossa-Richardson and Margaret Meserve, 313–337. Leiden: Brill.
———. 2018b. Touch me not: Sense and sensibility in early modern botany. *Early Science and Medicine* 23: 420–443.
———. 2018c. The temperament and the imagination in Elijah Montalto's *Archipathologia* (1614): A Galenic account of mental illness in the Renaissance. In *Dor, sofrimento e saúde mental na* Arquipatologia *de Filipe Montalto*, ed. Adelino Cardoso and Nuno Miguel Proença, 85–127. Ribeirão (Vila Nova de Famalicão): Edições Húmus.
Goclenius, Rudolph, the Elder. 1590. ΨΥΧΟΛΟΓΙΑ: *hoc est, De hominis perfectione, animo et in primis ortu huius, commentationes ac disputationes quorundam theologorum et philosophorum nostrae aetatis*. Marburg: Paul Egenolph.
Granger, Herbert. 1985. The *scala naturae* and the continuity of kinds. *Phronesis* 30: 181–200.
Hartbecke, Karin. 2006. *Metaphysik und Naturphilosophie im 17. Jahrhundert: Francis Glissons Substanztheorie in ihrem ideengeschichtlichen Kontext*. Tübingen: Niemeyer.
Harvey, William. 1766 [1651]. *Exercitationes de generatione animalium, quibus accedunt quaedam De partu, De membranis ac humoribus uteri, et De conceptione*, 160–603, in *Opera omnia*. London: William Bowyer.
Henry, John. 1987. Medicine and pneumatology: Henry More, Richard Baxter, and Francis Glisson's *Treatise on the energetic nature of substance*. *Medical History* 31: 15–40.
Hirai, Hiro. 2005. *Le concept de semence dans les théories de la matière à la Renaissance: De Marsile Ficin à Pierre Gassendi*. Turnhout: Brepols.
———. 2007. The invisible hand of God in seeds: Jacob Schegk's theory of plastic faculty. *Early Science and Medicine* 12: 377–404.
Kepler, Johannes. 1937–. *Gesammelte Werke*, ed. Walther von Dyck, et al., 19 vols. Munich: Beck.
Kohn, Eduardo. 2013. *How forests think: Toward an anthropology beyond the human*. Berkeley/Los Angeles: University of California Press.
Lloyd, G.E.R. 1996. *Aristotelian explorations*. Cambridge: Cambridge University Press.
Nutton, Vivian. 1987. Galen's philosophical testament: 'On my own opinions'. In *Aristoteles – Werk un Wirkung. Band II: Kommentierung, Überlieferung, Nachleben*, ed. Jürgen Wiesner, 27–51. Berlin: De Gruyter.
———. 2003. A translation of Galen's *De substantia virtutum naturalium* by Niccolò da Reggio. In *Medizin in Geschichte, Philologie und Ethnologie: Festschrift für Gundolf Keil*, ed. Dominik Groß and Monika Reininger, 321–332. Königshausen & Neumann.
Pagel, Walter. 1967. *William Harvey's biological ideas: Selected aspects and historical background*. Basel: Karger.
———. 1976. *New light on William Harvey*. Basel: Karger.
Petit, Caroline C.L. 2007. Gadaldini's library. *Mnemosyne: A Journal of Classical Studies* 60: 132–138.
Philoponus. 2005. *On Aristotle: On the Soul, 2.1–6*. Trans. William Charlton. London: Duckworth.
Plotinus. 1966–1988. *Enneads*, ed. A. H. Armstrong, 7 vols. London/Cambridge, MA: William Heinemann/Harvard University Press.

Regier, Jonathan. 2014. Kepler's theory of force and his medical sources. *Early Science and Medicine* 19: 1–27.

Roe, Shirley A. 1981. *Matter, life, and generation: Eighteenth-century embryology and the Haller-Wolff debate*. Cambridge: Cambridge University Press.

Saffrey, Henri-Dominique. 1996. Florence 1492: The reappearance of Plotinus. *Renaissance Quarterly* 49: 488–508.

Skemp, J.B. 1947. Plants in Plato's *Timaeus*. *The Classical Quarterly* 41: 53–60.

Sprague, Rosamond Kent. 1991. Plants as Aristotelian substances. *Illinois Classical Studies* 162: 221–229.

Temkin, Owsei. 1964. The classical roots of Glisson's doctrine of irritation. *Bulletin of the History of Medicine* 38: 297–328.

Vidal, Fernando. 2016 [2006]. *The sciences of the soul: The early modern origins of psychology*. Trans. Saskia Brown. Chicago: The University of Chicago Press.

Wildberg, Christian. 2009. A world of thoughts: Plotinus on nature and contemplation (*Enn*. III.8 [30] 1–6). In *Physics and philosophy of nature in Greek Neoplatonism: Proceedings of the European Science Foundation Exploratory Workshop*, ed. Riccardo Chiaradonna and Franco Trabattoni, 121–143. Leiden.

Wolff, Christian. 1732. *Psychologia empirica methodo scientifica pertractata*. Frankfurt/Leipzig: Officina libraria Rengeriana.

———. 1734. *Psychologia rationalis methodo scientifica pertractata*. Frankfurt/Leipzig: Officina libraria Rengeriana.

Chapter 20
"Vegetative Epistemology": Francis Glisson on the Self-Referential Nature of Life

Dániel Schmal

Abstract The aim of this paper is to examine Francis Glisson's theory of perception insofar as it concerns the lowest class of living beings: plants. Plants have a special status, they are located between inanimate objects and animals in the hierarchy of being. Unlike the former, they are organic, but unlike the latter they are unconscious. Peculiar to Glisson is the claim that vegetative organization requires *self-referential perception*. In light of traditional epistemology, this claim may sound puzzling, because we tend to associate self-representation with conscious thought. In order to make sense of Glisson's self-referential, but unconscious perceptions, the first part of the paper investigates the role of perception in the vegetative process of organization. The second part is a comparative study on two ways of drawing the distinction between unconscious perceptions and conscious sensations. The first one is the Cartesian approach as outlined by Antoine Arnauld in the early 1680s, the second one is Glisson's theory. In light of the comparison it is argued that Glisson's is a higher order theory of sensation in which the phenomenon of consciousness requires the delimitation of the all-inclusive, self-referential nature of perception.

Francis Glisson's medical and philosophical work has been the subject of in-depth examinations over the last half century or more. Thanks to the researches of Walter Pagel, John Henry, Guido Giglioni, François Duchesneau and others, much of the intellectual background of his thought and its impact on the history of science has

D. Schmal (✉)
Peter Pazmany Catholic University, Budapest, Hungary

Research Center for the Humanities, Institute of Philosophy
(Eötvös Loránd Research Network), Budapest, Hungary
e-mail: schmal.daniel@btk.ppke.hu

F. Baldassarri, A. Blank (eds.), *Vegetative Powers*, International Archives of the History of Ideas Archives internationales d'histoire des idées 234,
https://doi.org/10.1007/978-3-030-69709-9_20

been clarified.[1] One of the remaining points where something can be added to the picture, concerns the connection between Glisson's vitalism and the late scholastic philosophy of nature. In this respect, Glisson's ideas are more or less unique in the history of vitalism because of the extraordinary number of references to the scholastic doctrines of his age. It seems as if he attempted to put vitalism on a more solid footing by grafting the Helmontian medicine onto the late scholastic Aristotelian tradition.[2] One of the key issues, where this cross-fertilization of traditions is particularly marked, is the problem of perception. Glisson, in keeping with his vitalist commitments, attributes non-conscious perception to inanimate substances, plants and animal tissues, and once it comes to spelling out the epistemological details of this doctrine, he heavily draws on scholastic resources.

Guidio Giglioni, in his article on Glisson's reception by Albrecht von Haller in the eighteenth century, puts his finger on a difficulty in Glisson's arguments, namely that natural perceptions in his view are both *self-representative* and *unconscious*. He claims that in virtue of this tension "Glisson's vindication of natural perception verges on self-contradiction" (Giglioni 2008, 480). There is indeed a hint of contradiction in Glisson's theory of perception. On the one hand, he maintains that natural perception is a perception of *sui ipsius,* that is perceptions are always self-representative, on the other hand, Glisson declares that inanimate beings are devoid of self-awareness which would be necessary for consciousness. That is why Giglioni raises the question if it is "really plausible to advocate a notion of an unconscious self-perception" (*Ibid*). Taking Albrecht von Haller's point of view, the answer is negative: "[t]he fundamental problem with Glisson's theory of irritability is that it rests on a contradiction" (Giglioni 2008, 488).[3]

The aim of this paper is to examine Glisson's 'vegetative epistemology', in other words his theory of perception insofar as it concerns the lowest class of living beings: plants. Plants have a special status: in the hierarchy of being they are located between inanimate objects and animals. Unlike the former, they are organic, but in contrast to the latter they are unconscious. For these reasons, they form a link between these two domains, organic life and unconscious perceptions. If we are to make sense of Glisson's self-representative perceptions, we need to turn to this intermediate realm and address the question of how plants organize themselves in virtue of their perceptions, even in the absence of conscious sensations. To deepen this question, in the first part of my paper, I concentrate on perceptions as they contribute to the process of vegetative organization. The second part (divided into subsections) is a comparative study focused on two different ways of drawing the

[1] On Glisson's philosophical and medical views, just to mention a few seminal studies, see Duchesneau 1989, 183–196; Giglioni 2002a, b; Hartbecke 2012; Henry 1987; Hutton 2015; Pagel 1953; Pagel 1967; Rey 2014; Thomson 2008, ch. 3. On the impact Glisson made on the history of vitalism and vital reactivity see Giglioni 2008 and Wolfe 2013. For further items see the bibliographies in Henry 1987, 17. fn. 6; Giglioni 2002b, 175. sqq. and Hartbecke 2012, 273. sqq.

[2] On the Helmontian roots of Glisson's vitalism see Giglioni 2002b.

[3] In a personal conversation, Guido Giglioni emphasized that his verdict reflected Albrecht von Haller's perspective.

distinction between unconscious perceptions in plants and conscious sensations in animals. The first path is the Cartesian approach, taken by Antoine Arnauld in the early 1680s. The second is Glisson's theory. By comparing them, I hope to show that Glisson's theory of unconscious perception in plants and other inanimate beings may be implausible by our contemporary standards, but is not based on overt contradiction.

20.1 Perceptions and Vegetative Life

To begin, let us focus on how, according to Glisson, plants develop *mira organisationis varietate* as complex vegetative substances (*Ad lectorem,* 9., Glisson 1672, b2r). In general, he agrees with the traditional view that it is characteristic of all organisms that they preserve more or less the same corporeal structure while the underlying particles that build up or instantiate this structure continue to change and are replaced by new ones (IV. 32., Glisson 1672, 49). This is why the essence of an organization consists in what Glisson calls *confederation (confoederatio)*, a process which consists in bringing together and uniting more or less number of different particles which come to join the body of the plant. The plant's ability to recruit new matter is based on its nutritive power. The underlying process is – to use the Aristotelian term – generation *(generatio),* a substantial change which accounts for the fact that particles coming from the outside assimilate to and become real components of the plant's substance.[4] In the process of assimilation Glisson distinguishes two phases. The first consists in motions external to the plant. It is defined by the fact that the substantial change the new particles have to undergo to integrate into the organic body of the plant is initiated by some external causes. Thus the causes triggering the substantial change, are, at the beginning, *foreign* to the nature of the affected body. In contrast, the second phase of the process is internal. It is an *immanent* activity which, at the end of the generative process, gives rise to a recent form. Thus, while the first phase leads to new dispositions in the material stuff created by the *transitive* activity of the agent *(actio),* the new form itself (which is the accomplishment of the substantial change) emerges from the material stuff by its *immanent* activity *(actus).* This is how plants appropriate new matter through nutrition. Nutrition is a 'motion' in the Scholastic sense of the word which presupposes both external and internal factors.

At an interesting point in his treatise, Glisson illustrates the whole process by comparing it to a military operation.[5] According to him, the way in which the integrity of a substantial form – i.e. the form defining the current identity of a body – is threatened by the chemical change of its underlying particles is analogous to the

[4] Although Glisson uses the traditional language of hylomorphism, in his conception individual substances (such as plants or animals) are only particular *modifications* of the energetic nature of matter. See *Ad lectorem* 12. Glisson 1672, b3r–v.

[5] Glisson seems to be fond of military images. See Giglioni 2002a, 253.

dissolution of a state under the pressure of an invading enemy. As the passage highlights the nature of the process involved, I quote it at length:

> The attack against the present form can be compared to an invasion of a kingdom by an enemy. *Hostile movement* is like the troops of the invading army building fortresses and strongholds for themselves in the kingdom under attack. [...] Their *movements* and *preternatural effects* jointly produce the *idea* of the invading cause, and raise it high up as a flag. This *idea* is no less hostile to the present form than the causes which produced it, because it [i.e. the new idea] urges matter to dismiss the current form and embrace it [in its stead]. To this end it relaxes and weakens the union of the matter with the form. If [external] causes prevail, they gradually drive out the reigning form, but if the current form happens to gain the upper hand, they will be expelled in turn. *Preparatory dispositions* become more and more pronounced until they rise to the point of banning the existing form. It is in that instant that the preparation of the material reaches its climax. Since in the following moment the old form is expelled and the new one is ushered in, it is clear that matter had the determination to take up the [new] form beforehand, otherwise the effect would not have been immediate. (Glisson 1672, 156. *Emphasis added.*)[6]

Let me divide this long and enigmatic passage into three successive parts corresponding to three successive stages in the process. Whereas the last one matches what we have called the *internal* phase of the substantial change, the first two stages relate to its *external* phase, and highlight it in more detail. The first step marks the beginning of the physical interaction between a substance (say the nutrient in the soil) and an external cause (say the organic body of a plant). The latter launches an 'attack' against the former and attempts to destroy its current nature. What external causes do, is to modify the material disposition underlying the ruling form. By affecting the chemical composition of matter which supports the present *eidos*, the destructive factors change the due proportion of the elements, disturbing or blocking their natural activity. Fortresses and strongholds stand for the points of attack where this chemical change begins.

The second step is the most intriguing one. It begins with the *conception of a new idea* in the substance under attack. In speaking of *ideas*, Glisson reasserts his central claim, that matter is not an inert mass, but is full of activity and has true perceptions. The object of bodily perceptions, as we will see in the next section, is first and foremost the actual state of the body itself. It is in perceiving its own present state, that matter comes to conceive a new form, a new nature which would be more appropriate to its dramatically changing physical dispositions. Accordingly, new 'desires' arise in the original stuff, so that the body 'wishes' to become something else and to

[6] X. 29.: *Motus hostilis est hostis invadentis exercitus, hic in regno alieno arces et propugnacula sibi exstruit [...]. Porro, motus et affectiones praeternaturales conjunctim, loco vexilli, ideam causae invadentis erigunt. Hac idea non minus inimica est praesenti formae, quam sunt ejus causae: materiam enim ad suos amplexus, et ad dimittendam formam praesentem sollicitat ideoque illius cum hac unionem quodammodo laxat et debilitat. Hae causae, si praevaleant, tractu temporis formam praesentem expugnant, sin contra haec triumphet, illas foras deturbat. Praeviae dispositiones praevalentes sensim intenduntur, donec ad momentum quo forma prasens expellitur deventum fit. Eo enim instanti determinatio materiae praeparatoria completa fit. Cum enim in proximo momento forma vetus expellatur, et nova educatur, clare constat materiam fuisse in antecessum ad eam formam satis determinatam: aliter enim effectus non immediate consecutus fuerat.*

'take up' or 'instantiate' a new form or *eidos*. The idea of the new form, lurking behind the recently taken dispositions, is compared to the banner of the invaders rising over the land. The third and last step in the process of the substantial change is the *eductio formae,* the bringing forth of a new substantial form which is better suited to the new nature of the matter affected. This step – in contrast to the physical and chemical modifications described in the previous two stages – is a *metaphysical* change, within the substance. It gives birth to a new, triumphant form.

The shift from the physico-chemical level to a metaphysical one reveals that Glisson's account, unlike mechanical explanations of the time, appeals to something more than matter in motion. The new form is not induced externally, as a result of some external impact, but comes about as engendered by the *internal* potentialities of matter. Overall, these three steps clarify the process of organization and correspond to the three basic features that Glisson attributes to matter: *perception, appetite,* and *motion*. Plants have natural perceptions,[7] tendencies and motions which, in the chemical process of nutrition, support and foster organization by raising the 'idea' of the reigning form (the substantial form of the plant) in the pieces of matter recruited from the environment.

By attributing mentality to the whole nature, Glisson extends his model of change to matter in general and provides it with universal scope. At the same time, he does his best to avoid the pitfalls of pan-psychism. He attributes perception to matter, but having stated that matter has *some kind* of perception, he qualifies his claim by drawing a line between *natural* and *animal* perceptions (*Ad lectorem,* 10., Glisson 1672, b2v. sqq). On the one hand, natural perception is just a simple comprehension of some state of affairs, a bare *intentional* act which, in Glisson's view, is directed at itself.[8] Natural perception is self-representative in such a way that the subject and the object – the perceiving body and the body perceived – coincide. In animals, on the other hand, we find *sensations* or *animal perceptions*. The definition of the latter refers back to its natural counterpart, for it is nothing else but a natural perception accompanied by further, higher order acts whose role is to represent the former. What higher order operations represent, to be more precise, is the act of the original perception along with its first order content. In other words, the higher order act represents the fact that *a certain object is represented to the subject*. For Glisson this higher order perception demands a special physical organ (that of the *sensus communis*) which is one of the necessary conditions for sensation. Plants, representing

[7]Guido Giglioni warns the reader not to misconceive Glisson's notion of perception. In order to avoid anthropomorphic connotations he suggests that perception is "not a cognitive function, but, more fundamentally, a vital act". My take on this issue is that cognition, existence and power go parallel in Glisson (much in the same ways as in Spinoza). In light of this parallelism, I agree with Guido Giglioni that animism and anthropomorphism are to be avoided, but I suggest one should resist antropomorphism not by denying cognitive functions to the simple forms of material life but by degrading the nature of cognition in them. Elementary vital acts involve *elementary* cognition (Giglioni 2002a, 254.)

[8]See Glisson's emphatic claim in chapter XV.: *illius objectum enim est entitas propria quae repraesentat se, suas causas et effectus, item omnes influentias aliarum rerum, confoederationes, cooperationes, consensus et dissensus, etc.* (XV.10, Glisson 1672, 215.)

the lowest rank in the order of organic beings, cannot reach that level because they do not have the appropriate structural requirements in their bodies to retain first order perceptions. Without a principle of retention first order perceptions cannot become the objects of higher order acts.

Here we can return to the paradox presented at the outset, because now we are able to articulate it more clearly: according to Glisson, natural perceptions are directed at themselves, but as long as they stand by themselves, they remain unconscious. Without being accompanied by other higher order perceptions, they do not reach the level of animal perception. To get closer to a possible solution to the original puzzle, now we go to the other side of the line that separates plants from the animal kingdom and see how certain perceptions reach the threshold of sensation by transcending their vegetative foundations.

20.2 Arnauld and Glisson on Self-Referentiality

In our contemporary terms, Glisson's conception of consciousness refers to a higher order phenomenon, while representation in itself appears at the lowest level of perception. What needs to be explained in this constellation is how the perception's self-referential nature and consciousness get divorced in this account. To spell out the details of the problem, I am going to set Glisson's views against the Cartesian theory of consciousness developed more than a decade later by Antoine Arnauld in his 1683 treatise *On True and False Ideas (Des vraies et des fausses idées).* What makes this comparative approach promising is that Arnauld, one of the leading Cartesian thinkers of the time, provided us with the most elaborated theory of self-referentiality and reflexion in the late seventeenth century. Nevertheless, as no direct historical link can be established between the two authors, this argumentative move needs some justification. When instead of turning to the vexed question of the relationship between Glisson and Leibniz,[9] or Glisson and Spinoza,[10] I propose to focus on Arnauld, the idea is that the comparison of Glisson's insights with the leading Cartesian theory of his age may bring out similarities and divergences which help clarify Glisson's ideas.[11] Though structural analysis says nothing about the formation of one's ideas, it can reveal some patterns of thought which were commonly available to the protagonists.

Parallel readings of Glisson and Arnauld may show a number of convergent features as well as important differences between their respective approaches to the problem of cognition. For both of them, consciousness is closely related to some

[9] On the history of the speculations about the Glisson-Leibniz connection see Hartbecke 2012, 22.

[10] Glisson and Sinoza were associated as having common tendencies by Henry More already in the late seventeenth century, see Henry 1987, 28–29.

[11] From the vast scholarly literature on the Cartesian theory of consciousness and reflexion allow me to mention just a few items which I have made particular use of: Wilson 1978/2005, ch. IV. 2; Lähteenmäki 2007; Thiel 2011, 43–54; Simmons 2012; Alanen 2003; Nadler 2017.

kind of self-referenciality. The closest term in Arnauld's vocabulary, when talking about self-representation and self-referenciality, is *réflexion*. Although Glisson uses the corresponding Latin form in a much more restricted sense, for the purposes of our comparative study, I will adhere to Arnauld's terminology as a common ground. Arnauld famously maintained that there are two types of reflexion, one 'virtual' and one'express' as he calls them.[12] This is how he set apart the two subclasses:[13]

> Our thought or perception is essentially reflexive upon itself: or, as it is said rather better in Latin, *est sui conscia*. For I do not think without knowing that I think; I do not know a square without knowing that I know it... The second point is that, as well as this virtual reflexion *[qu'on peut appeler virtuelle]* which accompanies all our perceptions, there is also something explicit *[une autre plus expresse]*, which occurs when we examine our perception by means of another perception. (Arnauld 1990, 71.)[14]

Elsewhere I have argued that Arnauld's understanding of the Cartesian *cogitatio* in terms of virtual reflexion and express reflexion goes back to the Scholastic philosophy of his time.[15] This can be relevant to the issue at hand, as Francisco Suárez, who used these terms widely in the late sixteenth and early seventeenth centuries, was also a standard point of reference for Glisson in the *Tractatus de natura substantiae energetica* (he mentions the Spanish thinker more than 200 times in this treatise). Although no direct historical connection is known between the English physician and the French divine, the common link might have been Suárez, who had an enormous impact on the philosophical curriculum both in England and on the Continent. Perhaps the terminological similarities can be explained by the fact that they both were acquainted with Suárez's impressive *oeuvre*.

The meaning of *virtual reflexion* can be highlighted by presenting an example of how Suárez used the term. One of the most telling occurrences can be found in the 25th book of the *Disputationes metaphysicae* where the *doctor eximius* gives a detailed analysis of the way in which *ideas* (*exemplars* that exist in the mind of an artist) are supposed to govern the process of artistic production:

> It must be added, it seems to me, that there is no need for formal reflexion during the time the art-governed external operation is evolving. This can be proved, I believe, in a satisfactory manner, both by reason and experience, because direct representation or mental intu-

[12] Arnauld expressed his view that all modes of the *res cogitans* contain a reflexive element as early as his 1648 correspondence with Descartes. Here he speaks of "simple reflexion which is intrinsic to all thoughts" (*simplici reflexione omni cogitationi intrinseca*, July 1648, in Descartes 1897–1913. V, 213.)

[13] On the two forms of reflexion in Arnauld the basic piece of literature is still Nadler's seminal book on Arnauld (Nadler 1989, 120. sqq). Another important contribution focusing on Arnauld's debate with Malebranche is Moreau 1999. See also Thiel 2011, 52–53.

[14] *[N]otre pensée ou perception est essentiellement réfléchissante sur elle-même: ou, ce qui se dit plus heueresement en latin,* est sui conscia. *Car je ne pense point que je ne sache que je pense. Je ne connais point un carré que je ne sache que je le connais [...] La seconde est qu'outre cette réflexion qu'on peut appeler* virtuelle, *qui se rencontre dans toutes nos perceptions, il y en a une autre plus* expresse, *par laquelle nous examinons notre perception par une autre perception* (Arnauld 2011, 73.) I have slightly modified Stephen Gaukroger's translation.

[15] Schmal 2020.

> ition of the thing, or of the action to be produced, suffices to guide the operation. There is no need to know the exemplar as a formal object *(objectum quod);* it is enough to cognize it somehow, implicitly, in its tendency towards the object, through a latent and virtual reflexion which is included in each and every mental act as something that makes it known to itself not as an object but as the instrument by means of which the cognition is being performed. (Suárez 1856–1878, XXV, 909b–910a.)[16]

What Suárez is emphasizing in this passage is that exemplars do not need to enter the process of artistic production as objects of explicit knowledge – most of the time it is enough that they operate in the background. The reason is that working according to an internal plan is quite different from working after an external model. Consider the case of painting a portrait. In the process of painting the external model must be carefully observed by the artist whose eyes keep moving between the model and the canvas in an attempt to produce a faithful copy. Working after an internal plan, on the other hand, does not require a similar comparison. When the artist follows her imagination, she does not have to observe her mental images or compare the lines drawn on the canvas with her ideas serving as a model. In this case, the agent focuses exclusively on the thing to be produced, while the ideas 'in her head' operate just as 'guidelines' as it were in the process of making. Although they are known *somehow* even in this case, implicitly or as "tendencies driving towards the object", they do not function as explicit models, but remain in the background while the attention is focused on the object. Suárez's point is that although the mental act tending towards the object is a direct one, it is also present to itself 'virtually', and in this sense the act of cognition tending towards the object serves as its *idea*.

Arnauld seems to apply the point to all instances of thought in general. While our thoughts, in most cases, are first and foremost directed towards an ordinary object which is external to the mind, any act of thought is also presented to itself in an implicit or virtual manner. To a certain extent, Glisson's line of thought runs parallel with this reasoning. He also believes that each act of perception has a self-representative component in virtue of which it is, in Arnauld's words, *sui conscia*. In Glisson's phrasing the point is this:

> [I]n virtue of its natural perception, substantial nature perceives itself, its faculties, their operations, and finally the influence produced by external objects in it, the changes they undergo, along with the connection and confederation with them. When perceiving itself, the substantial nature also perceives all these things as affecting it in a certain way. (Glisson 1672, 213.)[17]

---

[16] *Disputationes metaphysicae,* 25. s. 1. n. 39.: *Addendum tamen existimo non toto tempore quo durat operatio exterior procedens ab arte necessariam esse illam reflexionem formalem; quod mihi quidem satis probare videtur experientia et ratio, quia ipsa directa repraesentatio, et quasi intuitus mentis circa rem vel operationem faciendam, sufficit ad dirigendam operationem. Tunc ergo non est necesse ut exemplar cognoscatur tanquam obiectum quod, sed satis est ut implicite et per modum tendentiae ad obiectum aliquo modo cognoscatur, ea tantum implicita et virtuali reflexione, quam quilibet actus mentis in se includit, ratione cuius dicitur cognosci seipso ut quo, quamvis non cognoscatur ut quod.*

[17] XV. 7.: *Natura igitur substantialis sua naturali perceptione 1. percipit se, 2. facultates suas, 3. operationes harum, 4. et ultimo, influxus, alterationes rerum aliarum extra se, necnon connexus, confoederationes, etc. cum iisdem. Percipiendo autem se, alia haec omnia una percipit, nempe ut se aliquo modo afficientia.*

As the passage shows, Glisson also believes that natural perception is self-referential by its very nature: each act of perception has two different objects, itself and the external thing. But unlike Arnauld, he makes it clear that even though the natural perception is self-referential by its nature, *consciousness cannot depend on* self-referentiality. What consciousness does depend on, Glisson claims, is the representation of a first order perception by a different one, a second order perception. Thus his view is that although natural perception represents itself, so long as it remains isolated from other perceptions of the required type, it falls short of full-blown *animal* perception. As he puts the point: "I admit that the natural perception acquires certain knowledge of its own acts *in some way*, but this does not happen through a different act of perception, it happens by the same act with which it perceives the object" (Glisson 1672, 213.)[18] This admission clearly indicates that Glisson, although aware of the problem, sees no tension between the self-referential character of the perceptions and the lack of consciousness. He believes that first order self-representation does not amount to true sensation, because true sensations, as we have seen, demand something more: a new act of cognition directed at the first as retained in the brain. In modern terminology, Glisson's view is a second (or higher) order theory of consciousness. The reason why plants have perception but lack consciousness is that the self-referential nature of their perceptions is not accompanied by higher-order thoughts.

Another point to consider is that although they would agree to ascribe a sort of self-referentiality to even the most rudimentary forms of perception, Glisson and Arnauld emphasize different aspects of the phenomenon. For Glisson the primary feature of a simple idea is its self-relatedness. For him the primordial form of cognition is self-knowledge, and all other types of cognition are just derivative of this as its modifications. Arnauld, by contrast, seems to give priority to the idea's external relation when he states that "I know myself in knowing [all] other things" (Arnauld 1990, 53; 2011, 53.)[19] This difference is not without significance.

The focus on the self-referential element indicates a strong ontological commitment on Glisson's part. Individual substances differ from their environment in virtue of their basic self-relatedness. Giglioni highlights this point succinctly:

> Viewed in this way, the individuation of substance is a kind of centripetal motion. A process of contraction stems from the substantial nature, such that it excludes all that is not immanent to substance. The individuum is a substance which has become thoroughly immanent, by turning within itself and internalizing any link with the outside. This is the end (which is also the terminus) where the movement of substance finds full satisfaction: 'The end of suppositalitas is the enjoyment of its own substantial nature as a complete individual substance' (Giglioni 2002a, 246. See also Glisson 1672, 50.)

Seen from this perspective, the self-representative character of a perceptual act is just the epistemological counterpart of the fundamental nature of a substance, *viz.* its tendency to 'become thoroughly immanent' and 'internalize any link with the

[18] *Fateor itaque perceptionem naturalem suos actus aliquo modo cognoscere, sed non per alium percipiendi actum, atqui per eundem quo percipit objectum. (Ibid.)*

[19] *Je me connais donc moi-même en connaissant toutes les autres choses.* In *Des vraies et des fausses idées,* ch. II.

outside'. In Glisson's view substantiality means self-preservation, self-expression and self-development in this thoroughly internalized way. This fact is captured by Glisson's basic claim that the essence of a substance is *energy (natura energetica).* In contrast to the Neoplatonic wisdom in Cudworth and others who deprive matter of intelligence on the grounds that matter cannot return to itself to make a reflexion, Glisson considers self-relatedness as essential to matter. Not only do certain vital phenomena (such as irritability) arise from the self-referential nature of matter, but the tendency to become self-contained and immanent also manifests itself in its rudimentary cognitive capacities. Hence, according to Glisson, self-representation is the most fundamental characteristic of all perception.

By contrast, in Arnauld self-relatedness is just a virtual aspect of the ordinary perception, even if reflexion accompanies the direct act simultaneously, and in an indispensable way, and cannot be torn from it, because they form a single whole. In Arnauld this complicated structure, integral to each act of perception, also reflects his metaphysical commitments about the thinking substance. He argues that representation, that is the idea's relatedness to something external is the distinctive mark of the *cogitatio,* the fundamental operation of the *res cogitans.* Famously, he maintained against Malebranche that this feature of the perception cannot be reduced to anything more fundamental since this constitutes its *form,* its essential nature, which had been created by the *fiat* of the infinite power of God.[20] Given that object-directed representations are always accompanied by virtual reflexion, perceptual content is not limited to the external object of which they are the representations, because virtual reflexion also displays the fact that the subject is perceiving the external object.

The result is that even though both Glisson and Arnauld take self-relatedness to be necessary even for the most basic form of perception, in Glisson's more internalist account self-representation is not just a virtual companion of the intentional act, but it forms the basic structure of the perception. For the sake of convenience, let me refer to this as 'immanent first order self-representation' as opposed to Arnauld's 'transient first order self-representation'.

Another revealing difference, which immediately goes to the roots of the problem at issue, is that according to Glisson first order self-representation does not imply that the subject is aware of the content of the perception with full consciousness. Arnauld, as we have seen, maintains that in addition to the perception of the external object virtual reflexion provides a kind of 'tacit' knowledge about the subject's perception of the object. On this account, the fact that these two pieces of information are so closely interwoven, seems to be responsible for consciousness. For Glisson, consciousness is not linked to the self-referential nature of perception, rather it is a question of complexity that transcends primary perception. It arises from the cooperation between *different* acts of perception, because for a being to be aware of something, a certain distance must be kept between the cognitive act and

[20] *quand on est arrivé jusqu'à connaître la nature d'une chose, on n'a plus rien à chercher ni à demander quant à la cause formelle* (Arnauld 2011, 52.) See also Arnauld's fifth methodological rule: Arnauld 2011, 50.

its object, so that the act of consciousness gains a certain perspective on the object cognized. Sensation requires that its content be *judged* from a distance. This is how Glisson justifies this claim:

> [S]ensation contains a kind of implicit judgement on the perceived object. For to be aware of one's perceiving *this or that* object – seeing *this* horse, hearing *that* man speaking – is to make a tacit assertion, as it were, about the object perceived (Glisson 1672, 214.)[21]

Consciousness requires judgement, not on an object, but on one's perception of an object. What is going here is that the very function which Arnauld ascribes to the virtual reflexion at the first level, is added to the perception by a separate act at the second level. It would be tempting to say that Glisson attributes to express reflexion the role which, according to Arnauld, is played by virtual reflexion. Nevertheless, this temptation must be resisted. By opting for a second level theory of sensation, Glisson does not attribute the role of Arnauld's virtual reflexion to its 'express' counterpart for that simple reason that, unlike Arnauld, he does not think that second order perceptions just bring into light what has been present in the first-level act 'virtually'. According to the theory of *natura energetica,* it is not the first order reflexion which comes to be developed into a second order content at a higher level.

This last aspect of the comparison brings us to the heart of the problem, so it needs to be explained in more detail. But before moving on to this point, we need a caveat. In light of the above quotation which defines Glissonian sensation as judgement, we must beware of putting too much weight on modern meanings of *consciousness*. Glisson, in fact, does not use the word consciousness and its relatives.[22] When he speaks of sensations, he simply contrasts basic perception and higher cognition; by calling the latter 'tacit assertion', he makes it clear that sensation is a propositional affair. Obviously, it has nothing to do with 'phenomenal consiousness' or 'what-it-is likeness' in our contemporary philosophy of mind. But be it as it may – whether natural perception is conscious or not by our current standards – the main point to focus on is that according to Glisson, primary (or natural) perception cannot make a judgement on itself to produce animal sensation.

20.3 Glisson on Sensory Awareness

In Glisson's *Tractatus* sensory awareness is based on *partial representation*. What is essential to sensation, he says, is precisely its partial character, the fact that it offers a limited point of view. Sensations do not represent the whole process of

---

[21] XV. 8.: *[S]ensus includit quasi implicitum quoddam judicium de re percepta. Est enim quasi tacita assertio de percepto objecto, se sibi conscium esse hoc vel illud se percipere, hunc equum se videre, hunc hominem loquentem se audire.*

[22] This is not surprising since the term *consciousness*, as is known, was not commonly used in its modern philosophical sense during the seventeenth-century. According to Thiel 1991, 79., Cudworth was the first in the English-speaking world to make extensive use of the word in its distinctively philosophical (as opposed to moral) sense, at the time of Glisson.

perception, but only its efficient cause, the external object. Thus the enigma of sensory awareness, the animal's acquaintance with its environment, is closely connected to the birth of this limitation. This needs a bit of unpacking.

In Glisson's view, as is mentioned, an idea is nothing else but a *motion* induced in a substance by an external cause. When talking about motion what he has in mind is not the effect of the impact *(motus factus)*, but the process of change unfolding under the impact of the cause *(motus fiens)*. Because *motus fiens* (the change in the making) is not identical to the effect which begins to exist only when the process of generation is over, what the idea represents is not the final product either, but the *actio moventis,* the action of the cause evolving in the object moved. Thus, at the most basic level of the natural perceptions, an idea is merely the cognitive side of the change which, seen from a physical viewpoint, is nothing but matter in motion. In Glisson's scholastic jargon, ideas represent the whole *ratio objectiva* of the material change. Because of this identity, perceptions encompass everything which is present in the motion: in plants they capture the transition produced by the cause and tending towards a certain goal in the most energetic way (*in actu flargantissimo*).

Not unlike *ideas* in Spinoza, the Glissonian *idea* refers to itself when representing the motion of which it is the idea (XI. 6., Glisson 1672, 167). Thanks to this self-referential character, its content is rich in information:[23] alongside the movement itself, it also extends over the whole dynamism of change, representing, albeit inadequately, both the cause and the target of the process. Although Glisson is not entirely explicit in this regard, the content may be indefinitely wide because the current impression made by the proximate cause is only the final link in a long chain of physical events. By implication, natural perceptions not only encompass all that they contain in themselves, but they grasp external things too to the extent that external beings enter the circle of the perception's self-knowledge, that is to say, inasmuch as external things have an impact on the susbtance.[24] Likewise, each act of natural perception extends to the effects of its own operation to the extent that they depend on its own activity.[25] "By virtue of perceiving itself, substantial nature perceives all these other things as somehow having an effect on itself" (Glisson 1672, 213.)[26]

[23] As far as the scope of the natural perception is concerned, Glisson makes it clear that this kind of self-knowledge is free from any ignorance and error. It seems to cover everything contained in the act itself: *Primo, perceptio naturalis integrum objectum suum, et quicquid in eo continetur, et quo ordine singula se respiciunt, perfecte, hoc est, absque ignorantia aut errore, noscit* (*Ad lectorem* 14., Glisson 1672, c1r).

[24] *Secundo, causas se afficientes in effectis, quamdiu hi sibi inhaerent, intuetur. (Ibid.)*

[25] *Tertio, omnes suas dispositiones seu aptitudines ad formam, ad quam a generante aut alias ordinata est, accurate percipit; ipsiusque ideae formae futurae, methodique qua materia ad operationes ejus accurate accommodetur, exactam notitiam habet.* [...] *Extenditur enim, directe, tantum ad subjectum percipiens. At oblique suam objectivam rationem ampliat: et non tantum ad alterationes in se factas earumque causas, sed et ad effectus a se oriundos, qui in eorum causis, nempe in seipso, relucent, quibusque referas ideam foetus formandi, extenditur* (*Ibid.*, Glisson 1672, c1r–v.)

[26] XV. 7.: *Percipiendo autem se, alia haec omnia una percipit, nempe ut se aliquo modo afficientia.*

By contrast, animal sensation has a narrow content.[27] It is not about itself, it is directed at an external object *as* external. In Glisson's view this fact calls for explanation. As he makes clear, animal sensations are limited, they have a limited point of view upon the elements contained in natural perceptions, because they bracket off certain parts of their content. The limitation is described as a material process. The possibility of developing animal sensations is contingent upon a special organization of matter which enables the animal to preserve past motions and make them the target of some second-order perceptions in the organ of the *sensus communis*.[28] According to all indications, the pivotal point in Glisson's reasoning is that certain representations, present at the level of first order perceptions in the external senses, get lost or blocked at the second level due to the limited capacity of the nerves to transmit information. What this means, on the cognitive side of the story, is that *consciousness depends on the fact that higher order perceptions leave behind self-referenciality*. Only this limitation explains that ordinary sensations are not about themselves or the sense-organs which receive the impact of external bodies. As Glisson says, "this is why animals perceive only the objects, and fail to discern the motion in their organs" (Glisson 1672, 168).[29]

For the sake of more clarity, let us consider the same results from another vantage point. By putting a body into motion, the *motor*, as we have seen, propagates its idea to the body moved.[30] What the receptive faculty of the latter perceives in this process, is the causation of the mover, and not the effect of the collision (not the the motion generated in the body). Glisson's point is that as long as the motion is in progress, the corresponding idea is that of the motor, since "action is a kind of assimilation, therefore it bears the image *(simulacrum)* or the idea of the efficient cause". As motion exists in the moved body, so the idea of the motor is also present in it: "The perceptive faculty of the passive matter necessarily perceives this idea because of its intimate presence in it..." (Glisson 1672, 167).[31] At this point we need to come back to the aforementioned distinction between action and effect, the *motus fiens* and the *motus factus*. This division is crucial for Glisson, because he assumes that the contrast between simple perception and animal sensation depends on the different intensity of these two types of motion. While the unfolding motion is vivid

[27] By *narrow content*, as we will see soon, I mean the content of an act of natural perception, because sense-perception as a whole gains a lot by limiting the information conveyed about itself. This process allows the animal to transgress the narrow limits of its own existence and to become familiar with the world and its inexhaustible richness.

[28] In contrast, natural perception does not require organization: *Dico, naturalem perceptionem ad rationem suam non requirere materiae organizationem, sed consistere posse vel cum, vel sine, eadem* (XV. 6., Glisson 1672, 211.)

[29] XI. 6.: *Haec causa est cur animalia percipiunt sola objecta, at motum in organo factum non discernunt.* See also chapter XV. 10., Glisson 1672, 215–216.: *Sensus enim entitatem propriam organi percipientis, hoc est ipsius partis, qua percipit, non percipit, sed objectum perceptum ut quid extra se contemplatur.*

[30] XI. 6.: *movens motu suam ideam in materia mota absque alia actione satis efficaciter imprimat...*

[31] *Ibid.*: *quam facultas perceptiva materiae passae, propter intimam sibi praesentiam, necessario percipit.*

and energetic (cf. *in actu flagrantissimo,* Glisson 1672, 168), the completed motion is weak (*segnem ideam praebet, ibid*). Accordingly, only the first reaches the 'animal faculty', the center of sense-perception in the animal. Because in the *actus flagrantissimus* the idea of the cause asserts itself in a more dramatic way than the self-knowledge of the perception, animals grasp the idea of the external cause and not that of the motion produced in the sense-organ. What we see again is that the self-referential part of the original perceptual content is fading to the point that only the idea of the external cause remains available to the cognizer.

At first glance, sensation's *distance* and *limitedness* might appear as handicaps as compared to the *transparency* and *immediacy* which characterize natural perceptions. But this appearance is misleading. Glisson makes it clear that sensations, despite or perhaps due to their limited content, take priority over simple perceptions for a number of reasons. While first order perceptions fail to extend directly over anything else but themselves, animal sensations are open to the world. Owing to the fact that they do not give direct access to themselves, they refer to something else which brings the subject beyond the limit of its own existence. What we gain by loosing large parts of the original content is, to use the modern terminology, *intentionality.* Let me underline the double character of this change. On the one hand animals get intentional sensations by transgressing the border of self-knowledge. On the other hand, they cannot do that without taking a distance and limiting the scope of the perceptual content. By linking these two aspects, what we get is that conscious perception can only be born by transcending or eclipsing original self-referentiality.

In Glisson's view, this adds a special perfection to animal *sensation,* a perfection which natural perceptions lack, precisely because of their immanency and interiority. Plants are unable to put their environment at a distance, that is why "natural perception, in relation to its objects, does not reach the perfection of sensation" (Glisson 1672, c1v).[32] The perfection Glisson has in mind arises from an exchange: the limitation of the self-related nature of the self-knowledge opens sensations to an infinite number of objects external to the observer. Leaving behind the *angustia objectiva* (representative narrowness, see *ibid.,* Glisson 1672, c1v) of self-knowledge, animals gain extended perceptual relationship with their environment.[33] A further advantage of sensations over simple perceptions is that, on account of the distance between cognition and its object, the subject is able to suspend its perceiving activity and to turn to other objects. Without distance and limitation there is no free direction of attention.

To sum up, original perceptions belonging to the energetic nature of matter extend to everything that exists or happens in the body of the plant they build up, and exercise control over the organization through the unconscious process of self-knowledge. At the same time – and exactly for the same reason – natural

[32] *Ad lectorem* 14.: *Quare naturalis perceptio objective a ratione sensuum plurimum deficit.*

[33] In Giglioni's version What is gained in the "public" knowledge of the senses is lost in the "private" knowledge of nature. Giglioni 2008, 470.

perceptions remain self-contained, and self-enclosed. They represent other causes as well, but only in relation to themselves. By contrast, animal sensations are perceptions of perceptions. They give the subject a certain freedom *vis-à-vis* the objects perceived, because they do not coincide with them.

20.4 Conclusion

In light of the above, we are in a position to advance an answer to our initial question as to the nature of unconscious perception. The first point to be emphasized is that the extremely rich content which belongs to the rudimentary forms of self-knowledge is, as Glisson frequently points out, always *simple*. Glisson's stress on the simplicity of natural perception suggests a fundamental distinction between the self-referential nature of a perception and its reflexivity. Simplicity does not even allow a virtual distinction between a perception and the act of knowing it. Like the Aristotelian *noesis noeseos,* natural perception is directed towards itself, but its self-related character does not imply any complexity in its essence. In other words, elementary self-knowledge is not rooted in anything like the first-level reflexion in Arnauld, but results from the act's essential transparency to itself.

At first glance, Glisson's natural perceptions have the very same structure as their counterparts in Arnauld, because both Glisson and Arnauld appeal to first order acts which fold back on themselves. Closer scrutiny, however, reveals important differences. One of the most revealing points is that for Glisson, self-representation is a more fundamental characteristic of perception than its relation to the object. If Arnauld, following Suárez's lead, speaks of virtual reflexion, Glisson's approach could be qualified as as 'virtual object-relatedness', that is 'virtual relation to something external'. This kind of virtuality can only be actualized through second order perceptions which represent the initial content without its self-representative features, 'from the outside' as it were. Although natural perception, being identical with its object, is not open to error, it represents the lowest degree of cognition exactly because self-relatedness is essential to it. In the realm of natural perceptions, there are no objects *opposed to, distinguished from* or *placed in front of* the observer.

As far as the 'vegetative epistemology' is concerned, virtual object-relatedness explains why plants remain caught in the immediacy of obscure and indistinct perceptions. Although natural perceptions are transparent in themselves and for themselves, they are not transparent for the sentient subject which, in animals, emerges from the *natura energetica* of matter in a way that allows the animal to distance itself from its own material basis and to stand in opposition to the rest of the world. These considerations suggest that whether one chooses to stick to the Arnaldian term 'reflexion' or replaces it with 'self-relatedness', unconscious self-representation points to a sound epistemological principle: consciousness presupposes intentionality. Without limitation, perspective and a distance between the subject and the object known, there is no consciousness or openness to the world.

Acknowledgments The work on this papaer was supported by National Research, Development and Innovation Office (Hungary): [Grant Number K 116234, K 120375, K 123839, K 125012]. I am particularly indebted to Guido Giglioni for his comments on an earlier version of my chapter.

References

Alanen, Lilli. 2003. *Descartes's Concept of Mind*. Cambridge, MA/London: Harvard University Press.

Arnauld, Antoine. 1990. *On True and False Ideas*, trans. and ed. S. Gaukroger. Manchester/New York: Manchester University Press.

———. 2011. *Des vraies et des fausses idées*, ed. Denis Moreau. Paris: Vrin.

Descartes, René. 1897–1913. *Œuvres complètes*, ed. Charles Adam and Paul Tannery, 12 vols. Paris: Vrin.

Duchesneau, François. 1989. *Les modèles du vivant de Descartes à Leibniz*. Paris: Vrin.

Giglioni, Guido. 2002a. Francis Glisson's Notion of *confoederatio naturae* in the Context of Hylozoistic Corpuscularianism. *Revue d'histoire des Sciences* 55: 239–262.

———. 2002b. *The Genesis of Francis Glisson's Philosophy of Life*. Maryland (PhD Dissertation): Baltimore.

———. 2008. What Ever Happened to Francis Glisson? Albrecht Haller and the Fate of Eighteenth-Century Irritability. *Science in Context* 21: 465–493.

Glisson, Francis. 1672. *Tractatus de natura substantiae energetica seu de vita naturae ejusque tribus primis facultatibus, perceptiva, appetitiva et motiva, naturalibus...* London: E. Flesher.

Hartbecke, Karin. 2012 [2006]. *Metaphysik und Naturphilosophie im 17. Jahrhundert: Francis Glissons Substanztheorie in ihrem ideengeschichtlichen Kontext*. Berlin: Walter de Gruyter (First edition: Tübingen: Niemeyer).

Henry, John. 1987. Medicine and Pneumatology: Henry More, Richard Baxter, and Francis Glisson's Treatise on the Energetic Nature of Substance. *Medical History* 31: 15–40.

Hutton, Sarah. 2015. *British Philosophy in the Seventeenth Century*. Oxford: Oxford University Press.

Lähteenmäki, Vili. 2007. Orders of Consciousness and Forms of Reflexivity in Descartes. In *Consciousness: From Perception to Reflexion in the History of Philosophy*, ed. Sara Heinamaa, Vili Lähteenmäki and Paulina Remes, 177–201. Dordrecht: Springer.

Moreau, Denis. 1999. *Deux cartésiens: La polémique entre Antoine Arnauld et Nicolas Malebranche*. Paris: Vrin.

Nadler, Steven. 1989. *Arnauld and the Cartesian Philosophy of Ideas*. Manchester: Manchester University Press.

———. 2017. Consciousness. In *The Routledge Companion to Seventeenth Century Philosophy*, ed. Dan Kaufman, 310–333. London/New York: Routledge.

Pagel, Walter. 1953. The Reaction to Aristotle in Seventeenth-Century Biological Thought: Campanella, Van Helmont, Glanvill, Charleton, Harvey, Glisson, Descartes. In *Science, Medicine and History. Essays on the Evolution of Scientific Thought and Medical Practice Written in Honour of Charles Singer*, ed. E. Ashworth Underwood, vol. I, 489–509. London/New York: Oxford University Press.

———. 1967. Harvey and Glisson on Irritability: With a Note on van Helmont. *Bulletin of the History of Medicine* 41: 497–514.

Rey, Anne-Lise. 2014. Metaphysical Problems in Francis Glisson's Theory of Irritability. In *The Life Sciences in Early Modern Philosophy*, ed. Ohad Nachtomy and Justin E.H. Smith, 85–97. Oxford: Oxford University Press.

Schmal, Dániel. 2020. Virtual Reflexion: Antoine Arnauld on Descartes' Concept of *conscientia*. *British Journal for the History of Philosophy* 4: 714–734.

Simmons, Alison. 2012. Cartesian Consciousness Reconsidered. *Philosophers' Imprint* 12: 1–21.

Suárez, Francisco. 1856–1878. *Opera Omnia*. Paris: Louis Vivès.

Thiel, Udo. 1991. Cudworth and Seventeenth-Century Theories of Consciousness. In *The Uses of Antiquity*, ed. Stephen Gaukroger, 79–99. Dordrecht: Kluwer.

———. 2011. *The Early Modern Subject: Self-Consciousness and Personal Identity from Descartes to Hume*. Oxford: Oxford University Press.

Thomson, Ann. 2008. *Bodies of Thought: Science, Religion, and the Soul in the Early Enlightenment*. Oxford: Oxford University Press.

Wilson, Margaret Dauler. 1978/2005. *Descartes,* New York: Routledge.

Wolfe, Charles T. 2013. Sensibility as Vital Force or as Property of Matter in Mid-Eighteenth-Century Debates. In *The Discourse of Sensibility: The Knowing Body in the Enlightenment*, ed. Henry Martyn Lloyd, 147–170. Dordrecht: Springer.

Chapter 21
Life in the Dark: Corals, Sponges, and Gravitation in Late Seventeenth Natural Philosophy

Raphaële Andrault

> *And for the essence of vital principles, we are yet more in the dark.*
>
> *As having little more knowledge hereof, than of their operations (Grew 1701, 54)*

Abstract This chapter examines how the borderline cases pointed out by English naturalists and philosophers in the second half of the seventeenth-century call into doubt the common notion of life as a vegetative power. In the first part of this chapter, I focus on Nehemiah Grew's notions of life and living beings by comparing his plant anatomy, in which he examines the cases of sponges and corals, with his physico-theology. In the second part, I confront Grew's views on life to those of John Ray and Ralph Cudworth. My aim is threefold. First, I want to show that there was then no agreement on what a "vital phenomenon" is. Second, I want to emphasize the fact that what we would classify today as physical phenomena, such as gravitation, were then regarded as living phenomena. Third, and accordingly, I would like to challenge the view that the English naturalists and philosophers who then developed a physico-theology simply used empirical vital phenomena in order to refute Descartes' metaphysics. In my view, there were no such things as "empirical vital phenomena." Indeed, the very identification and description of the phenomena considered as "vital" was a matter of interpretation, and those different authors did not share the same set of assumptions regarding what kinds of beings are alive. At the very least, they do not share the set of assumptions that define what *we* regard as "vital."

R. Andrault (✉)
CNRS IHRIM-ENS Lyon UMR5317, Lyon, France
e-mail: raphaele.andrault@ens-lyon.fr

F. Baldassarri, A. Blank (eds.), *Vegetative Powers*, International Archives of the History of Ideas Archives internationales d'histoire des idées 234,
https://doi.org/10.1007/978-3-030-69709-9_21

21.1 Introduction

In the *Essay concerning human understanding*, Locke claimed that we have no "clear distinct settled idea" of what "life" is:

> Life is a term, none more familiar. Any one almost would take it for an affront to be asked what he meant by it. And yet if it comes in question, whether a plant, that lies ready formed in the seed, have life: Whether the embryo in an egg before incubation, or a man in a swoon without sense or motion, be alive or no; it is easy to perceive that a clear distinct settled idea does not always accompany the use of so known a word as that of life is (Locke 1979, III, X, §22).

Locke was not the only one interested in the natural phenomena that questioned our common understanding of the differences between living and non-living things. At the end of the seventeenth century, medical, anatomical, and natural observations were full of borderline cases which called into question the boundaries between animate and inanimate, between plants and animals: the description of corals, sensitive plants, or unconscious patients challenged the usual taxonomy of beings. For example, in 1670, the physician Theodor Kerckring wondered if a patient, a young woman who could no longer feed herself, was still alive: he spoke then about a "*diuturna mors*", a durable death, rather than life (1670, 156). In the same decade, Johannes Swammerdam questioned the similarity between the contractility of muscular fibers in animal bodies and the spontaneous motions of the plant called "impatiens of Dodona" (1758, 592). Those cases challenged the very notion of life, which seems at first sight very simple, but turns to be, after careful examination, dreadfully difficult to delimit and define.

This chapter examines Nehemiah Grew's natural history, which offers an interesting perspective on the taxonomy of living beings in the late seventeenth century. Grew first developed in the 1670s a plant anatomy underlain by the mechanistic idea that there is a strong continuity between physical and living phenomena; he then wrote a catalog of "natural rarities," such as sponges and corals, that agrees with the morphological and chemical approach he endorsed in his plant anatomy. None of these works contains considerations on life or on the boundaries of the animate world. It was only 20 years later, in his apologetic book entitled *Cosmologia sacra,* that he introduced the thesis that there is in nature a vital world that exceeds the powers of matter. Cases of "vegetation", whether in plants or animals, thus demonstrate for Grew the existence of a "vegetative life," distinguished from a "sensible" one. And yet, these notions raise more issues than they resolve. How do they fit the empirical criteria that Grew uses in his natural history? How do the borderline cases with which he deals in his catalog of rarities fall into these different species of life? By addressing these questions, my purposes are twofold: first, I want to question the very notion of "vital empirical phenomena" (Garret 2003, 74); second, I want to emphasize the fact that the notion of life had then no taxonomic value. In the last part of this chapter, I confront my analysis of Grew's natural philosophy to

Cudworth's and Ray's notion of "plastic life," which raises the same kind of problems.[1]

21.2 Nehemiah Grew's Natural History

Nehemiah Grew (1641–1712) was a Calvinist physician, who graduated from Cambridge in 1661, and obtained a doctorate in medicine from Leiden in 1671. From 1661 until 1671, Grew likely practiced medicine in Coventry, before moving to London to become a fellow of the Royal Society thanks to his works on the anatomy of plants. In 1673, he published an *Idea of a Phytologic History,* and in 1682, an *Anatomy of Plants,* which collected previous publications communicated to the Royal Society. Between 1677 and 1680, he was the secretary of the Society, for which he wrote a catalogue of the rarities preserved at Gresham College (1681; see Hunter 1981). After this, he resumed his medical practice in London and Coventry. His last published book, the *Cosmologia Sacra, or Discourse of the Universe as it is the Creature and Kingdom of God* (1701), was not a work of natural history as such, but rather a text of physico-theology which aimed at "demonstrating the Truth and Excellency of the Bible," against the Anti-Scripturists," like "Spinoza and some others." The book is divided into five parts dealing respectively with the corporeal world and "nature's geometry;" the existence of a vital world; Providence and God's government; the "Hebrew Code;" and, lastly, the New Testament. In the first and second books, Grew thus placed his knowledge of natural history at the service of an apologetic undertaking. And yet, the principles that he defended in the *Cosmologia Sacra,* and particularly the three kinds of incorporeal "principles of life," did not fit the natural history he had proposed 20 years earlier.

Indeed, in the *Anatomy of Plants,* which was regarded as a pioneering work, Grew explained plant organism in terms of the mechanical interactions between particles (Arber 1941, 1942, 16; Hunter 1981, 189). According to Grew, "vegetation," that is, the growth and reproduction of plants, is due to operations such as filtration and fermentation, which mainly depend on the size and shape of the organs involved in the process – particularly the shape of the fibers that compose them–, and on their chemical nature. The proportion of the various chemical principles, i.e. nitrous, acid, alkaline and marine salts, is what explains the various faculties and powers of plants. As for these chemical principles themselves, they come from the microscopic "geometry" of nature, that is, from the respective position, shape, and motion of the small invisible particles that compose the various parts of plants. But "Grew's adherence to the mechanical theory of matter did not bring about his rejection of chemical principles" (Clericuzio 2000, 162; see also Roos 2007, 87–8): chemical principles simply correspond with the relevant level of explanation when

[1]On the notion of plastic life in Cambridge Neo-Platonists and on Cudworth, see Hutton in this volume.

the powers of plants, for example the purgative character of rhubarb, are at stake. To illustrate this anti-reductionist position, Grew used the example of the clock: to explain the functioning of a clock, it is not enough to describe its various preexistent parts; one should also describe the exact combination of parts that specifies a clock:

> So the *Principles* of the Purgative Parts of a *Root,* as of *Rhubarb*, although we should suppose them to be existent in the surrounding *Earth,* yet we cannot say, That *that Earth,* or the *Principles* therein contained, are Purgative; but only that they are such, as by being combined together, in such a peculiar way, may become *so.* So the several parts of a *Clock,* although they are and must be all prae-existent to it [...]; yet is it the *setting together* of such *Parts,* and in such a way only, that makes them a *Clock* (1682, "An Idea of a Philosophical History of Plant" §55, 21).

It is only when Grew addresses the question of the formation of roots that he speculates on the invisible components that ultimately account for the actions of chemical principles:

> [The] first Intention of *Nature* is, That some of those *Parts* be disposed to *Rest* [...] And the *Particles* of *Marine Salt,* being *Cubick;* and so, with respect to their *Figure,* of greater *Bulk* than those of any other *Salt;* they will hereby be most and first of all disposed to *Rest;* and so become, as it were, the Foundation of the following *Superstructure.*
>
> The Second Intention of Nature is, That the *Particles* be brought to *Rest,* in a certain *Position,* agreeable to the *Figure* of the *Parts* which are to be formed. And therefore in the next place, all those *Parts* of a *Plant* which are truly *Lignous,* by the *Marine Salt,* with the assistance of the *Alkaline,* but especially of the *Nitrous,* are made to shoot out in Length, or into an innumerable company of small *Cylindrick Fibres* [...]
>
> The next Intention is, That these *Fibres,* at the same time in which they are formed, may likewise receive such a *Posture* as will best answer the indented *shape* of the *Leaf.* (Grew 1682, IV:6 §10–12, 159)

In all cases, the interactions between particles that underlie the formation of the "anatomy of plants" first described by Grew are mechanical. A few paragraphs after the one just cited, Grew took up the Galilean notion that nature is written in a mathematical language: "Thus doth *Nature* everywhere *γεωμετρειν.* For what She appears in Her *Works,* She must needs be also in their *Causes*" (Grew 1682, IV:6 §19, 160). Beyond his work in plant anatomy, Grew adopted a conception of the universe as a "Great Engine," which he deems entirely compatible with "Religion, and the Sacred Scriptures:"

> So *God* may be truly the *Cause* of *This Effect,* although a Thousand other *Causes* should be supposed to intervene: For all *Nature* is as one Great *Engine,* made by, and held in His Hand. And as it is the Watch-makers *Art,* that the *Hand* moves regularly, from hour to hour, although he put no his Finger still to it: So is it the Demonstration of *Divine Wisdome,* that the Parts of *Nature* are so harmoniously contrived and set together; as to conspire to all kind of Natural Motions and Effects, without the Extraordinary and Immediate Influence of the *Author* of it (Grew 1682 II §5, 80).

The harmonious combination of the different parts of nature thus guarantees the regularity and productivity of the physical processes at stake – in this instance, the formation and growth of roots. Grew does not admit any vital principle that would be fundamentally irreducible to the physical interactions of plant particles.

Actually, Grew's experimental studies do not include any reflection on the general criteria for life, nor on the notion of a specifically vegetative life: in the *Anatomy of Plants,* there is nearly no mention of "life", "vitality," or soul of any kind, except in the expression "vital sap," or "vital part," for example, which mean the essential parts of a plant, or the kinds of parts that a certain species of plant has to have (1682, I:2 §23, 15; 1682 I:6 §4, 41). In addition, his plant anatomy is partly grounded in analogies with animals. For example, the description of the ascent and distribution of the "sap," or "vital sap," is supported by analogies with blood and nervous spirit in animals:

> So that as in an *Animal Body* there is no instauration or growth of Parts made by the *Bloud* only, but the *Nervous Spirit* is also thereunto assistant; so is it here: the *Sap* prepared in the *Cortical Body,* is as the *Bloud,* and that part thereof prepared by the *Lignous,* is as the *Nervous Spirit* (Grew 1682, I:2 §23, 15).

In other words, Grew does not regard plants as specific living beings endowed with a vegetative soul or a principle of vegetative life: since they are natural "engines" among others, there is no reason to fundamentally distinguish them from minerals or animals.

In the *Musaeum Regalis Societatis, or a Catalogue and a Description of the Natural and Artificial Rarities Belonging to the Royal Society, and Preserved at Gresham College* (1681), Grew described several natural objects that questioned the limits of plant life: sponges, corals, kermes berries, and cochineal.[2] Regarding sponges, Grew first related the common view according to which a thing is considered as an animal if it is endowed with sense-perception. Grew did not express his disagreement with this common view, but rather questioned the criteria according to which a thing is regarded as endowed with sense. According to Grew, the contractile movement produced by the sponge when it is touched cannot be used as an indication of its animal nature:

> It is the opinion of some, that *Sponges* have sense, because said to shrink, if they are pluck'd; and are therefore reckon'd amongst *Zoophyta.* But of that property I doubt very much. For a *Sponge* being a springy Body, and so extensible, and yielding a little to one that plucks at it; so soon as he lets his hold go, it will, from its elasticity, shrink up again. Which motion of restitution, some probably, have mistaken for the effect of a Cap-sense (Grew 1682, 252).

Elasticity, a property that belongs to every kind of body, suffices to explain the reactive motions of the sponge. Those movements are purely mechanical, and do not presuppose that the sponge has the kind of sense-perception which we observe in animals. The presence of reactive motions as such is not a proof of animalness.[3]

---

[2] Grew describes also different kinds of shells and sea-porcupine, but their animal nature is apparently obvious for him and for his contemporaries.

[3] On this last criterion, see for example Digby: "For speaking rigorously, I cannot allow Plants to have Life. They are not Se Moventia, They have not a principle of motion within them" (1660, 70). Regarding sensitive plant, Webster argues: "the greatest obstacle to the recognition of plant sensitivity was the generally accepted distinction between plants and animals, which had been accepted since the writings of Aristotle and Theophrastus – that plants were characterized by an insensitive

Besides, the structure of the sponge suggests that it is a plant, and not a "zoophyte," or a plant-animal:

> No *Sponge* hath any Lignous Fibers, but is wholly compressed of those which make the Pith and all the pithy parts of a Plant. Yet vastly thicker, and their Texture much more rare or open, so as to be visible to a good eye, especially assisted with an ordinary Glass. So that a *Sponge,* instead of being a *Zoophyton,* is but the one half of a Plant (Grew 1682, 252).

Similarly, Grew mentions the "sensitive plant" without calling into question its inclusion in the plant kingdom. This plant "takes its Name (as is commonly known) from its Imitation of sense or Animal motion. For so soon as you touch the Leaves, they presently fall, till they lie upon the ground. After a while, they rise again; but being touched, fall as before" (1682, 234; on the sensitive plant, see for example Hooke 1665, 124; Webster 1966; Giglioni 2018). However, as it is the case for sponges, the motility of sensitive plants is not a proof that they are animals and endowed with sense-perception. According to Grew, the unconscious and automatic motion of intestines and muscles "is imitated in a fainter degree, in those Plants, which are commonly, tho' unaptly, called sensitive" (*Cosmologia sacra,* 1701, 35). The criteria of color and structure, rather than motility, are equally decisive in the case of the "sea-blobber," that is, of what we call today bladder-wrack, a sea-plant that could well be an animal, owing to its small brownish bladders:

> 'Tis a Cluster of Small roundish Bladers, almonst in the shape of little *Oystershells*; of a light brown colour, all over veined with Fibers, like the uter Cover of a *Plumpstone.* Which makes it the more doubtful, whether it be an Animal Body, or a Vegetable. Which soever, it is supposed the *Matrix* of a *Sea-Insect* (Grew 1682, 250).

Generally speaking, the three natural kingdoms do not seem to play a decisive role in Grew's classification and descriptions. With regard to corals, Grew simply explains: "These having also a resemblance unto plants, and a near analogy unto those stones, last described in the precedent chapter [i.e. petrified plants], they may therefore not unaptly be here subjoyn'd" (275). Nothing here implies that Grew really considers corals as stones. Actually, Grew does *not* think that corals are minerals, since he relates Boyle's hypotheses about their reproduction.[4] But in the *Musaeum,* he did not at all address the question of whether corals are animals or plants. First, the corals of the museum of Gresham College are by definition dead corals, or skeletons of corals belonging to the mineral kingdom. Second, an apparent resemblance between two kinds of natural objects, as well as a common opinion about their nature, suffices to justify the order Grew follows in this catalog. For example, he regards "kermes berries" as animals, because of their fetid smell when

---

Vegetative soul, while animals had a sensitive soul. Animals made active responses to external stimuli, while plants were passive, having only the faculties of nutrition and reproduction. This principle had such authority that the seventeenth century proponents of plant sensitivity won only limited approval" (1966, 6).

[4] "On the Nature and Generation of *Coral,* it is affirmed by the Honourable Mr. *Boyle,* That whilst it grows it is often found soft and succulent, and propagates its *Species*" (Grew 1682, 278).

they are burnt, but describes them in the chapter given over to mosses and mushrooms "in compliance with the Vulgar Opinion":

> [As] the Pulp or Powder, so called, is a Cluster of small Animals: so the Husk itself is Animal Body, as it were grafted on the Stock or Leaf, whereon it grows; and so converteth all the nourishment it deriveth thence (as *Bread* eaten is turned to Flesh), into its own Animal Nature. And that the said Husk is really an Animal Body, appears by that *fetid* scent it gives, like that of Horns, Hair, and the like, upon its being burnt. A property, which I find belonging to no Plant whatsoever, except to some *Sea-Plants* [...]. So that, though in compliance with the Vulgar Opinion, I have placed it here, yet ought it to be treated of amongst Animals (1682, 240).

Grew uses again this chemical criterion of animality for the cochineal:

> To me, thus much seems evident, That' tis neither a Vegetable Excrescence, as some surmise; nor an *Insect,* as others: yet an Animal Body, as the *Kermes Berry,* by some *Insect* affixed to a Plant; and thence nourished for a time, but gather'd before it be fill'd with *Mites* or *Maggots*. For being held, as the *Kermes Berry,* in the flame of a *Candle*; it usually huffs and swells, but always stinks, like Hair or Horn when they are burnt (Grew 1682, 241).

And yet, as Grew indicates about kermes berries, sea-plants also have a fetid scent when they are burnt, even though they belong to the plant kingdom: "All Sponges stink, more or less, upon burning, as the Horny Sea-Shrubs. So that it is a property belonging to most of the Vegetable Productions in the Sea" (1682, 251–252). We have thus the following situation: in most cases, a "fetid scent" is a criterion of animalness, but in some cases, it can also suggest that the object under examination is a sea-plant. Grew's chemical criterion dividing plants from animals is not very systematic.

To put it briefly, in his natural history, Grew does not seem to attach a lot of importance to the question of whether the natural object he describes is a mineral, a plant, or an animal. In the rare cases where he raises this question, his criteria, which are merely morphological and chemical, seem neither systematic nor robust.

21.3 Nehemiah Grew's Physico-Theology

At the time when he wrote the *Anatomy of Plants,* Grew did not distinguish between minerals, plants, and animals, through the criteria of motion or sense-perception. In addition, he never mentioned the neo-Aristotelian tripartition between reason, sense-perception, and vegetative soul. His natural history dealt only with *bodies*, whether mineral, vegetal, or animal. Grew's apologetic book, the *Cosmologia Sacra,* which was published more than 20 years later and aimed at refuting materialism, is partly in accordance with his earlier views regarding the boundaries and the functioning of the plant kingdom. In addition, in the first book of the *Cosmologia Sacra*, Grew uses the very idea of a "geometry of nature" that he first presented in the *Anatomy of Plants.* His new notion of the "organism" of a body then summarizes Grew's anatomical research on this natural geometry (Cheung 2006, 319; Andrault 2014a, b, 37):

> How admirable also is the natural Structure or Organism of Bodies? The whole Body of a Plant, whether Herb, Shrup, or Tree, is composed of two Species of Fibers, so artificially managed; that all the Parts, from the Root to the Seed, are distinguished one from another, only by different Position, Proportion, and other Relations and Properties, of those two sorts of Fibers: As I have made to appear, in the Anatomy of Plants (Grew 1701, 18).

However, in the first chapter of the second book of the *Cosmologia Sacra,* entitled "Of Life," Grew introduces the usual tripartition of the principles of life which was completely absent from his earlier works. According to this text, there is a "vital" and incorporeal world that completes the corporeal world and partly accounts for it:

> Life and Motion, being, as is said, the two instruments of Commerce, between the Vital and the Corporeal worlds. Whence also the union of Soul and Body, and of all things Vital and Corporeal, is nothing else, but the Congruity between the Life and Motion, which they either have, or are Capable of. Now the several Species of Life, seem to be reducible unto these Threes; viz. Vegetable Life, Sense, and Thought. The lowest Species of Life, whereof we can have any conception, is such a sort of Life, as is without Sense (Grew 1701, 34).

Grew endorses a dualistic conception of nature in which there are two kinds of beings: beings that are corporeal, mechanical, or more generally body-like, on the one hand, and those that are "incorporeal," "vital," or more generally soul-like, on the other hand. Grew articulates this dualistic conception of nature with a tripartition of the incorporeal principles, which are divided into three different species of life, according to an Aristotelian framework. "Vegetable life", "sense" and "thought" are thus the main components of the vital kingdom, in contrast to the corporeal world. Generally speaking, there is nothing original in that. We can find similar views in Ralph Cudworth's *True Intellectual System of the Universe* (1678), which also aimed at the refutation of the so-called atheistic tendency of Descartes' mechanical philosophy, in the same manner as Henry More's *Antidote against Atheism* (Duchesneau 1998, 169–181; Hutton 2015, 139, 142–3, and 161). Like Cudworth, Grew adopts two different conceptions of life (Andrault 2014a, b, 37). On the one hand, according to a generic conception of life, what is vital is incorporeal, and vice versa. For example, Grew argues in the very beginning of the second book that "there is a Vital Substance in Nature, distinct from a Body. [...] God, who as he is the Cause of all other Beings, is the most Substantial Being: is Himself a Substance Vital and Incorporeal" (1701, 31). In this sense, life is an "incorporeal" power that may account for a large range of natural capacities, from growth and reproduction to sensorial perception to reason: life explains what is not explicable physically, i.e. in terms of the mechanical motions and shapes of invisible corpuscles. Grew's notion of life is clearly opposed to Descartes' view, according to which life is a mechanical process, guaranteed by the continual movements of blood and nervous fluid (1701, 32–3). On the other hand, according to its specific conception, life is sometimes identified with vegetative life, that is to say, with the lowest common denominator of all the things that are regarded as living beings.[5] According to

[5] On the connection between plants and vegetative soul, see Baldassari (2018, 513–4).

a common view, this lowest species of life, which is the only one which animates plants, is characterized negatively, by the lack of sense-perception:

> [It] is the same Vegetable Life, whereby these Operations are performed in Animals, as well as in Plants. And which are still going on, while we are asleep, and all our Senses locked up. [...] There are Sundry motions, both in Plants and Animals, depending upon this Vegetable Life. The motion of restitution, in the Bladders of the Barque and Pith, for the expressing of their Liquors, may hereby be promoted. [...] A motion, whereof we have no perception, as depending on the vegetable life in an animal. As also doth the Peristaltick Motion of the Arteries. [...] As then, Hairs and Feathers, are Plants growing upon an Animal; so these are vegetable motions, in an Animal. And therefore, though Vital, yet have nothing to do with Sense (Grew 1701, 35).

Classically enough, vegetative life is also partly responsible for the generation of bodies, because "for the performance of this Work, a Vital or Directive Principle, seemeth of necessity to be assistant to the Corporeal." Most notably, the vegetative principle of life produces the specific "mixtures" that determine the actions of the corporeal principles. In both cases, according to the general sense of life, which refers to incorporeal principles, and the specific sense of life, which refers specifically to vegetative power, "life" is sharply distinguished from the "corporeal," i.e. the mechanical world.

What is original in Grew's conception of life is rather the fact that he refuses to identify it with a source of motion, or a dynamic power. Indeed, according to the Cartesian conception of matter, the quantity of motion remains constant in the universe.[6] Consequently, the different species of life cannot produce new motion; they only transfer it in a certain manner:

> Yet neither by this Life, nor the Subject of it, do I mean, a Principle of Motion. The Universal Stock of Motion, as that of Matter, being neither encreased, nor diminished, but only transfer'd. But I mean, a certain Power to determine the manner of its being transfer'd; or of returning an Impression upon Bodies, suitable unto that which it receives (Grew 1701, 35).

Vegetative life transfers motion immediately, without retaining it, in contrast to animal life. It was rather new not to define vegetative life by its passivity, or lack of motion, but rather by the manner in which it transfers motions:

> The difference between Vegetable and Sensible Life, seemeth to be this; that in the former, the Impression arising from the Motion, is entirely reflected, and lost in an Instant. In the latter, it is Immersed and retein'd.

[6] Grew's mechanical conception of matter is somehow difficult to combine with his rehabilitation of vital principles. On this point, see Garret: "Bayle ends with a sentiment that Grew would have found congenial: like most Cartesians Grew denies that matter can have the power of self-movement. But Grew has not realized that the postulation of a vital principle allows the atheist to assert the coherence of self-moving or directive matter after all [...] But as Bayle points out, the postulation of a vital principle, naturally compatible with matter, and capable of directing matter, leaves the atheist with the conceptual resources to explain teleological phenomena without God. The atheist threat of active self-moving matter is not lessened by the appeal to Grew's vital principles and this irony was not lost on the theist Pierre Bayle" (2003, 81).

Does this difference between "vegetable life" and "sensible life" fit the empirical criteria Grew used in his natural history? Does this conceptual difference agree with the descriptions of borderline cases which Grew presents in his *Cosmologia Sacra*? In other words, is there a discrepancy between the conceptual tripartition of the species of life and the empirical identification of a thing as a plant or an animal? At first sight, there is no contradiction between them. For instance, Grew regards the *urtica-marina*, i.e. the sea-nettle, as well as the *aurelias*, i.e. insect nymphs, as endowed with sense-perception; he does not reduce them to plant life:

> The next Species of Life, above the Vegetable, is that of Sense. Wherewith some of those Productions, which we call Plant-Animals, as the *Urtica Marina*, and the like, are endowed, as their highest Species of Life. In which rank, we may also place the *Aurelias* of all Creeping Insects (Grew 1701, 37).

Even if Grew does not explain this view, and does not try to connect it with his definitions of vegetative life, one can consider that the sea-nettle and the nymphs are endowed with sense-perception because they do not reflect the motion they receive immediately, in contrast to the sensitive-plant which immediately contracts its leaves when it is touched. However, it remains difficult to explain Grew's notion of vegetative life by means of all the examples he includes in the *Cosmologia Sacra*. Indeed, Grew regards the motions of which we have no perception, and particularly the organic motions that are performed in our body when we are asleep, as "depending upon the Vegetable Life" (1701, 35):

> whereby the Vegetable Life, is made so far independent on the Sensible: that notwithstanding the perpetual Motion and Clutter in all the rooms of the House, *viz.* our bodies; yet we take no notice at all of it, nor are in the least, hinder'd or diverted, from any thing we have to do, as Animals, or as Men (Grew 1701, 36).

But in what sense are the internal motions of an animal or a man asleep immediately "reflected" – and not "retained and immersed" –, to take up Grew's own definition of "vegetable life"? According to Grew, the development of feathers and hairs depend also on vegetative life. But is it really possible to identify the slow development of hairs with a vegetative movement understood as the immediate "reflection" of a motion received? What is the difference between the inner motions of the insects in the chrysalid and the organic motions of a man asleep? Why is the first regarded as depending on a "sensible life," while the second is said to be produced by the vegetative life? In the case of a man asleep, the specificity of the vegetative life seems to consist only in the lack of awareness during the performance of the automatic movements. But if this is the case, how is it possible to claim that the nymph and the sea-nettle have more than a vegetative life, that is to say, that they are endowed with sense-perception and awareness? Is it even possible to ascribe sense-perception to a natural being by means of merely observational criteria? In addition, Grew denied, in the *Anatomy of Plants,* that the sensitive plant is an animal by arguing that its movement could simply be explained mechanically, by the mere elasticity of matter. Why, then, in the *Cosmologia Sacra,* where he still considers the "sensitive plant" as a plant, does he regard more generally the automatic "motion of restitution, in the Bladders of the Barque and Pith" of plants as the effects of an incorporeal vegetative principle (1701, 35)?

The difficulty of Grew's notion of "vegetable life" is not only due to the discrepancy between, on the one hand, the empirical approach to plant phenomena he adopted in his *Anatomy of Plants* and *Musaeum Regalis Societatis*, and, on the other hand, the apologetic use of those same phenomena he developed in the *Cosmologia Sacra*. The problem is more general, and concerns the difficulty in making conceptual distinctions between different species of life that agree with anatomical investigations. Grew combines a dualism, which aims at refuting materialism, with a neo-Aristotelian tripartition of the species of life that was common, and very intuitive, but does not easily fit into his mechanical approach to physical phenomena. Such a mechanical conception implies, on the contrary, a certain continuity within nature, particularly between simpler and more complex phenomena, which allows the naturalist both to use analogies between species and to give microscopic causes to macroscopic phenomena.

21.4 Life Is Everywhere… and Nowhere?

In his article "Vitalism and Teleology in the Natural Philosophy of Nehemiah Grew," Brian Garrett claims that "[according] to the vitalist Grew, matter, active or otherwise, is insufficient to explain the empirical vital phenomena, and thus an incorporeal principle is required" (2003, 74). He adds that Grew's "sensitivity to empirical phenomena reveals, as it did for many of his fellow Englishmen, the poverty of Cartesian explanations of living phenomena" (81). It is true that Grew considers Descartes' natural philosophy as insufficient to explain vital phenomena. And it is not impossible that Grew's empirical research eventually convinced him of this insufficiency – even if nothing in Grew's natural history departs from mechanical philosophy. Nevertheless, the very notion of "*empirical vital phenomena,*" as well as the idea of the "poverty of Cartesian explanations of *living phenomena,*" may be misleading. They could suggest that a simple statement of fact suffices to claim the existence of living phenomena and to experimentally delimit the range of phenomena that are "vital," in contrast to non-vital phenomena. And yet, this is not the case for three major reasons. First, acknowledging the existence of a vital world is a doctrinal option which is grounded in the sharp criticism of a merely physical approach to nature, both from an epistemological and a religious point of view. Second, according to Grew, life and vitality are not in themselves phenomena, but rather the causal principles of natural phenomena. Third, and more importantly, the limits of the phenomena that are only explicable by referring to vital principles are particularly uncertain. On the one hand, what Grew regards as endowed with life does not correspond to what we would regard today as a "vital phenomenon." On the other hand, the boundaries of the non-vital world are not clearly drawn; it is not even sure that there are natural phenomena that are not partly caused by vital principles: it is not impossible that all the natural things require in one way or another the action of a vital principle – that is to say, that all phenomena are "vital" in some sense.

Indeed, Grew regards gravitation, "magnetic inclination," and even colors, as the effects of a vital principle:

> By virtue of this Power [Life], I suppose it is, That all Bodies have their Sphere of Activity, whereby they operate one upon another, more or less. That there are Dark Rays, as well as Light ones [...]. This aforesaid Power, from whence ariseth the Speer [*sic*] of Activity, is more conspicuous, in all the sorts of Magnetick Inclination, and in the Gravitation of all Bodies. But it is that also, wherewith every Corporeal Principle may probably be Endow'd: Or to speak properly, may be animated (Grew 1701, 35).

In this paragraph, Grew contemplates the possibility that every corporeal principle is endowed with a principle of life. A few lines below, his statement is more affirmative:

> On the Directive Power of the former [vital principle], and the Regularity of the latter [corporeal principle], whereby it is capable of Direction; depends the Generation of all Bodies. The said Power, being one and the same vegetable Life, infused into all the Parts of Corporeal Nature; but more remarquably into Plants and Animals (Grew 1701, 35).

First, the principle of life does not limit its action to the things that Aristotle regards as endowed with life (*De anima* 412a, 2011, 48). Second, it is not a principle of individuation. Indeed, the principle of vegetative life is neither proper to an *individual* thing, since it is "*infused* into all the parts of corporeal nature", nor proper to a *living* thing, in contrast to a non-living thing, since it is a principle with which every corporeal principle is said to be endowed. For this very reason, it is not easy to understand the exact link between the "vital" (i.e. incorporeal) and the mechanical (i.e. corporeal). Generally speaking, they seem to be conceived of as two principles of explanation, if not two causal powers; but occasionally, they are rather presented as two "worlds" or two different kinds of things. In the first case, they can be mixed and can cooperate to produce one single phenomenon; in the second case, if the corporeal and the incorporeal refer to two kinds of things, it is more difficult to conceive of how "every Corporeal Principle" is maybe "endowed" with life (35).

In any case, we must acknowledge that there is no place in Grew's physico-theology for an inanimate world with which we can compare the vital world. Indeed, if at first sight the "vital world" refers to the set of plants, animals, and human beings, it actually appears after scrutiny that vegetative life animates also bodies that are usually considered as inanimate. It is as though there was no part of the world where the effects of "life" were not mixed with the effects of the corporeal principles. In this sense, the notions of life and vitality do not have any particular "taxonomic" value (Foucault 1966, 173): they cannot help to classify phenomena or objects as either animate or inanimate.

One could object that Grew's physico-theology lacks coherence, or that Grew is not representative of the thought of the other English philosophers that introduced into their natural philosophy some incorporeal principles of life so as to refute Descartes' mechanism. But we can find similar statements in the theological works of the botanist John Ray, as well as in Ralph Cudworth's *True Intellectual System of the Universe,* which inspired Ray's physico-theology. To be sure, Grew, Ray, and Cudworth did not defend the same metaphysics. For example, Cudworth did not admit the Cartesian principle according to which the quantity of matter and the quantity of motion remain constant in the world. Accordingly, the principles of life that he admitted, and most notably "the plastic nature" and "the plastic life of nature," are dynamic powers that give new motions to a dead matter (Andrault 2014a, b, 36).[7] However, despite their differences and the different notions that they use to mean life and vital principles, they share with Grew the following set of assumptions. First, they endorse a clear dualism between physical entities and soul-like, or vital entities, which serves an apologetic undertaking. Second, they rehabilitate Aristotle's tripartition of souls, which is presented as a common and intuitive opinion: they accept a vegetative life, responsible for generation, growth, and unconscious movements in living beings, and they deem that this vegetative life is distinct from an animal life, endowed with sense-perception and consciousness, or *synesthesia* (Cudworth 1678, I:3, §37, 159). Third, the rehabilitation of Aristotle's tripartition is combined with semantic innovations and metaphysical indecisions that do not allow us to strictly identify Ray's and Cudworth's vegetative powers with a principle of individuation that operates in living beings only. Indeed, in both cases, it is not clear if "life" and vital principles are a kind of soul, a specific power of the soul, or a principle of another kind. It is not clear either if there is a soul for each individual, or rather a common soul or spirit for the whole nature. Lastly, exactly like More's "spirit of nature," those plastic or vegetative lives eventually "extend [their] operative domain to furnish explanations for all natural phenomena," and not specifically for vital phenomena (Gabbey 1990, 28).

I will now turn to these two authors before concluding.

21.5 A Brief Comparison with John Ray and Ralph Cudworth

John Ray (1627–1705), an English naturalist who studied at Cambridge, was also a fellow of the Royal Society from 1667 on.[8] His works include a huge *Historia Generalis Plantarum* (1686–1704), which proposed a system of classification, as well as a book entitled *The Wisdom of God* (1691; 1714[6]), which uses Ray's plant

[7] For this reason, I do not think that Grew simply took up Henry More's and Ralph Cudworth's "vitalistic theories" (see Hutton 2015, 150). More generally, on the complexity of the mechanical philosophy in England before Newton, see Henry 1986.

[8] Incidentally, exactly like Nehemiah Grew's father, John Ray lost his living in refusing to comply with the Uniformity act of 1662 (even though Ray continued to belong to the Church of England as a layman, and was not a Presbyterian like Obadiah and Nehemiah Grew). Their nonconformist religious opinions, which they shared with Ralph Cudworth, also educated in Cambridge, led them to develop their own theological views in the form of a physico-theology, closely related to their scientific knowledge of nature.

and animal anatomy in order to demonstrate the necessity of an omniscient Creator. According to him "the formation and the organization of the Bodies of animals" cannot be explained by matter only, and require a "mental and final cause" (1714, 44). Those vegetative functions of formation, growth, and reproduction of plants and animals are said to depend on some "vital principles." But for Ray, the actions of these principles go far beyond the vegetal and animal kingdoms:

> So are there many *Phaenomena* in Nature, which being partly above the force of these *mechanick Powers*, and partly contrary to the same, can therefore never be salv'd by them, nor without *final Causes* and some *vital Principles*; As for example, that of Gravity or the Tendency of Bodies downward, the motion of the *Diaphragm* in Respiration, the *Systole* and *Diastole* of the Heart, which is nothing but a Muscular Constriction and Relaxation, and therefore not *mechanical*, but *vital*. We might also add, among many others, the Intersection of the Plains of the Equator and Ecliptick or the Earth's diurnal motion upon an axis not parallel to that of the Ecliptick, nor perpendicular to the Plain thereof (Ray 1714, 43).

The range of "phenomena in Nature" that require the action of vital principles thus includes the gravity of bodies and the earth's orbital motion. We have the same situation as in Grew's *Cosmologia Sacra*: the reasons for admitting vital principles are not restricted to what we would call today "vital phenomena." They include everything that appears to go beyond the explicative power of mechanism. Exactly like Grew whom he sometimes cites (1714, 59), Ray does not even make a distinction between gravity and animal movements. In addition, the motions of the heart are said "vital," but Ray does not explain why it is the case.

> This Pulse of the Heart Dr. Cudworth would have to be no Mechanical, but a Vital Motion, which to me seems probable, because it is not under the command of the will [...]. What directs and moderates the Motions of the Spirits [i.e. the animal spirits that are in the blood]? They being but stupid and senseless matter, cannot of themselves continue any regular and constant motion, without the guidance and regulation of some intelligent Being [...]
>
> I should rather attribute them to a *Plastick Nature* or Vital Principle, as the Vegetation of Plants must also be (Ray 1714, 46–47).

One thing is certain: matter is conceived of as purely mechanical; conversely, what is irreducible to the mechanical motions of a "stupid matter" is perceived as dependent upon vital principles. Those are thus immaterial. Because of these semantic equivalences, the vital principles could be produced by a vegetative soul distinct from an animal soul. But Ray prefers taking up Cudworth's notion of a single immaterial plastic nature that would be responsible for the formation of all living beings. Indeed, Aristotle's vegetative soul is supposed to account for the formation and growth of plants. But how, then, are we to explain the growth of cuttings? Is this vegetative soul divisible? The "vegetative soul, if that were the Architect, would be divisible, and consequently no spiritual or intelligent Being" (1714, 52). A "single Agent" must thus "preside over the whole Oeconomy of the Plant" (*id.*):

> I therefore incline to Dr. Cudworth's Opinion, that God uses for these Effects the subordinate Ministry of some inferior *Plastick Nature;* as in his Works of Providence he doth of Angels (Ray 1714, 52).

Ray thus endorses Cudworth's thesis, according to which it is not God himself that causes the formation and growth of living beings, but an incorporeal "plastic nature" – i.e. a formative nature – that it is an intermediary between God and nature.[9]

The Cambridge Platonist Ralph Cudworth (1617–1688) published in 1678 *The True Intellectual System of the Universe: the First Part, wherein all the Reason and Philosophy of Atheism is Confuted and its Impossibility Demonstrated,* in which he propounded the theory that "a plastic nature," or "the plastic life of nature," causes the phenomena which he considers to go beyond the explicative power of mechanism.

> The Occasion of which Miscarriage [of Descartes] hath been already intimated, namely from the acknowledging only *Two Heads* of Being, *Extended* and *Cogitative*, and making *the Essence of Cogitation* to consist in *Express Consciousness*; from whence it follows, that there could be no Plastick Nature, and therefore either all things must be done by *Fortuitous Mechanism*, or else God himself be brought Immediately upon the stage, for the solving of all *Phaenomena*. Which Latter Absurdity, our Philosopher being over careful to avoid, cast himself upon the Former, the banishing of all *Final* and *Mental* Causality quite out of the World (Cudworth 1678, 1:3, §38, 175).

Life as such, that is vegetative life, lacks sense-perception and consciousness, as the well-known case of the sensitive plant shows:

> clear and express consciousness is not essential to life. There is some appearance of life and vital sympathy in certain vegetables and plants, which however called sensitive plants and plant-animals, cannot well be supposed to have animal sense and fancy (Cudworth 1678, I:3 §37, 160).

However, while lacking sense-perception, the plastic nature or the vegetative life is still something incorporeal:

> But though the Plastick Nature be the Lowest of all Lives, nevertheless, since it is a *Life*, it must needs be *Incorporeal*; all Life being such (Cudworth 1678, 1:3 §37, 163).

According to Cudworth, as long as these vital principles are conceived of as incorporeal and as the "servants" or "drudging executioners" of God – and not as the immanent powers of the nature itself–, they are compatible with Christianity. But Cudworth is not firm on the metaphysical nature, or substratum, of these vital principles:

> though it be not reasonable to think, that every Plant, Herb, and Pile of Grass, hath a Particular Plastic Life, or Vegetative Soul of its own [...]; yet there may possibly be, for ought we know, one Plastick Nature or Life belonging to the whole Terrestrial globe, by which all Plants and Vegetables, continuous with it, may be differently formed, according to their different Seeds, as also Minerals and other Bodies framed, and whatsoever else is above the Power of Fortuitous Mechanism effected (Cudworth 1678, 1:3, §37, 171).

---

[9] Ray targets Boyle's *Free Inquiry into the Notion of Nature* here: "we [Ray and Boyle] are in the main agreed, differing chiefly about the Agent that executes those Laws, which he holds to be God himself immediately, we a Plastick Nature" (1714, 51). One of Ray's arguments is the existence of monsters: if God were immediately responsible for the formations of bodies, the existence of monsters would remain inexplicable. Similarly, the "flow and gradual process that is in the generation of things" would be a "vain and idle pomp or trifling formality, if the Agent were omnipotent."

Cudworth adds in the following page:

> [Life lower than the Animal] is either a Lower Faculty of some Conscious Soul, or else an Inferior kind of *Life* or *Soul* by itself; but essentially depending upon a Higher Intellect (Cudworth 1678, 1:3, §37, 172).

But what is the exact link between the plastic life of the "terrestrial globe" which he presented first, and this vegetative faculty or soul of which he now speaks? Are there different kinds of plastic natures – that is to say, either the plastic natures which are souls of their own, or the single plastic nature which is "infused" into the whole earth to account for the growth of every pile of grass?

Regarding the metaphysical substrata of those vital principles, Ray has no definite claims either. Indeed, if he first endorsed Cudworth's notion of a general plastic life of nature endowed with vegetative powers, in contradistinction to a vegetative soul, in other passages he seems simply to take up Aristotle's vegetative soul:

> Bodies are either Animate or Inanimate [...]. Animate bodies are either such as are endued with a vegetative soul, as plants; or a sensitive soul, as the bodies of animals, birds, beasts, fishes, and insects; or a rational soul, as the body of man (Ray 1714, 59; see also 100)

Ray takes up the common tripartition of animate beings (plants, animals, and human beings), which are respectively endowed with their corresponding soul, without endeavoring to articulate this tripartition with his notion of a plastic life. In addition, strikingly enough, Ray appeals neither to Cudworth's plastic life of nature, nor to Aristotle's vegetative soul when borderline cases of vegetation are at stake. Indeed, in his *Historia Plantarum,* he addresses the debated question of whether corals and other "lithophytes" are stones or plants. According to him, they are plants. However, the reason he puts forward has nothing to do with the definitional features of the vegetative soul. It is only the physical structure of the branches that suggests that corals grow by a process of intussusception, i.e. by the penetration of the nutritive fluid into the inner parts, in contrast to the process of the apposition of new particles by which stones are shaped (Ray 1704, Supplementum 2).

21.6 Conclusion

As we carefully read the apologetic books written by Cudworth's, Ray's, and Grew's – all bestsellers in the early eighteenth century–, we realize the vagueness of the notions of life and vitality. Those authors did not clearly delimit the exact operations produced by vegetative life, the exact metaphysical substratum of vital operations, and the exact range of bodies that have a vegetative soul. First, they refused to be dogmatic in this matter: they rather intended to articulate Christian philosophical tradition with common or "popular" views on living things, on their taxonomy, and their principles of action. Second, their "principles of life" were first and foremost a polemical tool against Descartes' mechanism. Accordingly, they insisted on what Descartes' mechanical notion of life failed to explain – not on what their own notions of an immaterial life should explain. They considered that a certain number of motions such as gravitation were "above" mechanism, and, for this very reason, were obviously "vital" and "incorporeal." But they did not explain in what sense the growth of plants and the gravitation of bodies depend on the same vegetative power, or on the same kind of incorporeal principles of life. Are the so-called inanimate bodies, the sponges, and the corals endowed with the same vegetative life? And for the same reasons? Regarding gravitation, it is the notion of a directed force that leads Cudworth, Grew, and Ray to attribute to it vegetative life. Regarding sponges and corals, which, of course, do not manifest the same obvious force as gravitation, Grew and Ray rather appeal to their observable structure and chemical constitution as sufficient criteria for classifying them among plant kingdom. Moreover, when Grew addresses the case of the sensitive-plant or of the sponge, he prefers not to retain the criterion of spontaneous movement as a proof of animalness. Third, and most importantly, all three, in keeping with the practice of the period, do not assign to the notion of life a taxonomic role. On the one hand, they do not use the neo-Aristotelian tripartition of souls and powers to classify the objects under examination. As soon as they address borderline cases between minerals and plants, or between plants and animals, they rather use distinguishing criteria that are less systematic and more adapted to their own empirical approach. On the other hand, they show no concern for the boundaries between animate and inanimate in their empirical works.[10] They do not have the same interest for the limits of the living world as the physiologists of the nineteenth century and the biologists of the early twentieth century. Grew and Ray do not endeavor to defend the specificity of their subject matter; their works naturally include considerations on minerals and corporeal principles. Actually, there was in the seventeenth century no clear-cut line of distinction between mathematics, physics, natural history, and anatomy. For these reasons, we should not project onto these works our own conception of the boundaries of biology or of the distinction between the living and the non-living.

References

Andrault, Raphaële. 2014a. *La vie selon la raison. Physiologie et métaphysique chez Spinoza et Leibniz*. Paris: Champion.

———. 2014b. *What is Life?* A Comparative Study of Ralph Cudworth and Nehemiah Grew. In *The Life Sciences in Early Modern Philosophy*, ed. Ohad Nachtomy and Justin Smith, 29–46. Oxford: Oxford University Press.

Arber, Agnes. 1941. The Tercentenary of Nehemiah Grew (1641-1712). *Nature* 147 (3734): 630–632.

———. 1942. Nehemiah Grew (1641-1712) and Marcello Malpighi (1628-1694): An Essay in Comparison. *Isis* 34: 7–16.

Aristotle. 2011. *De Anima. Of the Soul,* trans. M. Shiffman. Newburyport: R. Pullins.

[10] See also Foucault 1966, 174: « Le naturaliste, c'est l'homme du visible structuré et de la dénomination caractéristique. Non de la vie. »

Baldassarri, Fabrizio. 2018. Descartes' Bio-Medical Study of Plants: Vegetative Activities, Soul, and Power. *Early Science and Medicine* 23 (5–6): 509–529.

Cheung, Tobias. 2006. From the Organism of a Body to the Body of an Organism: Occurrence and Meaning of the Word 'Organism' from the Seventeenth to the Nineteenth Centuries. *The British Journal for the History of Science* 39 (3): 319–339.

Clericuzio, Antonio. 2000. *Elements, Principles, and Corpuscles. A Study of Atomism and Chemistry in the Seventeenth-Century*. Dordrecht: Kluwer Academic Publishers.

Cudworth, Ralph. 1678. The True Intellectual System of the Universe: the first part, wherein all the reason and philosophy of atheism is confuted and its impossibility demonstrated. London.

Digby, Kenelm. 1660. *A Discourse Concerning the Vegetation of Plants*. London: Printed by J.C. for John Dakins.

Duchesneau, François. 1998. *Les modèles du vivant de Descartes à Leibniz*. Paris: Vrin.

Foucault, Michel. 1966. *Les mots et les choses. Une archéologie du savoir*. Paris: Gallimard.

Gabbey, Alan. 1990. Henry More and the Limits of Mechanism. In *Henry More (1614–1687). Tercentenary Studies*, ed. S. Hutton, 19–35. Dordrecht: Kluwer.

Garret, Brian. 2003. Vitalism and Teleology in the Natural Philosophy of Nehemiah Grew. *The British Journal for the History of Science* 36/1 (1): 63–81.

Giglioni, Guido. 2018. Touch Me not: Sense and Sensibility in Early Modern Botany. *Early Science and Medicine* 23 (5–6): 420–443.

Grew, Nehemiah. 1682. *The Anatomy of Plants with an Idea of a Philosophical History of Plants, and Several other Lectures, read Before the Royal Society*. London: Rawlins.

———. 1701. *Cosmologia Sacra, or Discourse of the Universe as it is the Creature and Kingdom of God*. London.

Henry, John. 1986. Occult qualities and the experimental philosophy: Active principles in pre-Newtonian matter theory. *History of Science* 24 (365): 335–381.

Hooke, Robert.1665. Micrographia. London: J. Martyn and J. Allestry.

Hunter, Michael. 1981. Early Problems in Professionalizing Scientific Research: Nehemiah Grew (1641-1712) and the Royal Society, with an Unpublished Letter to Henry Oldenburg. *Notes and Records of the Royal Society* 36 (1981–2): 189–209.

Hutton, Sarah. 2015. *British Philosophy in the Seventeenth Century*. Oxford: Oxford University Press.

Kerckring, Theodor. 1670. *Spicilegium Anatomicum*. Amsterdam.

LOCKE, John.1979. An Essay concerning Human Understanding, ed. Peter H. Nidditch. Oxford: Oxford University Press.

Ray, John. 1686. *Historiae plantarum tomus primus*. London.

———. 1704. *Historiae plantarum tomus tertius*. London.

———. 1714 [1691]. Wisdom of God Manifested in the Works of the Creation. In *The Sixth Edition Corrected*. London: William Innys.

Roos, Anna Marie. 2007. *The Salt of the Earth: Natural Philosophy, Medicine, and Chymistry in England, 1650–1750*. Leiden: Brill.

Swammerdam, Johannes. 1758. Collections académiques. *Tome V de la Partie étrangère et le second volume de l'Histoire naturelle séparée. Contenant les observations de Jean Swammerdam sur les insectes*. Dijon: Desventes.

Webster, Charles. 1966. The Recognition of Plant Sensitivity by English Botanists in the Seventeenth Century. *Isis* 57 (1): 5–23.

Chapter 22
Vegetable Life: Applications, Implications, and Transformations of a Classical Concept (1500–1700)

Fabrizio Bigotti

Abstract Throughout the pre-modern era, plants, animals, and humans shared a deep ontological continuity as different levels in the hierarchical organisation of life. Popularised by Lovejoy as 'the great chain of being' (*scala naturae*), such continuity relied on extensive use of metaphors, as well as on functional and psychological analogies in which the vegetable realm was understood as the primary and most fundamental manifestation of life. An appeal to the vegetable powers and to the plant-like functions of animals continued even during the golden period of mechanical philosophy and was actually revived as part of different theoretical strategies, stretching from Renaissance naturalism to eighteenth-century vitalism and materialism. From this standpoint, the survival of a classical model poses questions as to the kind of applications, implications, and transformations the vegetal imagery was susceptible to. In this paper I take a *longue-durée* approach to the notion of vegetable life in order to show the complexity of the theme and its variations in medical and philosophical sources as well as in its technological applications and visual renderings. Notably, in the first part, I consider the analogical assumptions underlying the use of vegetable metaphors in ordering, classifying, and visualising of the *materia medica*. In the second, I analyse the survival of the notion of vegetable soul in the seventeenth century and its interaction with various trends proper to the new mechanical philosophy. Finally, in the third part, I dwell on the anthropological implications of 'the vegetable' as a means to highlight the 'material' and 'natural' side of human existence, an approach that runs counter to but parallels the dualist accounts of mind and body in Descartes and his epigones.

F. Bigotti (✉)
Institut für Geschichte der Medizin, Julius Maximilians Universität Würzburg, Würzburg, Germany

Centre for Medical History (CMH), University of Exeter, Exeter, UK
e-mail: fabizio.bigotti@uni-wuerzburg.de; f.bigotti@exeter.ac.uk

F. Baldassarri, A. Blank (eds.), *Vegetative Powers*, International Archives of the History of Ideas Archives internationales d'histoire des idées 234,
https://doi.org/10.1007/978-3-030-69709-9_22

Scholars of the early modern period are familiar with the notion of vegetative soul.[1] As a concept that enjoys the *longue dureé* proper to the most relevant topics in intellectual history, it crops up regularly in medical, philosophical, and theological texts, from the classical age well through the Enlightenment. If, as a formal term, 'vegetative soul' pertains to the domain of Aristotelian philosophy, where it points to a characteristic power or capacity proper to plants, informally, it could be found anywhere plants were available, studied and regarded as living beings rather than machines. As Brooke Holmes has shown in a recent contribution, one of the reasons why the vegetative soul was invested with such an important role is that it made 'manifest', by comparison, features and processes which in the animal body remained 'obscure'.

22.1 Vegetable Analogies

Although this was neither the only nor the most important use 'the vegetable' was susceptible of, its adoption was significant and widespread especially in botany and pharmacology. Comparing and equating by means of analogy, in fact, allowed not only the classification of various plants into the progressive order of species and genera, but could indicate subtler similarities between the structures and functions of plants and those of animals and humans. The likening of trees to humans, in particular, grew out of a learned tradition, stretching back at least to the time of Plato,[2] but it was equally present in the agrarian culture of the early Romans.[3] With the assimilation of the classical tradition, the comparison gained wider currency in European culture and, by the end of the twelfth century, it became part of the very definition of man, as the French theologian Alain de Lille (ca. 1128–1203) poignantly recalls in *Distinctiones monasticae*:

> 'Tree' means 'man', and rightly so. In Greek, in fact, man is said ἀνθρωπος which means inverted tree (*arbor versa*). And indeed, while a man has his mouth in the upper part of his body, a tree has it in its lower part. (Chambers 1961, 295)

Not unlike theology, the medical literature is rife with examples of analogical reasoning, in which plants feature prominently as a term of comparison. This development is proper to the late medieval and early modern botanical repertoire, where plants are classified according to their medical use and suitability to the treatment of the different parts of the human body.

---

[1] Lovejoy 1933; Kent Sprague 1991; Giglioni 1996; Ingensiep 2001; Perler 2008.

[2] See Plato, *Timaeus*, 90A; *Cratylus*, 399c on which Chambers 1961, 292–293.

[3] Onians 1951, 124–125: 'That the head was believed by the early Romans to contain, to be the source of, the seed will also explain, as nothing else can, why in the vegetable kingdom not only the blossom or fruit containing the seed but also, in old Latin, that from which a plant sprung, its root, was called *caput,1* and why in nature at large, from Plautus onwards, the source of a stream was called its 'head'.

A case in point is that of medieval herbals, which made us familiar with the *mandragora* rendered as an anthropomorphized plant. Much as it borrowed from ancient sources, significant differences occur within the medieval tradition and its ancient counterpart in Galen and Dioscorides. These consist chiefly in the fact that, while for the latter, similarities between the anatomical parts of man and plants point to an underlying similarity in their 'chemical' make-up, late medieval and early modern botanical treatises rely on a universal symbolism, by virtue of which common *notae* (i.e. 'notes', 'signs', 'characteristics') allow the doctor to discover in plants the suitable remedy for each affected part in the human body (Maclean 2011).

These *notae* are regarded as universal signs insofar as they share a common origin from a single *exemplar* consistent with the design of the world originally outlined by God.[4] As such, it is the *exemplar*, not the material substratum, that underpins the very process of grasping and classifying similarities, seen as pale reflections of wider and deeper natural connections.

As is known, this conception forms the base for the part-to-part analogy which presides over the Paracelsian doctrine of signatures (*signatura rerum*), a doctrine which, between the end of the sixteenth and early seventeenth century, finds its most prominent representatives in Giovanni Battista Della Porta (1535–1615) and Jakob Böhme (1575–1624).[5] Both by reason of its being published later and because it did not deal specifically with plants, the *De signatura rerum* by Böhme (1622) did not enjoy as much popularity as Della Porta's *Phytognomonica* (1588), which was reprinted several times up to 1655. As to its title and much of its content, the *Phytognomonica* mirrors Della Porta's celebrated *De humana physiognomonia*, published a few years earlier (1583), from which it develops the idea that plants' virtues can be spotted at a glance (Della Porta 1588, 16).[6] The image of a lynx discovering a plant growing on the top of a rock with the motto *aspicit et inspicit* ("she looks at and into") engraved on the frontispiece of the work [Fig. 22.1], testifies to this approach and highlights the author's intent to use analogy to discover the inner essence of things.

Compliant with the tenets of physiognomics, plants' external shape deserves particular attention. Plants represent the "simplest" (*simplicia*), not only in the traditional sense in which they form the basic elements of the *materia medica* (*simplicia medicamenta*) that are combined by the physician to obtain more powerful drugs (*composita medicamenta*), but also in the sense that, within the reciprocities of the

[4] Della Porta 1588, 16–17: 'Erat enim in humana Physiognomonia per apparentem morem diiudicandi modus, ut quibus moribus se quisquam exercendo praefiguraret, tales ei mores inesse pronunciabat. [...] Haud secus existimandum est in hac Phytognomonica, quod plantae internos mores frondium, fructum, radicum stirpium, actionum, morum, aliorumq[ue] quoad possunt conspicua allusione, et consimilitudine demonstrant. *Est quippe per similitudinem demonstrandi modus, quo saepissime summus rerum opifex divinas et occultas res solet patefacere, ut supremam idearum similitudinem referrent*'. Italics added.

[5] See Paracelsus 1553. For an analysis of this work and the meaning of the Paracelsian doctrine of signatures see Porter 2005; Findlen 1990 and, more extensively, Bianchi 1987.

[6] On the common logic of Della Porta's physiognomics and phytognomics and their sources see Ferrini 2012.

Fig. 22.1 Giovanni Battista Della Porta, *Phytognomonica* (Della Porta 1588), Frontispiece, Upper String

analogical comparison, they stand for the simplest 'type' in which a given animal or human characteristic can be found in nature. Not unlike animals and humans, in fact, plants are endowed with their *ingenium*, an innate characteristic which changes itself at the alteration of the plant's soil, climate, and structure. As a consequence, odours, colours and other such characteristics are seen as secondary with respect to the structure and innate quality proper to the plant (Della Porta 1588, 12–14).

If difficult to disentangle from the magical, hermetical, and occult practices which were thriving in the late Renaissance, Della Porta's and Bohme's attempts are more generally consistent with the very logic of the *historia naturae*, whereby botanical species are classified by selecting a number of relevant *notae*, listing their properties and counting their variation in other *specimina* pertaining to the same genera, so as to the determine their degree of intensity, similarity and/or affinity.[7] At the same time, this method of combining signs shares common grounds with that of the *mathesis universalis*. In fact, while, according to the precepts of the *signatura rerum*, one property could be transferred from one species to another by virtue of an observed similarity in their external characteristics, but without the necessity to find the same structure in both, in those of the *mathesis universalis* each being could be converted into a series of characters expressing its prominent features which were then reproduced in any given circumstances by simply combining and re-arranging the number of salient signs according to some formal rules (*ars combinatoria*). The two approaches take the scholastic interest in logical relations, classification and description of signs to extreme lengths, with a view to reducing nature to an easily graspable language of signs.[8] It was thus consistent with these premises that trees, seen as articulated unities, were taken as the *exemplar* of any 'natural' method of

[7] Findlen 2001.

[8] Maclean 2011, 276: 'The physiognomonic treatises (broadly understood) inherited from the ancient world and the Middle Ages by the Renaissance, while for the most part made up of inventories of signs, refer to a variety of semiological and logical problems, all of which are appropriated and repeated by their later emulators.'

division and composition between the different classes of *notae*. The *exemplar* was instantiated in the various 'trees of knowledge' (*arbores scientiarum*) developing out of the famous 'Porphyrean tree' (*arbor Porphyriana*), where each branch of the tree points to a specific discipline and to its position in the hierarchical progression of knowledge. As a specific tradition of it, the so-called 'genealogical trees' (*arbores consanguineitatis*) deserve particular attention, whereby not only is each branch/member of the family localized in respect to the totality of the tree by reason of its proximity and degree of affinity to another member, but a close underlying analogy between blood and sap presides over the application of the whole scheme [Figs. 22.2, 22.3 and 22.4].

22.2 Vegetal Anatomies

Another way in which continuity between humans, animals, and plants was witnessed in nature rested on their being functionally related. If the tree of consanguinity entails a functional analogy between sap and blood, in medicine the analogy goes even deeper and plants are assumed as a model to understand and explain the development of the embryo, at least insofar as nutrition and growth are concerned. The works of Hippocrates and Galen, likening the spreading (*disseminatio*) of veins, arteries, and nerves to the 'branches' (*rami*) and the 'ramifications' (*ramificationes*) of a tree, are certainly revealing of such an analogy,[9] which reached its peak at the end of sixteenth century, with the anatomical plates of Adriaan van den Spiegel (1578–1625) providing possibly the most vivid representation of the ramification of the umbilical veins [Fig. 22.5].

However, Spiegel's virtuosity in the anatomical rendering has probably less to do with a desire for accurate description than with an interest in the underlying functions underpinning generation. Whereas, the vegetable analogies by Hippocrates and Galen still make up – in part at least – for the necessity to get a glimpse of the characteristics of the specimen in the absence of anatomical illustrations, Spiegel utilises his plates to show that structures in plants and animals resemble each other by virtue of their developing out of the same underlying process.[10] Girolamo Fabrizi da Aquapendente's (1533–1619) plates in *De venarum ostiolis* (1603) represent another good example of this approach, insofar as the comparison of the valves of the veins to the knots in the branches and stem of plants (*ac nodi in plantarum*

[9] Hippocrates, *De natura pueri*, 27. 1, in Littré 7.528; Galen, *De foetum formatione*, 3.15 (K 5: 666); *De semine*, 1.8.4 (K 4:540).

[10] Fabrizi d' Aquapendente 1600, 3, 112: 'Unde ex iis quae necessaria plantis sunt, dicebat Galenus, licet cognoscere, quot et qualia sunt ea, quibus foetus indiget quousq[ue] una anima, perinde ac plantae, regitur et gubernatur [..] Neque ulla alia inter vasorum umbilicalium, et plantarum radices conspicitur differentia, nisi quod vasa duum sunt genera, venae scilicet, et arteriae, plantarum autem radices unius sunt naturae, et substantiae.' Spiegel 1626, 21, using Aristotle's opinion on the functional compatibility of plants and umbilical veins to support his own thesis of the latter's function.

Fig. 22.2 "Tree of sciences" (*arbor scientiarum*). From Ramon Lull (1232–1316), *Arbor Scientiae Raimundi* (Lull 1515)

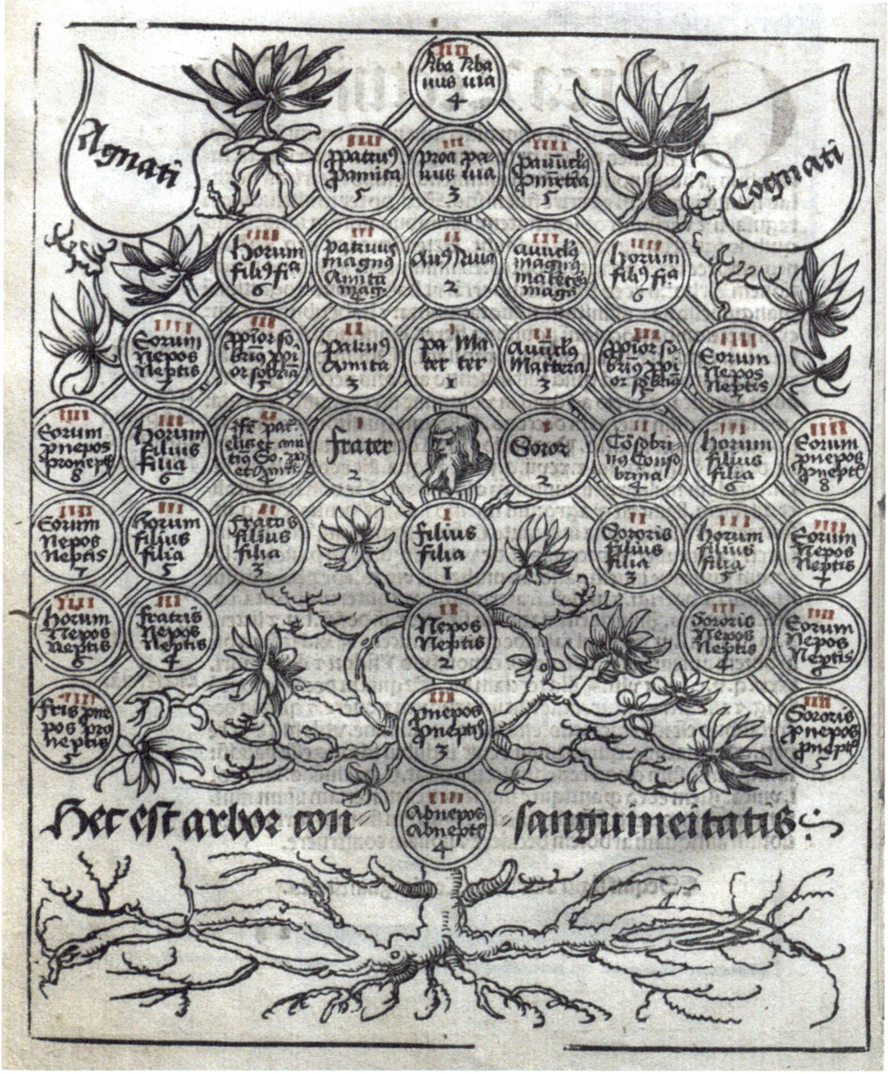

Fig. 22.3 "Tree of consanguineity" (*arbor consanguineitatis*). From Giovanni D'Andrea (1270–1348), *Arbor consanguinitatis cum suis enigmatibus* (D'Andrea 1506). Yale University Library

Fig. 22.4 An eighteenth-century engraving of the "Porphyrian tree" (*arbor prophyriana*). From Alexander Brunczwik, 'Tractatus in Aristotelis logicam' (1748) engraving by Martin Prokeš (fl. 1708–1733). Hungarian Franciscan Library, Item no. 017053, Budapest

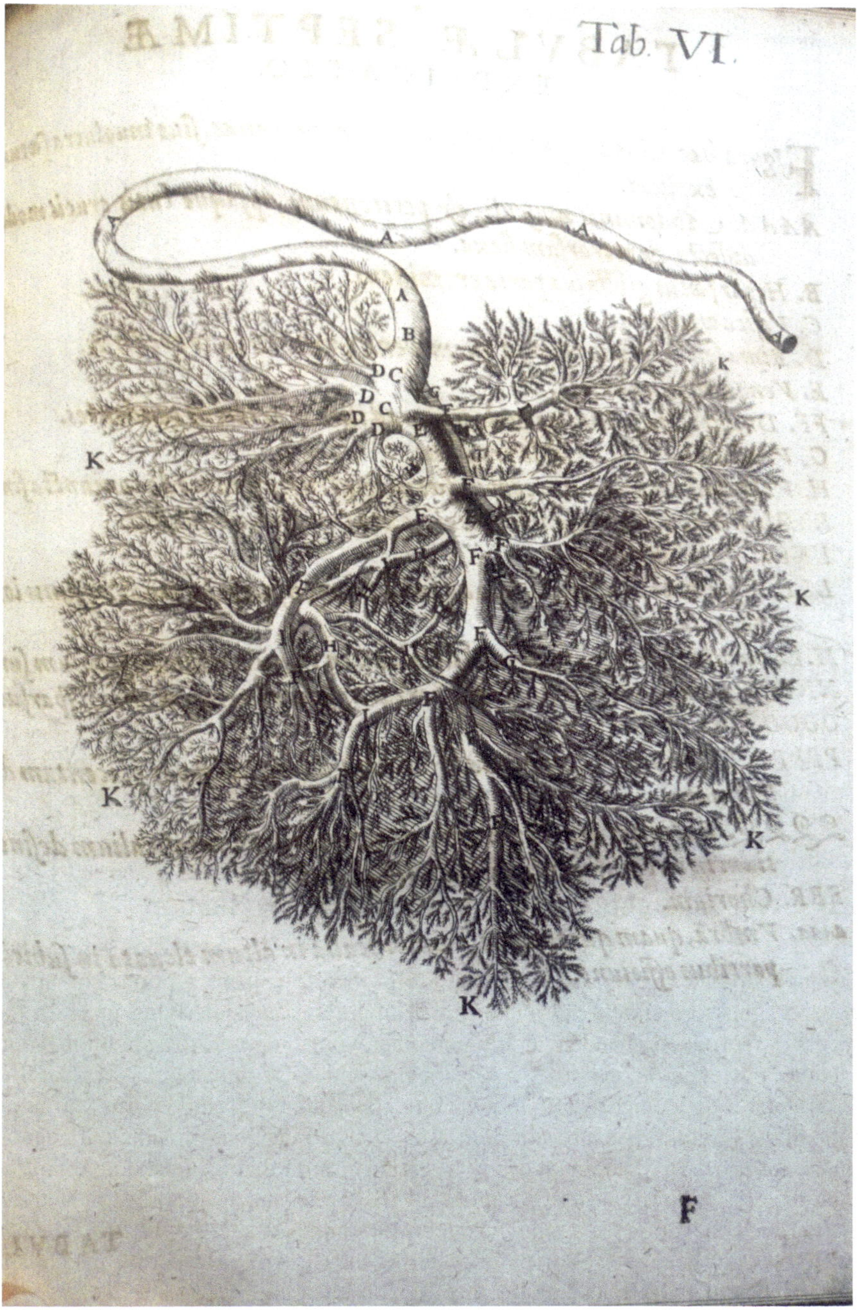

Fig. 22.5 Umbilical vessels (*vasa umbilicalia,* corresponding in modern anatomy to the vessels of the placenta) seen as ramifications of a tree, from Adriaan van den Spiegel, *De formato foetu liber singularis* (Frankfurt 1626), tabula sexta. From the author's private collection

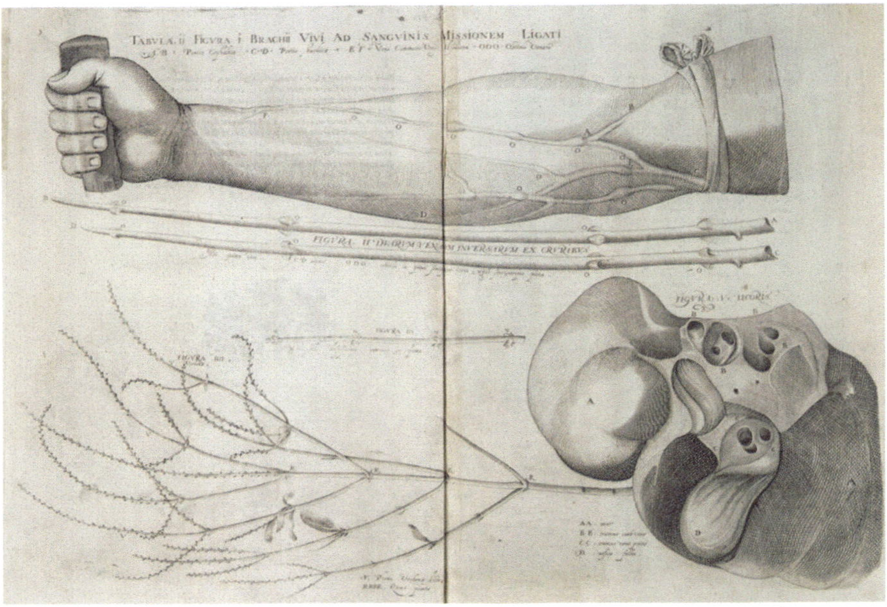

Fig. 22.6 Anatomical similarities between the valves of the veins and the knots in the branches of a plant, from Girolamo Fabrizi da Aquapendente, *De venarum ostiolis* (1603)

ramulis et caule) illustrates the precepts of Galenic physiology, with veins seen as suppliers of nutrients to the whole body, a function all too comparable to that of plants' branches distributing sap from the roots to the leaves [Fig. 22.6].

22.3 The Vegetable in Generation and Reproduction

The application of vegetable analogies shifts therefore from a pure comparison between signs to a functional analogy that highlights an underlying unity in the process of reproduction and generation; in other words, it points to a common 'soul' or natural power acting in both. Seen as the first and most fundamental stage of life, the "vegetable" becomes synonymous with *nature* or φύσις, namely 'the sprouting of the blade of grass'. As a result, processes such as nutrition, growth, and generation are inscribed in the domain of the '*vegetative soul*', that is in the domain of the 'natural', the 'simple' and the 'primordial'. It is relevant, in this regard, that the Hippocratic conception[11] likening the maternal womb to the fertile soil and the male semen to a plant seed grew in popularity during the early modern period. The image bore witness to the biblical motto 'be fruitful and multiply' (Genesis 9:7) with

[11] Hippocrates, *De Natura Pueri*, 27 in Littré 7.528.

LIBER PRIMVS DE GENERATIONE HOMINIS.

CAPVT I.

De genitali femine, quid fit, vnde & quomodo nafcatur.

PRocreationem naturalem hominis, talem omnino obferuamus, qualem omnis generis ftirpium generationem animaduertimus. Vt enim illæ ex femine fui quæque generis telluri commiffo, germinant, augentur, & ad perfectam fuæ naturæ formam naturaliter excrefcunt: ita homo que-

Fig. 22.7 The tree of life and death from Jakob Rueff, *De conceptu et generatione hominis* (3d ed., 1587), Bavarian State Library, Munich

which Jacob Rueff (1500–1558) opens his chapter on generation in the *De generatione et conceptu hominis* (3d edition, Frankfurt 1587) by engraving Adam and Eve beside the tree of life and death [Fig. 22.7].

On the other hand, however, the image was strictly adherent to practices of agrarian culture, i.e. tilling the earth, sowing, harvesting, cooking/baking the products of the earth and re-using the food waste so as to fertilize the soil and restart the cycle. Nutrition, generation and reproduction become so strictly intertwined and so tightly dependent upon vegetable analogies that, in Rueff's own description, they all become interlocked stages of an ideal 'vegetation cycle':

> We observe that the natural generation of man is like the generation of all the different species of plants. Indeed, as plants sprout out of their different seed that has been left in the ground, and then grow and naturally reach their natural form, in a similar manner man, though a rational creature, as to the nature of his body, takes his natural beginning from the semen of the male that has been ejaculated into the female womb as into a ground. However, that very semen, which we call also sperma, as for its origin and nature is only a superfluous humor – a nourishment's residuum, as it were – and what remains of the third bodily concoction. This, brought from the main parts of the body to the genitals through tiny organs, is made available for reproduction. (Rueff 1587, cc. 1r-v).

The continuity between the vegetable and human realm extends also to those actions by means of which plants select and absorb their nutrients, i.e. *attraction*, *retention*, *alteration* and *expulsion*, known as "natural faculties" (φυσικαί δυνάμεις), after Galen's definition (Galen, *Nat. Fac*, 12, K II, 26–27). Foetuses were regarded as the quintessential embodiment of these actions. In particular, the attribution to them of a plant-like life developed from an Aristotelian intuition. According to Aristotle, each life form in order to survive needs at least to nurture, reproduce and grow, thus presupposing the very functions proper to plants, their 'vegetative soul'. This communality was particularly clear in chicken eggs and was then extended to all animals, whose embryos 'grow by means of the umbilicus in the same way as a plant by its root' (Aristotle, GA II 4, 740b9). Regarded as the cause of growth, nutrition is revealed as pivotal to understanding how continuity unravels in nature. However, while reproduction was portrayed as a *cycle* (pursuant with the idea of the 'circle of life'), nutrition enjoyed a different representation which might be termed as a form of *encapsulation*. By virtue of this *encapsulation*, the vegetable becomes synonymous with food, the primary source of sustenance for animals and humans, and thus the equivalent to matter for all other living creatures.

Acquapendente's unpublished *Methodus anatomica*, a series of lecture notes taken by one of his German students in late 1579, which have been recently rediscovered by Michael Stolberg,[12] reveal how the encapsulation model could work, along the lines of the more classical articulation of matter and form. Taken as food, plants represent the key to unravel the strict interdependence of life forms in the *scala naturae* from the simplest to the most complex and back again, up to reach the *materia prima*:

[12] See Stolberg 2018, 391.

> The origin of all plants derive from the four elements, and plants then provide the aliment for sheep, pigs, goats and other animals; these <animals>, on the other hand, become in turn the aliment to men. Out of this food, blood is produced in men, along with phlegm, and the two kinds of bile [...]. And this blood is matter of our generation which, although it consists of four humours, nevertheless takes its name from the prevalence of blood.[13] [...] Now, matter is not one and the same in all bodies, but matter of animals are, as we have said, the four humours. Matter of humours, on the other hand, are the plants themselves, for humours are combined together by plants, while matter of plants are the four elements: fire, air, water, and heart. Indeed, plants originate immediately out of these elements. But then again, the matter of the four <elements> is inform prime matter which, as Galen says in the *Quod animi mores*, chapter 3, is devoid of any quality and can only be grasped by a mental act. (Fabrizi d'Acquapendente 1579, ff. 226v-227r)[14]

Like all forms of continuity, the *scala naturae* implies a certain blending of categories and hence a disquieting blurring of the distinction between different realms of nature. Intermediate forms of life between plants and animals were not only seen as possible on a theoretical level, but in fact predicted. Known as *zoophyta*, these 'animal-plants' were remarkable for their capacity of absorbing nutrients from the ground, like plants, while moving and perceiving, like animals. With a Latin word which translates and reverses the order of the Greek terminology, Jean Bodin (1530–1596) names them as 'plantanimal', listing amongst them the famous *mimosa pudica* or 'touch-me-not', because of the way it reacts to tactile stimuli (Bodin 1597, BK III, 297). By displaying both plants' and animals' abilities, in fact, *zoophyta* on the one hand justified Aristotle's embryological tenets, which required embryos to 'recapitulate' the chain of being by living their primordial stage as plants, while, on the other, they opened up the possibility that both plants and animals could share certain feature of sensibility (Aristotle GA, II 4, 736b 20ff.). And here we find possibly another model in which the vegetable is referred to in the early modern period, that is the *recapitulation* model, which mirrors the *encapsulation* seen at work in Fabrizi d'Acquapendente.

Even though Aristotle had denied plants any capacity of sense perception, the issue of sensibility had become a contentious one already in late antiquity,

[13] Fabrizi d'Acquapendente 1579, f. 221r: 'ex 4 el[emen]tis plantarum omnium ortus existit, quae postea ovibus, suib[us], capris, aliisque a[n]i[m]alibus al[imen]tum p[rae]stant, at hominibus haec ipsa animalia alimentum sunt. Ex quibus q[ui]d[em] cibis sanguis in illis gignitur, et pituita, et geminae biles, sed sanguis plurim[us] solus in venis apparet, alii aut[em] pauciores. Atque is ipse sanguis e[st] nostrae g[e]n[erati]o[n]is materia, qui ex 4 q[ui]dem humoribus constat a dominio tamen sanguis d[ici]t[u]r.' The present transcription and translation have been realised in my capacity as a research fellow on Michael Stolberg's project *Anatomische Lehre und Sektionpraxis in Padua* (1540–1600) funded for three years (2018–2021) by the Deutsche Forschungsgemeinschaft (DFG).

[14] The original Latin is: 'M[a]t[eri]a vero non o[mn]ib[us] corporibus una et eadem est, sed m[a]t[eri]a a[n]i[m]alium sunt sicut diximus quattuor humores. M[a]t[eri]a vero quattuor humor[um] sunt ipsae plantae ex plantis [e]n[im] compositis fiunt quattuor humores, plantar[um] aut[em] m[a]t[eri]ae sunt quattuor el[emen]ta ignis aer aqua terra. [P]lantae [e]n[im] p[ro]xime ex his oriunt. El[emen]torum aut[em] quattuor m[a]t[eri]a est m[a]t[eri]a prima informis q[uae] ex Gal[eni] s[e]n[ten]tia in lib[ro] q[uod] a[nim]i mores 3 ea. est expers q[u]alitatis et q[uae] cogit[ati]o[n]e t[antu]m deprendit[ur] [...].'

occupying Galen's mind in his last writings. In the recently rediscovered *De propriis placitis* ("On my own opinions"),[15] circulating in the early modern period as a mutilated tract under the name *Fragmentum de substantia facultatuum naturalium,* Galen concedes that, in selecting their nutrients, plants exert a capacity that is akin to perception in animals.[16] This somewhat allows them to feel pleasure and pain with regards to what sustains or hinders their existence. It was in keeping with this conception of the vegetative soul as encompassing also the domain of pain and pleasure that Renaissance thinkers such as Juan Huarte (1529–1588), Bernardino Telesio (1509–1588) and Eustachio Rudio (1548–1612) worked out the outlines of pure naturalist ethics.[17]

22.4 Vegetative Soul and Sexual Drive

Such attempts were geared around the idea that animals and humans are driven by a fundamental concern for self-preservation (*cura sui*), most notably consisting in nutrition and reproduction. Hence, mechanisms that improve nutrition by means of selecting the aliments in a healthy diet as well as practices that allow the control and rationalisation of sexual intercourse were given special attention. The theme features prominently in Galen's *Quod animi mores*, and Renaissance physicians were keen to develop such a suggestion to extreme lengths.

Huarte's *Examen de ingenios para la sciencias* (Baeza 1575), for instance, is presented by the author himself as a way to allow the sovereign to select and direct to better employment the different capacities (*ingenios*) operating at all levels of the social scale (Huarte 1594: Dedication to Philip II). The means to operate in this direction are provided by medicine and consist in proposing a grand eugenic project, by means of which new offspring could be generated safely, healthily and, above all, intelligently. This last feature is indeed paramount in understanding Huarte's programme. Intelligence and knowledge, in fact, are considered as organic functions that are embodied and, therefore, inheritable from parents to children. Huarte's basic claim is that we are a consequence of our parents' choices and our abilities can be improved or hindered by our sexual and nutritional habits. To prove this point, he scornfully rejects metaphors and analogies between the brain and the sexual organs but dwells on the effects deriving from the castration of animals. Regarded as the deliberate removal of the organs that preside over the generation of new life, in fact, castration represents in itself a kind of inverse proof which, by suppressing its manifestation, reveals the true power that vegetative soul holds over the human psyche:

---

[15] Baudon-Millot and Pietrobelli 2005.

[16] Baudon-Millot and Pietrobelli 2005, 189–190, 212.

[17] See Bigotti 2019: 137–183.

> The temperature of the four first qualities (which we heretofore termed nature) hath so great force, to cause that (of plants, brute beasts, and man) each one set himselfe to performe those workes which are proper to his kind, that they arrive to that utmost bound of perfection which may be attained, suddenly and without any others teaching them; the plants know how to forme roots under ground, and by way of them draw nourishment to retain it, to digest it, and to drive foorth the excrements [...] And if you ask who hath taught the brute beasts to doe these works, (which breed us such marvell) and men to discourse with reason; Hippocrates answereth, it is the natures of them all without any teacher: as if he should say, the faculties or the temperature of which they consist, are all given them without being taught by any other. Which is clearly discerned, if they passe on to consider the workes of the soule vegetative [...]. We see by experience, that if mans seed consist of an evill substance, and enjoy not a temperature convenient, the vegetative soul runs into a thousend disorders: for if the same be cold and moist more than is requisit, Hippocrates sayth, that the men proove Eunuches, or hermofrodites; and if it be very hote and drie, Aristotle sayth, that it makes them curle, paled, crook-legged, and flat nosed as are the *Aethiopians*, and if it be moist, the same Galen sayth, that they grow long and lithie and if it be drie, low of stature. [...] [B]ut the member which most partaketh the alterations of the belly, all Phisitions say, is the brain, though they have not set down the reason wheron they ground this correspondencie. True it is Galen prooveth by experience, that by speying a sow, she becommeth faire and fat, and her flesh verie savory: and if she have her cods, she tasteth little better than dogs flesh. Wherby we conceive, that the belly and the cods carrie great efficacie, to communicat their temperature to all the other parts of the body; especially to the brain, for that the same is cold and moist like themselves. (Huarte 1594, 34–35)

In applying the same experience to man, Huarte defends the thesis that genitals are even more important than the heart itself in granting human health and ingenuity:

> and because we sayd, that from the wit and manners of a man we conjecture the temperature of his cods, it is requisit that we take notice of a notable point, mentioned by *Galen*, namely, that to make us understand the great vertue which a mans cods possesse, to give firmnesse and temperature to all the parts of the body, he affirmeth that they are of more importance than the heart: and he rendereth a reason, saying, that this member is the beginning of life, and nought else, but the cods are the beginning of living soundly and without infirmities. How much it endammageth a man to be deprived of those parts (though so small) there need not many reasons to proove, seeing we see by experience, that forthwith the haire and the beard pill away, and the big and shrill voice becommeth small, and herewithall a man leeseth his forces and naturall heat, and resteth in far woorse and more miserable condition than if he had bene a woman. But the matter most worth the noting is, that if a man before his gelding had much wit and habilitie, so soone as his stones be cut away, he groweth to leese the same, so far foorth as if he had received some notable dammage in his very braine. And this is a manifest token, that the cods give and reave the temperature from all the other parts of the body. (Huarte 1594, 279)

As it appears, by using the effects of castration and abnormal generation, Huarte backs up his claims about the naturality of the human soul in its entirety and, with it, the materiality of mind. For all purposes, materiality here becomes synonymous with sexual drive which must be harnessed and manipulated, aiming at compensating – on the political and ethical level – for the fact that the instincts of the vegetative soul are irrational and, as such, irrepressible. And yet, the case for the materiality of the soul becomes even tighter by postulating that sexual reproduction depends, as a specific case of it, on spontaneous generation (*generatio ex putri*), for, in that case, the link to the vegetable and the natural becomes all the stronger.

In the *De diluviis*, an extract of his commentary to Aristotle's *Meterologica* (II.6), Avicenna (980–1037) had already upheld the notion of spontaneous generation of man.[18] The thesis remained in any case controversial, as orthodox Aristotelians – let alone Platonic philosophers or Christian theologians – were unwilling to concede that so complex a creature as a human being, could be generated out of the heat produced by putrefying organic matter. Physicians and naturalist philosophers were somewhat more positively inclined to accept and develop the consequences of Avicenna's claim, which found followers in Telesio and Rudio.

Bernardino Telesio explained the emergence of life as an effect of a primordial heat, acting as a living force in any part of the cosmos.[19] Originally endowed with a special sensibility and inclination (*ingenium*) to select what is proper and improper for the survival of life, heat shaped all living organisms which no longer differed in essence – and soul – but in degree, according to the intensity of that heat and its reacting in each material substratum with different intensities.[20] In this model, plants are primordial animals with a rudimental form of sensibility, which displays itself in their capacity of selecting and absorbing nutrients. A similar model holds in medicine, with prominent Paduan physicians such as Eustachio Rudio, a colleague of Fabrizi d'Acquapendente and one of Harvey's teachers in Padua, conceiving of sensibility as the result of material alterations.[21] Unlike Telesio, however, Rudio denies the need to ascribe a rudimental sensibility to matter and considers animal intelligence as an outcome of the transformations occurring in matter through spontaneous generation. Both Telesio and Rudio regarded the distinct souls and faculties required by traditional natural philosophy as effects of a unified process which links organic and inorganic life, a model which supersedes in many ways the traditional account of living beings as superintended by different faculties and souls. Particular attention should be given to the fact that both Telesio and Rudio insist on the importance of disentangling the consideration of animal functions from the necessity to postulate a rational and transcendent purpose, a model for which the vegetable realm provided plenty of examples. Only when a clear break between the rational and the vegetable-animal continuum had been made was it possible, in fact, to distinguish the mind from the body and to ask nature to respond to the abstract laws of mechanics.

[18] Throughout the middle ages, and still for much of the early modern period, this chapter of Avicenna's physics devoted to meteorology circulated independently as a Latin text under the name *Libellus de diluviis* see Hasse and Büttner 2018, 313–370.

[19] Hirai 2012, 71–83.

[20] Bigotti 2019, 149–169.

[21] Bigotti 2019, 169–183.

22.5 Vegetable Machines

From this standpoint, it is usually presumed that the spreading of mechanical reasoning, along with the mechanization of life's functions and the rejection of intentionality to the entire non-human sphere, led equally to the rejection of the whole apparatus of Aristotle's natural philosophy with its essences, souls, and virtues. A closer look at the philosophies and theories of the seventeenth and eighteenth century, however, reveals that this presupposition is only partially true. An appeal to vegetable metaphors, analogies, powers and, to a certain extent at least, even to the notion of a vegetative soul was in fact retained. As a result of the impact that Cartesian philosophy exerted on seventeenth-century science, a substantial rift between mechanical philosophy and natural history was introduced in the early modern disciplines of knowledge, due to the former's inability to make sense of the simplest functions in living beings, with the sole exception of the movement of muscles and fibres. As a consequence, the notion of a vegetative soul survived in natural history where it could be used to substantiate, analogically, claims about the natural continuity of living beings ('*natura non facit saltus*'). And yet, in a sense, the notion lived on even in mechanical philosophy, if only as an attempt to link the complex functions in animals to the simplest natural operations, by means of a systematic study of plants' functions.

An example of the former kind of survival can be found in seventeenth-century ovism, most usually considered a direct rebuttal of the model of 'spontaneous generation'. Indeed, Harvey's thesis that all living beings ought to develop out of an egg (*omne vivum ex ovo*) rests firmly on the idea that eggs, much like the plant's seeds, are the minimal and most fundamental elements necessary to the propagation of the species:

> We, however, maintain (and shall take care to show that it is so), that all animals whatsoever, even the viviparous, and man himself not excepted, are produced from ova; that the first conception, from which the foetus proceeds in all, is an ovum of one description or another, as well as the seeds of all kinds of plants. Empedocles, therefore, spoke not improperly of the oviparum genus arboreum, "the egg-bearing race of trees". The history of the egg is therefore of the widest scope, inasmuch as it illustrates generation of every description. (Harvey 1651: 2) […] The egg, moreover, and this is especially to be noted, corresponds in its proportions with the seeds of plants, and has all the same conditions as these, so that it is to be regarded, not without reason, as the seed or sperma of the common fowl, in the same way as the seeds of plants are justly entitled their eggs, not only as being the matter or that from which, but the efficient or that by which the pullet is engendered. In which finally no part of the future offspring exists *de facto*, but in which all parts inhere *in potentia*. (Harvey 1651, 76)

Unlike generation, whose mechanical explanation fell short of carrying conviction even amongst the mechanical philosophers themselves, nutrition and growth became subjects of investigation in their own right, with mechanical models applied especially to vegetation. Most of them relied on newly available technology, such as the instruments invented by Santorio Santori (1561–1636) and adopted by his followers.

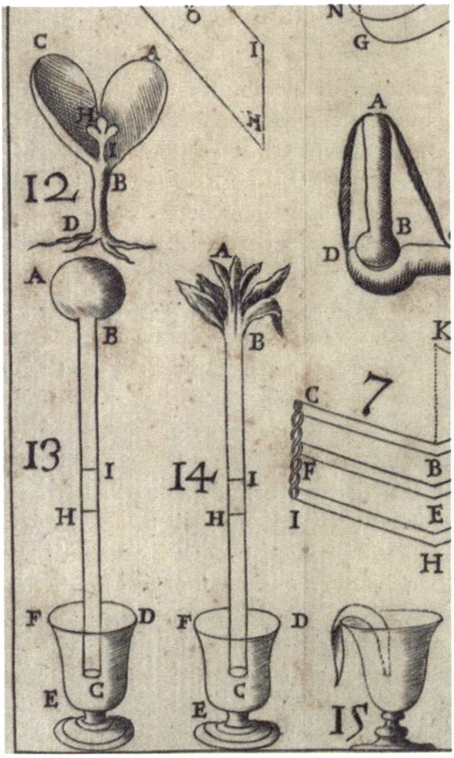

Fig. 22.8 Germination and growth of a plant seed explained as a result of the rarefaction and condensation process occurring in a thermometer: 12) plant seed, 13) Santorio's thermometer, 14) growth of a plant. From Giovanni Alfonso Borelli, *De Motu Animalium* (Pars II, 1685), Tabula XVIII, detail

Giovanni Alfonso Borrelli (1608–1689), for one, borrowed Santorio's thermometer to explain mechanically the germination and nutrition of plants by means of the action of rarefaction and condensation occurring within the stem (Borelli 1685: 260–263) [Fig. 22.8],[22] while Stephen Hales (1677–1761) applied the methods of Santorio and his followers in his *Vegetable staticks* (London 1727) to understand how plants were embued with their nutrients from the soil, what effects air had on their vegetation and how sap was maintained and distributed throughout the stem and the leaves. If Santorio's studies on metabolism (*Ars de statica medicina*, 1614) were originally conceived as a means to measure how the human body keeps the balance between intake and output by constantly discharging a specific quantity of insensible matter (*perspiratio insensibilis*), Hales retraces this mechanism back to its very origin, first in the vegetable realm, and then in animals (*Haemestaticks*, 1733). Notably, by virtue of the very functional analogy between blood and sap, he explores how plants maintain a homeostatic balance of their fluids which allows them, by decreasing or increasing the quantity of sap at any given moment, to select and absorb different nutrients from the ground (Hales 1727, iii) a procedure which

[22] Borelli 1685, *Propositiones* CLXXV–CLXXVII, 260–263.

is then transferred to the human blood, whose pressure Hales quantified by inserting fine tubes of glass into animals' arteries and then measuring the height to which the column of blood rose (Hales 1733, 1–3).[23]

Although motivated by different purposes from those of Harvey, both Borelli's and Hales' experiments transferred to the vegetable realm either the simplest mechanisms of physics or the basic functions by means of which the human body was meant to sustain itself. Furthermore, as Fabrizio Baldassarri has recently argued, Descartes himself shared similar preoccupations with plants for, although he did not deal with botany in any depth and was unclear as to whether to consider plants as inanimate or living beings, his studies developed along the same lines as those of seventeenth-century naturalists, aiming at linking the study of plants with the simplest functions of the body and both to the general laws of mechanics (Baldassarri 2018, 511; Baldassarri 2019, 48).

22.6 Vegetable and Animal Together: Organisms

Another way in which the vegetative soul survived the development of mechanical philosophy was by morphing into 'organism'. Seen as the instruments of a *soul*, which provided them with the ability to preserve themselves, organisms were nevertheless understood mechanically as well-organised parts put in motion by a material agent. Plants made an exceptional case in this direction, as they represented easily available specimens ready to dissect and yet, at the same time, capable of resisting the most adverse conditions.

This resistance to decay was something that was neither necessary nor relevant for mechanical philosophers. In providing new conceptual and scientific instruments for the study of nature, in fact, Descartes and his followers had not been successful in explaining how living beings could resist decay or what allowed them to adapt to the changes occurring in the environment. The elementary laws of mechanics were reckoned as insufficient to explain this aspect of life, and the very fact that Descartes' subjective idealism remained unconcerned with the material processes underpinning the emergence of thought and life – in that it strongly denied any such connections – provided new stimulus to re-thinking vegetable powers within a different conceptual framework. Hence, when the organism resurfaces in late seventeenth-century natural philosophy, first with Friedrich Hoffmann (1660–1742) and then within the Leibniz-Stahl controversy around the nature of organisms (1709 but published posthumously in 1720),[24] it did so by bringing with it a new understanding of vegetation, as most notably regarding functions such as irritability and self-preservation, besides the most common nutrition and generation.

[23] Booth 1977, 794.

[24] Duchesneau 2014, 100. See also Duchesneau's contribution to this volume.

Georg Ernst Stahl's (1659–1734) concept of organism provides a good example of this turn. Objecting to a purely mechanical explanation of life on the grounds of his chemical theory, Stahl admitted a single vital agent that selected nutrients and allowed their chemical combination to be constantly kept under control, in order for the body not to be destroyed by the increased acidity of the resulting compound.[25] Stahl's model of reasoning hinges so largely on a concept of life coinciding with plants' capacity to select their nutrients and to grow, that Leibniz in his reply to him (*Animadversion* 9) could claim: "it is in the author's power to call 'life' what others call 'vegetation'". As François Duchesneau has noted, in fact, "For Stahl the physiology of generation was an analytical extension of that of nutrition" (François and Smith 2016, xxxii).

22.7 "The Plant-Man"

If on a very different level compared to Stahl's vitalism, the vegetative soul as a model of organisation underpins also the undertakings of physicians such Julien Offroy De La Mettrie (1709–1751). By drawing on the formulaic imagery of the vegetable-animal continuum, La Mettrie's *L'homme plante* (1748) uses plants as an anthropological model which inscribes man in the unity and materiality of life processes:

> In order to judge the analogy which we find between the two main kingdoms <of nature>, it is necessary to compare the parts of plants with those of man and apply to animals what I say with regards to man. In our species like in plants, there is a principal root and some capillaries roots. One is formed by the reservoir of the lumbar region and the thoracic canal, and the other by the lacteal veins. Everywhere we see the same uses and the same functions. Food is carried by these roots throughout the whole of the organised body. (de La Mettrie 1748, 6)

L'homme plante reminds us of one of the main reasons why vegetable powers and analogies were yet again taken seriously in the eighteenth century: they allowed the rehabilitation of the materialistic doctrines that link man to the lowest forms of life, in a way that is opposed, but indeed more powerful, than the mechanical model.

---

[25] Duchesneau and Smith 2016, xxvii: 'Stahl's concept of life stemmed from his inquiries into the formal, material, and final causes of physiological processes in their relation to the structures and composition of the body. It was based on empirical evidence that the organic body is intrinsically corruptible. The structure of the organs on which vital acts depend results from an unstable chemical mixture that is prone to fermentation and dissolution while the persistence of these vital acts is at stake. The right option for the physician was to make life depend on a power that preserved the complex and fragile living aggregate: "The conservation of such a corruptible thing, so that it is not corrupted by its very act, is precisely what should be understood by the current term 'life.' And this is the relation by which the body, as a mere mixture, is to be contrasted with and distinguished from the body as living.'

Seen as the common ground out of which all natural powers arise, in fact, mechanism falls short of explaining how continuity unravels in nature, and how lower forms of life could evolve into the highest, an explanation that fitted perfectly the model of vegetation, due to its double meaning of 'primordial' and 'material'. As such, the metamorphosis which turns man into a plant has nothing to do with poetical images of Ovidian ascendance: if anything it reveals at once the commonalities existing between the vegetable and human realms, insofar, at least, as nature proceeds 'uniformly' in her processes distributing the same characteristics simultaneously in all her products (de La Mettrie 1748, Préface). The project undertaken in *L'homme plante* goes therefore in the direction of strengthening such a connection even further, by reducing man to its primordial instincts of nutrition and reproduction:

> We can consider the virgin, or rather non-pregnant, womb or - if one prefers - the ovary, to be like a seed not yet fertilised. The woman's stylus is her vagina; her vulva and her *mons veneris*, with the odour given off by the glands in these parts, correspond to the stigma; and these things – i.e. the womb, the vagina and the vulva – form the pistil, which is the name that modern botanists give to all the female parts of plants. I compare the pericarp to the womb in a state of pregnancy because its function is to cover the foetus. We have our seed, like the plants, and it is sometimes very abundant (de La Mettrie 1748, 12–13).

It is important to highlight that La Mettrie's remarks were not isolated but part of an ongoing debate between promoters and detractors of the application of mechanical philosophy to life, certainly including Paul-Henri Thiry D'Holbach (1723–1789) and Claude Adrien Helvétius (1715–1771). As Kathleen Welmann has noticed (Wellmann 1992, 3), these authors have been the object of much neglect for their materialism which seemed to fall back to dogmatic assumptions. In this sense, Welmann's quotation of Cassirer's contention that the materialists sign a "retrogression into the dogmatic mode of thinking which the leading scientific minds of the eighteenth century oppose and endeavour to eliminate"[26] might help understanding why the position of La Mettrie, Helvetius and D'Holbach can be taken in countertendency to the scientific spirit of the Enlightenment. Indeed, one of the reasons for reclaiming the plant-animal continuity and the vegetative soul was the fact that mechanical philosophy relied directly on a dualistic account of mind and matter, a presupposition consistent with religious and moral concerns. It was probably such presuppositions that convinced La Mettrie, who had tried to carve a mechanism to fit his materialistic purposes in *L'Homme Machine* (1747), to withdraw provisionally from that endeavour one year later. The separation of mind and body presupposed by Descartes' mechanism entailed the possibility of subtracting the mind from the very possibility of analysis that mechanical philosophy itself had successfully inaugurated (de La Mettrie 1753: XXXVIII). By claiming back the model of 'continuity' and 'uniformity' proper to the vegetable understood as the most fundamental manifestation of nature, La Mettrie attempted to counter dualism, an undertaking that, in his *Système d'Épicure* (Berlin 1750, Amsterdam 1753), morphed natural continuity into a blind evolution from inert matter:

[26] The original reference is to Cassirer 1954, 124.

> Can all animals, and thus man, whom no wise man even thought of removing from that category, be truly the sons of the earth, as the fable says of the giants? As the sea perhaps originally covered the whole surface of our world, could it not have been itself the floating cradle of all the beings eternally enclosed in its breast? That is the theory of the author of Telliamed which comes down more or less to that of Lucretius; for it still needs the sea, absorbed by the earth's pores and consumed little by little by the sun's heat over an infinite period of time, to have been forced on withdrawing to leave the human egg beached on the shore, as it sometimes leaves fish. As a result of which, without any other incubation than the sun, man or any other animal could come out of its shell, as some still hatch out today in hot countries, and as chickens do in warm dung thanks to the naturalist's art. (de La Mettrie 1753, XXXII, transl. in de La Mettrie 1996)

The notion of evolution had the possibility to neutralize any necessity to claim intentionality or supernatural powers in the development of life. Thus, the likening of men to plants, and plants to proto-humans ushered in a reductionism and materialism which could be adopted successfully also at a political level, first and foremost against religion: the eighteenth-century attempt at rewriting the natural history of the soul by focusing on the powers of nature works, therefore, also in this direction. And yet, it is a direction that is neither new nor original. It takes on the tradition of the *historia naturae* in ways that are both traditional and radical, and eventually reframes them, so as to counter the path that goes from the Cartesian to the Kantian idealism. The involuntary epigones of this very movement, were – much unknowing – Carl von Linné (1707–1778) and Johan Wolfgan von Goethe (1743–1832). The conceptual remnants of the vegetative soul were still sufficient to Linné to elaborate a system of botanical classification whose foundations rested on reproduction and therefore on the morphological analysis of the sexual organs of flowers, while the vegetable realms manifested themselves to Goethe as the essential clue to crack the laws of metamorphosis.

In this last respect, it may be seen as an utter joke that the vegetable world, which has historically been denied both motion and perception, came finally to be regarded as the very embodiment of both. And yet this outcome is perhaps less surprising when we consider, as Goethe suggests us, that natural change is most apparent in all its flamboyant richness precisely in the vegetable world:

> *Gaze on them as they grow, see how the plant*
> *Burgeons by stages into flower and fruit,*
> *Bursts from the seed so soon as the fertile earth*
> *Sends it to life from her sweet bosom, and*
> *Commends the unfolding of the delicate leaf*
> *To the sacred goad of ever-moving light!* (Goethe 2009, 1)

In the end, and by way of conclusion, it is important to remind ourselves that the vitality of the imagery associated with the vegetative life is not only an evolutionary truism but has deep roots in our anthropological desire to feel part of the living world, from which our rationality somewhat puts us apart. And yet, such a desire is indeed difficult to recapture in a single philosophical tradition or formula. Eventually, it rests on the observation that the functions by means of which plants grow, reproduce, and die, make us – who share these very functions with them – something more than thinking monads.

References

Baldassarri, Fabrizio. 2018. Descartes' Bio-Medical Study of Plants: Vegetative Activities, Soul, and Power. *Early Science and Medicine* 23 (5–6): 511–529.

Baldassarri, Fabrizio. 2019. The Mechanical life of Plants: Descartes on Botany. *British Journal for the History of Science* 52 (1): 41–63.

Baudon-Millot, Véronique, and Antoine Pietrobelli. 2005. Galien Ressuscité: édition princeps du texte grec du 'De propriis placitis'. *Revue des Études Grecques* 118 (1): 168–213.

Bianchi, Massimo Luigi. 1987. *'Signatura Rerum'. Segni, Natura e Conoscenza da Paracelso a Leibniz.* Rome: Edizioni dell'Ateneo.

Bigotti, Fabrizio. 2019. *Physiology of the Soul. Mind, Body and Matter in the Galenic Tradition of Late Renaissance (1550–1630).* Turnhout: Brepols.

Bodin, Jean. 1596. *Universae naturae theatrum.* Lyon: Roussin.

Booth, Jeremy. 1977. A Short History of Blood Pressure Measurement. *Journal of the Royal Society of Medicine* 70 (11): 793-799.

Borelli, Giovanni Alfonso. 1685. *De motu animalium.* Leiden: Van Der Aa, Boutesteyn, De Vivie, Gaesbeeck.

Brunczwik, Alexander. 1748. *Tractatus in Aristotelis logicam.* Hungarian Franciscan Library, Budapest. Item no. 017053.

Cassirer, Ernst. 1954. *The Philosophy of the Enlightenment.* Boston: Beacon Press.

Chambers, Alexander B. 1961. 'I was but an Inverted tree': Notes toward the History of an Idea. *Studies in the Renaissance* 8 (1): 291-299.

de La Mettrie, Jean Offray. 1748. *L'Homme plante.* Voss: Potsdam.

———. 1753. Système D' Épicure. In *Oeuvres philosophiques de Mr. De La Mettrie.* Amsterdam

———. 1996. *Machine Man and Other Writings*, trans and ed. Thomson, Ann, Glasgow: Cambridge University Press.

Della Porta, Giovanni Battista. 1588. *Phytognomonica, octo libris contenta.* Naples: Salviano.

Duchesneau, François. 2014. The Organism-Mechanism Relationship. An Issue in the Leibniz-Stahl Controversy. In *The Life Sciences in Early Modern Philosophy*, ed. Ohad Nachtomy and Justin E.H. Smith, 98–114. New York: Oxford University Press.

Duchesneau, François and Justin Smith, eds. 2016. *The Leibniz-Stahl Controversy.* New Haven\ London: Yale University Press.

D'Andrea, Giovanni. 1506. *Arbor consanguinitatis cum suis enigmatibus.* Nuremberg: Hieronymus Hoelzel.

Fabrizi d'Acquapendente, Girolamo. 1579. Methodus Anatomica (Ms). *Forschungsbibliothek Gotha, Chart. A* 628: ff. 221r–ff. 239r.

———. 1600. *De formato foetu.* Venice: Bolzetta.

Ferrini, Maria Fernanda. 2012. Animi a corporis motibus non sunt impassibilies'. I 'Physiognomica' del 'Corpus Aristotelicum' nel 'De humana physiognomia' di Giovann Battista Della Porta. *Rivista di Cultura Classica e Medioevale* 54 (2): 333–361.

Findlen, Paula. 1990. Empty Signs? Reading the Book of Nature in Renaissance Science. *Studies in History and Philosophy of Science, Part A* 21 (3): 511–518.

———. 2001. Building the house of knowledge. In *The Structure of Knowledge: Classifications of Science and Learning since the Renaissance*, ed. Tore Frängsmir, 5–52. Berkley: University of California.

Giglioni, Guido. 1996. Campanella e Glisson. Motivi ilozoistici nella filosofia inglese della seconda metà del Seicento. *Bruniana & Campanelliana* 2 (1/2): 237–224.

Goethe, Johann Wolfgang. 2009. *The Metamorphosis of Plants* (transl. Miller, Gordon L.). Cambridge, MA\London: The MIT Press.

Hales, Stephen. 1727. *Vegetable Staticks or an Account of Some Statical Experiments on the Sap in Vegetables being an Essay on the Natural History of Vegetation.* London: Innys.

———. 1733. *Statical Essays: Containing Haemestatics.* London: Innys.

Harvey, William. 1651. *Exercitationes de generatione animalium.* London: Du-Gardianis.

Hasse, Dag Nikolaus, and Büttner Andreas. 2018. Notes on Anonymous Twelfth-Century Translations of Philosophical Texts from Arabic into Latin in the Iberian Peninsula. In *The Arabic, Hebrew and Latin Reception of Avicenna's Physics and Cosmology*, ed. Hasse Nikolaus Dag and Bertolacci Amos, 313–370. Berlin/Boston: De Gruyter.

Hirai, Hiro. 2012. Il calore cosmico di Telesio fra il *De generatione animalium* di Aristotele e il *De carnibus* di Ippocrate. In *Bernardino Telesio tra filosofia naturale e scienza moderna*, ed Mocchi Giuliana, Plastina Sandra, Sergio Emilio, 71–83. Roma: Fabrizio Serra. Translated into English in Hirai, Hiro. 2019. Telesio, Aristotle, and Hippocrates on Cosmic Heat. In *Bernardino Telesio and the Natural Sciences in the Renaissance*, ed. Pietro Daniel Omodeo, 51–65. Leiden: Brill.

Holmes, Brooke. 2017. Pure life: the Limits of the Vegetal Analogy in the Hippocratics and Galen. In *The Comparable Body – Analogy and Metaphor in Ancient Mesopotamian, Egyptian, and Greco-Roman Medicine*, ed. Z.J. Wee, 358–386. Leiden/Boston: Brill.

Huarte, Juan de San Juan. 1594. *Examen de Ingenios. The Examination of Mens Wit* (Trans. R.C. Esquire, after Camillo Camilli 1582). London: Adam Flip.

Ingensiep, Hans Werner. 2001. *Geschichte der Pflanzenseele. Philosophische und biologische Entwürfe von der Antike bis zur Gegenwart*. Stuttgart: Kröner.

Kent Sprague, R. 1991. Plants as Aristotelian Substances. *Illinois Classical Studies* 16 (1/2): 221–229.

Lovejoy, Arthur O. [1933] 2011. *The Great Chain of Being. A Study of the History of an Idea*. Cambridge, MA/London: Taylor and Francis.

Lull, Ramon. 1515. *Arbor Scientiae Raimundi*. Lyon: Badius.

Maclean, Ian. 2011. The Logic of Physiognomy in the Late Renaissance. *Early Science and Medicine* 16 (4): 275–295.

Onians, Richard Broxton. 1951. *The Origins of European Thought. About the Body, the Mind, the Soul, the World Time, and Fate*. Cambridge: Cambridge University Press.

Paracelsus. 1553. *Labyrithus medicorum errantium Teophrasti Paracelsi*. Vischer: Nuremberg.

Perler, Dominik, ed. 2008. *Transformations of the Soul, Aristotelian Psychology 1250–1650*. Leiden: Brill.

Porter, Martin, ed. 2005. *Windows of the Soul: The Art of Physiognomy in European Culture 1470–1780*. Oxford: Clarendon Press.

Rueff, Jakob. 1587. *De conceptu et generatione hominis de matrice et eius partibus*. Frankfurt: Libri sex.

Santori, Santorio. 1614. *Ars Sanctorii Sanctorii de statica medicina*. Venice: Polo.

Stolberg, Michael. 2018. Learning Anatomy in Late Sixteenth-Century Padua. *History of Science* 56 (4): 381–402.

van den Spiegel, Adriaan. 1626. *De formato foetu liber singularis*. Frankfurt: Meriani.

Wellmann, Kathleen. 1992. *La Mettrie. Medicine, Philosophy and Enlightenment*. Durham/London: Duke University Press.

Chapter 23
The Notion of Vegetative Soul in the Leibniz-Stahl Controversy

François Duchesneau

Abstract In the controversy that arose between Gottfried Wilhelm Leibniz and Georg Ernst Stahl following the publication of the latter's *Theoria medica vera* (Stahl 1708), both theoreticians strove to distance themselves from what they took to be the classical conception of the vegetative soul, while relating and partly reducing their opponent's doctrine to it. I shall first attempt to establish their respective meanings for such a soul and the reasons why they would challenge the relevance of that notion for physiology. I shall then try to show how they nevertheless integrated explanatory functions equivalent to those attributed to that metaphysical entity into the theoretical frameworks they proposed and favoured. As a consequence of alternative interpretations of this kind, eighteenth-century physiologists will tend to link vegetative life either with "organic mechanism" or with various kinds of "vital principle," akin to Stahl's *anima*.

As a proponent of a physiology that linked every organic operation in the living body with a soul's activity and direction, Georg Ernst Stahl (1659–1734), the famous author of the *Theoria medica vera* (1708), would be expected to sympathize with traditional views on the vegetative soul. Instead, he proved to be a staunch critic of such views. As for Gottfried Wilhelm Leibniz (1646–1716), his physiological theory entailed a substantive link between monads and organic bodies, with plain monads' perceptions and appetites affording sufficient reasons for so-called vegetative processes. But Leibniz as well seems to have held a critical stance towards appeals to vegetative souls for causal accounts of organic processes. Although in the controversy that they fueled against each other they argued for adverse conceptions of the soul-body relationship, they both tended to disqualify the explanatory role that the concept of vegetative soul might have represented for former physiologists. They would dilute those conceptions into reshaped doctrines

F. Duchesneau (✉)
Département de Philosophie, Université de Montréal,
C.P. 6128 succursale Centre-ville, Montréal, Québec H3C 3J7, Canada
e-mail: francois.duchesneau@umontreal.ca

F. Baldassarri, A. Blank (eds.), *Vegetative Powers*, International Archives of the History of Ideas Archives internationales d'histoire des idées 234,
https://doi.org/10.1007/978-3-030-69709-9_23

they repelled, and in so doing they seem to have influenced the removal of vegetative souls from the lexicon of experience-based physiology. But, at the same time, they achieved some transposition of the operations to which vegetative souls had been previously dedicated, and thus provided new models for analyzing vegetative functions. In the famous controversy that took place between the Halle medical theorist and the Hanover scientist and philosopher between 1709 and 1711, the direct confrontation between animism and organic mechanism entailed antagonistic explanatory schemes about the basic processes of life, which they would connect differently with mental cognitive operations.

23.1 Stahl's Critical Stance About the Vegetative Soul[1]

What does one find concerning the vegetative soul in Stahl's *Theoria medica vera* (1708)?[2] In the first part of the first section of the *Theoria* devoted to the scope and end of the body, Stahl precisely referred to ancient conceptions of the vegetative soul. He criticized them for appealing to agents that differed from the unique and real soul and would exert vital functions outside of the latter's jurisdiction. According to Stahl, more recent doctrines upheld similar erroneous views, but the root of such deviances had to be sought in Galen and his disciples, even if the ancient physiologist's name was not mentioned.

> A more ancient opinion which attributed other agents to the human body besides the rational soul, conceived of them as other souls and proposed them for these functions, to wit a vegetative soul and a sensitive one. To the vegetative one they referred the functions of life and nutrition; to the sensitive one the operations of sense and motion. At the exclusion of those souls, they made the sole strictly human soul capable of intellection. However, they granted the vegetative and sensitive ones a knowledge (*gnosin*) according to which they administered their whole business in able, orderly and appropriate fashion. (Stahl 1737: 206)

Another version of the same doctrine that should probably be linked with the Aristotelian tradition was accounted for in the following article:

> Others, formerly as it appears, thought that a unique human soul, termed rational because of its specific and thus superior energy (*energiam*), would own also these less noble and thus lower energies, in virtue of a quite plausible argument, that what is capable of more would also be of less. However, by an inveterate tendency to daringly multiply abstract concepts, they employed this otherwise quite good principle for introducing and inconveniently framing up diverse faculties of that soul, *qua* substantial energies. And they would amplify and augment this framing-up by their very use of terms, since they did not seem to speak of a plain capability anymore, but, by a nomenclature of natural types, to set onto stage an absolutely real power to act, active and positively efficient, which, under the auspices of the soul, does and achieves what they used to attribute to the latter. (Stahl 1737: 206–207)

---

[1] On Stahl's physiological and pathological theories, see de Ceglia (2000); de Ceglia (2009); Duchesneau (1998); Duchesneau (2012); Geyer-Kordesch (2000).

[2] I shall quote from the second and revised edition of Stahl's *Theoria medica vera*: Stahl (1737). The translations from Latin to English are mine.

In the *Negotium otiosum* (Stahl 1720), in which the Leibniz-Stahl controversy was first edited, similarly grounded statements entailed the rejection of the vegetative, as well as the sensitive, soul according to both interpretations. For instance in Stahl's *Summary of the principal points of doubts* (*Conspectus*), one reads about Galen and others of that trend:

> The frivolous nonsense that even Galen himself, along with the more ancient Sophists, oppose to this truth [that the soul should be attributed the energy to directly excite the motions appropriate for the body's conservation and functions] by the threefold order of souls—the vegetative, sensitive, and rational—is known in the Schools. In truth, the physicians, even if many of them have related something on this matter, attribute these three faculties to a single soul, with only the spirits being called in for support, lest the soul should appear to act immediately on the body. But, by this very fact, they brought on the suspicion of tacitly acknowledging the reason, which was deemed to constitute the foundation of the adverse opinion, namely, that the incorporeal soul could have no immediate interaction with the body. Thus, because of that hypothesis, they remained constantly exposed to the objections, far from absurd, of their adversaries. (Duchesneau and Smith 2016: 223)

An analogous argument found its way into Stahl's reply to Leibniz's exception XXI:

> This disagreement [the presumed drastic separation between that which thinks and that which moves] is not, as the most ill-informed imagine, something new, and still less something due to innovators who jokingly run them down, since it is best noted that Galen himself, already for this reason, is found guilty of a most delusory levity, since he subordinated to the dreams of philosophers (certainly the Sophists) the familiar opinion received from physicians from much earlier times, which attributed to one and the same human soul (according to the hackneyed rule: whoever can do more can do less) the faculty of understanding as well as of sensing and moving, and finally of nourishing, when he attributed to one and the same human body three plainly distinct souls: the rational, the sensitive, and the vegetative. Of which none has any right of dominion in the others' spheres, etc. etc. (Duchesneau and Smith 2016: 361)

These statements need first to be interpreted in light of Stahl's doctrine of the soul-body interrelation and his mechanism-organism distinction. Then we shall be able to turn to Stahl's opinion on the consequences that followed from the ancient doctrine of vegetative soul as resumed in late seventeenth and early eighteenth century physiological mechanism.

Stahl's physiology seeks to explain how life is preserved and restored from inner disturbances in a body that is a complex and unstable aggregate of chemical compounds, ever subject to imminent corruption.

> The conservation of such a corruptible thing, so that it is not corrupted by its very act, is precisely what should be understood by the current term 'life.' And this is the relation, according to which the body, as a mere mixture, is to be contrasted with and distinguished from the body as living. (Stahl 1737: 200)

In fact, the composition and conformation of these heterogeneous organic aggregates require a formative and preservative principle that cannot be reduced to any mechanical apparatus, however sophisticated. The integration of the various organic parts so that they yield their specific functions demands a hegemonic agent that can actualize and maintain this combination through inner motions. Failing this agent's constant intervention for harmonizing compounds and mechanisms, the living body would instantly corrupt and dissolve. Considering the circulatory processes of vital fluids in particular suggested a relation of structural-functional order dependent on a dynamics of vital motions, which in turn meant that a single operative agent would coordinate the entire system of functional acts. Hence Stahl's strict opposition to doctrines that would appeal to several soul-like agents to account for the constitution and operations of organic devices as means to compensate for the shortcomings of mechanist explanations. Vegetative souls would enter that category, but Stahl extended it to fit alternative notions as well, like Joan Baptista van Helmont's *archaeus*.[3] Stahl indeed favoured the hegemony of a single soul that formed, organized, operated and preserved the global economy, what he called the "organism," of the body's heterogeneous components and processes.

On the other hand, the requirement for constituting and preserving this organism is that the agent responsible for that task own appropriate intentional capacities, which involve relevant mental aptitudes. In the chapter of *Theoria medica vera* devoted to the ends and finality of the body, countering purely mechanist interpretations, Stahl appealed to the inherent teleology of physiological processes, that is, on the material side, to the instrumental destination of organs, and on the formal one, to adequate cognitive aptitude for achieving such end results. (Stahl 1737: 202–210) He argued that the functions attributed to the soul imply that they directly elicit intentional corporeal motions following the acts of that psychic principle, a function that surpasses the potential of any concurring material causes. For instance, Stahl insisted on the inherent finality of organic processes like secretions and excretions:

> [From the analysis of vital motions] one draws the true reason of a properly medical physiology, for thence one understands not only that the whole process of secretions and excretions is grounded in the proportion of matter to be moved, rather than in any other necessity arising from the organs, but that in addition the true and firm foundation for a motile determination affecting this type of matter is properly and simply based on a goal, and to be more precise, on an end pre-set by way of estimation and intention, namely, that the integrity of the body be preserved not as being organic at present, but as being destined to remain organic. (Stahl 1737: 430)

A vital agent is needed in order to control the formation of organic mixtures as a means for achieving and preserving the constitution of the living bodies of plants and animals. Further to this organic-genetic activity, the soul has to assume the actualization and direction of the diverse specialized functions that the body yields. Thus, organism, that is the goal-directed structuring and animating of the complex organic machine that would naturally dissolve otherwise, exists to provide two types of functional acts[4]: acts of vital conservation on the one hand, which include all formative, nutritive, and developmental processes; on the other hand, specialized processes of motility connected to what later physiology will characterize as

[3] On Joan Baptista van Helmont (1577–1644), see Pagel (1982); Duchesneau (1998): 18–29.

[4] On the invention of the concept of organism and its original meaning, see Cheung (2010); Duchesneau (2014); Duchesneau (2018): 29–49.

relational or animal life, which more directly involves sensitivity and various modes of cognition. This physiology conceived both types of acts to depend on a single agent: this agent intervenes at once in the body that it builds and in which it fulfils vital functions, and it produces functional mental acts, which correspond to the so-called animal functions; but both types of acts flow from basic cognitive operations, which represent this agent's essential nature and are accomplished by and through motions. And motion itself is taken to be something incorporeal that is added to bodies and is generically of a similar nature and "universal disposition" (*universae habitudinis*) with the soul itself. (Stahl 1737: 208)

Above all, Stahl objected to modern physiologists who reduced motions taking place in living bodies to local motions, that is to say, to modes of material reality, necessarily accomplished without a final destination, on which the mechanical analysis of organic processes should be based. He felt that organic motions were to be considered as end-directed and functionally adapted, which made them soul-dependent. He thus professed:

> Vital motions are directed and executed by the soul's action alone: they are truly organic acts instituted by a superior active cause in the corporeal instruments in order to produce certain effects, not only generically determined and specifically necessary, but also in a most special way, particularly and subtly proportioned to fit the requirements flowing from various temporal circumstances and accidental external causes. (Stahl 1737: 428–429)

As vital motion is the generic mode of all acts performed by the animating principle in living beings, no reason can be found why we should attribute lower-level functions to a soul distinct from others which would be endowed with either sensitive or cognitive dispositions. Neither would there be any justification for drastically segregating vegetative from sensitive-motive and intellective faculties within a presumed hegemonic soul. Indeed, Stahl had to counter the objection that because there are many organic processes that seem to fall short of cognitive awareness and whose relationship to an intellective soul would not be empirically supported, we should favour a set of heterogeneous soul-like agents as grounding principles for essentially diverse functions. This is why he brought in his distinction between *logos* (*ratio*), a kind of rational instinct underpinning vegetative and spontaneous processes, and *logismos* (*ratiocinatio*), a power that implies sensory representations and their assimilation within discursive and voluntary processes. (Stahl 1737: 38) And there would be strict analogy and continuous transition between both forms of reason as they yield all operations that qualify life through its manifold expressions.

A further argument for rejecting several distinct animating principles or powers ruling over the organic body stems from the latter's individuality and unity, and the synergies of motions that ground and warrant its constitution. There could not be but a unique form for this integrative whole of material parts and processes, which represents the true destination of the soul in all its operations. Stahl thus recommended that the physiologists

> learn to rightly assess the nature of motion as well as that of the immediate agents of motion, the presence of the soul in the body, that is, in a single individual, and nowhere else, and finally the fact that the full set of actions of this soul, from what we know and truly perceive, are exerted only in connection with, in, and by corporeal things. (Stahl 1737: 378)

Interestingly enough, Stahl accuses the doctrine of vegetative souls for having fostered abstruse consequences, which have ultimately brought support to the materialists, characteristically represented by the mechanist natural philosophers, those modern disciples of Democritus and Epicurus. Thus the multiplication of causal agents has favoured appealing to spirits for the explanation of several physiological processes, and consequently, to corpuscular clusters endowed with dynamic dispositions. According to Stahl's interpretation, as it was presumed that no direct exchange could take place between immaterial and material entities, it seemed that the standing and formal powers of the rational soul were endangered by the supposition of inferior souls or animating faculties. So it came to be presumed in late scholastic philosophy that the vegetative and similar operations might be achieved by undetermined "spirits" of various kinds, subservient to the true soul as its instruments. Those spirits, which will later be interpreted in micro-corpuscular fashion, would own no spontaneous powers, but would be activated, delivered and distributed through the structure by the soul, which alone knows what is necessary or useful for the various parts of the structure in such and such circumstances. However, according to Stahl, this doctrine was modified so as to integrate a distinction between influent and inherent spirits, the former provided by surrounding bodies, the latter possessing an intense active capacity of their own which would endogenously intervene for transmitting and steering up reactions throughout the organic body. The incredible speed of these presumed nervous transmissions suggested to some an analogy with the phenomena of light, and others would even consider the spirits involved to be of the same nature as the ingredients of light.

The transfer of dynamic and intentional powers of animation to spirits, that is, to special organic constituents, seemed to Stahl to have occasioned van Helmont's typical system. Hence Stahl's interpretive statement:

> As there are diverse consequences of doctrines on matters that suffer from inherent difficulties, the ill-considered implications of these fictions generated out of these alternatives a new confused doctrine concerning the *archaeus*, which, according to all other descriptions, is the Ancients' vegetative soul. Thus, from the hypothesis of spirits, in Helmont's, the fiction of inherent and influent spirits, and even of inherent ones proper to each organ and plainly endowed with a specific nature, spread out: at places, the Ancients' vegetative soul is really described, but it is referred to by the name of spirit, sometimes called vital, sometimes animal. Meanwhile, to both archaeus and spirit, a perfect awareness of their action (*gnosis agendi*) is attributed, otherwise said, a somewhat impressed and most regular power to act: and this absolute power to act is not even such that it is in and by itself without assistance from the soul, but also such that the soul has been so excluded from archaeus and spirit that it cannot even concur with them. (Stahl 1737: 208)

Stahl's story about the transition from the Ancients' vegetative soul to the late scholastic theory of spirits and from the latter to van Helmont's vital medicine can indeed be considered a fiction of his own making. But my point is the following. Stahl had to repudiate from his physiology any allegation that he had resumed elements from those former doctrines. As I occasionally pointed out, his conception of the organic body is entirely mechanistic. (Duchesneau 2010: 135–153; Duchesneau 2012: 25–67) It is a clock of some kind, except that, left to its purely material properties, this clock, far from preserving its structures and functional operations, would destroy itself, or rather be destroyed by the very physical-chemical processes flowing from its own constitution in interaction with the physical milieu. The maintenance of life for an integrative organic body can only depend on the action of a single immaterial soul endowed with rational power. This power would conceive and frame up appropriate motions for producing, preserving, and operating the living being's "organism" against the mechanical processes that flow from its material constituents, once deprived of the soul's dominion. Obviously, the operations and functions identified with the vegetative soul still require to be grounded in a non-material agent, but this agent is endowed with a rational capacity of its own that suffices to account for the purposive design of all relevant organic motions, including those that were referred to the operations of higher-level souls or powers. This hegemonic vital agent is an organic mind, and the concept of it will provide the pattern for most notions of vital principle in later physiology.

23.2 Leibniz's Interpretation of the Vegetative Soul[5]

Now what about Leibniz's critical assessment of the Ancients' vegetative soul? In a way, there is much that evokes this notion in his natural philosophy. But, in the writings of his controversy with Stahl, the main allusion to this subject is indirect. Thus at the end of the preamble to his Animadversions on Stahl's *Theoria medica vera*, Leibniz writes:

> It appears that formerly the ancients also accommodated the observation of ends to the utility of medicine, but in this instance the more recent authors of what is called vital medicine, from whose ranks Paracelsus and van Helmont emerge, have relied on a special kind of reason, namely a certain Archaeus, by whose appetite, irritation, quickness, and laziness the actions of the body would be directed. But whatever is true in this agrees with our view; and it is not necessary to conceive of something in body other than what contains, what is contained, and what produces impetus, nor to invoke other appetites than those of the soul. So, therefore, the Archaeus should not be sought after except in the soul and in the corporeal spirits that are in agreement with it, nor do we need some plastic or hylarchic principle, nor various princes or little kings distributed in organs like the cardianax, gastrianax, and other such ones. (Duchesneau and Smith 2016: 29)

And Leibniz ends up including Stahl as the latest proponent of this vital medicine based on admitting various animistic explanatory principles. Now this assimilation of Stahl's theory with a vital medicine which would have resumed the ancients' doctrine of the vegetative soul involved certain interpretive shifts that need to be qualified. Commenting on the preliminary essays of the *Theoria medica vera*, Leibniz was convinced that Stahl took life to depend on a power of a non-material kind that would rule over the structures and operations of organic bodies by

[5] On Leibniz's conception of the living being and his views on physiology, see Andrault (2014); Duchesneau (1998); Duchesneau (2010); Duchesneau (2018); Nachtomy (2009); Roland (2012); Smith (2011).

countering the mechanical and chemical dispositions of their component mixtures to dissolve and return to a non-vital condition. The so-called vegetative functions—organic framing-up, growth, nutrition by intussusception, secretions and excretions—would be materially enacted, not by a pure psychic entity analogous to a perceiving mind, but by some spontaneous non-reflective agent, unfolding blind designs of self-constitution and preservation for its material components. And the image of the archaeus would come immediately to the fore as an incarnation of such a vegetative power, along with Henry More's hylarchic principle or spirit of nature,[6] (More 1659) or Ralph Cudworth's plastic nature. (Cudworth 1678) For Leibniz, on the contrary, all operations identified with vegetative life would be monad-dependent in terms of their formal raison d'être, but mechanism-dependent in terms of their material delivery and unfolding, and this meant that, insofar as they characterize corporeal substances and their specific properties, they are to be accounted for as resulting from complex mechanical devices. This is a clear implication of his Animadversions 8 and 9 on Stahl's *Theoria medica vera*:

> [Stahl] rightly urges at various times a great distinction between living beings and other mixtures. I am in the habit of locating life in perception and appetite. The celebrated author instead places it in that very ability of guarding one's own body against the propensity for destruction, since otherwise the bodies of living beings would be so perfectly fluent that life would resemble salt, as was said in jest of the soul of a pig. (Duchesneau and Smith 2016: 33–35)

> I have referred this back to the force of vegetating, by which the living body perfects, nourishes, repairs, and propagates itself, and this, I maintain, follows from the very structure of the machine, even if everywhere the soul is conspiring. And we see a certain analogue to vegetation in a body that is so very subtle, but so little living, to wit, the flame, which nourishes and propagates itself, and when its aliment begins to be depleted, dashes about in wonderful motions, acting in order to preserve itself. Nor should God have been prevented from giving to bodies much more enduring than ours organs and sense, by which they should at any rate be deemed living beings. But I do not wish to quarrel over words. It is in the author's power to call 'life' what others call 'vegetation.' (Duchesneau and Smith 2016: 35)

From Stahl's reply, it can be gathered that there arises a major difference between the two theories concerning the operations that can be ascribed to the "force of vegetation (*vis vegetandi*) by which the body perfects, feeds, repairs, and propagates itself." (Duchesneau and Smith 2016: 99) But the problem lies with identifying the origin and ontological foundation of this force, not with describing the vegetative functions that are found in all living beings. For Leibniz, it is clearly the case that these operations should be accounted for by what constitutes organic mechanism, that is, conspiring motions in the structures involved, which reflect the perceptive-appetitive states of the dominant monad, as principle of unity of the organic body, at whichever stage of development it is taken to be. From that viewpoint, nothing in the living being may be attributed to a faculty or non-material entity that acts directly in the world of bodies so as to causally determine vegetative processes. Indeed,

[6] For a reedition of *The Immortality of the Soul* by Henry More, see More (1987).

minds, souls, monads form a continuum of substantive principles that underpin those mechanisms, but they do not introduce any formal heterogeneity in the laws that rule over bodily phenomena such that life should be categorized as entailing different natures according to whether it be found in plants, animals, or humans. As a matter of fact, all corporeal substances are monad-endowed organic bodies, but those bodies are entirely dependent on mechanical processes for yielding vital and animal motions. For Leibniz, there cannot be any possible location for vegetative souls or powers in the inner constitution of any category of bodies. This is not to say that one could not identify categories of vital processes that should be accounted for through specific organic processes, but these processes would boil down to sets of mechanisms whose constitutive law reflects the inner sequences of perceptions-appetites in the corresponding monads. The substantive and phenomenal levels should be kept separate for the sake of explaining specific processes, which indeed was not the case with the traditional physiological theories that appealed to the notion of vegetative soul as a causal agent operating in living bodies.

If there could be some apparent equivocation between Stahl's animism and the ancients' vegetative soul doctrine—an equivocation that I have tried to dispel—a similar condition is less easily to be found in Leibniz's conception of the strict correspondence between vital mechanisms in organic bodies and entailment of internal perceptual states in monads. My view is that Leibniz's criticisms against plastic natures which were strictly analogous to those he proffered against Stahl's physiological soul had been targeted to include vegetative souls. In 1705, Leibniz had published his *Considérations sur les principes de vie et sur les natures plastiques par l'auteur du système de l'harmonie préétablie.* (Leibniz 1965: GP VI 539–546) He had been previously challenged by Pierre Bayle to differentiate his views from Cudworth's, whose doctrine of *The True Intellectual System of the Universe* (1678) Jean Le Clerc had brought back to the fore in a series of journal articles against the mechanist conception of living beings, published in the *Bibliothèque choisie* from 1703 to 1706.[7] Leibniz rejected the supposition that plastic natures would intervene to impose, organize and preserve organic processes, as if these could not be explained according to the plain laws of mechanics. But would those laws suffice to account for the end-directness and apparent spontaneity living beings display in the course of physiological phenomena? In the article *Rorarius* of his *Dictionnaire historique et critique* (Bayle 1697), Bayle had presumed that Leibniz's monad-governed organic bodies represented neo-Aristotelian substances whose formal components would actuate functions and endow individual material entities with self-organizing powers—a supposition that would make them equivalent to plastic natures. According to Cudworth, "[Plastic] nature is, either a lower Part or Faculty of some Conscious Soul; or else an Inferior kind of Life by itself, which is not without Soul, but Subordinate to it and dependent on it." (Cudworth 1678: 166) Here, it should be pointed that Cudworth's plastic nature and More's hylarchic principle

[7] See Bayle (1704) in which the latter pointed to the similarity between Leibniz's views on monads and organic beings and those identified with the doctrine of plastic natures. In both theories he denounced a kind of altered neo-Aristotelianism (*péripatétisme rectifié*). (Bayle 1704: 394)

were first meant to characterize a general principle inherent in all of nature that imposed organizing and regulative laws onto physical phenomena which, strictly enacted by efficient mechanical causes, would generate disorderly inorganic effects. Of this general argument, an excellent illustration is afforded in More's *Enchiridion metaphysicum* (1671).[8] But, relative to organic beings, this general plastic nature, or hylarchic principle, or "spirit of nature," would be substituted by souls of diverse capacities holding responsibility for vital, animal, and higher functions. In this case, individual plastic natures or equivalent souls would constitute immanent principles of life within created substances, yielding non-mechanical, goal-directed processes over and above what would result from their bodily make-up.

Leibniz counteracted by dissociating all souls from the condition of what he designated as formal plastic natures. He extended the essence of higher souls, namely perceptive-appetitive capacity to all monads from minds to souls, and from souls to "plain monads" (*monades brutes*). Correlatively, the organogenetic function of plastic natures or vegetative souls would disappear. This organogenetic function and the operative processes deriving therefrom would be transferred to the organic body that develops or unfolds its structures in strict subordination to physical laws, even though the complexity of phenomena may prevent our awareness of the fact *pro tempore*. Preformation provides the sufficient reason for a primordial organization of organic bodies whose temporal delivery yields such a functional ordering of processes as corresponds to the descriptive features of vegetation, severed from any psychic causation. Leibniz's pre-established harmony implies that all changes occurring in the motions of organic bodies can only depend on the interaction of those bodies and the laws ruling over it. And the same applies *mutatis mutandis* to the series of monadic states. Leibniz refers to divine preformation the causation of these series as well the entailment of the laws they deploy, and these laws, mechanical or psychological, suffice to account for the detail of what changes respectively in body and soul. What is truly excluded is the interaction of substantial form and phenomenal matter at the physical level, which seemed to be the case with vegetative souls that would operate directly in and on their organic bodies.

What did the preformation of living beings imply in the context of Leibniz's critique of plastic natures? It implied first that the original production or formation of living beings *qua* machines of nature could not be explained by reference to the laws of mechanism as they apply to inorganic bodies. For instance, Leibniz would laud Cudworth for having argued that vital organization as such could not result from combinations of material parts endowed with corpuscular motions. But, if this applies to the original production of beings, there is no question of allowing incorporeal substances or powers to causally determine the operations and transformations of actual organic bodies, as if those substances could act directly on bodies independently of the laws of mechanism. This objection could be raised against Cudworth's "immaterial plastic principles" as well as against the structuring souls of Aristotelians like Julius Scaliger, Helmont's archaeus, or Stahl's physiological

[8] For a translation of that work, see Jacob (1995).

soul. Hence, secondly, it would be necessary to admit divine preformation as the sole sufficient reason for the initial formation of living beings and the programmed correspondence between the soul's and the organic body's respective developmental laws. Hence, also, Leibniz's critical appraisal of speculative entities similar to vegetative souls:

> To this I can say *non mi bisogna, e non mi basta*, because this preformation and this infinitely complex organism provide me with material plastic natures that meet the need. Immaterial plastic natures, by contrast, are as unnecessary as they are incapable of satisfying it. For since animals are never formed naturally from an inorganic mass, the mechanism, though incapable of producing their infinitely varied organs anew, can at least draw them out of pre-existing organic bodies by a process of development and transformation. (Leibniz 1969: 589)

23.3 Conclusion

What may one conclude concerning physiological theory from this critical appraisal of the notion of vegetative soul by both Leibniz and Stahl? At the time, that concept clearly symbolized a kind of abstract entity representing the unknown cause of a vaguely circumscribed set of organic processes taking place in plants and animals and involving formation, growth, nutrition, secretion, excretion, and similar functions. To support this speculative view, one had to presume that these processes could neither be reducible, on the one hand, to purely physical ones, nor, on the other hand, to mentally-induced operations, more or less tied with sensitive awareness. To a given point, even limited allegiance to the mechanist ideal of the moderns' *philosophia naturalis* required that that concept be either rejected or rescinded. Either one could propose a mechanist reduction of those phenomena and processes over which vegetative souls had been called upon to rule, or an inclusion of the same into the field of reference of a vital principle endowed with efficient and regulative powers of control over mechanisms. The latter option was what Stahl's programmatic animism boils down to. The former one represented the main methodological consequence for physiology to be drawn from Leibniz's theory of the organism as involving preformation and monad–organic body correspondence. What I found astonishing is that both models kept influencing theory-framing in Enlightenment physiology, with at times attempts at producing conceptual mixtures such that one would seem to be faced with resurrected vegetative souls under other names. (Duchesneau 2018) For instance, this seems to be the case with the *vis essentialis* of Caspar Friedrich Wolff as a principle of epigenetic formation in plants and animals, (Wolff 1759) and, quite differently, with the *force de vegetation* of John Turberville Needham as a principle of organic structuration in micro-organisms. (Needham 1750).

References

Andrault, Raphaële. 2014. *La vie selon la raison: Physiologie et métaphysique chez Spinoza et Leibniz*. Paris: Honoré Champion.

Bayle, Pierre. 1697. *Dictionnaire historique et critique*. Rotterdam: Chez Reinier Leers.

———. 1704. Mémoire communiqué par Mr. Bayle. *Histoire des ouvrages des sçavans* 20: 369–396.

de Ceglia, Francesco Paolo. 2000. *Introduzione alla fisiologia di Georg Ernst Stahl*. Lecce: Pensa Multimedia.

———. 2009. *I fari di Halle: Georg Ernst Stahl, Friedrich Hoffmann e la medicina europea del primo settecento*. Bologna: Il Mulino.

Cheung, Tobias. 2010. What is an "organism"? On the occurrence of a new term and its conceptual transformations 1680-1850. *History and Philosophy of the Life Sciences* 32: 155–194.

Cudworth, Ralph. 1678. *The True Intellectual System of the Universe*. London: Richard Royston.

Duchesneau, François. 1998. *Les Modèles du vivant de Descartes à Leibniz*. Paris: Vrin.

———. 2010. *Leibniz. Le vivant et l'organisme*. Paris: Vrin.

———. 2012. *La Physiologie des Lumières: Empirisme, modèles et théories*. Paris: Classiques Garnier.

———. 2014. The organism-mechanism relationship: An issue in the Leibniz-Stahl controversy. In *The Life Sciences in Early Modern Philosophy*, ed. J. Smith and O. Nachtomy, 98–114. New York: Oxford University Press.

———. 2018. *Organisme et corps organique de Leibniz à Kant*. Paris: Vrin.

Duchesneau, François, and Justin Smith. 2016. *The Leibniz – Stahl Controversy*. New Haven: Yale University Press.

Geyer-Kordesch, Johanna. 2000. *Pietismus, Medizin und Aufklärung in Preussen im 18. Jahrhundert: Das Leben und Werk Georg Ernst Stahls*. Tübingen: Max Niemeyer.

Jacob, Alexander. 1995. *Henry More's Manual of Metaphysics*. Hildesheim: Georg Olms.

Leibniz, Gottfried Wilhelm. 1965. In *Die philosophischen Schriften (GP)*, ed. C.J. Gerhardt. Georg Olms: Hildesheim.

———. 1969. In *Philosophical Papers and Letters*, ed. L.E. Loemker. Dordrecht: Reidel.

More, Henry. 1987. *The Immortality of the Soul*. Dordrecht: Martinus Nijhoff.

Nachtomy, Ohad. 2009. Leibniz and the *Logic of Life*. *Studia Leibnitiana* 41: 1–20.

Needham, John Turberville. 1750. *Nouvelles Observations microscopiques avec des découvertes intéressantes sur la composition et la décomposition des corps organisés*. Paris: Chez Louis-Étienne Ganeau.

Pagel, Walter. 1982. *Joan Baptista van Helmont: Reformer of Science and Medicine*. Cambridge: Cambridge University Press.

Roland, Jeanne. 2012. *Leibniz et l'individualité organique*. Montréal/Paris: Presses de l'Université de Montréal/Vrin.

Smith, Justin. 2011. *Divine Machines: Leibniz and the Sciences of Life*. Princeton: Princeton University Press.

Stahl, Georg Ernst. 1737. *Theoria medica vera*. Halle: Impensis Orphanotrophei.

———. 1720. *Negotium otiosum*. Halle: Litteris et Impensis Orphanotrophei.

Wolff, Caspar Friedrich. 1759. *Theoria generationis*. Halle: Litteris Hendelianis.

Chapter 24
Vegetation and Life from Wolff to Hanov

Matteo Favaretti Camposampiero

Abstract A committed supporter of mechanism, Christian Wolff mentions the vegetative soul as a notable example of an "empty term" or meaningless expression. In spite of this unconditional rejection of the vegetative soul, his major disciple in the field of natural philosophy felt the need to reintroduce something similar. In the 1760s, Michael Christoph Hanov revised Wolff's account of vegetation and life by claiming that life is the product of a vegetative force and not of pure mechanism. After reconstructing both Wolff's and Hanov's accounts, this paper explores the reasons and implications of the latter's revival of vegetative powers and argues for its relevance to the early history of biology.

Christian Wolff was not only the author of lengthy volumes on metaphysics, practical philosophy, and mathematics. Especially in the early and middle stages of his academic career, he composed several works on natural philosophy and experimental science, a certain number of which concern what we would now call the life sciences. A staunch supporter of the application of quantitative and experimental methods to the study of living bodies, Wolff by no means neglected the vegetable kingdom. On the contrary, he cultivated agronomy and botany, devising experiments[1] and devoting short treatises and entire sections of his systematic works to the anatomy and physiology of plants.[2]

As with the other parts of Wolff's philosophy of nature, his philosophy of plants is based on the mechanistic view that bodies are machines. All entities that do not fit into the mechanistic picture should be eliminated from the ontology of physics. Accordingly, forces and faculties of the traditional philosophy of nature must be

[1] See Wolff's letter to Leibniz, 15 March 1716 (Leibniz and Wolff 1860: 182).

[2] See especially Wolff 1718a; Wolff 1719; Wolff 1723b: §§384–407; Wolff 1725a; and Wolff 1725b: §§214–266. For an overview, see Mühlpfordt 1989.

M. Favaretti Camposampiero (✉)
Department of Philosophy and Cultural Heritage – Ca' Foscari University of Venice, Venezia, Italy
e-mail: matteo.favaretti@unive.it

F. Baldassarri, A. Blank (eds.), *Vegetative Powers*, International Archives of the History of Ideas Archives internationales d'histoire des idées 234,
https://doi.org/10.1007/978-3-030-69709-9_24

banished as occult qualities, since they provide merely alleged causes that fail to respect the explanatory patterns of mechanical philosophy. A case in point is the vegetative soul of plants, which according to Wolff should be simply proscribed. However, in spite of Wolff's unconditional rejection of the vegetative soul, his major disciple in the field of natural sciences felt the need to reintroduce something similar. Michael Christoph Hanov tacitly transgressed the prohibition of irreducibly non-mechanical explanations by assuming that life is the product of a vegetative force and not of pure mechanism. He saw no other way to account for the most basic functions of living beings than to restore a non-mechanical power. Thus, Hanov's foundations of the life sciences actually turn out to be a hybrid of Wolffianism and vitalism. As this is the context in which the term *biologia* first appeared, exploring the background and implications of Hanov's theory of life casts light on the eighteenth-century origins of biology itself.

24.1 The Mechanical Life of Plants

Wolff's interest in the sciences of nature and life conformed to his teaching duties at the University of Halle. A couple of years after he was appointed professor of mathematics in 1707, Wolff was allowed to teach experimental and (shortly thereafter) theoretical physics as a substitute for Friedrich Hoffman, who had moved to Berlin to serve as royal physician (Wolff 1841: 146). From 1715 Wolff held the professorship of physics (Wolff 1841: 149; see Mühlpfordt 1989: 79–80). His physics lectures included the life sciences, a subject that Wolff described as encompassing the "external ornament of the Earth, in which I include vegetables, animals, and the human being" (Wolff 1718b: sect. 2, ch. 5, §27). Accordingly, he divided his exposition into two parts: "I treat of vegetables separately, but I expound the doctrine of animals together with the doctrine of humans, since the reason for the phenomena of the human body is the same as for the phenomena of the animal body" (Wolff 1718b: sect. 2, ch. 5, §27). This implies that the phenomena of vegetable bodies abide by a different *ratio* and therefore need a separate account. Wolff's treatises on physics (Wolff 1723b) and physiology (Wolff 1725b) adopt the same twofold division of the domain of living bodies.

The lectures on plants treated four main topics: "Concerning plants, I consider their vegetation, nutrition, growth, and seed generation" (Wolff 1718b: sect. 2, ch. 5, §28). These are the basic operations of plants, the functions that every vegetal body performs. Wolff's method for studying life conforms to his metaphysical principle that functions depend on structures; in order to explain a certain function, one must first consider the material structure that makes it possible. In the case of animals and humans, this means that anatomy, the description of the structure of the living body (Wolff 1728: §746), is preparatory to physiology or the science of its functions (see Favaretti Camposampiero 2016b: 89). The same order of priority applies to botany as well: "To attain limpid truth in these things, I begin with explaining the structure of plants on the basis of the discoveries by Malpighi, Grew, and others. Then I investigate the nourishment of plants and the causes of vegetation, and I use these manifest principles to interpret the aforementioned phenomena" (Wolff 1718b: sect. 2, ch. 5, §28).

Knowledge of the anatomy of plants can only be based on observations. Wolff repeatedly places his own contributions to the study of plants – along with the work on leaves by his pupil Ludwig Philipp Thümmig (1721) – in the tradition of the great micro-anatomists of the seventeenth century: Malphighi, Grew, and Leeuwenhoek (see, e.g., Wolff 1726: §184). In Germany, Wolff and Thümmig were among the first to apply microscopes extensively to the study of plants, which earned them a place in the history of phytotomy (see Wolff 1723a: §§90–96; Sachs 1875: 266–267). However, the quotation above shows that Wolff valued phytotomy mainly for the sake of physiology. Anatomical knowledge is necessary to interpret the phenomena of vegetation; that is, to discover their causes. Wolff's so-called *German Physics* uses the metaphysical concept of essence to explain "why we must investigate the structure of plants":

> We find that all plants are composed of certain parts and each part consists again of other smaller parts. As the essence of a body consists in the manner of its composition (Wolff 1720: §606), we know the essence of plants if we know the parts of which they are composed and also the parts of which their parts are composed, and if we understand how they have been composed of their parts and how these, in turn, have been composed of the other smaller parts. If we want to indicate the correct reason for the properties of plants and thus also the reason why plants nourish themselves and grow, live for a time then die, and generate the likes of them, we must deal with their essence. (Wolff 1723b: §385).

The manner of composition of an organic body is what Wolff calls its structure (see Wolff 1726: §83; Wolff 1731: §§275–276). Thus, the passage above concludes that "we must first examine the structure of plants" (Wolff 1723b: §385). Explaining physical phenomena consists in deriving the actual behaviour of bodies from their structures and the laws of motion. This is the core idea of Wolff's notion of mechanical philosophy. The proper role of the mechanical philosopher is to "explain the effects according to the true laws of nature" and to make "the possibility [of those effects] conceivable to the mind" (Wolff 1710: 17). But if this possibility depends on the structure of bodies, then bodies are not different from machines, and mechanics is the foundation of the science of nature: "We call machines all the bodies in the universe whose operations follow from their structures by virtue of the eternal and indispensable laws of nature, since in art we call machine anything to which this notion can be applied" (Wolff 1710: 17–18). Mechanics, which was originally the science of artificial machines, has thus become the foundation of the science of nature (Wolff 1710: 16).

Basically, Wolff thinks that modern advancements in the study of plants and life in general depend on the progress made by scientists in the knowledge of vegetable and generally organic structures. To know the structure of an organic body is to know its essence and to have a real or genetic definition of that body:

> A genetic or real definition explains the manner in which a thing can be made. But from a distinct notion of the structure of an organic body we understand how it can be composed. Thus, we find the real definition of an organic body if we search for a distinct notion of its structure. Needless to say, microscopes can often be of use in this, namely when through the microscope one sees things that escape unaided sight. (Wolff 1728: §686).

Wolff adds that definitions of this kind can be found not only in treatises on human and animal anatomy but also in Grew's and Malpighi's investigations of the anatomy of plants (Wolff 1728: §686n).

As early as 1709, in a dissertation discussed by his pupil Georg Remus, Wolff credits Malpighi and Grew with "showing very clearly that even trees and plants are composed of mere pipes [*tubulis*]" (Wolff 1709: sect. 2, §32). Wolff's idea is to use this tubular model for both vegetable and animal structures so as to account for the effects of winter on all living bodies. He considers that modern anatomy and physiology have proved that life and health depend on the circulation of fluids in the body; but since fluidity requires heat, "cold causes the stagnation of fluids" (Wolff 1709: sect. 2, §32). On the other hand, the experiments reported by Mariotte (1679: 63ff.) and Perrault (1680) show that some fluids also circulate in the tubular structures of plants. Hence, concludes Wolff, "what has been said about the force of winter on the bodies of animals can be applied even to trees and plants" (Wolff 1709: sect. 2, §32). This provides a good illustration of the explanatory power of mechanical structures: from the similarity of structures between animals and plants it is possible to infer a similarity of functioning and then to conclude that even the life and state of health of plants depend on the optimal circulation of the "nutritious juice" or sap through their pipes (Wolff 1709: sect. 2, §32; see also Wolff 1723b: §§403–405).

Conversely, the neglect of anatomical structures explains the failures of ancient botanists: "Since the ancients did not care about [examining the structure of plants], it is no wonder that in this matter they put forward nothing but empty words" (Wolff 1723b: §385). This is the *pars destruens* of Wolff's mechanism. Every account of natural phenomena that does not conform to the mechanist explanatory pattern is dismissed as mere verbiage or meaningless speech. In epistemological contexts, Wolff regularly evokes the vegetative soul along with the attractive force to illustrate his views about verbalism and his related criticism of both scholastic and Newtonian physics. The vegetative soul of plants or *wachsendmachende Seele* is supposed to be "a spiritual being that makes plants capable of growing" (Wolff 1965: ch. 2, §7), which is an example of how we can utter (seemingly) intelligible speech about nothing. Although each single word in the description is perfectly meaningful, the entire phrase is meaningless, for there is no idea that we can associate with it (Wolff 1965: ch. 2, §7 and §10). The same also holds for the interpretation of botanic phenomena in terms of sympathy and antipathy or natural love and hate (Wolff 1965: ch. 2, §7; Wolff 1718b: sect. 2, ch. 7, §7).

Furthermore, Wolff refers to the vegetative soul as an example of alleging a spurious cause, which consists in forging a name to denote the cause of certain perceived effects without having any notion of that cause other than the very notion of the effects. The name thus introduced is semantically empty, and those who use it fail to discriminate between words and things:

> For instance, the ancient scholastics noticed several effects in plants, namely, that plants nourish themselves, grow, and generate the likes of them. As they did not understand the anatomy of plants, that from which these effects derive they called *anima vegetativa* or *wachsendmachende Seele*. As many diverse effects they came across in plants, as many

> powers they ascribed to this soul, like a power to attract sap through the roots, a power to divide it throughout the plant for its nourishment, a power to form new leaves and twigs, a power to form a new plantlet in the seed. But as soon as the effects known from experience are removed, there remains nothing but empty words. Thus, they passed words off as things. (Wolff 1965, ch. 9, §3).

According to Wolff, there is no genuine concept of the vegetative soul that could be used to account for natural phenomena. We have no idea of what the vegetative soul should be apart from the effects that it should account for (see Wolff 1725a: 107, 133). As the passage above points out, the reason why the vegetative soul was introduced was ignorance of the anatomy of plants. Accordingly, the mechanistic alternative to such spurious explanations consists in deriving (physiological) effects from (anatomical) structures and not from fictional entities. Descartes leads the way in this endeavour: "However, after Descartes successfully subverted this verbiage [*Wörter-Kram*], nowadays judicious people are ashamed of adducing the soul as cause of the effects in plants and in the bodies of animals and humans" (Wolff 1965: ch. 9, §3).[3]

On the other hand, "if we want to indicate occult causes, in ancient physics we find hardly anything but empty words of this kind" (Wolff 1965: ch. 9, §3). Wolff's rejection of the vegetative soul is part of his campaign against occult causes, which is also instrumental in his and Leibniz's polemics against Newtonian forces and vitalistic medicine. All vital processes can and must be explained in terms of physical laws, structures, and related functions.

24.2 Biology and the System of Sciences

Wolff's German treatises on natural philosophy have no counterpart among his Latin works. After the early 1720s, Wolff ceased to develop a systematic philosophy of nature, although he conducted ground-breaking investigations into the epistemological and metaphysical foundations of medicine (Favaretti Camposampiero 2016a, 2016b).

After Wolff's death, the task of completing his system in the various fields of natural science was undertaken by Michael Christoph Hanov (1695–1773). During the 1760s, Hanov published a monumental *Philosophia naturalis sive physica dogmatica* in four volumes that ranged from general physics to anthropology. The phrase "dogmatic physics" had been used by Wolff himself (Wolff 1730b: §509n) to denote the whole of natural philosophy (including physics, teleology, and physiology) as distinct from natural history (which includes the description of experiments), for "dogmatic" is a book that expounds universal truths, whereas historical

[3] Wolff (1718: sect. 2, ch. 5, §39) mentions the "obsolete hypothesis of the scholastics concerning the soul" as an example of a physical hypothesis that conceals ignorance, hinders scientific progress, and should be banished from both medicine and physics. See also Wolff's 1722 Preface to Sturm's *Physica electiva* (in Wolff 1755: sect. 3, 143) and Favaretti Camposampiero (2012).

books describe mere facts (Wolff 1728: §§744–752). To confirm the Wolffian imprint, each of Hanov's volumes bears the following subtitle: "As a continuation of Christian Wolff's philosophical system". In the preface to the first volume, Hanov forestalls the objection that he is not always faithful to Wolff by pointing out that Wolff himself did not hesitate to enhance or even correct his earlier doctrines in the light of later discoveries (Hanov 1762: Preface, unpaginated). Thus, Hanov tries to cloak his departures from the master by presenting them as a necessary effort to update and refresh the Wolffian system.

The title of the third volume of *Philosophia naturalis sive physica dogmatica* refers to a discipline called *biologia* (Hanov 1766). This title is perhaps the only part of Hanov's work that has received any attention in the contemporary historiography of science, for the reason that it contains what might be the first occurrence of the word "biology" to designate the science of life in general (McLaughlin 2002: 2; see also Toepfer 2011: vol. 1, 254) and thus disproves the widely received dating of this term to the early 1800s.[4] Nevertheless, scholars still tend to minimize the relevance of Hanov's coinage to the history of the life sciences. Some defend a qualified version of the received dating of the word by distinguishing between "the word itself", which indisputably occurs in Hanov's work, and the word "in its modern sense" that first appears in the works of Burdach, Treviranus, and Lamarck (Ben-Menahem 2009: 1541n). Unfortunately, it is not clear precisely what sense (other than the "modern" one) should be ascribed to Hanov's use of the word. Others have suggested that in Hanov's volume "biology seems to have designated zoology" (Richards 2003: 16n). In fact, as Peter McLaughlin (2002: 2) has noted, Hanov's biology also includes phytology or the science of plants and has therefore a general character that certainly transcends zoology. Even McLaughlin, however, concludes that what prompted Hanov's linguistic innovation was not a new concept of the life sciences but merely the typically Wolffian, "almost pathological penchant for subdividing, classifying, and giving Greek names to things" (McLaughlin 2002: 3). The new term "seems to have been an afterthought" that "was merely grafted on to an already completed compendium" and thus "plays no real role in the content of Hanov's own philosophy" (McLaughlin 2002: 2–3). A closer look at Hanov's work makes it possible to reassess the significance of his coinage by clarifying its meaning in relation to Hanov's concept of life.

The word *biologia* appears not only in the title of Hanov's volume but also in the preface and the running title of the first chapter of the biological section. Moreover, the overview (*Conspectus*) of the volume introduces the adjectival form; the second section of the third part of physics is called "The Biological Section", which contains "the science of living things" (Hanov 1766: unpaginated). Biology as the science of all living things includes four main disciplines: general biology, phytology (divided into general phytology, dendrology or the science of trees, and physical botanology or the science of herbs), zoology (divided into general zoology, first theriology or the science of inferior beasts, and second theriology or the science of

[4] See, e.g., Schiller (1978: 1). There is no mention of Hanov in Zammito (2018).

higher beasts), and anthropology (divided into general anthropology, natural physiology, and physical pathology: Hanov 1768).

Hanov's classification of sciences is clearly inspired by the traditional scholastic partition whereby each discipline includes a general part and one or more special parts, each of which may in turn replicate the same arrangement (for instance, the science of trees is divided into general and special dendrology).[5] But it is also – indeed, primarily – inspired by the exposition of the parts of philosophy in Wolff's *Discursus praeliminaris de philosophia in genere*, which subsumes two main life sciences under physics: phytology or the science of plants, further divided into general phytology, botanology or the science of herbs, and dendrology or the science of trees and bushes (Wolff 1728: §83); and physiology or "the science of the animate [i.e. ensouled] body", which appears to subsume the physics of both animals ("animal economy") and humans, thereby including anthropology or the "physical examination of the human being in particular", physiology or "the science of the healthy human body", and physical pathology or "the science of the diseased body as such" (Wolff 1728: §84).

In Wolff's classification, the primary distinction is between vegetable bodies, which are the object of phytology, and animate or ensouled bodies, which are the object of physiology (see Hatfield 2018: 65). Wolff presents anthropology and animal economy (or zoology, to use Hanov's term) as parts of a single general discipline, physiology, whereas Hanov favours the tripartite phytology, zoology, and anthropology. However, the most substantive difference from Hanov's classification is that Wolff introduces no intermediate level between general physics or the science of bodies on the one hand, and the special disciplines of phytology and physiology on the other hand. In Wolff's classification there seems to be no biology as the science of all living bodies in general, as if vegetable bodies and ensouled bodies were not subsumed under the general category of the living. Hanov, by contrast, explicitly considers that the division into phytology or the science of *vegetabilia* and zoology or the science of *animantia* arises from the "distribution of living beings into *animata* and *inanimata*" (Hanov 1766: Preface, VI).

This difference in nomenclature reveals a change in the manner of defining the very object of the life sciences. In the preface to the third volume of *Physica dogmatica*, Hanov introduces biology as follows: "In the Second Section, Biology, we move from the things devoid of life properly so called [*vita proprie dicta*] to the things endowed with life, by analysing first, in Chapter 1, life's own nature and character [*propria vitae natura et indole*]" (Hanov 1766: Preface, V; see also §425). Indeed, Hanov's biological section begins with the chapter "On Corporeal Life" (*De vita corporea*), whose running title reads "On the General Biology of Bodies" (*De biologia corporum generali*). Thus, the task of general biology is to analyse the concept of life so as to discover its essential character.

---

[5] See also the further occurrence of "biology" in the fourth volume of *Physica dogmatica*: "Quae universis singulisque animantibus communia sunt, tradenda sunt in Zoologia generali, uti fecimus in Biologia et phytologia generali" (Hanov 1768: §141). It is incorrect to argue from this passage, with McLaughlin (2002: 3), that Hanov's actual use of "biology" refers only to general biology.

Consider, however, that the biological section featuring this chapter on general biology is in fact only the initial part of biology, which is specifically devoted to vegetable bodies.[6] The reason for treating life in general along with vegetative life is that Hanov considers vegetation to be "the lowest degree of life" (Hanov 1766: Preface, V). The life of vegetable bodies is closer to bare life than the life of animals, since plants are not ensouled. In plants we observe life at its most basic, which means that the necessary and sufficient conditions for vegetative life are the necessary and sufficient conditions for life in general. Thus, biology begins with vegetation in order to investigate the boundary between the living and the non-living.

24.3 Organic Bodies

In eighteenth-century Germany, the inclusion of plants among living beings was not to be taken for granted. Theologians, for instance, argued that in Scripture plants are not said to live. To these scruples, Hanov replies that the issue is merely terminological. Although the books of Moses use the term "life" in the narrow sense of "ensouled life", they nevertheless recognize that plants enjoy properties like birth, germination, growth, vegetation, reproduction, disease, and death, all of which are typical of inanimate or soulless (*inanimis*) life but are actually common to living bodies in general (Hanov 1766: §434n). Hanov also invokes the authority of the (pseudo-)Aristotelian *De plantis* (815a), which recognizes the presence of "occult" life in plants (Hanov 1766: §450n). Moreover, Hanov quotes Linnaeus's famous saying that "stones grow, plants grow and *live*" (Hanov 1766: §450n).[7]

Although Hanov assumes the existence of both living and non-living bodies as empirical evidence (1766: §425), he thinks that the rationale for this distinction requires philosophical analysis. In the preface, he already states the necessary and sufficient conditions for something to be alive:

> Life, when it subsists in the lowest degree that we call vegetation, requires an organic structure of the body and a vital force of forming and nourishing itself [*vim vitalem se formandi et nutriendi*], which is in some way analogous to sense in the obscure perception of the vegetating principle and observes a *sui generis* rule insofar as nothing hinders it. (Hanov 1766: Preface, V).

The first condition for life concerns the body's structure, which must be organic. Hanov's discussion of this topic appears *prima facie* faithfully Wolffian. It draws first on Wolff's characterization of the organic body as "that *body* which in virtue of

[6] See the Overview: "Sectio II. Biologica s. rerum viventium scientia, cujus initium: de vita vegetantium corporum./ Cap. I. de vita corporea./ Cap. II. Phytologicum generale sive de regno vegetabili in genere./ Cap. III. De arboribus in genere, sive Dendrologia generalis./ Cap. IV. de Dendrologia speciali" (Hanov 1766: unpaginated).

[7] Emphasis added by Hanov. See Linnaeus (1751: 1): "Lapides crescunt. Vegetabilia crescunt et vivunt. Animalia crescunt, vivunt, et sentiunt".

its composition is apt for a certain specific action" (Wolff 1731: §274),[8] and second on Wolff's distinction between "the *simple organic* body [*organicum simplex*], which is not composed of any other organic parts, and the *composite organic* body [*organicum compositum*], the parts of which are organic" (Wolff 1731: §274; see also Hanov 1766: §440).

Concerning the first point, Hanov correctly interprets the concept of "aptitude for a certain action" in functional terms: "We call *functions* the actions that are particular [*peculiares*] and therefore different from the others and characteristic. The Greeks call them *διοικησεις*, dispensations and, as it were, administrations" (Hanov 1766: §440). Bodies are organic insofar as they are made to serve specific functions. Moreover, Hanov is faithful to Wolff's understanding of the relation between structure and function; the structure of a certain organic body makes it apt for a specific function, which in turn contains the reason why that body has precisely that organic structure.[9] Thus, organic structures are inherently functional.

Concerning the second point, Hanov actually merges Wolff's distinction between simple and composite organic bodies (Wolff 1731: §274; see Favaretti Camposampiero 2019) with Wolff's distinction between monofunctional and polyfunctional machines (Wolff 1730a: §3), for he assumes that a plurality of functions can only result from a plurality of functional parts.[10] This is a minor difference, however, since Wolff indeed used both distinctions to discriminate between natural machines, which are composite and polyfunctional at all levels, and artificial machines, which are ultimately composed of simple and monofunctional machines (Favaretti Camposampiero 2016b: 84). The major difference is that Hanov makes clear from the outset that his discussion concerns only "natural organs",[11] whereas Wolff had included all artificial instruments in the category of organic bodies.[12] Thus, the distinction that Hanov aims to establish by means of the two criteria above is not between natural and artificial bodies (as with Leibniz and Wolff) but rather between living and non-living natural organic bodies. Only the former should be called organic in the narrow sense of the term: "In the stricter sense or primarily, we call organic those composites or bodies the most parts of which (or different single

[8] See Hanov (1766: §440): "We call *organic* the *bodies* that by their essence and nature are apt for particular actions".

[9] See Hanov's multiple references to Wolff concerning the structure of organs: "The manner of composition of an organ, hence its essence and nature, is called *structure* (Wolff 1731: §275ff.). This structure entails such an artful and ingenious fabric that it constitutes an instrument that is by its own force suitable for a particular function (Wolff 1731: §278), which contains the reason why the forces and parts are shaped and connected in this manner rather than otherwise (Wolff 1731: §279)" (Hanov 1766: §441).

[10] The simple organic body "is only apt for one single action characteristic to it; hence it has no organic parts but consists of corpuscles", whereas the composite organic body "is apt for several different actions at one time, for even its parts are organic" (Hanov 1766: §440). Hanov ascribes this distinction to Wolff (1731: §274).

[11] See Hanov (1766: §440): "The *organs* in general are simply the instruments of the agents. Here, we are concerned with natural organs".

[12] See Wolff (1731: §276n): "All the instruments that are used in art are organic bodies".

parts of which) are organic, that is, living organs, destined for specific functions" (Hanov 1766: §440).

Of course, Hanov recognizes the distinction between artificial and natural bodies. For instance, the amazing functions performed by a pocket watch are merely the effect of a spring that has been wound on its mechanical structure: "Thus, the mechanical or made organ [*organum mechanicum s. factum*] differs from the native or natural one [*natum seu naturale*]" (Hanov 1766: §448n). However, this distinction is not so essential for him as the distinction between living and non-living bodies. Non-living organic bodies include both human and natural artefacts. Examples of the latter are spiderwebs, bird nests, and honeycombs,[13] whose structures are clearly designed for certain functions. On a larger scale, every cosmic machine can be described as a non-living divine organic body. Indeed, both human and natural artefacts lack the characteristic powers of life, including of course the reproductive power:

> Although divine organic bodies like air, aether, water, ordinary earth, and the bodies made of these, like mountains, clouds, and the stars themselves, surpass to that extent human [artefacts] by all their nature and very useful character, they are nevertheless themselves merely cosmic machines [*machinae mundanae*], always provided with only one kind of motive force and by no means apt to generate, multiply, and propagate themselves. As this is doubtless, it remains that even bare natural organs cannot be called living, for the reason that they have only the ordinary motive force and do not produce the effects of living beings. (Hanov 1766: §448n).

The existence of organic bodies devoid of life shows that simply having an organic structure – a structure so organized as to obtain certain purposes – is not a sufficient condition for being alive. As Hanov puts it, "the organic structure of bodies that regularly conspires [*consentiens*] to obtain their purpose is by no means sufficient to explain the life of organic bodies" (Hanov 1766: §448). In this formulation, the claim is ambiguous between an epistemological and a metaphysical interpretation. According to the former, we are not (yet) able to reduce life to structure. According to the latter, life is in itself irreducible to structure.

Hanov actually endorses both claims. His argument for the epistemological claim draws on Wolff's distinction between physical and mechanical explanations, along with the related concepts of corpuscular philosophy and mechanical philosophy. Basically, mechanical and physical explanations correspond to two different levels of explanatory reduction. According to both Wolff and Hanov, the properties and behaviours of bodies can be explained by adducing either mechanical or physical reasons. The former are those that are fully reducible to the four mechanical principles: figure, size, motion, and position, while the latter include all the observable qualities of bodies that are not or cannot be explained mechanically, since the perceptions we have of them are inherently confused. Whereas mechanical principles are realities – they really do belong to bodies – physical reasons have a merely phenomenal status, just like the phenomena they have to explain (Wolff 1731: §237; Hanov 1766: §445). As the corpuscles that compose bodies have both mechanical

[13] See Hanov (1766, §442n, § 448, §452n).

and physical qualities, the corpuscular philosophy transgresses the boundaries of the mechanical philosophy in that it relies not only on mechanical but also on physical patterns of explanation (see Wolff 1731: §241).

On this basis, Hanov argues that all organic bodies but especially living ones are out of the reach of mechanical philosophy. Concerning organic bodies in general, he maintains that "no one can philosophize mechanically about those organs whose structure cannot be explained yet on the basis of the inmost character of the parts, their figure, position, number or magnitude, and motion or motive force" (Hanov 1766, §446). Hanov's view of our epistemic limits is more drastic than Wolff's. Not only is the "inmost structure" of bodies presently unknown to us, but there is also no hope that it might cease to be inscrutable, which rules out a fortiori[14] the possibility of an entirely mechanical explanation of life: "Corporeal essence or life cannot be explained by mechanical reasons alone, but one must use physical reasons above all" (Hanov 1766: §447).

Although Hanov is more resolute in stressing the explanatory limits of mechanism, his view thus far is not incompatible with Wolff's own position. His real departure from genuine Wolffianism consists in shifting from the epistemological claim that we cannot explain life mechanically to the metaphysical claim that life does not depend on purely mechanical principles. Whereas the epistemological claim leaves open the possibility that living organic bodies are merely more complex than non-living bodies, the metaphysical claim states that no level of structural complexity is in itself sufficient to produce life. As noted above, there is a further necessary condition for a body to be living. In addition to the organic structure of the body, life requires "a vital force of forming and nourishing itself" (Hanov 1766: Preface, V). The next section elaborates on this vital force, which is expected to make the difference between the living and the non-living.

24.4 Vegetative Force and Vegetative Soul

Like all organic bodies, plants are machines, specifically hydraulic machines. But whereas artificial machines are devoid of life, plants are "living hydraulic machines" (Hanov 1766: §515). The difference has to do with force. In order to perform its functions, every organic body needs a force that, as a principle of motion, activates the machine. Wolff had observed that human artefacts and all machines produced by finite agents generally are not really self-moving but require an external force to be applied to them in order to move. By contrast, the machines of nature

[14] The a fortiori step of the argument is made explicit in the scholium: "Indeed, not even minerals could be so far accounted for on the basis of mechanical principles; much less will there ever be hope to do so concerning living beings" (Hanov 1766: §447n). Elsewhere in the same work, however, Hanov restricts the necessity of physical reasons to living bodies alone: the "natural change" of an organic body "can be explained on the basis of its structure through the rules of motion, if it is devoid of life; but in the case of living beings one further needs physical principles" (§516).

have a principle of motion that belongs to the machines themselves (Wolff 1720: §643); thus, they do not need to be wound up like clocks. Hanov takes this inherent force to be distinctive not of natural beings in general but of living bodies in particular. Moreover, he specifies the inner character of this force by tracing its origin back to the "elements" or simple substances (see Hanov 1762: §46; Hanov 1765: §103n) that ultimately compose the body:

> Artefacts or the organs generally that are produced by a finite force lack an inherent [*propria*] force to do what can be effected by them with an external force, as long as this force that is added or included in them is able to act [...]. By contrast, the organs that are produced and destined by the divine force, which are also called natural, originate and grow from the elements, which are endowed with an intrinsic constant (essential) force [...]. *These organs*, endowed with a manifold essential force, are called *living organs*, because we perceive that they act with such a constant effectiveness that they deserve to be termed artful and active by themselves [*per se daedala et actuosa*]. For the same reason, they differ a great deal from bare organs that are only passive. (Hanov 1766, §449).

If the vital force belongs primarily to the elementary components of the body, the difference between the living and the non-living must depend on a fundamental difference between those elements. There are elements whose nature consists in a vital force and elements whose nature consists in a non-vital force, according to God's will. For the divine force does not directly produce the living organs; rather, it makes life possible by bestowing a vital nature upon some elements so that they can form a living body. Thus, Hanov's view is that life is not an emergent property of complex organic structures but rather an essential property of each single simple substance that composes the organic structure of a single living being. This primitive character of divine gift makes the vital force and life itself not further explainable.[15]

Several passages describe the vital principle that accounts for the most basic functions of living bodies as a vegetative force. Hanov introduces this term when claiming that even vegetation, the lowest level of corporeal life, is irreducible to mechanism and must spring from a dynamic source, which he calls *vis vegeta*, *vis vegetans*, or *vis vegetandi*.[16] Thus, the question arises as to how this concept relates to the traditional doctrine of the vegetative soul. If the former turns out to be merely an updated version of the latter, we should conclude that Hanov's purported continuation of Wolff's system actually and openly revives a doctrine that Wolff had relegated to the figments of human language and imagination.

In fact, however, Hanov's strategy is subtle enough to accommodate Wolff's criticism of scholastic physics. In the chapter on general phytology, Hanov adopts the empiricist stance that the principles of vegetation must be investigated a posteriori (Hanov 1766: §513), since finite human minds cannot fathom "the intimate forces and differences of the elements" (Hanov 1766: §512). Any attempt by humans

[15] See Hanov (1766, §450n): "No life can be conceived or exist without an essence endowed with a vital nature, which only God can give".

[16] See Hanov (1766, §450): "Thus, *corporeal life*, the *lowest* of all, consists in *vegetative force* [*in* vi vegeta *(vegetante)*] and continuous vegetation, but not in bare mechanism".

to demonstrate such principles a priori results in fact in the unwarranted invention of things that humans cannot know, like the vegetative soul and the *anima mundi*:

> Related to this are the things affirmed by the ancients about the vegetative soul, by Tommaso Campanella about the *sensus rerum*, which he wrongly attributes even to plants, and by Andreas Rüdiger (1716: book 1, ch. 4, §32, §37), who maintains that plants possess not only some kind of sense but also an intellect or an intelligent principle [...]. Likewise, Malpighi (1675: 32 [in fact, 15]) thinks that plants occupy the lowest rank of animals and have tracheas instead of lungs. Others invented the soul of the world and ascribed to it, among other tasks, the formation of plants. (Hanov 1766, §512).

However, Hanov rejects the vegetative soul for different reasons than Wolff. Hanov's point is not that soul-based explanations of physical phenomena do not fit into the mechanistic model and are therefore pseudo-causal and unintelligible. Rather, he argues that only animals have souls, for the soul is the principle of sense perception and motion; but plants neither sense nor move and hence are not ensouled. Once the soul (*anima*) is characterized as "a substance endowed with the force of (perceiving at least clearly, i.e.) sensing, appointed or destined to locally move and govern its own living body" (Hanov 1766: §465), it appears indeed inappropriate to talk of plants having souls. Another passage reinforces the impression that Hanov's objection to the vegetative soul is more terminological than substantive:

> Aristotle ascribes to [plants] a nutritious or vegetative soul and says that whatever lives has it (*De anima*, III, c. 12). However, since according to the ordinary use of language the soul is at least provided with sense, whereas plants [*vegetantia*] are deprived of it, here it is preferable to retain the bare entelechy as the internal principle that governs vegetation and is called *vegetamen* in Latin. Among the moderns, Tommaso Campanella in the book *De sensu rerum*, Francesco Redi in *De generatione insectorum*, and Rüdiger in *Physica divina* attributed sense to plants, but different from the sense of animals. (Hanov 1766: §451n).

Whereas Wolff attacks vitalism in general, Hanov's targets are restricted to animism (the view that everything has a soul) and hylozoism (the view that matter is alive). Against the latter, he stresses the distinction between the living and the non-living, arguing that both ancient and modern hylozoists have conflated the concept of life and the more universal concept of force.[17] Against the animistic account of life, he invokes the distinction between *animatum* and *inanimatum*. Living beings are either ensouled or soulless:

> With respect to their essence and nature, living beings divide into *animata* and *inanimata*. Those are called *animata* which possess not only an organic body capable of moving but also a sentient soul that governs that living body. Those are called *inanimata* which have no soul, no sensation, no locomotive faculty; thus, those which only vegetate. (Hanov 1766: §466).[18]

---

[17] See the appendix "De vi materiali vel hylozoismo" in the second volume of *Physica dogmatica* (Hanov 1765: 760–768).

[18] See Hatfield (2018: 66): "Here, 'animate' clearly means 'cognitively ensouled' and not simply 'living', and 'inanimate' means those living beings that lack the cognitive powers of an ensouled being".

The former are generally called *vegetabilia*, "from the vegetative life [*vita vegeta*] that is manifest in all of them" (Hanov 1766: §468); the latter are called *animalia* or *animantes* (Hanov 1766: §470). Just as there is no animal without a soul,[19] there is no plant endowed with a soul. Aristotle was right in dismissing Anaxagoras's and Empedocles's belief that plants have sense and appetite, but he was wrong in conceding that plants are nevertheless animated by vegetative souls which effect their motions: "He gratuitously supposes that there are ensouled things [*res animatae*] that are not animals" (Hanov 1766: §468n).

Despite the claim that plants are soulless, it would be hasty to conclude that Hanov simply rejects the vegetative soul. Although what he calls vegetative force cannot be described as a soul proper, it nevertheless performs quasi-mental functions analogous to those of the soul. Hanov considers both the formation and conservation of living organs to consist in the union of tiny organic particles called *primordia*. In order for these *primordia* to join together so as to form a living structure, they must be able to seek, attract, and select one another. According to Hanov, these basic vital operations have a quasi-cognitive character that cannot be explained away in terms of purely mechanical interactions of particles. On the contrary, they must be performed by a force that has perceptual powers and is identical with the vital force of the elements or simple substances that compose the *primordia* of all living organs:

> And this force, admirably artful [*daedala*] and variously *organ-forming* [*organopoea*], cannot be conceived without a perception at least momentarily clear or without a certain desire for uniting and forming oneself, and it therefore deserves to be called *generative* [*genetica*] *and formative* [*formatrix*]. [...] At the same time, the living force [*vis viva*] is deemed to be *sustaining or nourishing*. (Hanov 1766: §450).

Life requires more than generic perceptual powers; it requires a perception that is "at least momentarily clear". According to the Leibnizian and Wolffian accounts of perceptual clarity, clear perceptions are salient enough to be conscious, whereas obscure perceptions are below the level of consciousness. As clarity characterizes sense perception (sensations are essentially clear-confused perceptions), the vegetative force that makes life possible must be to some extent similar to the sensory faculty of animal souls. In order to find and discern assimilable particles, that force must be able to quasi-sense them by what Hanov calls "a sort of simulacrum of touch or taste".[20] This does not amount to a sensation proper; the momentary clarity of vegetal perception is but one step above the obscurity of mineral perception:

> Since the *vis vegetandi* avidly receives other things and distributes things that are similar or assimilated to itself so as to form its organs, it must be able to somehow discern, by its own

[19] See Hanov (1766: §467n): "Thus, it is foolish to imagine, against §466, an animal lacking every sense and touch but alive, or intermediate between plants and animals".

[20] Hanov (1766: §452): "If the organs of living beings develop naturally without miracle, and if naturally contingent things cannot develop by chance or fortuitously (Wolff 1931: §96) but rather depend on their own structure and the laws of motion (Wolff 1931: §513f.), their vegetative force and *vital principle* must *discern naturally*, by a sort of simulacrum of touch or taste, *the things that are useful to it from those that are useless* [...]".

effort, the things that are beneficial to it from those that are noxious, and to be brought to add the former to itself and to avoid and remove the latter no matter how. As this cannot be done by blind impetus without a perception sufficient to it (thus, in an obscurity unaware of itself, of course, but superior to the one that occurs in the elements of minerals and other gross materials, and capable of discerning the things that are brought in contact with it) or without the bare appetite [...] and aversion that originate from that perception, in the obscurity of elemental perceptions there must be higher degrees such as required for corporeal life. (Hanov 1766: §451).

In the same vein as Leibniz (*Monadologie*: §26, in Leibniz 1885: 611) and Wolff (1734: §765) had ascribed an *analogon rationis* to animals, Hanov ascribes an analogue of sensation ("sensus vel clarae perceptionis analogum", Hanov 1766: §451) to plants, an analogue of vegetation to minerals, an analogue of organization to mixtures (Hanov 1766: §441), and so on throughout the scale of beings.[21] What varies from one step of the scale to the other is the degree of perceptual clarity that a certain class of beings (or better, the elements that form their bodies) can reach, for although soulless living beings are deprived of motion and sensation proper, they are all endowed with a perceptive force or *vis percipiendi* (see Hanov 1766: §465, §467n). In this respect, Hanov's rejection of animism and hylozoism needs to be qualified. What Hanov actually refuses is to call *anima* the perceiving and dynamic principle of substances lower than animals. In the appendix on hylozoism, he agrees that even the elements of corporeal things possess "a never idle force, however not life properly so called, but either a principle or an analogue of it, like a shadow or a simulacrum" (Hanov 1765: 767).

Needless to say, such a universal attribution of quasi-perception and quasi-life draws powerful inspiration from Leibniz's monadology. Whereas Wolff refrained from characterizing the force of the elements of material things in perceptual or representational terms (see Wolff 1720: §§598–599; Wolff 1724: §13; and Wolff 1734: §712n), Hanov is openly committed to Leibniz's tenet that all kinds of simple substances are in a respect similar to minds, in that every monad is endowed with (or even individuated by) an inner force that produces two sorts of internal changes, perceptions and appetitions, so that what distinguishes lower monads from animal souls and these souls from rational spirits is their respective degree of perceptual clarity.[22] Furthermore, Hanov draws on Leibniz's doctrine of the dominating monad to account for the teleological organisation of living bodies:

As this force tends towards a common use and purpose with wonderful agreement, it contains the primary perfection of the elements, which Aristotle and Leibniz after him have called entelechy. As in every living being all the organs are formed and directed towards one single end, a single principle appears to rule all the others [...], and every living being has a single entelechy, a single *monas vegeta*, that governs corporeal life and incessantly strives for the goal destined by God, unless prevented by a greater force. (Hanov 1766: §451).

[21] See Hanov (1766: §451n): "Just as the bodies that have regular structure and figure, like salts, etc., are closer to vegetable bodies, etc., than the others, and just as below we will show that there is something similar to reason in beasts, so it is not in the least contrary to reason that there may be something similar to sense in the lowest genus of living beings".

[22] Hanov (1766: §451) refers to Leibniz's *Monadologie*, §18ff. See Leibniz (1885: 609–611).

24.5 Natural Instinct and Biological Laws

Thus, it might be tempting to interpret Hanov's vital force as a belated attempt to rescue Leibnizian monadology from Wolff's dismissal. At the same time, there should be little doubt that Hanov in fact distorts the genuine Leibnizian doctrine for his own purposes. His use of monadology to provide biological explanations blurs Leibniz's distinction between physics and metaphysics, and his claim that monads govern corporeal life contradicts Leibniz's denial of causal interaction between the simple substances and the physical world, since it implies that the monad's inner activity influences physiological processes.[23] Whereas Leibniz strove to keep monadology compatible with mechanistic physics, Hanov turns monadology against modern mechanism.

Indeed, it is precisely at this point that Hanov finally admits his disagreement with Wolff. In *Psychologia rationalis*, Wolff pointed out that pre-established harmony requires a full mechanization of the body's activity, whereby all bodily motions, including all the voluntary motions that compose rational and free human behaviour, must result from a "mechanical production" (Wolff 1734: §615). According to Wolff (1734: §112ff.), the brain is equipped with a system of material representations – *ideae materiales in cerebro* – which makes it possible for the nervous fibres to translate mental contents and processes into corporeal entities and mechanical motions. Human thoughts are mostly verbalized within the mind itself, but every piece of mental language has a physical counterpart available in the brain, the material idea of the words that compose it. Insofar as rational thoughts and wills are usually expressed in words, they can be represented in the brain and can thus govern the body's behaviour in purely mechanical fashion (see Wolff 1734: §§395–397, §403, §§414–418, §§615–616; Favaretti Camposampiero 2009: 512–608). In this way, all bodily phenomena can be derived from purely mechanical causes without any reliance on non-physical, mental causation, simply by assuming that God has designed the brain so that it constantly represents (or harmonizes with) the present mental state.

Hanov's assessment of this doctrine is explicit enough. He writes that he "cannot share" Wolff's account of voluntary motions, because it entails that organic bodies are entirely governed by blind mechanisms and thus makes the soul – and indeed any non-physical entity – completely irrelevant to the explanation of physical phenomena[24]:

---

[23] Like several Wolffians, Hanov was highly critical of pre-established harmony and a partisan of physical influx; see Favaretti Camposampiero (2017).

[24] See Wolff (1734: §616n): the body would perform exactly the same voluntary motions (as well as all its other activities) even if the soul did not exist. Materialists are right in recognizing "that the motions that we call voluntary follow from the motions that pre-exist in the brain by virtue of mechanism alone" (Wolff 1734: §616n); they are wrong only in inferring from this that the soul *does not* exist. In his anthropology, Hanov criticizes Wolff's theory of material ideas as a materialistic doctrine; see Hanov (1768: §503n) and Favaretti Camposampiero (2017: 86–88).

> To imagine or dream that the written works of Homer, Virgil, and Cicero, or the mechanical organs and burning glasses of Archimedes, etc., coalesced according to the laws of motion without any perception would also be patently absurd but still much easier than to imagine or dream that the organic bodies form and act only mechanically. Thus, I cannot share what can be read in *Psychologia rationalis*, §615ff. (Hanov 1766: §452n).

Contrary to mechanical causes, Hanov's vegetative force is no blind agent. Against the idea that highly organized structures may result from mechanical interactions, he evokes the old adage that the works of nature (including both non-living artefacts like honeycombs and the living bodies of plants and animals) are works of intelligence. "Natural instinct" is the expression he uses to describe the sort of pre-cognitive intelligence that he ascribes to nature. Organic bodies – and living bodies a fortiori – are not the products of random collisions of particles but the offspring of the natural instinct that God instils in the vegetative elements:

> Honeycombs conform to all the rules of the most exquisite mechanics. Bees cannot understand these rules, but by intervention of the divinely infused instinct they act exactly as though they understood them and had learnt to conform their actions to them. If waking bees do these things by natural instinct [...] then even the unconscious vegetative elements, in constructing their organs, can perform by natural instinct the things to which they are destined by God [...] (Hanov 1766: §452n).

This comprehensive consideration of the natural world is nevertheless far from reducing the gap between the living and the non-living. Whereas honeycombs simply conform to the laws of mechanics, living structures form and act according to a law of their own, the *lex vegetationis*.[25] Crucial to Hanov's idea of biology is the recognition that life has its own laws. Although all living bodies qua bodies obey the general laws of motion, they also obey, qua living beings, a set of more specific laws which only apply to whatever has a vital nature: "The general laws of living beings must be elicited secondly from their vital nature, firstly from God's nature. Just as the common laws of motions are derived from the common nature of bodies, so the particular laws can be learned only from the bodies' own nature, which in living beings is life qua vital nature" (Hanov 1766: §476). This is the domain of the sub-discipline called bionomy (*bionomia*), further divided into general bionomy, which investigates the common laws that derive from the common nature of all living things, and the special bionomies of phytology and zoology, which investigate the laws of vegetal and animal life respectively.[26] Hanov deplores the current neglect of biological laws and presents his own exposition of bionomy as ground-breaking, which in his view justifies its possible flaws.[27]

---

[25] See Hanov (1766, §452n).

[26] See Hanov (1766, §476): "As even the nature of living beings is partly common to all living beings, partly a property of their different genera and individuals, here we will only have to treat the common laws of living beings or general bionomy. Special [*propria*] bionomy will have its own reserved places in phytology and zoology".

[27] See Hanov (1766, §476n): "So far there is profound silence in physics concerning the laws of living beings or bionomy. Thus, if in breaking this ice many things will be left to posterity to be completed and corrected, this will have to be ascribed to the common fate of new truths".

The general conclusion to be drawn is that Hanov's account of vegetation, especially his doctrine of the vegetative force, bears clear marks of vitalism. Despite its author's claim of offering an updated sequel to the Wolffian system, the *Physica dogmatica* in fact reveals serious departures from Wolff's mechanistic approach to living phenomena. The above investigation has shown that what was really at stake in this late revival of vegetative powers was the very concept of life and the boundary between the living and the non-living. In light of this, the first naming of biology as a unitary and self-standing discipline turns out to be an understandable consequence of the (paradoxically) anti-Wolffian developments of Wolffianism itself, and the fact that it was Hanov and not Wolff[28] who first introduced the word *biologia* no longer appears to be mere chance.

References

Ben-Menahem, Ari. 2009. *Historical Encyclopedia of Natural and Mathematical Sciences*. Berlin [etc.]: Springer.

Favaretti Camposampiero, Matteo. 2009. *Conoscenza simbolica: Pensiero e linguaggio in Christian Wolff e nella prima età moderna*. Hildesheim: Olms.

———. 2012. Dividing Fiction from Reality: Existence and Nature in Christian Wolff's Metaphysics. In *Existence and Nature: New Perspectives*, ed. Matteo Favaretti Camposampiero and Matteo Plebani, 65–97. Heusenstamm: Ontos.

———. 2016a. Bodies of Inference: Christian Wolff's Epistemology of the Life Sciences and Medicine. *Perspectives on Science* 24: 361–379.

———. 2016b. Christian Wolff's Philosophy of Medicine: An Early Functional Analysis of Health and Disease. *Quaestio* 16: 75–94.

———. 2017. De l'automate à l'homme entier: Anthropologie et médecine chez Wolff et Hanov. In *L'Homme entier: Conceptions anthropologiques classiques et contemporaines*, ed. Faustino Fabbianelli and Jean-François Goubet, 71–90. Paris: Classiques Garnier.

———. 2019. Machines of Nature and Machines of Art: Christian Wolff's Reception of Leibniz. *Rivista di storia della filosofia* 74: 431–452.

Hanov, Michael Christoph. 1762. *Philosophiae naturalis sive physicae dogmaticae tomus I. continens physicam generalem, coelestem et aetheream tanquam continuationem systematis philosophici Christiani L.B. de Wolff*. Halle: Renger.

———. 1765. *Philosophiae naturalis sive physicae dogmaticae tomus II. continens aërologiam et hydrologiam vel scientiam aëris et aquae. Tanquam continuationem systematis philosophici Christiani L.B. de Wolff*. Halle: Renger.

———. 1766. *Philosophiae naturalis sive physicae dogmaticae tomus III. continens geologiam, biologiam, phytologiam generalem et dendrologiam vel terrae, rerum viventium et vegetantium in genere, atque arborum scientiam. Tanquam continuationem systematis philosophici Christiani L.B. de Wolff*. Halle: Renger.

———. 1768. *Philosophiæ naturalis sive physicæ dogmaticæ tomus IV. et ultimus, continens botanologiam physicam, zoologiam, anthropologiam generalem, physiologiam et pathologiam physicam, hoc est herbarum, animalium atque hominis scientiam. Tanquam continuationem systematis philosophici Christiani L.B. de Wolff*. Halle: Renger.

[28] By contrast, Mühlpfordt (1989: 103) emphasizes Wolff's role as a forerunner of the unity of biology.

Hatfield, Gary. 2018. Baumgarten, Wolff, Descartes, and the Origins of Psychology. In *Baumgarten and Kant on Metaphysics*, ed. Courtney D. Fugate and John Hymers, 61–77. Oxford: Oxford University Press.
Leibniz, Gottfried Wilhelm. 1885. *Die philosophischen Schriften*, ed. Carl Immanuel Gerhardt, vol. 6. Berlin: Weidmann.
Leibniz, Gottfried Wilhelm, and Christian Wolff. 1860. *Briefwechsel zwischen Leibniz und Christian Wolff*, ed. Carl Immanuel Gerhardt. Schmidt: Halle.
Linnaeus, Carolus. 1751. *Philosophia botanica*. Stockholm: Kiesewetter.
Malpighi, Marcello. 1675. *Anatome plantarum [Pars prima]*. London: Martyn.
Mariotte, Edme. 1679. *Essais de physique. Premier essai: De la végétation des plantes*. Paris: Michallet.
McLaughlin, Peter. 2002. Naming Biology. *Journal of the History of Biology* 35: 1–4.
Mühlpfordt, Günter. 1989. Die organischen Naturwissenschaften in Wolffs empiriorationalistischer Enzyklopädistik. *Il cannocchiale* 2–3: 77–106.
Perrault, Claude. 1680. De la circulation de la sève des plantes. In *Essais de physique, ou recueil de plusieurs traités touchant les choses naturelles, Tome I*, 173–304. Paris: Coignard.
Richards, Robert J. 2003. Biology. In *From Natural Philosophy to the Sciences. Writing the History of Nineteenth-Century Science*, ed. David Cahan, 16–48. Chicago: The University of Chicago Press.
Rüdiger, Andreas. 1716. *Physica divina*. Frankfurt: Andrea.
Sachs, Julius. 1875. *Geschichte der Botanik vom 16. Jahrhundert bis 1860*. München: Oldenbourg.
Schiller, Joseph. 1978. *La notion d'organisation dans l'histoire de la biologie*. Paris: Maloine.
Thümmig, Ludwig Philipp. 1721. *Experimentum singulare de arboribus ex folio educatis ad rationes physicas revocatum*. Halle: Hilliger.
Toepfer, Georg. 2011. *Historisches Wörterbuch der Biologie*. 3 vols. Stuttgart and Weimar: Metzler.
Wolff, Christian. 1709. *Consideratio physico-mathematica hiemis proxime praeterlapsae*. Halle: Zeitler.
———. 1710. Epistola gratulatoria in qua vera philosophiae mechanicae notio explicatur. In *Wolff* (1755: sect. III, 15–19).
———. 1718a. *Entdeckung der wahren Ursache von der wunderbahren Vermehrung des Getreydes*. Halle: Renger.
———. 1718b. *Ratio praelectionum wolfianarum in mathesin et philosophiam universam*. Halle: Renger.
———. 1719. *Erläuterung der Entdeckung der wahren Ursache von der wunderbahren Vermehrung des Getreydes*. Halle: Renger.
———. 1720. *Vernünfftige Gedancken von Gott, der Welt und der Seele des Menschen, auch allen Dingen überhaupt*. Halle: Renger.
———. 1723a. *Allerhand nützliche Versuche*. Vol. 3. Halle: Renger.
———. 1723b. *Vernünfftige Gedancken von den Würckungen der Natur*. Halle: Renger.
———. 1724. *Monitum ad Commentationem luculentam de differentia nexus rerum sapientis et fatalis necessitatis*. Halle: Renger.
———. 1725a. Anhang von der Gewißheit der Erfindung von der Vermehrung des Getreydes. In *Entdeckung der wahren Ursache von der wunderbahren Vermehrung des Getreydes*, 2nd ed., 65–136. Halle: Renger.
———. 1725b. *Vernünfftige Gedancken von dem Gebrauche der Theile in Menschen, Thieren und Pflantzen*. Frankfurt and Leipzig: Renger.
———. 1726. *Ausführliche Nachricht von seinen eigenen Schrifften*. Frankfurt: Andreä and Hort.
———. 1728. *Philosophia rationalis sive Logica, methodo scientifica pertractata*. Frankfurt and Leipzig: Renger.
———. 1730a. De notione sanitatis. In *Horae subsecivae marburgenses anni 1729. Trimestre aestivum*, 318–355. Frankfurt and Leipzig: Renger.
———. 1730b. *Philosophia prima, sive Ontologia, methodo scientifica pertractata*. Frankfurt and Leipzig: Renger.

———. 1731. *Cosmologia generalis, methodo scientifica pertractata*. Frankfurt and Leipzig: Renger.
———. 1734. *Psychologia rationalis methodo scientifica pertractata*. Frankfurt and Leipzig: Renger.
———. 1755. *Meletemata mathematico-philosophica*. Halle: Renger.
———. 1841. *Eigene Lebensbeschreibung*. Ed. Heinrich Wuttke. Leipzig: Weidmann.
———. 1965. *Vernünfftige Gedancken von den Kräfften des menschlichen Verstandes und ihrem richtigen Gebrauche in Erkäntniß der Wahrheit*, ed. Hans Werner Arndt. Hildesheim: Olms.
Zammito, John H. 2018. *The Gestation of German Biology: Philosophy and Physiology from Stahl to Schelling*. Chicago: The University of Chicago Press.

Chapter 25
Bichat's Two Lives

Tobias Cheung

Abstract In *Recherches physiologiques sur la vie et la mort* (1800), Xavier Bichat, a medical doctor at the Hôtel-Dieu in Paris, distinguishes between the "organic" and the "animal life" of humans and animals. The organic life is directed towards the inside of their bodies. It is composed of organs which maintain cycles of assimilation, consumption and excretion. The animal life is directed towards their outside. It is composed of sense organs, nervous systems, brains and locomotor muscles. In this essay, I will focus on four aspects of Bichat's distinction: first, on the role of both lives within the "animal series" from the first animal, the zoophyte, to the last animal, the human, second, on the difference between animals and plants, third, on various forms of interactions or "communications" between both lives, and, fourth, on their relation to "social orders." Further on, I will discuss the position of Bichat's thesis of two lives within the discursive field between medicine, anatomy, physiology, philosophies of human and animal subjectivity and anthropology.

In his lectures on natural philosophy, published in 1817, Hegel distinguishes three aspects of animal subjectivity: its inward orientation, its outward orientation and the unity of both orientations in an "independent individual" with relations to other individuals.[1] Hegel's distinction relies on the work of Xavier Bichat, a medical doctor at the Hôtel-Dieu in Paris.[2] In *Recherches physiologiques sur la vie et la mort* (1800) and *Anatomie générale appliquée à la physiologie et à la médecine* (1801), Bichat examines the "general mode of existence" of organized bodies in focusing on different types of agents that maintain their "lives" in resisting against their

[1] Hegel refers himself to Bichat's physiology; see Hegel 1999, 354–356 (*Encyclopädie der philosophischen Wissenschaften im Grundrisse*, §§ 353–356).

[2] For bio- and bibliographical references, see Arène 1911; Entralgo 1948; Haigh 1975; Rey 1997; and Huneman 1998. Unless otherwise indicated, all translations in this article are the author's.

T. Cheung (✉)
Humboldt University, Berlin, Germany
e-mail: tobias.cheung@staff.hu-berlin.de

F. Baldassarri, A. Blank (eds.), *Vegetative Powers*, International Archives of the History of Ideas Archives internationales d'histoire des idées 234,
https://doi.org/10.1007/978-3-030-69709-9_25

"death."[3] Among these agents, he is mainly interested in animals and humans whose "individual lives" are composed of "two lives": one directed towards their inside and one directed towards their outside. He calls the first life "inner" or "organic life" and the second "outer" or "animal life." The orders of both lives are based on the activities and interactions of two different sets of organs within the same body. While the first set is composed of digestive, respiratory and excretory organs, the second set includes sense organs, nerves, brains and locomotor muscles. The organic life maintains the order of an organized body within its "borders" through "cycles of assimilation and dissimilation" (*cercles d'assimilation et de désassimilation*), or "cycles of creation and destruction" (*cercles de création et de destruction*), and the animal life supports it through "all connections that bind our existence to everything that surrounds us" (*tous les liens qui attachent notre existence à tout ce qui nous entoure*).[4] The different forms of interaction between both lives determine the "general mode of existence" of "living individuals" (*individus vivants*), but not their "generative reproduction" as individuals of the same "species". Bichat's physiology is a science of the reproduction of life-maintaining processes within individuals from their birth to their death – and not of the reproduction of this reproduction from one birth to another.[5]

Like Théophile de Bordeu and other medical doctors of the school of Montpellier, Bichat assumes that physical forces and their "mechanistic" orders are not sufficient to explain the existence of self-active and self-maintaining "systems" (*systèmes*) of animals and humans. Rather, in his view, it is necessary to refer to special "vital forces" that are only active within "living bodies." As variations of just one single force, "vital forces" play for the foundation of Bichat's physiology as an experimental science the same role as the gravitational force for Newton's physics.[6]

[3] Cf. Bichat 1802, 1: "*La vie est l'ensemble des fonctions qui résistent à la mort*;" and Bichat 1802, 307, footnote 1. For the general history of experimental physiology in the eighteenth and nineteenth centuries, see Canguilhem 1965; Rothschuh 1968; Delaporte 1979; and Lesch 1984.

[4] Bichat 1802, 34, 125 and 145. In the supplements to *Die Welt als Wille und Vorstellung*, first published in 1844, Schopenhauer compares his notion of a "blind will to life" and its relation to reflexive faculties with Bichat's difference between the organic and the animal life. Cf. Schopenhauer 1986, 337–343; and Janet 1880. For Bichat's distinction of two lives, see Rey 1997, 190–194; Rey 2000, 354–358; and Stahnisch 2003, 91–94.

[5] Cf. Bichat 1802, 3–4: "La génération n'entre point dans la série des phénomènes de ces deux vies, qui ont rapport à l'individu, tandis qu'elle ne regarde que l'espèce: aussi ne tient-elle que par des liens indirects à la plupart des autres. Elle ne commence à s'exercer que lorsque les autres fonctions sont depuis longtemps en exercice; elle s'éteint bien avant qu'elles ne finissent." Unlike Lamarck, Geoffroy Saint-Hilaire and Georges Cuvier, Bichat does not discuss the problem of the fixism and the transformism of species. For the historical context of this debate, see Appel 1987; Laurent 1987; and Corsi 1988.

[6] For the wider historical and philosophical context of concepts of vital forces, see Lenoir 1980 and 1982; Monti 1997; Williams 2003; Steinke 2005; Schmitt 2006; Cheung 2008a and 2008b; Wolfe 2010 and 2017; Duchesneau 2017 and 2018; and Zammito 2018. In his *Cours de philosophie positive* (1830–1842), Comte criticizes Bichat's concept of "vital forces" as a theory of "metaphysical" entities that would hinder scientific progress. Cf. Comte 1830–1842, vol. 3, 143: "La prétendue indépendance des corps vivants envers les lois générales, si hautement proclamée encore, au

However, Bichat is also aware of the problem that the principles and laws of "vital forces" are difficult to define, because their effects are related to very variable phenomena. In general, "vital forces" represent "energetic" dispositions that generate activities through organ-specifc stimuli-reaction-schemes or "sensibilities". The amount or "dosis" of the dispositions is not directly "measurable", but their expression depends on physical and chemical properties of organic "membranes" (*membranes*) or "tissues" (*tissus*), whose modifications have again an influence on the activity of organs.[7]

Through partial destructions or deteriorations of "regular" or "normal states" (*états réguliers, normaux*) of tissues, Bichat examines the negative influences of modified organic material on the processual framework of organized bodies as causes of "illnesses" or "death" and infers from these influences their positive role in "healthy" bodies. His experimental physiology is a science of living, partially dead, almost dead and dead bodies. Like the vivisectionists William Cruikshank, Jean César Legallois and Franz Joseph Gall, Bichat reconstructs the order of life through the way how it ends in focusing on the effects of the "loss of action" (*défaut d'action*) of single organs within the order of all organs.[8] His medical research on the properties of "normal" and "modified tissues" combines, similarly to the works of James Carmichael Smyth, John Hunter and Philippe Pinel, the physiology of animals and humans with their anatomies and pathologies.[9] However, Bichat always looks from both ends of the life of organized bodies into the order of their two lives. He is not only interested in the enchained processes of their "partial deaths" (*morts partielles*), but also in the development of their embryonic life into that of an adult body.[10]

Bichat's physiology finally includes, like those of Gottfried Reinhold Treviranus, Georges Cabanis and the French Ideologues, a science of man or an anthropology based on a sensationist theory of the production of knowledge and of "intellectual phenomena", whose transformations reach from Locke to Étienne Bonnot de

---

commencement de ce siècle, par le grand Bichat lui-même, n'est plus désormais directement soutenue, en principe, que par les seuls métaphysiciens;" and Comte 1830–1842, vol. 3, 153–154. For Comte's model of animal agents and its relation to Bichat's physiology, see McCormick 1976; Cherni 1999; and Cheung 2011 and 2014, 249–277.

[7] Cf. Bichat 1802, 89, footnote 1: "Ces expressions dose, somme, quantité de sensibilité, sont inexactes en ce qu'elles présentent cette faculté vitale sous le même point de vue que les forces physiques, que l'attraction, par exemple; en ce qu'elles nous la montrent comme susceptible d'être calculée, etc. Mais, faute de mots crées pour une science, il faut bien, afin de se faire entendre, en emprunter dans les autres sciences;" and Bichat 1802, 90–91.

[8] The "general death" of an organized body always depends on the "partial death" of one of the "centers" of the organic and the animal life: the heart and the brain. Cf. Bichat 1802, 189–197 and 316–330. For the historical context of vivisectionist experiments at the turn of the eighteenth century, see Rupke 1987; Hagner 1997; and Cheung 2013.

[9] For the discursive field between anatomy, physiology and medicine around 1800, see Foucault 1963; Canguilhem 1966; Keel 1979, 1982 and 2001; and Schmitt 2009.

[10] Duden & Schlumbohm (2002), Schmitt (2005) and Wellmann (2010) focus on different aspects of the history of embryologies and of their relation to physiologies.

Condillac's *Traité des sensations* (1754) and Paul Thiry d'Holbach's *Système de la nature* (1770).[11] The special position of man is again related to differences between the lives of plants, animals and humans, and yet these lives resemble each other. For Bichat, higher animals basically possess the same set of organs and the same processual structure as humans, but they are less "perfect" in their ability to access and control the world that surrounds them. At the same time, all animals are more "perfect" or "successful" than plants in these activities, although some of them seem to exist like a plant. In a similar way as in Charles Bonnet's great chain of beings, the first animal of Bichat's "animal series" (*série animale*) resembles a plant, and it is within this series that the "two lives" become what they are in humans.[12] In this sense, there is, in Bichat's concept of the order of life among organized beings, an animal within the human and a plant within the animal.

25.1 The Difference of Two Lives, or How One Life Becomes Two Lives

For Bichat, the life of plants is a life "within itself." Their existence is "bound to the soil that received their seed," and they have no inner organs, but just one organized inner space, in which different liquids circulate within interconnected canals. These liquids have been passively absorbed from the outside, while others evaporated. By contrast, higher animals possess inner organs, they can move their bodies, and they combine their "inner life" with a self-regulated "outer life" composed of specialized organs for this activity.[13] Plants thus only have one life and animals two lives. However, some animals seem to have more than one and less than two lives. These animals possess just one inner organ. They are, so to speak, on the way to become an animal without being a plant. Their two lives are just recognizable as "obscure traces" of what they are in higher animals. However, their second, inner "border" makes a major difference within the general mode of existence of an organized body. It produces not only a second, sensible skin, but also a new, enfolded entity with a life of its own – a cavity within a cavity that is able to actively absorb and reject particles, because its sensibility regulates the opening and closing movements of its wholes. The new type of agent is an agent that folds itself into itself, thus a double agent with a secret life within its body.

At this point of transition, the last plant becomes a first "sketch" (*ébauche*) and a "knitting pattern" or "robe" (*canevas*) of the first animal at the very lower end of the

[11] For the wider historical context of anthropologies and the movement of the French Ideologues, see Moravia 1966 and 1974; Yolton 1991; Cipollini 1998; Guichet 2010; and Cheung 2017.

[12] For Bonnet's great chain of being, see Lovejoy 1936; Anderson 1976; and Cheung 2005a and 2005b. Around 1800, the research about the order of life in plants depended on the emerging experimental physiologies of animals. Cf. Delaporte 1979, 11: "Dans l'histoire de la physiologie végétale, en général, c'est l'histoire de la physiologie animale qui fournit un modèle d'intelligibilité."

[13] Cf. Bichat 1799, 379–380; and 1802, 2.

animal series; an animal that folds this robe into itself without producing an "external apparatus of organs," but just sensible and contracting wholes in its skin. Bichat calls the first animals "plant-animals" (*animaux-plantes*) or "zoophytes" (*zoophytes*).[14] In a similar perspective, his colleague at the Hôpital Saint-Louis in Paris, Anthèlme Richerand, refers in his *Nouveaux élémens de physiologie* (1802) to zoophytes as the first "animal organisms" that could absorb and digest outer particles through an "inner surface" (*surface intérieure*) forming a single "cavity" that folds itself into many organs in higher animals:

> Every animal may be considered, in abstraction, as a nutritive tube, open at its extremities; the whole existence of the polypus seems reduced to the act of nutrition, as its whole substance is employed in the formation of an alimentary tube, of which the soft parietes, extremely sensible and contractile, appropriate to themselves, by a sort of absorption, the substances which are brought into it. ... It is in the thickness of the parietes of this animated tube, between the mucous membrane that lines it inwardly, and the skin with which this membrane is continuous, that all the organs are placed, which serve for the transmission and elaboration of fluids, together with the nerves, the muscles, in short, all that serves for the maintenance and the conservation of life.[15]

Within Bichat's "animal series", the folding process, that results into the first stomach-like cavity, gets very complicated in higher animals. From lower to higher animals, this growing complexity relies on a process through which the plant body is not simply redoubled into inner cavities of the same type. Rather, a new plurality emerges within the inside of organized bodies, a plurality that cannot be seen from their differently shaped outer appearances: a plurality of organs with a "proper life" and associated in different sets of organs. The relation between these organs and the body to which they belong represents both a unity-in-variety of one and the same folded skin and a unity-in-plurality of interacting organs with different sensibilities. Leibniz's principles of the most perfect theo-ontological order, that constitute the

---

[14] Cf. Bichat 1802, 2–3; and 1801, vol. 1, Considérations générales, xlii. After Abraham Trembley's experiments on the regeneration of entire freshwater polyps that were cut into many pieces in the 1740s, preformationists and their epigenetic critics discussed the regenerative properties of zoophytes. These debates reached from Bonnet's observations on the parthenogenesis of aphids to the embryological and morphological works of Albrecht von Haller, Johannes Friedrich Blumenbach and Caspar Friedrich Wolff. For the role of zoophytes in the natural history of the eighteenth century, see Querner & Jahn 2003; Gibson 2012; and, with a focus on Charles Bonnet, Cheung 2016a. Roger (1993), Sloan (2002) and Cheung (2006) reconstruct different aspects of preformationism and its critique in the eighteenth century.

[15] Richerand 1817, vol. 1, 16–17: "Tout animal peut être réduit par la pensée à un tube nutritif ouvert par ses extrémités; toute l'existence du polype paroît réduite à l'acte nutritif, comme toute sa substance employée à la formation d'un sac alimentaire dont les parois molles, très-sensibles et contractiles, travaillent à s'approprier par une sorte d'imbibition les substances qui y sont attirées. ... C'est dans l'épaisseur des parois de ce tube animé, entre la membrane muqueuse qui revêt son intérieur, et la peau avec laquelle cette membrane se continue, que se trouvent tous les organes qui servent au transport et à l'élaboration des humeurs, les muscles, les nerfs, en un mot, tout ce qui sert à l'entretien et à la conservation de la vie." For Richerand's physiology, see Cheung 2016b.

foundation of Bonnet's great chain of beings, thus reappear in the order of organs within living bodies.[16] Bichat is mainly interested in this order.

The plurality of organs in Bichat's animal series is not a playful manifold of possible variations. Organs fulfill a certain "role" or "function" in the "system" to which they belong in order to get from the "system" what they need for their own "lives". There is not only a new plurality of organs in higher animals, but also a plurality of new relations between them. Organs co-exist, and their relations belong to different orders of interdependence and interconnectedness, of interactions and "communications" and of mutual exchanges and transformations. The stomach, the gall bladder, the liver and the kidneys, they all fulfill different functions, these functions rely on different sensibilities and different, organ-specific cycles of production, consumption and reproduction accomplished by different organic materials or "textures", while all functions, sensibilities and textures are related to each other within a single "system". In animals, life becomes complicated, but better organized than in plants.

Animal bodies exist as fusions of inner forces and textures of organic fibers. Their textures form the "particular machines" through which vital forces express themselves, and the "particular machines" are again parts of the "general machine" of the entire "network" (*réseau*) of interconnected textures of organized bodies. Each operational phase of the stimuli-reaction-schemes of organs – their "irritability", "contractility" and "extensibility" – corresponds to certain "properties" (*propriétés*) of interconnected textures, although the "dosis" of the energetic disposition of their sensibility is not itself determined through this material constellation.[17]

To analyze the physical properties of organic textures, Bichat uses dead organic material. He cuts it out, dries and heats it and treats it with alcohol, acids, bases and water. Further on, he examines specific combinations of properties and textures and classifies them as different types of "tissues". These tissues represent the anatomical "elements" of the processual framework of organized bodies:

> All animals are an assemblage of different organs, which, executing each a function, concur in their own manner, to the preservation of the whole. It is several separate machines in a general one, that constitutes the individual. Now these separate machines are themselves formed by many textures of a very different nature, which really compose the elements of these organs. Chemistry has its simple bodies, which form, by the combinations of which they are susceptible, the compound bodies; such are caloric, light, hydrogen, oxygen, carbon, azote, phosphorus, etc. In the same way anatomy has its simple textures, which, by their combinations four with four, six with six, eight with eight, etc., make the organs.[18]

[16] For Leibniz's and Bonnet's construction of the chain of beings, see Rieppel 1988; Fichant 2003; Cheung 2009 and 2016a; and Duchesneau 2010.

[17] Cf. Bichat 1801, vol. 1, Considérations générales, lxxviii-lxxix: "Dans les causes qui mettent en jeu la contractilité, les unes appartiennent donc à la vie; les autres en sont indépendantes; elles ne tiennent qu'à l'organisation. Tous les organes sont essentiellement contractiles; mais chacune des causes qui les font contracter, n'agit que sur tel ou tel tissu."; and 1802, 104–114.

[18] Bichat 1801, vol. 1, Considérations générales, lxxix: "Tous les animaux sont un assemblage de divers organes qui, exécutant chacun une fonction, concourent, chacun à leur manière, à la conser-

The more complex the "network" and the processual framework of interactions get in the ascending animal series, the more specialized become the relations between tissues and vital forces with different sensibilities in the organic life. At the same time, however, the "central organ" of the animal life becomes a seemingly unstructured "pulpy mass". This is for Bichat an indication that the somehow "higher" sensibility of the organs of the cerebral apparatus depends less on physical properties than that of the "inner organs." Conversely, from higher to lower animals and from brains to hearts, the activities of organs rely more on mechanic reactions of physical textures than on the sensibilities of vital forces. The inner activities of plants "resemble" circulatory, nutritive and secretory phenomena, but they are not performed by specialized organs.[19] They still possess some kind of force-related sensibility, although their movements seem to rely only on textures and "physical laws." For Bichat, most of the things that happen in plants are still very "obscure."[20]

Besides the relation between vital forces and tissues, the processual framework of organized bodies depends on a certain imperfection of the "cycle of assimilation and dissimilation." As elements of a twofold nutritive process, that occurs both inside and outside of themselves, organs perform acts, and this activity represents "work" (*travail*) – for example the "work of digestion" – resulting into the deterioration and modification of the involved materials and their energetic dispositions. These materials and dispositions have to be regenerated or reproduced. Through their self-maintaining activity, organs and their organized bodies thus constantly re-produce their needs for reproduction, and yet they cannot satisfy these needs within themselves through a recycling process of formerly assimilated, "worn-out material." Parts or liquids, that became useless, have to be excreted and to be replaced by assimilated, "foreign" material taken from the outer world. This transfer from the outside to the inside is a selective process. It has to be as selective as the difference between the needs of the "inner organs" and their limited capacities to assimilate different "foreign bodies."

Folded into organs that are able to detect and grasp suitable "foreign bodies," the "outer skin" has to have a sensibility that is not only directed towards the inside of its body, but also towards the things that surround it. Higher animals possess various organs and organic apparatuses that interact with the bodies and the changing circumstances of the surrounding world. Their sensibility is different from that of the "inner organs." The plurality of organs therefore relies on a difference between the

---

vation du tout. Ce sont autant de machines particulières dans la machine générale qui constitue l'individu. Or ces machines particulières sont elles-mêmes formées par plusieurs tissus de nature très-différente, et qui forment véritablement les élémens de ces organes. La chimie a ses corps simples, qui forment, par les combinaisons diverses dont ils sont susceptibles, les corps composés: tels sont le calorique, la lumière, l'hydrogène, l'oxigène, le carbone, l'azote, le phosphore, etc. De même l'anatomie a ses tissus simples; qui, par leurs combinaisons quattre à quatre, six à six; huit à huit, etc., forment les organes;" translated by George Hayward (in Bichat 1822a, 44), modified by T. Cheung.

[19] Cf. Bichat 1802, 85–86 and 99–104.

[20] Cf. Bichat 1802, 99–104.

"inner organs" and other, co-existing organs that mediate between them and the outer world. This difference is the difference between Bichat's two lives. One life, called organic, represents the order of "inner organs," and the other, called animal, the order of the "outer organs." Unlike single organs and similar to apparatuses of organs, the organic and the animal lives are, as meta-apparatuses, composed of many organs and apparatuses. There are thus the "proper lives" of many organs and the two lives of one organized body that exists as a "living individual":

> ... the functions of the animal are of two very different classes. The functions of the first class are composed of a habitual succession of assimilation and excretion; through these functions, the animal lives within itself, transforms into its proper substance the particles of bodies that are next to it, and afterwards rejects them when they become heterogeneous. By the other class, it lives externally, is the inhabitant of the world, and not as the vegetable of the place where it was born.[21]

25.2 Forms of Interaction, or How Two Lives Become One Life

For Bichat, all interactions or "communications" between self-active agents rely on stimuli-reaction-schemes with a certain sensibility. His "systems" of "mutual dependencies" and "influences" of multiple agents include seven variations or forms of "communications": contractile, sympathetic, nervous, perceptive, reflexive, regulative and social "communications". While all "inner organs" "communicate" with each other through contractile and sympathetic interactions, the forms of interaction between the organs of the animal life are not only different from those of the "inner organs," but also among themselves. Through their contact with "foreign bodies," the sense organs produce "sensations" (*sensations*) and send these sensations through specific nerve-"channels" and their interconnected tissues directly to an organ that has no direct access to the outer world: the brain. Its activity regulates and controls again the "volontary muscles" that move the common body of all organs.[22] The brain is thus, like the heart, a "central organ," with which all the other organs of

[21] Bichat 1802, 3: "... les fonctions de l'animal forment deux classes très-distinctes. Les unes se composent d'une succession habituelle d'assimilation et d'excrétion; par elles il transforme sans cesse en sa propre substance les molécules des corps voisins, et rejette ensuite ces molécules lors qu'elles lui sont devenues hétérogènes. Il ne vit qu'en lui, par cette classe de fonctions; par l'autre il existe hors de lui; il est l'habitant du monde, et non comme le végétal du lieu qui le vit naître."

[22] Cf. Bichat 1801, 1: 115–116; and Bichat 1802, 85–100; Bichat 1802, 124: "Dans la vie organique, la sensibilité est la faculté de recevoir une impression; dans la vie animale, c'est la faculté de recevoir une impression, plus, de la rapporter à un centre commun;" and Bichat 1802, 166. Through their contact with and their reaction to bodies that are "foreign to our proper substance", the "mucous membranes" (*membranes muqueuses*) of "inner organs" represent for Bichat the "inner sense of touch" (*tact interne*) of the organic life. Cf. Bichat 1802, 49: "... les membranes muqueuses, espèces d'organes qui, toujours en rapport avec des corps étrangers à notre propre substance, sont le siége d'un tact interne, analogue en tout au tact extérieur de la peau sur les corps qui nous entourent."

the animal life "communicate." However, it does neither, like the "inner organs", contract itself nor, like the sense organs, produce sensations of "foreign bodies.".[23]

The brain possesses a complex inner processual framework. In a first operational phase, it gets "affected" by the sensations of the sense organs and reproduces them within itself as "perceptions," which it uses for "comparative operations" that establish "memories" and "imaginations." These classified perceptions become again the stimuli of an agent-like intracerebral entity. As if another organ would be folded into the brain, Bichat assumes that this entity is equipped with a proper stimulus-reaction-interface for which perceptions play the same role as the sensations of the sense organs.[24] He calls such an agent within the brain a "soul" – without clarifying its ontological status or anatomical position.[25] Its sensibility, however, seems to depend even less on organic textures than that of the other parts of the brain. Stimulated by perceptions, the reflexive processes of the soul produce their own perceptions, on the basis of which it performs volitional acts and judgements. These acts and judgements are "communicated" as sensation-like stimuli, mediated by nerves, to different apparatuses of organs of "voluntary muscles," through which the brain regulates the spatial relation and the access of the entire body to and its exchange processes and "communications" with "foreign bodies."[26] Through this operational framework, the brain is also able to adapt and eventually optimize its own reactions to the different "states", "needs" and "pathological" modifications of "inner organs" as well as to the changing "outer circumstances."

Equipped with a brain, the animal life seems to be the ruler of an organized body that is produced and reproduced by a different life to which it is only attached. However, both lives perfect each other, they only exist and die together, and they interact with each other, although each life influences and changes the other life in its own way and according to its own rules. Through their interactions, they become one single "individual life." Each "function" of this life belongs to either the organic or the animal life, and yet all "functions" are the "functions" of the same life.[27] The "ordinary" interactions between the animal and the organic life concern the access to and the "communication" with the outer world, the delivery and the incorporation of outer "foreign bodies" (in different states of aggregation) – for example through the mouth and the nose – and the excretion of "wasted" or "used material."

The brain "presides" (*préside*) over the movements of the "volontary muscles" of the eyes, of mastication, of locomotion and of the lower abdomen.[28] It also "controls" the intercostal muscles and the diaphragm as "intermediary agents" (*agens intermédiares*) and boundary phenomena that are "nearly situated between" the order of

---

[23] The heart contracts, like all other "inner organs", "with necessity" immediately after the "arrival" of a stimulus. Cf. Bichat 1802, 98.

[24] Cf. Bichat 1802, 23: "le cerveau est à l'ame ce que les sens sont au cerveau;" Bichat 1802, 24–25; 53.

[25] For the historical debates about a soul-organ, see Figlio 1975 and Hagner 1997.

[26] Cf. Bichat 1802, 187.

[27] Cf. Bichat 1802, 185: "Toutes les fonctions appartiennent ou à la vie animale, ou à l'organique. De là deux classes très-distinctes entr'elles."

[28] Bichat 1802, 189–190.

both lives, because their movements dilate and contract the lungs.[29] Every "function" thus belongs either to the organic or the animal life, and some "functions" are more influenced by both lives than others. Whatever lies between the supply and the excretion of "foreign" and "used material," participates at the "cycle of assimilation and dissimilation" of the organic life. The brain cannot directly "react upon" this cycle.[30] "Inner organs" have their own "lives," possess their own sensibilities and ways of "communication," produce their own products and maintain not only the "lives" of each other, but also those of the organs of the animal life, whose products are not digestible for the lives that nourish them.[31]

Even a short interruption of the "twofold movement of assimilation and dissimilation" would terminate the lives of all organs, while the organs of the animal life show signs of "fatigue" and need to refill their energetic dispositions through "periodic intermittences" or a significant reduction of their activities.[32] "Inner organs" never sleep, and their activities constantly modify each other. The permanent variations of their activities and "states of excitation" contrast for Bichat with those of the sense organs, which rely on the influence of relatively "constant" and "rhythmically" changing physical phenomena of the outer world.[33] If the regular "irregularity" (*irrégularité*) of the order of interaction of the "inner organs" with

[29] Cf. Bichat 1802, 48–49 and 330.

[30] Cf. Bichat 1802, 321.

[31] Like the "life" of all other organs, the one of the brain relies on its supply with red blood. At the same time, it regulates the movements of the muscles that are necessary for the influx of fresh air into the lungs. After the "partial death" of the brain, the red color of the blood changes to black. However, Bichat observes that the heart still beats, although slower and slower. The heart thus dies after the brain, and yet it cannot live longer than just a couple of minutes without it. Cf. Bichat 1802, 171: "Il est facile de prouver que le mouvement du sang, en se communiquant au cerveau, entretient son action et sa vie.", and Bichat 1802, 363: "... la vie organique ne cesse pas immédiatement par la cessation de l'action cérébrale, il y a donc des agens intermédiaires qui déterminent, par leur mort, cette cessation. Or ces agens sont principalement, comme dans la mort du cœur par celle du cerveau, les organes mécaniques de la respiration. Voici la série des phénomènes qui arrivent alors. 1°. Interruption des fonctions cérébrales. 2°. Cessation des fonctions mécaniques du poumon. 3°. Anéantissement de ses fonctions chimiques. 4°. Circulation du sang noir dans toutes les parties. 5°. Affaiblissement du mouvement du cœur et de l'action de tous les organes. 6°. Suspension de ce mouvement et de cette action." This series of "partial deaths" is different in "lower animals." Cf. Bichat 1802, 147 and 331.

[32] Cf. Bichat 1802, 34–36.

[33] Cf. Bichat 1802, 32–33: "Les forces vitales, et les excitants qui les mettent en jeu, sans cesse variables dans l'estomac, les reins, le foie, les poumons, le coeur, etc., y déterminent une instabilité constante dans les phénomènes. Mille causes peuvent à chaque instant doubler, tripler l'activité de la circulation et de la respiration, accroître ou diminuer la quantité de bile, d'urine, de salive sécrétées, suspendre ou accélérer la nutrition d'une partie; la faim, les aliments, le sommeil, le mouvement, le repos, les passions, etc., impriment à ces fonctions une mobilité telle, qu'elles passent chaque jour par cent degrés divers de force ou de foiblesse. Tout, au contraire, est constant, uniforme, régulier dans la vie animale. Les forces vitales des sens ne peuvent, de même que les forces intérieures, éprouver ces alternatives de modifications, ou du moins à un degré aussi marqué. En effet, un rapport habituel les unit aux forces physiques qui régissent les corps extérieurs: or, celles-ci restant les mêmes dans leurs variations, chacune de ces variations anéantirait le rapport, et alors les fonctions cesseraient."

their "thousand modifications" would be "communicated" to the brain through inner senses and sensations that are as distinct as those of the outer senses, the animal life would become "overexcited", "confused" and "dysfunctional".[34] There are thus no inner eyes of the brain, but only somehow blurred representations of generalized (ordinary or pathological) "states of excitation" and activities of "inner organs", "communicated" through alterations of the blood current, "sympathetic movements" and particular ramifications of cerebral nerves into the organic life. Bichat also mentions a kind of a second nervous system composed of "ganglions." In difference to cerebral nerves, ganglions are not related to a "single center," but to a "network" with many centers distributed among the "inner organs." Its "aggregations" and "branches" constitute the "spinal cord" and diffuse into "volontary muscles," while some cerebral nerves are attached to "glands" and "involuntary muscles." However, the role of ganglions in the organic life is for Bichat still too "obscure" to speculate about it.[35]

High "states of excitation" and intense activities of single organs often affect multiple organs that are not part of the same apparatus or functional ensemble. The simultaneous and "concurrent activities" of co-affected organs reinforce again themselves and expand their influence on the "ordinary activity" of other organs. Bichat calls this phenomenon "sympathetic communication" (*communication sympathique*).[36] Besides the bloodstream and nervous "communications", it represents the third form of interaction between the two lives. The brain normally perceives the effects of the "concurrent activities" of "sympathetic affections" as an "inner sentiment" (*sentiment intérieur*) of a "region" (*région*) or an "area" (*foyer*) of its body or just of the "point" (*point*), where they come into contact with cerebral nerves. However, on their way from the "inner organs" to the brain and its nervous system, the "sympathetic communications" are themselves modified through the "proper lives" of affected organs.[37] The "inner sentiment" thus only gives a vague and sometimes erroneous topological indication of the "place" (*lieu*), from which the movements originated. Further on, very strong sympathetic affections can have a direct effect on "voluntary muscles" and cause "spasmic movements."[38]

---

[34] Cf. Bichat 1802, 32–36 and 80–81.

[35] Cf. Bichat 1801, 1: 115–116; and 1802, 70–73.

[36] Cf. Bichat 1802, 65–66.

[37] Cf. Bichat 1802, 67–69 and 72–74: "... dans les organes internes, le sentiment né de l'affection d'une partie, est toujours un indice infidèle du siége et de l'étendue de cette affection: par exemple, la faim porte son influence sur la totalité de l'estomac, et cependant le cardia semble seul nous en transmettre la sensation. Une large surface enflammée dans la plèvre ou le poumon, ne donne lieu le plus souvent qu'à une douleur concentrée sur un point. Combien de fois à la tête, à l'abdomen, etc. une douleur fixe et occupant un petit espace, ne coïncide-t-elle pas avec une affection largement disséminée, et ayant même un siége tout différent de celui que nous présumons! Il ne faut donc jamais considérer le lieu où nous rapportons le sentiment, comme le sûr indice du lieu précis qu'occupe l'affection, mais seulement comme un signe qu'elle se trouve là, ou dans le voisinage;" and Bichat 1802, 75. For Lamarck's concept of the inner sentiment and its role in the animal series, see Gissis 2010.

[38] Cf. Bichat 1802, 67.

There are high "states of excitation" of "inner organs" – indicated, for example, by a high heartbeat, fast breathing and sweating –, whose effects are similar to those of "pathological states" without being caused by them. They belong to "sympathetic phenomena" that cause specific "disturbances" or "troubles" (*troubles*) between both lives. Leaving the explanation of their ontological status and their differences from "emotions" and other "affections" to "metaphysicians", Bichat calls them "passions." Based on observations and experiences, they can be classified according to specific patterns of "states of excitation" and their effects. In the "common language", these states have various names, like "anger", "hate", "joy" and "anxiety". While the "seat" of the passions is only within the organic life, their "natural stimulants" (*excitants naturels*) come from the sense organs.[39] The "external sensitive apparatus" gives the "inner organs" "opportunities" (*occasions*) to reinforce or weaken their activities. These modified activities can be again influenced by "pathological states."[40] Since plants possess neither inner organs nor sense organs, they can also show no signs of passions.[41]

Passions only occur between "inner organs," and they operate like amplifiers of their activities.[42] However, the sympathetic effects of passions influence not only the activity of "inner organs," but also that of the animal life.[43] While "anger" accelerates the bloodflow, which could cause an "overexcitation" of the cerebral apparatus,

[39] Like Franz Joseph Gall and Johann Spurzheim, Comte criticizes Bichat's localization of the siege of the passions within the organic life. Cf. Comte 1830–1842, vol. 3, 423.

[40] Cf. Bichat 1802, 53; Bichat 1802, 60: "Non-seulement les passions portent essentiellement sur les fonctions organiques, en affectant leurs viscères d'une manière spéciale, mais l'état de ces viscères, leurs lésions, les variations de leurs forces concourent, d'une manière marquée, à la production des passions."

[41] Cf. Bichat 1802, 62: "On demandera sans doute ici comment les végétaux qui vivent organiquement, ne nous en présentent aucun vestige? c'est que, outre qu'ils manquent de l'excitant naturel des passions, savoir, de l'appareil sensitif extérieur, ils sont dépourvus des organes internes qui concourent plus spécialement à leur production, tel que l'appareil digestif, celui de la circulation générale, celui des grandes secrétions, que nous remarquons chez les animaux; ils respirent par trachées, et non par un foyer concentré, etc. Voilà pourquoi les passions sont si obscures, et même presque nulles dans le genre des zoophytes, dans les vers, etc. pourquoi, à mesure que dans la série des animaux, la vie organique se simplifie davantage, perd tous ses organes importans, les passions décroissent proportionnellement."

[42] Cf. Bichat 1802, 62: "Tout tend donc à prouver que la vie organique est le terme où aboutissent, et le centre d'où partent les passions;" and Bichat 1802, 152.

[43] Cf. Bichat 1802, 54: "Il est sans doute étonnant que les passions qui entrent essentiellement dans nos relations avec les êtres placés autour de nous, qui modifient à chaque instant ces relations, sans qui la vie animale ne seroit qu'une froide série de phénomènes intellectuels, qui animent, agrandissent, exaltent sans cesse tous les phénomènes de cette vie; il est, dis-je, étonnant que les passions n'aient jamais leur terme ni leur origine dans ses divers organes, qu'au contraire les parties servant aux fonctions internes, soient constamment affectées par elles, et même les déterminent suivant l'état où elles se trouvent. Tel est cependant ce que la stricte observation nous prouve."; and Bichat 1802, 321, footnote 1: "Ces mots *passion, émotion, affection, etc.* prétentent, je le sais, des différences très-réelles dans la langue des métaphysiciens; mais comme l'effet général des sentimens qu'ils expriment est toujours le même sur la vie organique, comme cet effet général m'intéresse seul, et que les phénomènes secondaires m'importent peu, j'emploie indifféremment ces mots les

other passions, like "anxiety," reduce it, but have sympathetic effects that could result into "spasmic movements" of "voluntary muscles."[44] The brain is able to regain control over its muscles.[45] However, the passions are as active and as variable in their activities as the "inner organs." The systemic correction of the "troubles" that they produce would be nearly as troublesome for the brain as a distinct perception of the "ordinary activity" of the organic life.

"Living individuals" represent "modifications" of the processual framework of interacting organs of the "primitive organization" (*organisation primitive*) or the "original constitution" (*constitution originale*) of their "class" (*classe*). For Bichat, such modifications occur during their embryological development – caused, for example, through the specific composition of their nutritive substances –, but they are not related to the "usage" (*usage*) or the "exercise" (*exercice*) of their organs.[46] They result in physical "dispositions", in which some organs have a "predominant" influence on the activities of other organs and on the passions. Bichat relates these dispositions to different types of Hippocratic "temperaments" (*tempéraments*) – for instance, a "sanguine temperament" is based on a dominant "pulmonary apparatus" that reinforces the influx of red blood into the "circulatory system" –, which have again a permanent influence on cerebral activities.[47] Bichat calls this influence the "moral affection" (*affection morale*) of the animal life. It corresponds to what is "commonly" called the "character" (*caractère*) of an individual.[48] "Involuntarily," the animal life is thus entangled in the organic life and co-exists with it as an "individual life," that is, as a certain "mode of life" with "moral affections," also a "social life" in "societies."[49]

---

uns pour les autres." Arène (1911) compares Bichat's concept of passions with that of Descartes in *Les passions de l'âme* (1664).

[44] Cf. Bichat 1802, 63–69.

[45] Cf. Bichat 1802, 67: "Quand elles [les passions] sont très-fortes, l'affection très-vive des organes internes produit si impétueusement les mouvemens sympathiques des muscles, que l'action ordinaire du cerveau est absolument nulle sur eux. Mais la première impression étant passé, le mode ordinaire de locomotion revient;" and Bichat 1802, 321–323.

[46] Cf. Bichat 1802, 150–151.

[47] Cf. Bichat 1802, 60–61: "Qui ne sait que l'individu dont l'appareil pulmonaire est très-prononcé, dont le système circulatoire jouit de beaucoup d'énergie, qui est, comme on le dit, très-sanguin, a dans les affections une impétuosité qui le dispose sur-tout à la colère, à l'emportement, au courage; que là où prédomine le système bilieux, certaines passions sont plus développées, telles que l'envie, la haine, etc.; que les constitutions ou les fonctions des lymphatiques sont à un plus haut degré, impriment aux affections une lenteur opposée à l'impétuosité du tempérament sanguin. En général ce qui caractérise tel ou tel tempérament, c'est toujours telle ou telle modification, d'une part dans les passions, de l'autre part dans l'état des viscères de la vie organique et la prédominance de telle ou telle de ses fonctions. La vie animale est presque constamment étrangère aux attributs des tempéramens."

[48] Cf. Bichat 1802, 68 and 77–78.

[49] Haines (1969, 153-155) refers to various aspects of Bichat's notion of "social order". For the wider context of the relation between physiology and society at the end of the eighteenth and the beginning of the nineteenth centuries, see Sarasin & Tanner 1998.

While higher animals are mainly directed by innate "natural instincts" towards the outer world, humans have for Bichat the privilege to be able to "enlarge the narrow circle, in which the animals remain" (*agrandir le cercle étroit où restent les animaux*).[50] In humans, the animal life optimizes its support for the organic life in "perfecting itself." However, the self-perfecting process of individuals and the social orders, in which this process takes place, mutually influence each other. Social orders are, like organized beings, systemic orders of interactions, although they do not have a proper life, because "societies" are not themselves organized beings with embodied "individual lives," but orders, in which organized beings exist and interact with each other. Further on, they represent orders that "perfect" both the conditions of existence and the ability of humans to perfect these orders. At the very end of the animal series, interacting humans thus establish in the outer world – somehow detached from their bodies, and yet relying on the "performance" of their organs – a self-perfecting order that perfects what the first animal potentially already was: a self-perfecting agent. In contrast, the "inner organs" of human animals just do what they have always done as agents within cycles of assimilation and dissimilation. "Inner organs" interact, they possess "proper lives," but they do not have "social lives":

> The actual state of society is nothing but a more regular development, a more marked perfection of the performance (*excercice*) of the different functions of this [animal] life; for one of its greatest characters ... consists in its capability to unfold itself, while, in the organic life, there does not exist a part, which in the least degree may pass the limits which are set to it by nature. We live organically in as perfect, in as regular a way, when infants, as when men; but what is the animal life of the child compared with that of the man of thirty years of age?[51]

The history of the "development" of the animal life of a man of thirty years begins immediately after his birth. When the human fetus leaves the womb of its mother, where it lived "like a plant" with seemingly only one life, its sense organs are activated and a "new mode of existence" begins: its "individual life", composed of two lives.[52] Through the "use of the sense organs, the "locomotor apparatus" and the

[50] Cf. Bichat 1802, 51–52.

[51] Bichat 1802, 52: "La société actuelle n'est autre chose qu'un développement plus régulier, une perfection plus marquée dans l'exercice des diverses fonctions de cette vie [animale], lesquelles établissent nos rapports avec les êtres environnans; car ... c'est un de ses caractères majeurs de pouvoir s'étendre, se perfectionner, tandis que dans la vie organique chaque partie n'abandonne jamais les limites que la nature lui a posées. Nous vivons organiquement d'une manière tout aussi parfaite, tout aussi régulière dans le premier âge que dans l'âge adulte; mais comparez la vie animale du nouveau né à celle de l'homme de trente ans, et vous verrez la différence;" translated by George Hayward (in Bichat 1822b, 55), modified by T. Cheung.

[52] Cf. Bichat 1801, 1: 115: "S'il est une circonstance qui établisse une ligne réélle de démarcation entre les deux vies, c'est sans doute le mode et l'époque de leur origine. L'une, l'organique, est en activité dès les premiers instans de l'existence; l'autre, l'animale, n'entre en exercice qu'après la naissance, lorsque les objets-extérieurs offrent à l'individu qu'ils entourent, des moyens de rapport, de relation: car, sans excitans externes, cette vie est condamnée à une inaction nécessaire, comme sans les fluides ..., qui sont les excitans externes de la vie organique, celle-ci s'éteindrait.";

functions of the brain, the infant "learns" (*apprend*) how to use them. However, unlike other animals, humans have to be "educated" in order to learn how to "unfold" the "capability" of their animal life.[53] While "nature" is the only teacher of their "inner life", their "outer life relies on a kind of apprenticeship (*apprentissage*)," which is organized through "societies."[54] Conversely, the order of "societies" depends on interactions that are operated by educated agents whose activities fulfil, like the activities of organs, a certain "social" function. Bichat refers to these activities as "social habits" (*habitudes sociales*) and "professions" (*occupations*).[55]

While social habits are based on conventions about rhythmically and regularly performed "general movements" (*movements généraux*), which regulate interactions of the daily life, professions represent specialized "work"-processes.[56] The apprenticeship of social habits and professions depend on two enchained forms of education: the "natural education of the outer organs" (*éducation naturelle des organes extérieurs*), based on the principle that the "habit of acting optimizes the action" itself, and the "social" or "artificial education" (*éducation sociale, artificielle*), which relies on teachers and institutions.[57] Through the repetitive usage of its limbs, the infant "naturally" learns how to use them better or more efficiently. While repeated movements physically strengthen the "muscles" that perform them, repeated similar impressions lower the "vivacity" of the sensations of the sense organs, to which Bichat refers as a confusing mixture of painful and joyful excitements, and increase at the same time the "attention" to differences between them.[58] The higher "precision" of these differences, or "particular sensations" (*sensations particulières*), is the basis for reflexive processes in the brain, which are themselves necessary for a "social education" and the apprenticeship of a habit or a profession. On the contrary, the "inner organs" always "work" with the same "precision" and depend for this "precision" on the constant "vivacity" of their reactions, without

---

and Bichat 1801, 1: 125: "Le fœtus n'a, pour ainsi dire, rien dans ses phénomènes de ce qui caractérise spécialement l'animal; son existence est la même que celle du végétal; sa destruction ne porte que sur un être vivant, et non sur un être animé." Ingensiep (2006) reconstructs the usage of the term *vegetative* around 1800. For Cabanis's version of the developmental history of the life of the fetus and the infant, see Cheung 2014, 146–153. Bichat refers also to Condillac's model of an inner statue in the *Traité des sensations* (1754). Cf. Bichat 1802, 5.

[53] Bichat 1802, 130.

[54] Bichat 1802, 130.

[55] Cf. Bichat 1802, 131–132.

[56] Cf. Bichat 1802, 26–28.

[57] Cf. Bichat 1802, 142–143 and 151–252.

[58] Cf. Bichat 1802, 40–41 and 47: "L'enfant qui vient de naître, et pour qui tout est nouveau, ne sait encore percevoir dans ce qui frappe ses sens, que les impressions générales. En émoussant peu à peu ces impressions qui retiennent d'abord toute l'attention de l'enfant, l'habitude lui permet de saisir les attributs particuliers des corps; elle lui apprend ainsi insensiblement à voir, à entendre, à sentir, à goûter, à toucher, en le faisant successivement descendre dans chaque sensation, des notions confuses de l'ensemble, aux idées précises des détails. Tel est en effet un des grands caractères de la vie animale, qu'elle a besoin, comme nous le verrons, d'une véritable éducation."; and Bichat 1802, 88.

which the organic life would become dysfunctional.[59] The "education of the brain" thus relies on "acts" that "only gradually reach the level of precision for which they are made" (*n'aqiuièrent que graduellement le degré de précision auquel ils sont destinés*).[60] Such a gradual perfection of two enchained forms of education equally applies to the perfection of the "painter's eye", the "arm of the baker" and the cerebral processes of the philosopher.[61] As three different "classes" of professions, they improve the "natural" activities of the three organs or apparatuses of organs that constitute the animal life: the sense organs, the locomotor muscles and the brain.[62] Professions represent, like habits, perfected sets of activities that the infant and the child already perform. However, the "perfection" of the usage of an organ produces for Bichat an "inequality" *(inégalité)* of the distribution of "energy" or "force" within a body that only possesses a certain "sum" of this "force." The "force" that has been used for the "perfection" of one organ is not available for that of other organs. Good philosophers are, in this perspective, bad bakers.[63]

Bichat's ascending order of perfection within the animal series ends in a seemingly unlimited self-perfecting process of the animal life, and yet the performance of its organs depends on the "work" of organs that lie deep under the skin. These organs do not perfect themselves. Folded into the same body, the animal life would be, without their productivity and vivacity, a seemingly disembodied "cold series of intellectual phenomena" within a body that lost its life.[64] Humans are, in this sense,

[59] Cf. Bichat 1802, 48: "Rapprochons maintenant de ces phénomènes [de la vie animale], ceux de la vie organique; nous les verrons constamment soustraits à l'empire de l'habitude. La circulation, la respiration, l'exhalation, l'absorption, la nutrition, les secrétions ne sont jamais modifiées par elle. Mille causes menaceroient chaque jour l'existence, si ces fonctions essentielles pouvoient en recevoir l'influence;" and Bichat 1802, 127: "Nous remarquerons bientôt que les organes de la vie interne atteignent tout-à-coup la perfection; que dès l'instant où ils agissent, ils le font avec autant de précision que pendant tout le reste de leur activité."

[60] Bichat 1802, 128.

[61] Cf. Bichat 1802, 130–131.

[62] Cf. Bichat 1802, 131–133.

[63] Cf. Bichat 1802, 133: "Vous ne verrez presque jamais coïncider la perfection d'action des organes locomoteurs avec celle du cerveau ni des sens, et réciproquement il est très-rare que ceux-ci étant très-habiles à leurs fonctions respectives, les autres soient très-aptes aux leurs. ... Une somme déterminée de force a été répartie en général à cette vie: or cette somme doit rester toujours la même, soit que sa distribution ait lieu également, soit qu'elle se fasse avec inégalité; par conséquent l'activité d'un organe suppose nécessairement l'inaction des autres;" and Bichat 1802, 134–135: "Considérez dans la série des animaux la perfection relative de chaque organe, vous verrez que quand l'un excelle, les autres sont moins parfaits. L'aigle à l'œil perçant, n'a qu'un odorat obscur; le chien, que distingue la finesse de ce dernier sens, a le premier à un moindre degré; c'est l'ouïe qui domine chez la chouette, le lièvre, etc.; la chauve-souris est remarquable par la précision de son toucher; l'action du cerveau prédomine chez les singes, la vigueur de la locomotion chez les carnàssiers, etc. etc. Chaque espèce a donc une division de sa vie animale qui excelle sur les autres, celles-ci étant à proportion moins développée: vous n'en trouverez aucune où la perfection d'un organe ne semble s'être acquise aux dépens de celle des autres."

[64] Cf. Bichat 1802, 54.

human animals that never leave their bodies, and these bodies have two lives that co-exist as if they were one life.

25.3 Concluding Remarks

Bichat's *Recherches physiologiques sur la vie et la mort* is a treatise about a class of organized bodies that exist as self-active agents, called animals, their general difference from another class of organized bodies, called plants, and two specific types of agents located at both ends of the animal series: zoophytes, the first animals, and humans, the last animals. More precisely, it is a treatise about three aspects of agents: first, the processual framework that maintains a certain order, which represents their "life", second, pathological modifications that eventually result into their death, and, third, their ability to perfect themselves as beings that resist their own death. Plants also live and die, but their lives depend on rather passive processes of absorption and exhalation. Animals perform actions that incorporate, assimilate and excrete "foreign bodies". They produce their own needs, they produce movements to satisfy them, and they produce waste or material that they cannot use any longer. Animals also reproduce each other, but this activity and its results are not part of Bichat's physiology of the life and death of the "individual" of organized bodies.

The first animal possesses the first inner organ: an inner sack that results from the folding process of a plant-like body without inner organs. For Bichat, this cavity is the first agent within an organic body. As an agent within an agent, the folded cavity controls and regulates the influx and the efflux of particles, and this exchange process is necessary for the assimilation of "foreign bodies" and the excretion of "wasted material." Each animal has to replace what its actions "consumed", because it is not able to recycle the "useless material" that it produces. It is therefore not enough that an animal eats only one time in its life, if it should live longer than the time that it takes to consume what it ate. Life is, at its very basis, an order of repetitive movements that establish exchange processes between the insides and outsides of folded skins.

The advantage of the first animal over a plant without inner organs is that it possesses a sensible mechanism of contracting and expanding movements situated at the openings of its inner cavity. From now on, it is possible to improve the access to, the inner supply with and the excretion of "foreign bodies." Bichat refers to these improvements as perfections of perfectible and self-perfecting animals. These perfections constitute the ascending order of the animal series from the first to the last animal. *La vie est ce qui résiste à la mort,* and some lives do it better than others.

The differentiation of the first animal results into a complex order of interactions between organs that are as self-active as they are sensible. Bichat calls these interactions "communications." Interactions rely, as every activity, not only on different sensibilities, but also on specific organic textures of interconnected tissues. Organs interact as embodied agents with other agents within the folded inner space of the tissues of a body whose skin surrounds them, and they interact through this skin with an outer world in order to maintain both their "proper lives" and the life of the ensemble of organs, united in a single body. Organs exist, because they interact and co-exist. The more differentiated their order of interactions becomes, the more the order of living bodies differs from the order of bodies that either have no organs, but live a different life, as plants, or simply never lived.

In his *Recherches*, Bichat examines the difference of two sets of organs and refers to it as a difference between two lives: the organic and the animal life. Each of these lives is composed of the life of many organs, and yet they only represent a difference whose "traces" are already present in the life of the first animal. One life that assimilates and digests, and one life that controls and regulates the influx and efflux of "foreign" and "wasted bodies." Through their two lives, animals live in two worlds: an inner world, enclosed by an outer skin, and an outer world, on the other side. In the inner world, they maintain an order, and in the outer world, they get what they need for it. Humans perfect the way how they get it. Their inner order of various forms of interaction creates an outer order, that optimizes, as a social order, the relation between the animal life and the organic life. Whatever the "inner organs" need, social orders are made to satisfy them. However, without red blood, the brain would immediately collapse. As long as the brain lives, its performance relies on the division of "labor" between interacting agents and on the different forms of "precision" between enchained actions that concern both the inner and the outer life of their bodies.

Like Georges Cuvier's and Henri Milne Edwards's comparative anatomies, but more focused on processual aspects of orders of interaction, Bichat's physiology finally joins, together with those of Cabanis and Anthèlme Richerand, the discursive constellation that leads to the social physiologies of Auguste Comte and the Saint-Simonians. For Bichat, however, the seemingly unlimited perfectibility of the animal life through the education of humans, which are all equal as exemplars of the same "class" of animals, has to be part of a social progress that respects and integrates the characters of its members as an expression of an inner life that is slightly modified in each individual, but remains the same from birth to death. Humans are perfectible animals that never cease to live like the plants that they have been in the wombs of their mothers.

References

Anderson, Lorin. 1976. Charles Bonnet's Taxonomy and Chain of Being. *Journal of the History of Ideas* 37: 45–58.

Appel, Toby A. 1987. *The Cuvier-Geoffroy Debate. French Biology in the Decades Before Darwin.* New York/Oxford: Oxford University Press.

Arène, A. 1911. Essai sur la physiologie de Xavier Bichat. *Archives d'anthropologie criminelle de médecine légale* 48: 753–825.

Bichat, Xavier. 1799. Mémoire sur les rapports qui existent entre les organes à forme symétrique et ceux à forme irrégulière. *Mémoires de la société médicale d'émulation* 2: 477–487.

———. 1801. *Anatomie générale appliquée à la physiologie et à la médecine*, 4 vols. Paris: Brosson & Sarrazin.
———. 1802. *Recherches physiologiques sur la vie et la mort. (1st ed. 1800)*. 2nd. ed. Paris: Brosson & Gabon.
———. 1822a. *General Anatomy, Applied to Physiology and Medicine*. Trans. George Hayward, vol. 1. Boston: Richardson and Lord.
———. 1822b. *Physiological Researches on Life and Death*. Trans. F. Gold, with notes by F. Magendie. George Hayward. Boston: Richardson and Lord.
———. 1829. *Anatomie descriptive*. 5 vols (1st ed. 1801–1803). New edition. Paris: Gabon.
Canguilhem, Georges. 1965. *La connaissance de la vie*. Paris: Vrin.
———. 1966. *Le normal et le pathologique*. Paris: Presses universitaires de France.
Cherni, Zeineb Ben Saïd. 1999. *Biologie, sociologie et histoire des sciences chez Auguste Comte*, 2 vols. Doctoral thesis. Paris: Université Paris VII-Denis Diderot.
Cheung, Tobias. 2005a. Charles Bonnets allgemeine Systemtheorie organismischer Ordnung. *History and Philosophy of the Life Sciences* 26: 177–207.
———. 2005b. *Charles Bonnets Systemtheorie und Philosophie des Organischen*. Frankfurt am Main: Harri Deutsch Verlag.
———. 2006. The Hidden Order of Preformation: Organized Bodies in the Writings of Louis Bourguet, Charles Bonnet and Georges Cuvier. *Early Science and Medicine* 11: 11–49.
———. 2008a. *Res vivens. Agentenmodelle organischer Ordnung 1600–1800*. Stuttgart: Rombach.
———. 2008b. Regulating Agents, Functional Interactions, and Stimulus-Reaction-Schemes: The Concept of "Organism" in the Organic System Theories of Stahl, Bordeu, and Barthez. *Science in Context* 21: 495–519.
———. 2009. System, Mikrooperator und Transformation: Leibniz' gemeinsames Ordnungsdispositiv der Monade und des Lebendigen im naturgeschichtlichen Kontext. In *Der Monadenbegriff im 17. und 18. Jahrhundert*, ed. Hanns-Peter Neumann, 143–201. Berlin/New York: Walter de Gruyter.
———. 2011. Comtes Milieutheorie Proleptisch-Reaktiver Organismen. *Laufener Spezialbeiträge* 1: 84–89.
———. 2013. Limits of Life and Death: Legallois's Decapitation Experiments. *Journal of the History of Biology* 46: 283–313.
———. 2014. *Organismen. Agenten zwischen Innen- und Außenwelten 1780–1860*. Bielefeld: Transcript.
———. 2016a. Einheit und Vielfalt in Bonnets Systemtheorie organisierter Körper. In *Wünschenswerte Vielheit. Diversität als Kategorie, Befund und Norm*, ed. Kristian Köchy and Thomas Kirchhoff, 151–169. Freiburg: Karl Alber.
———. 2016b. Inside and outside: The Animal Worlds of Cuvier's, Richerand's and Cabanis's Living Systems. In *Animals. New Essays*, ed. Andreas Blank, 181–201. München: Philosophia.
———. 2017. Cabanis and the Order of Interaction. In *Human and Animal Cognition in Early Modern Philosophy and Medicine*, ed. Stefanie Buchenau and Roberto Lo Presti, 236–245. Pittsburgh: Pittsburgh University Press.
Cipollini, Enrico Marco. 1998. *Analisi dei Rapports cabanisiani. Antropologia filsosofica*. Padova: Libraria Padovana Editrice.
Comte, Auguste. 1830–1842. *Cours de philosophie positive*, 6 vols. Paris: Bachelier.
Corsi, Pietro. 1988. *The Age of Lamarck*. Revised and updated. Trans. Jonathan Mandelbaum. Berkeley: University of California Press.
Delaporte, François. 1979. *Le second règne de la nature*. Paris: Flammarion.
Duchesneau, François. 2010. *Leibniz: le vivant et l'organisme*. Paris: Vrin.
———. 2017. Degrees and Forms of Sensibility in Haller's Physiology. In *Human and Animal Cognition in Early Modern Philosophy and Medicine*, ed. Stefanie Buchenau and Roberto Lo Presti, 201–220. Pittsburgh: Pittsburgh University Press.
———. 2018. *Organisme et corps organique de Leibniz à Kant*. Paris: Vrin.

Duden, Barbara, and Juergen Schlumbohm, eds. 2002. *Geschichte des Ungeborenen. Zur Erfahrungs- und Wissenschaftsgeschichte der Schwangerschaft. 17.-20. Jahrhundert.* Göttingen: Patrice Veit.
Entralgo, Pedro L. 1948. Sensualism and vitalism in Bichat's Anatomie Générale. *Journal of the History of Medicine and Allied Sciences* 5: 47–64.
Fichant, Michel. 2003. Leibniz et les machines de la nature. *Studia Leibnitiana* 35 (2003): 1–28.
Figlio, Karl M. 1975. Theories of Perception and the Physiology of Mind in the Late Eighteenth Century. *History of Science* 13: 177–212.
Foucault, Michel. 1963. *Naissance de la clinique: Une archéologie du regard médical.* Paris: Presses universitaires de France.
Gibson, Sussanah. 2012. On Being an Animal, or, the Eighteenth-Century Zoophyte Controversy in Britain. *History of Science* 50: 453–476.
Gissis, Snait. 2010. Lamarck on Feelings: From Worms to Humans. In *The Body as Object and Instrument of Knowledge*, ed. Ofer Gal and Charles Wolfe, 211–239. Heidelberg/London/New York: Springer.
Guichet, Jean-Luc. 2010. La question de l'animalité, pivot du matérialisme et de la définition de l'humain chez Cabanis. *Dix-huitième siècle* 42: 367–384.
Hagner, Michael. 1997. *Homo cerebralis. Der Wandel vom Seelenorgan zum Gehirn.* Berlin: Verlagsbeteiligungsgesellschaft.
Haigh, Elizabeth. 1975. Roots of the Vitalism of Xavier Bichat. *Bulletin of the History of Medicine* 49: 72–86.
Haines, Barbara Ann. 1969. *Henri de Saint-Simon and the Idea of Organism.* Doctoral thesis. Cardiff: University College of Wales.
Hegel, Georg Wilhelm Friedrich. 1999. *Hauptwerke in sechs Bänden*, 6 vols. Darmstadt: Wissenschaftliche Buchgesellschaft.
Huneman, Philippe. 1998. *Bichat, la vie et la mort.* Paris: Presses universitaires de France.
———. 2006. Leben zwischen ,"Vegetativ" und "Vegetieren". Zur historischen und ethischen Bedeutung der vegetativen Terminologie in der Wissenschafts- und Alltagssprache. *International Journal of History & Ethics of Natural Sciences, Technology & Medicine* 14: 65–76.
Janet, Paul. 1880. Schopenhauer et la physiologie française. Cabanis et Bichat. *Revue des deux mondes* 39: 35–59.
Keel, Othmar. 1979. *La généalogie de l'histopathologie. Une révision déchirante: Philippe Pinel, lecteur discret de James Carmichael Smyth (1741–1821).* Paris: Vrin.
———. 1982. Les conditions de la décomposition «analytique» de l'organisme: Haller, Hunter, Bichat. *Études philosophiques N. S.* 1: 37–62.
———. 2001. *L'avènement de la médecine clinique moderne en Europe: 1750–1815. Politiques, institutions et savoirs.* Montréal: Les presses de l'université de Montréal.
Laurent, Goulven. 1987. *Paléontologie et évolution en France de Cuvier et Lamarck à Darwin.* Paris: Éditions du Comité des travaux historiques et scientifiques.
Lenoir, Timothy. 1980. Kant, Blumenbach, and Vital Materialism in German Biology. *Isis* 71: 77–108.
———. 1982. The Strategy of Life. In *Teleology and Mechanics in Nineteenth-Century German Biology.* Chicago: Chicago University Press.
Lesch, John E. 1984. Science and Medicine in France. In *The Emergence of Experimental Physiology, 1790–1855.* Cambridge, MA/London: Harvard University Press.
Lovejoy, Arthur O. 1936. *The Great Chain of Being. A Study of the History of an Idea.* Cambridge, MA/London: Harvard University Press.
McCormick, Michael A. 1976. *The Biological Theory of Auguste Comte.* Doctoral thesis. Austin: The University of Texas at Austin.
Monti, Maria T. 1997. Les dynamismes du corps et les forces du vivant dans la physiologie de Haller. In *Vitalisms from Haller to the Cell Theory: Proceedings of the Zaragoza Symposium.*

XIXth International Congress of History of Science, 22–29 August 1993, ed. Guido Cimino and François Duchesneau, 41–66. Florence: Olschki.
Moravia, Sergio. 1966. Aspetti della 'science de l'homme' nella filosofia degli 'idéologues'. *Rivista critica della storia della filosofia* 21: 398–425.
———. 1974. Il pensiero degli idéologues. In *Scienza e filosofia in Francia (1780–1815)*. Florence: La Nuova Italia.
Querner, Hans, and Ilse Jahn, eds. 2003. *Christoph Gottfried Jacobi und die Süßwasserpolypen des Abraham Trembley*. Marburg: Basilisken-Presse.
Rey, Roselyne. 1997. Bichat au carrefour des vitalismes. In *Vitalisms from Haller to the cell theory: proceedings of the Zaragoza symposium. XIXth international congress of history of science, 22–29 August 1993*, ed. Guido Cimino and François Duchesneau, 175–204. Florence: Olschki.
———. 2000. *Naissance et développement du vitalisme en France de la deuxième moitié du 18e siècle à la fin du Premier Empire*. Oxford: Voltaire Foundation.
Richerand, Anthèlme. 1817. *Nouveaux élémens de physiologie* (1st ed. 1802), 7th ed, 2 vols. Paris: Crapart.
Rieppel, Olivier. 1988. The reception of Leibniz's philosophy in the writings of Charles Bonnet (1720–1793). *Journal of the History of Biology* 21: 119–145.
Roger, Jacques. 1993. *Les sciences de la vie dans la pensée française du XVIIIe siècle*. Paris: Albin Michel.
Rothschuh, Karl Eduard. 1968. *Physiologie. Der Wandel ihrer Konzepte, Probleme und Methoden vom 16. bis zum 19. Jahrhundert*. Freiburg/Munich: Karl Alber.
Rupke, Nicolaas Adrianus, ed. 1987. *Vivisection in historical perspective*. London: Croom Helm.
Sarasin, Philipp, and Jakob Tanner, eds. 1998. *Physiologie und industrielle Gesellschaft. Studien zur Verwissenschaftlichung des Körpers im 19. und 20. Jahrhundert*. Frankfurt am Main: Suhrkamp.
Schmitt, Stéphane. 2005. From eggs to fossils: epigenesis and transformation of species in Pander's biology. *International Journal of Developmental Biology* 49: 1–8.
———. 2006. *Les forces vitales et leur distribution dans la nature: Un essai de systématique physiologique*. Turnhout: Brepols Publishers.
———. 2009. From physiology to classification: Comparative anatomy and Vicq d'Azyr's plan of reform for life sciences and medicine (1774–1794). *Science in Context* 22: 145–193.
Schopenhauer, Arthur. 1986. *Sämtliche Werke*. 5 vols. Frankfurt am Main: Suhrkamp.
Sloan, Phillip. 2002. Preforming the Categories: Eighteenth-Century Generation Theory and the Biological Roots of Kant's a Priori. *Journal of the History of Philosophy* 40: 229–253.
Stahnisch, Frank. 2003. *Ideas in action: der Funktionsbegriff und seine methodologische Rolle im Forschungsprogramm des Experimentalphysiologen François Magendie (1783–1855)*. Münster: Lit.
Steinke, Hubert. 2005. *Irritating Experiments. Haller's Concept and the European Controversy on Irritability and Sensibility, 1750–90*. Amsterdam /New York: Rodolpi.
Wellmann, Janina. 2010. *Die Form des Werdens: eine Kulturgeschichte der Embryologie 1760–1830*. Göttingen: Wallstein.
Williams, Elizabeth A. 2003. *A Cultural History of Medical Vitalism in Enlightenment Montpellier*. Aldershot/Burlington: Ashgate.
Wolfe, Charles T. 2010. Endowed Molecules and Emergent Organization: The Maupertuis-Diderot Debate. *Early Science and Medicine* 15: 38–65.
———. 2017. Models of Organic Organization in Montpellier Vitalism. *Early Science and Medicine* 22: 229–252.
Yolton, John W. 1991. *Locke and French Materialism*. Oxford: Clarendon.
Zammito, John H. 2018. *The Gestation of German Biology: Philosophy and Physiology from Stahl to Schelling*. Chicago: University of Chicago Press.

The manufacturer's authorised representative in the EU is Springer Nature Customer Service Centre GmbH, Europaplatz 3, 69115 Heidelberg, Germany. If you have any concerns regarding our products, please contact ProductSafety@springernature.com

Printed and bound by CPI Group (UK) Ltd, Croydon, CR0 4YY
30/09/2026
02225694-0015